Doves, Diplomats, and Diabetes

Milind Watve

Doves, Diplomats, and Diabetes

A Darwinian Interpretation of Type 2 Diabetes and Related Disorders

Springer

Milind Watve
Indian Institute of Science Education
and Research Pune (IISER-P), India

ISBN 978-1-4614-4408-4 ISBN 978-1-4614-4409-1 (eBook)
DOI 10.1007/978-1-4614-4409-1
Springer New York Heidelberg Dordrecht London

Library of Congress Control Number: 2012943959

Printed on acid-free paper

Springer is part of Springer Science+Business Media (www.springer.com)

Prologue: Advantage Ignorance

A physicist, a chemist, and a biologist decided to collaborate for an ambitious experiment that needed expertise from all the three disciplines. Painstakingly they put all the experimental set up together. They performed the experiment and it appeared to work. They got some results in hand. Now they thought they wanted to prove the reproducibility of the results and so they did it again. Everything worked fine and they had the final readings again, just that it was somewhat different than the earlier one. To see which of the two results was correct; they performed it again and got a still different result. Pretty soon they realized that although they could draw some roundabout conclusions, the experiment was not quite precisely reproducible. Everyone started wondering what could be wrong. The physicist said that the experimental systems were perhaps not very stable. He started checking all the circuits, instruments and their errors, sensitivity to voltage fluctuations, temperature variations, etc. The chemist believed that they needed greater care and more precision in performing the experiment and thought they would achieve it by doing it again and again and he would be happy when they get three consecutive identical readings. The biologist appeared least worried and said "well, there is bound to be some intrinsic variation. Let us take an average!!" Leaving aside the naivety of the story, it illustrates how people trained in different disciplines of science think in different ways, although they might be addressing the same problem. It is difficult to predict whose approach would ultimately turn out to be fruitful. What is important is to try more than one way of thinking and that increases the chances of finding a solution. Until you have the solution you need to keep your mind open to all the different possible approaches. At times the solution comes from the most unexpected approach and perhaps the least authoritative person.

So many times in the history of science, a person crossing over the borders of disciplines has contributed novel concepts to science. This is natural as a person trained in one way of thinking crosses borders and goes to a different field of science he carries a different line of thinking to that field. Perhaps people trained in that field were trained to see things with a specific and limited viewpoint and therefore failed to see some important angle. A fresher might just happen to see those things owing entirely to his freshness. This is extremely important in today's science, particularly in the field of medicine where professional specialization is making the vision deeper but simultaneously narrower so that one seldom talks about his concepts, ideas, and visions to people from very different fields. Such talks would often be chaotic and fruitless but once in a while may give unexpected rewards. The ideas behind this book have arisen in precisely the same manner.

It is tempting to relate another story that I happened to hear sometime in my school days. A war broke out on the border of a country and it was urgent and desperate to move soldiers, arms, and ammunition to the front. Trucks loaded with ammunition were being transported through a village that had an ancient historical arc across the road. The first truck got stuck there since its height was just half an inch more than what the arch allowed. Looking at the urgency and desperation a quick solution was most imperative. Troop leaders and village elders came together and debated on the alternatives, whether to make an alternative road, whether to demolish the arch, or whether to dig up the road to reduce its level. Ultimately, a simpler and efficient solution came from a small illiterate village kid who suggested "why not deflate the tyres a little bit." This will reduce the height of the truck by an inch or so which could be enough to solve the problem. We were told this story in school to emphasize that wisdom can come from anyone and we need to keep our minds open for it.

My viewpoint to this book is that of this young boy who could see things differently. Wisdom lies in the ability to keep our minds open to a suggestion that can come from one of the most unauthoritative and unexpected persons. My writing a book on diabetes was a surprise to most of my friends who must have wondered how someone who never had any training in medicine or physiology can write a book on diabetes. People have spent their entire lives researching on some specialized aspects of diabetes adding inch by inch to the understanding of the complex disorder. Is it possible that some new and revealing light comes from someone completely naïve to the field?

The story narrated above is the only possible answer to the question. I claim no authority by knowledge. But when the emperor has no clothes only a child has the authority to expose him. I can certainly write with this authority. I did happen to see things differently and therefore in the true spirit of science it might be worth taking this point of view seriously.

I have one major advantage with which I can dare to write such a book. That is an advantage of ignorance. In today's science, and particularly in the field of medicine and biomedical research, expertise is highly valued, quite rightly so. But there is a negative side of expertise too. It builds castles of comfort zones around a person. A researcher is comfortable in his/her own area of research and generally avoids talking about things beyond this comfort zone. With increasing depth of research the comfort zones are becoming increasingly narrower. As a result there is no one left to take a wider perspective. Having a wider perspective is important in order to understand any complex system. But when a system is big and complex and the visions of insiders are narrow, an outsider is more likely to do a job complementing the insiders' work so that a coherent picture emerges.

Making a map of a large territory involves many compromises and trade-offs. The bigger the area the more difficult it is to fill in details. Too many details ultimately defeat the purpose of making a map. A map is always an oversimplified version of the territory. But every map has a purpose and how much one should simplify depends upon whether the purpose is being served.

When available data are very large and still growing exponentially, taking a birds' eye-view involves an inevitable trade-off with details. This is what I have to do and it is a difficult choice. I have tried to go into only as much depth as is needed for the expression of the central argument of the book but have avoided unnecessary details.

Besides both these limitations, that of my expertise and that of the inevitable trade-off between width and depth, as far as I know this is the first time that type 2 diabetes is being depicted on such a wide canvas. When the canvas becomes wide, the entire picture looks very different and it is this difference that I want to convey through this book.

Most books are written by experts in a field and are meant for general readers, students of the field, and other experts too. This is a book written by a general reader for experts, students in that field, and other general readers. I am an undergraduate teacher by profession and until recently I did not hold a job that expected me to do research. I am not writing this book as a researcher. As a teacher one has to face unexpected and provoking questions from students. It is the students' questions that have driven all my science so far and this book is no exception. Being a teacher solved many of my problems. I did not know how to write a scholarly book. I just did it as if I am talking to a group of students. Therefore this book is not to be treated as an authoritative piece of work. It is intended to raise questions and suggest alternative viewpoints in order to find solutions. I would be happy even if experts in this field do not agree with me on my solutions. They still need to answer the questions I have raised and that would inevitably lead to some good science.

Recently I happened to attend a talk by the 1989 Noble laureate Mike Bishop addressed to an audience of young scientists settling into research careers. Although no more young, I found his advice extremely useful for being able to write this book. "Dare to be wrong," he said "and follow your nose." A number of times I stopped and asked myself a troubling question. If what I see is very different from what everyone else says and believes, is it possible that there is something wrong in my vision itself? But I followed my nose and that helped me complete this book. I am open to the possibility that my thinking is completely wrong. But if the questions I raised are getting answered while proving me wrong, I would say it was worth being wrong after all.

The book is arranged in the following sequence. Most students of medicine and physiology are not really familiar with evolutionary theory beyond knowing that it is something about "survival of the fittest." Real insights into the subtleties of evolutionary theory started coming since 1960s. It is hard to give an adequate account of modern evolutionary theory in a small chapter, but I have attempted to do so in Chap. 1 restricting to the concepts that are used later in this book. As a counterpart, for readers who have not studied medicine or physiology Chap. 2 is devoted to an introduction to textbook picture of type 2 diabetes. The real stuff in the book begins at Chap. 3, which raises a number of questions and exposes a number of paradoxes in the classical theory of diabetes. This would sound like harsh criticism, which is quite against my nature, but it is only meant to raise questions and expose paradoxes only on the background of which the following chapters can be appreciated. Chapter 4 takes a quick review of the earlier attempts to explain evolutionary origins of obesity and diabetes. Here we also lay out a set of expectations from an evolutionary theory of diabetes. At the end of the book we need to return and see how much of it the new theory meets. Chapter 5 makes a fresh beginning by looking at some behavioral aspects of life in the wilderness that are relevant to physiology. The foundation of an alternative theory of insulin resistance is laid in this chapter itself, but not in the human context. Things are different in the human context and that difference is made clear in Chap. 6. Here the central argument of the new theory is stated, but it is still superficial, the details of which start flowing through the next several chapters. From Chaps. 7–13 all the complex pathophysiological mechanisms of diabetes are elaborated at some depth in the light of the new way of thinking. As we progressively build the arguments based on evidence, logic, and a little bit of mathematics, the existing theory of diabetes appears increasingly inadequate and perhaps completely wrong. Chapter 12 suggests a rather startling possibility that insulin resistance and relative insulin deficiency are not the cause of diabetic hyperglycemia. The alternative cause suggested appears more logical and evidence based. Chapter 13 leads to another fundamental deviation from classical thinking that insulin resistance is unlikely to be central to the pathophysiology of type 2 diabetes and that hyperglycemia may not be the root cause of diabetic complications. The reader needs to go stepwise to appreciate these possibilities. The chapters are arranged in a logical sequence to build the series of arguments. There are alternative pathways for the pathophysiology of diabetic complications. Chapter 14 broadens the theme and suggests how the same line of thinking may be useful in understanding a variety of other modern diseases. Chapter 15 takes a brief review as well as a critical look of the entire synthesis and suggests possible lines of research for the future.

I would request readers to begin reading with a fresh and open mind and a skeptical but unprejudiced viewpoint.

Acknowledgments

A number of people have been instrumental in persuading me to write this book. My interest in diabetes was triggered by Dr. Chittaranjan Yajnik and his research group at KEM hospital, Pune, about a decade ago. We wrote a couple of papers together. I would have liked him to coauthor this book since we developed some of the basic ideas together. But looking at his busy schedule, I suspect, making him coauthor would have perhaps delayed the book by at least another 10 years.

As the ideas were being developed, discussions with a number of students and coworkers were extremely crucial. I need to mention particularly Maithili Jog, Anagha Kale, Pramod Patil, Samit Watve, and Prajakta Belasare among several others. We had a weekly forum called "Science Katta" where free and random branching discussions on anything in science took place. That was the first outlet for these ideas and the first platform to debate on them. These heated debates in which mainly undergraduates participate have been the most important chaperones for my thinking. After having published over half a dozen papers, I found myself increasingly unhappy with research papers as a medium to express this set of new ideas. This was because a wide perspective was developing and it was necessary to paint that on a sufficiently wide canvas. Only a small part could be painted in one research paper at a time which was unlikely to give readers any idea about the broader picture. Coincidently, I happened to meet Janet Slobodien, the editor for Ecology and Evolution titles of Springer in August 2010 who convinced me that a book would be the right canvas.

While the book was not even half way, I had the opportunity to organize a meeting funded jointly by the Royal Society, London, and the Department of Science and Technology, India. Participants in the meeting were experts in widely different areas including epidemiologists, physiologists, psychiatrists, clinical diabetologists, neurobiologists, cell biologists, geneticists, epigeneticists, population ecologists, primatologists, ethologists, psychologists, and philosophers of science. This was just the right kind of audience to give an exposure to and debate on the upcoming ideas. This was an extremely enlightening experience for me. I was encouraged by observing that although not everyone agreed with my way of thinking; they thought it was worth taking seriously, debating, and putting it to test. During and following this meeting, detailed discussions with many participants were useful. They include Jonathan Wells, Jason Gill, Raj Bhopal, Daryl Shanley, Chittaranjan yajnik,

Giriraj Chandak, Neeta Deshpande, Nishikant Subhedar, Vidya Bhate, Palok Aich, Pranay Goel, Sanjeev Galande, Saroj Ghaskadbi, Sutirth Dey, Vidita Vaidya, Mili Bandopadhyay, Asmita Phadke, Ullas Kolthur, K.P. Mohannan, Kalpana Joshi, and Leena Phadke. A large number of questions and objections were raised that put me back on the thinking track. The critical thinking and dialogue that was initiated in this meeting helped me refine the upcoming hypotheses and put things in a better perspective. A few new proposals were generated and some new lines of research were also planned during the meeting.

The synthesis in this book is a reinterpretation of existing information that is based on a large body of research particularly over the last two decades. Therefore although I might be tempted to call it *my hypothesis* or *my theory*, I presume it is only a natural culmination of the rapid accumulation of new data throwing new light on old things. I am sure if I did not come up with a new interpretation, someone else would have, sooner or later. I view it as a natural culmination of research efforts by a large community of researchers.

In preparation of the book I was helped mainly by my current and former students including Maithili Jog, Anagh Purandare, Manawa Diwekar, Gauri Tendulkar, Sakshi Sharda, Anamika Chatterjee, Neelesh Dahanukar, Ulfat Baig, Uttara Lele, and Ashwini Keskar. When I was only expecting some information inputs, Pramod Patil wrote almost half of Chap. 2 and the tables thereof. Organizational help in research from Vinay Kolte, Charu Kumbhar, Prajakta Belsare, Archana Watve, and Mukta Watve was helpful. Selected chapters or the full text at a draft stage were read and commented by Jonathan Wells, Palok Aich, Sutirth Dey, Bill Sheehan, K.P. Mohanan, Chaitali Anand, Uttara Lele, and Pritee Oswal and their detailed comments were useful in shaping the book. Salil Lachke, a former student of mine who is an Assistant Professor at University of Delaware now, commented through a series of cartoons, most of which I have included in the book since they convey a meaning beyond words. The credits for other figures and illustrations go mainly to Neelesh Dahanukar, Manawa Diwekar, Anamika Chatterjee, Sakshi Sharda, Rama Hardeekar, and Ajey Hardeekar.

The research leading to this book did not have any project-specific funding, but institutional support by IISER, Pune; particularly the extraordinary academic freedom in this institute needs a special mention. Two Indian companies, LabIndia and Anujeeva Biosciences Pvt. Ltd., extended crucial financial support earlier during my difficult times without which the development of the concept would have been difficult.

Contents

1 A Darwinian Way of Thinking

Ernest Rutherford, a renowned physicist of the last century, is not only known for his discoveries but also for an often-quoted and apparently rude comment, "Only physics is science—all else is stamp collecting" [1, 2]. He was not wrong perhaps at least with reference to the mainstream biology of his times, although a conceptual revolution had already happened. Biology in its early days was nothing more than collecting information and classifying it, precisely what stamp collectors do with stamps instead of information. Physics on the other hand was able to predict things ahead of finding them based on sound theoretical foundation. For example the existence of Pluto, the ninth planet in the solar system, was predicted first and discovered eventually. So is the case with existence of some of the subatomic particles and phenomena which were predicted much before any experiments could demonstrate them. This hardly happened in biology at that time. Biology has come a long way since Rutherford's days. The number of examples where some phenomenon in biology is predicted first and experimentally demonstrated later is on a steep rise.

One can predict things in two distinct ways. One is by finding patterns, associations, or correlations. This is not new and this is what humans have been doing since ages. For example, the association between clouds and rain can enable us to make some short-term predictions about rain. We have been doing this much before any understanding of the physical forms of water, the energetics of vaporization and condensation, etc., was known. Similarly, by observations and associations, one can predict seasonal changes in different phenomena such as flowering of plants or bird migration even without knowing any science behind it. This is a trivial form of predictability. There is no insight into causality involved in these predictions. The other is by developing an understanding of the causal connections and the ability to make good predictions by insightful understanding. This insight earmarks high-quality science. This has started happening in biology in the recent decades, taking biology out from the "stamp collection" phase to that of a full-fledged science.

There are two main reasons for this. One is our attempt to understand biology in terms of its molecules, interactions between them, and organization of these molecules into a hierarchy of structures. At this level biology becomes a complex version of physics and chemistry. All biology is in fact physics and chemistry in the sense that all laws of physics and chemistry are applicable to biological systems, and the systems work just as physical systems would work. Apart from being much more complex than systems that classical physics has handled, there is nothing non-physical or metaphysical in biology. Therefore as the physical understanding of biological systems grows, there is better predictability in biology.

However, at the same time, a physical and chemical understanding of biology is highly inadequate. In addition to being complex physics, there is something more in biology which is not

M. Watve, *Doves, Diplomats, and Diabetes: A Darwinian Interpretation of Type 2 Diabetes and Related Disorders*, DOI 10.1007/978-1-4614-4409-1_1, © Springer Science+Business Media New York 2013

so much there in physics. An appropriate equation to represent biology would be

$$\text{Biology} = \text{physics} + \text{chemistry} + \text{history}$$

By history I mean the history of evolution. This is something unique to biology. There are certain parts of physics and chemistry where history does play a role. Astrophysics, geophysics, and geochemistry are good examples. But the role of history in shaping biology as a whole is not matched by any other branch of science. It is possible to study physics and chemistry with little reference to history and still get good insights into the subjects, but this is most unlikely to work in biology. Rutherford himself was perhaps aware that there is something in biology beyond physics and chemistry. Another of his quotes is almost equally famous, "When we have found how the nucleus of atoms is built up we shall have found the greatest secret of all—*except life*." One needs something more than physics and chemistry to understand biology and that something is nothing other than the history and the theory of evolution.

A large number of biologists do ignore evolution and try to interpret systems as they appear today. Typically research and education in almost all branches of medicine fall in this trap. A student of medicine would typically learn his courses in anatomy, physiology, biochemistry, and the like without worrying much about how the things he has been studying came about and why they are like that. Such a view is bound to be too myopic, and biologists that ignore evolutionary history miss the point just too often. Understanding how things evolved is an essential part of understanding how and why they are so. The question *why* in biology is just as important as *what* and *how*, and unless the question *why* is addressed, biology remains nothing more than stamp collection.

The Bridge on the River *Why*

The question "why" can be asked and answered at a variety of levels. Why did a chameleon change its color? The why in this question can be answered in three different ways, each of the answers being "correct" in its own way. (1) The first answer could be that the chameleon changed its color because the environment changed. This is a correct answer but it only talks about the stimulus that brought about the change. (2) It can also be said that the chameleon changed its color because some pigment cells under its skin contracted and some others expanded. This is also correct as a cause of the color change, but this answers *how* the change was mediated. In reality this question is more of a *how* question than a *why* question. (3) At the third level we can argue that since the ability to change colors has offered selective advantages over several generations, all chameleons are able to change their colors to suit their environments. This type of reasoning tries to go at the evolutionary roots of the phenomenon and is therefore called an ultimate cause in contrast with the first two that are called proximate causes. A proximate cause is unlikely to work in the absence of an ultimate cause having worked over a long time. The ultimate cause forms the link between the stimulus and the response. For any nonchameleon the same environmental stimulus will not be able to bring about the contraction or relaxation of pigment cells under the skin, because it has not evolved that way.

Although the distinction between proximate and ultimate reasoning is made in many fields, in biology, it gets a more definitive and stable meaning. In other fields, what is ultimate at one level may become proximate at another level. In biology, these levels are more or less stable since evolution is the only and the most robust foundation for ultimate causation. So a stimulus that brings about a change and the mechanism behind the change can be considered to be at a proximate level, and the evolutionary process by which the mechanism and its responsiveness to the stimulus came into being at the ultimate level. Unfortunately almost throughout the history of biology, different sets of people have worked on the two levels of reasoning. Even today researchers working at the proximate and ultimate level have different mindsets, and they may not even talk to each other and come out of their myopic vision, although there are some signs of the ice melting. Good biology is one where the two levels of reasoning are handled simultaneously and interactively, but so far, there are relatively few examples where the two have really been synergistic. Nevertheless, trying to bring the two together is an emerging trend in biology [3–7].

Medicine and physiology have worked mostly at the proximate level worrying little about why the systems are the way they are. Evolutionary ecologists, on the other hand, remain confined to the ultimate level most often. They would be interested in working out why animals behave the way they do, without worrying much about the mechanisms behind the behavior. Bridging the two is one of the biggest challenges of biology in near future. This book is such a bridging attempt. The attempt is limited to a specific set of questions, but the same questions are explored at both the ultimate and proximate levels simultaneously.

To make the book readable to both the proximate and ultimate biologists and to almost any science enthusiast, in Chaps. 1 and 2, I will introduce both the fields separately in the context of the book. Chapter 1 introduces the way of thinking of an evolutionary biologist and is intended to provide the necessary background for a reader who may be familiar with medicine and physiology but not with evolution and ecology. Chapter 2 tries to capture the classical proximate level thinking in diabetes and related areas of medicine and physiology, and readers acquainted with ecology and evolution but not with physiology and medicine should find it useful.

Describing the concept of evolution in its full elegance is beyond the scope of this book. I have made an attempt in this chapter to introduce those concepts of evolutionary biology which we will use further in trying to understand diabetes and related disorders. An elaborate account of evolutionary concepts may no more be needed since a number of authors including Richard Dawkins, Daniel Dennet, and Matt Ridley among others have already done a great job in a series of books [8–10] readable for a nonspecialist reader. At a little more technical level, comprehensive textbooks of evolutionary biology are available including those by Futuyma [11] or Strickberger [12]. Nevertheless we will get introduced briefly to some of the key concepts of evolutionary biology that we would be using in our arguments later in this book.

The Complex Face of Adaptation

Adapting to the environmental challenges is the major if not the only force driving evolution. There is generally no dispute about this. That animals adapt to their environments was known even prior to Darwin. But throughout the history of evolutionary theory, there are serious debates about the subtleties of the nature of challenges and the nature and mechanisms of adaptations. Some debates are ongoing and yet unresolved. Some simple examples of adaptations are long known and simple to understand. Questions like why gazelle or cheetah runs fast, why carnivores have sharp canines, or why polar bears have thick fur have very simple and obvious answers that almost everyone knows. Many others look very obvious but are not in fact that straightforward. For example, flowers make nectar to attract pollinators. A simple interpretation is that pollinators get benefitted by the nectar reward, and flowers get pollinated in turn making it a simple and ideal mutualistic relationship. The reality is not as simple as it looks. What will happen if a flower decides not to make nectar? Pollinators have no way to know whether or not there is nectar in a given flower. They will discover it only by entering or probing into the flower. Once they do so, there is a good chance that the pollination job might get done, and therefore even if they do not find any nectar, the flower has nothing to lose. It is possible therefore that flowers "cheat" pollinators by not making nectar, and cheaters may do better than the honest mutualists since the cheaters get the same reward at a lesser cost. This is not a speculation alone. In a large number of species of flowering plants, some proportion of flowers are empty [13–15]. If all flowers on a given plant are empty, it is possible that pollinators will learn to avoid this plant. Therefore there is an optimum proportion of empty flowers that would keep the pollinators lured but at the same time save some cost. What this optimum proportion is depends on ecological and social factors. If the plants grow gregariously, i.e., in large numbers growing tightly together, it is more difficult for the pollinator to remember individual cheaters and shun them. In such a situation cheating is more prevalent. On the other hand, if plants grow individually isolated, they can be easily remembered and punished by pollinators, and therefore solitary species of plants cheat less often as compared to gregarious ones. Although the flower–pollinator mutualism is known for a long time, the presence of cheaters and the complexities in the mutualistic relationship came to light very recently. This is quite representative. A layman's perception of evolution often does not go beyond what he feels Darwin must have said. In popular perception Darwin's was the last word in evolution, and he established evolutionary theory once for all. The reality is far from it. Although what Darwin said was a major quantum leap in biological thinking, today's evolutionary biology has progressed far beyond in terms of understanding the subtleties of the evolutionary process. The punch line of Darwinism—"survival of the fittest"—still remains grossly true, but we now understand that it is not as naïve as it looks. The cheater flower example is appropriate for demonstrating a number of important principles in evolution. There may not be a unique "good" character for a species as a whole. What is "good" for an individual and therefore what is selected by natural selection is highly contextual. For example if we assume that pollinators are good at learning, the optimum proportion of flowers a given plant should produce would depend upon what other plants are producing. If others are producing very few flowers with nectar, it makes sense for you to make more but just more enough for the pollinators to understand that you are more rewarding than others. By doing this you can attract more pollinators and increase your reproductive success as compared to others. But over several generations, this would lead to all plants making copious nectar and investing a lot in it. Somewhere along this axis, tables can turn once again. Since most plants are making copious nectar, pollinators need not worry much about learning and avoiding cheaters, and if this happens, a single or few cheaters can take advantage of it. This means that there may not be an all-time optimum strategy, and the optimum itself might undergo self-driven oscillations. This is quite contrary to a popular perception that evolution leads to "perfection." The definition of "perfect" is itself not universally stable, and the short-term "goal" to be achieved by evolution may itself keep on changing. Alternatively there can be more than one point of "perfection," and different genotypes in a single population may achieve different perfections. This is a very important concept for the arguments in this book, and we will elaborate on it once again later in this chapter when we learn about the theory of games. But before going there, we need to make our ideas clear about a few other issues.

Although the "optimum" is complex and contextual, individuals do not have to understand the complex context and make a conscious choice of the best option. Plants do not have a brain and therefore presumably do not understand what cheating is. They also cannot calculate the cost or benefit of a behavioral strategy, and yet the optimum behavioral strategy evolves. In this and several other examples, mathematical ecologists have calculated the optimum for a given situation, and the observed behavior of a species in the wild is very close to the calculated optimum. This apparent intelligence is built into the system by the process of evolution. So, even unintelligent organisms appear to have very wise strategies to face a variety of environmental situations and challenges, acquired through natural selection over generations.

There is no free lunch in evolution. The cheetah, for example, has evolved to be the fastest among ground mammals. But this does not come as a free lunch. There are costs associated with the running ability in terms of physiological and anatomical adjustments of the body needed to achieve the fastest speed. For aerodynamic reasons cheetahs need to have a small head. A large head like a lion will offer greater aerodynamic resistance. A smaller head necessitates a compromise in the power of the jaw muscles. Also as their survival depends on speed, cheetahs cannot afford to get injured. As a result, cheetahs are generally the losers if there is a conflict with other carnivore species on a kill. Similar to the cheetah, any organism has to bear some cost for achieving anything. In other words, if you want to gain something, it is often inevitable that you lose something else. Evolutionary biologists call it an evolutionary "trade-off." The cheetah's running speed is itself a good example of an evolutionary trade-off. There are umpteen examples of it in biology. Whether there are any achievements by any species in the history of evolution without involving any direct or indirect cost or a trade-off is doubtful.

Adaptations can take funny turns at times. Take the ability to type on a computer. When did we evolve this ability? Computers are so recent that not more than two generations have used them. Add the history of typewriters and a few more generations get added. But it takes several generations for any novel character to evolve. It is obvious that we did not evolve the ability to type on a keyboard. Our hands evolved to climb trees and then to make and use stone tools. But the hand structure evolved for making and using tools could be easily trained to type on a keyboard without any heritable biological change being involved. There is no doubt that some training and programming of neural circuits are necessary to be efficient in typing but no heritable change is involved. When an organ evolved because of one kind of selective pressure suddenly finds a new application and starts a new function it did not evolve for, the phenomenon is called exaptation [16]. Even an exaptation may have an associated trade-off. A hand making stone tools needs to have power and impact resistance. A skilled typing hand needs finer coordinated movements but hardly any power. If you keep on using your hands for typing alone for a long time, you will improve both speed and perfection in typing. At the same time, you are most likely to lose power and impact resistance, and if at all you have to make stone tools, you may find your hands entirely useless for this work. The modern human life is full of such exaptive changes, and we could be bearing quite some cost for it.

There is another limitation of the adaptive evolution paradigm that we need to understand well in order to understand the evolution of diabetes and related disorders. An adaptive character evolves by selective pressure faced by the population over a long time, in a way organisms evolve for their past, not for the present or the future. If there is a range of variation of an environmental parameter, organisms evolve to cope with that range, but they may fail to face the challenge if it goes beyond the range that it has evolved for. For example, every species has a range of temperatures that it can cope with. Some bacteria evolved in hot sulfur spring can grow at much higher temperatures demonstrating that life can exist at that temperature, but organisms not evolved for this range will not be able to tolerate that temperature. This meaning of an "out-of-range" stimulus is easy to understand. Situations can be funny if the environmental stimulus is out of range for what the species has evolved for but is not lethal.

Niko Tinbergen, one of the founders of ethology, the science of animal behavior, did an intriguing experiment several decades ago. While trying to answer the question "What features of an egg make a female gull recognize and incubate it?" Tinbergen made several versions of artificial eggs making serial changes in their shape, size, color, and markings. An interesting finding of these experiments was that females preferred to incubate large-sized eggs if the marking on them were similar to a real egg. This preference could be stretched to ridiculous levels. When offered a football-sized artificial egg, a female would typically try to sit on it ignoring her own real eggs. The intriguing observation was that although it was impossible to sit on the huge "egg," she preferred the supernormal "egg" over the real eggs [17] (Fig. 1.1).

Fig. 1.1 Response to a supernormal stimulus: A gull female prefers an unrealistically large artificial "egg," while she ignores her true eggs (Redrawn from Tinbergen) [19]

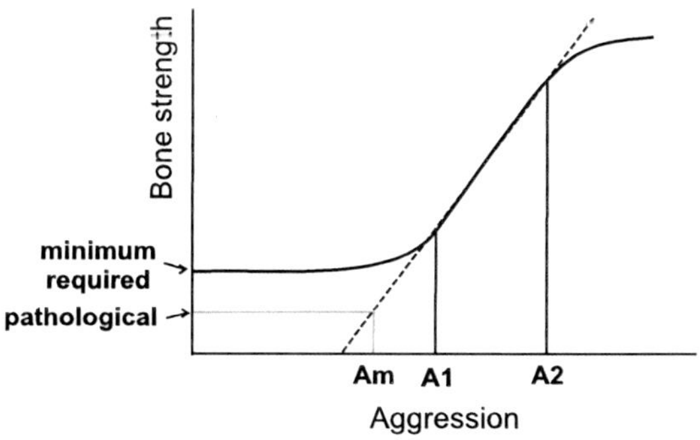

Fig. 1.2 How supernormal stimuli can lead to pathological effects: This is a conceptual diagram showing how bone strength might have evolved to respond to aggression. A minimal basal bone strength is needed by the body even at rest. Greater strength is needed during activity and particularly during aggressive encounters. Freestyle wrestlers, for example, need to have stronger bones. Evolution might have programmed the body to build stronger bones when facing frequent aggression. However, there is a physical upper limit to bone strength. The curve of aggression and bone strength is therefore sigmoid. If within the natural range of aggression (*A1* and *A2*) that an individual faces in the wild, the relationship is almost linear, then a linear response may evolve. If the evolved response is linear for an actually sigmoid curve, the response can turn pathological when the stimulus is out of the linear part of the curve. Here at *Am*, the abnormally low levels of aggression and physical activity in modern urban life, the bone density could become less than the minimum required

Stimuli such as a football-sized egg are called supernormal stimuli. Why does the mother gull give an exaggerated response to the supernormal stimulus is not very difficult to understand. Naturally there will be some variation in egg sizes. It has been shown in several species of birds that the parents preferentially feed the larger and stronger of the chicks. One with a wider gape and stronger begging behavior gets a larger share in food and parental care. This is adaptive since this way the parents ensure that their next generation is stronger and more competitive. Therefore, naturally, birds show a slightly higher preference for larger offspring while tendering or feeding. The same could be true for incubating eggs. Among the natural size variants, the mother gull could be showing slight preference for the larger ones. So the tendency that could have evolved is to prefer the larger egg. The question of how large a real egg could be did not arise during evolution since the natural size range was very narrow, and before the crazy ethologist, nobody gave them a football-sized egg. When the experimenters offered that, it worked as a supernormal stimulus and evoked a response directly proportional to its size. This is an example of how a supernormal stimulus can evolve a maladaptive response.

I will explain with one more example, and this time one that is directly relevant to the central argument in this book. Bones need to be strong not only in order to carry the weight of the body and support its mechanical operations but also to survive possible stresses and impacts. The higher the intensity of the impacts faced, the greater is the required bone strength. If an individual is physically more active and aggressive such as a freestyle wrestler, greater bone strength is needed, and the body has mechanisms to invest more calcium and other necessary components to build up the required bone strength. But if it is not used for a long time, following the principle of parsimony, the body also disinvests from bones. However, ideally, there should be a minimum bone strength needed even in the absence of aggression. Therefore the relationship of aggression and bone strength should be sigmoid as in Fig. 1.2. Naturally all wild animals are exposed to some range of aggressive interactions say between A1 and A2. If this range is confined to the nearly linear part of the sigmoid curve, a linear response may evolve although a sigmoid one should. If the evolved response is linear and as in modern human urban life, the level of aggression experienced is much below the range for which the species evolved, i.e., extreme left of the curve, the bone density may go below the minimum needed as in the figure. We would expect in such a case that a chronic lack of aggression can result into gradual depletion of bone density leading to osteopenia or osteoporosis. This is how supernormal stimuli are likely to work. When a species lives in a novel environment for which it has not evolved, it faces a number of such supernormal stimuli which reflect in abnormal behavior or physiology. Zookeepers are aware of a number of such behaviors and health problems in captive animals whose environment and behavioral needs are widely different from the ones for which they evolved. With a dramatically altered lifestyle, the human race is facing many such supernormal stimuli, and they may be the roots of many of the modern diseases.

Adaptation Is Not All

Although adaption is undoubtedly a major driving force of evolution, all evolution is not driven by adaptive changes. A large number of mutations are selectively neutral, i.e., they bring about neither deleterious nor beneficial phenotypic effects. This can happen either because a small change in DNA such as a single-base substitution does not change the amino acid sequence of the protein coded by that gene, because even after changing one amino acid the protein function does not change substantially, or because a visible phenotypic change may take place but is neither detrimental nor beneficial. Examples of different phenotypes that are selectively neutral are different eye colors or ABO blood groups in humans. Since people with different ABO types do not have different rates of survival or reproduction, natural selection does not act on these blood groups. Such characters are called neutral characters. The frequency of genes coding for such neutral characters does not change by natural selection. But it can change by chance alone or by what is called random drift. Sometimes certain alleles or variants of genes may become extinct or may drive other alleles to extinction purely by chance, a process called fixation. This is more likely to happen in a small population. Neutral mutations, genetic drift, and chance fixation are quite common and are responsible for molecular differences across different species. For example, human hemoglobin and chimpanzee hemoglobin differ by one amino acid [18]. Both the proteins work equally well. It is not that their function or their efficiency differs in the two species. But different variants have gone to fixation in the two species. Some researchers have claimed that genetic drift is important in the origin of obesity and diabetes, and we will see their argument more carefully in a later chapter, but the contribution of drift to evolution in general is certainly very important, and a number of patterns in molecular evolution are better explained by random drift than by adaptive evolution.

The Watchmaker Is Simpler than the Watch: Mechanisms of Evolution

The concept of evolution is fundamentally based on inheritance. Interestingly when the theory of evolution developed, i.e., the time of Lamarck, Darwin, Wallace, and others, the mechanisms of inheritance were completely unknown. It is awesome that these great minds perceived evolution in the absence of any idea about how biological inheritance works. But the absence of this knowledge left many weak points and voids in their arguments too, which with modern knowledge we hope to fill up in a better way. In the era of expanding molecular and genomic information, a number of paradoxes still remain unresolved, but our understanding has grown much beyond the Lamarck–Darwin era.

There has been an old debate about Lamarckian versus Darwinian mechanisms of inheritance. Lamarck was a predecessor of Darwin in the history of evolutionary theory and the first person to clearly come out with the idea of evolution on the platform of science. There are certain important differences in Lamarckian and Darwinian thinking though. Lamarck did not talk about natural selection. He thought that organisms adapt to their environment by suitably modifying themselves and that these modifications are inherited in the following generations. This is the central Lamarckian principle called inheritance of acquired characters. Both Lamarck and Darwin were aware of the fossilized remains of weird-looking animals, but while Darwin interpreted them as extinct species, Lamarck thought they were ancestral forms that kept on modifying gradually to become today's species. Since both Lamarck and Darwin were completely ignorant about the mechanisms of inheritance, the debate was more of a futile exercise until well-designed genetic experiments started addressing these questions toward the middle of the twentieth century. Darwin himself did not argue against Lamarckian thought. In fact he suggested a possible mechanism of inheritance speculatively which allowed Lamarckian inheritance. The Lamarckism–Darwinism debate caught fire after Darwin's death, and by this time, the difference in the two "isms" did not remain only in interpretation of the mechanisms of evolution. By this time Lamarckism had got a different connotation, and the debate transformed into a philosophical one about whether living organisms have an innate tendency to improve and strive to reach perfection as believed by the Lamarckian school or that evolution proceeds without any such intentions or foresights as the Darwinists thought.

In the 1940s and early 1950s, a series of experiments in bacterial genetics demonstrated that mutations took place without any foresight about their useful or detrimental effects. This supported the Darwinian school, and Lamarckism receded from mainstream biology. In the later part of the twentieth century, the role of nucleic acids in inheritance and the mechanism by which information in DNA sequence is used to build proteins were discovered, and it was clear that there is no mechanism for reverse flow of information from proteins to nucleic acids. This means that phenotypic adaptations cannot change genetic information, making inheritance of acquired characters mechanistically impossible. This gave rise to neo-Darwinism, essentially the Darwinian thought refined by the knowledge of twentieth-century molecular genetics. Neo-Darwinism has two essential components, namely, spontaneous mutation and natural selection.

Lamarckism however did not disappear from layman's perception of evolution. A large number of people still believe that just as our tail vanished when our ancestors became bipedal, our nails and eyebrows will vanish since we do not use them anymore now. This version of the principle of "use and disuse of organs" has a distinctly Lamarckian flavor. By Darwinian thinking our nails and eyebrows will vanish only if individuals without nails or eyebrows have a greater survival chance or reproductive success as compared to those having them. As we do not see this happening, we will not lose nails and eyebrows by natural selection. Interestingly the principle of "use and disuse of organs" works within the lifetime of an individual for most if not all organs. For example an extensively used muscle over a long time becomes stronger, and unused muscles become weak and also degenerate to a variable extent. This tendency called disuse atrophy is a useful adaptation. Since anyone's access to energy resources is finite, it makes sense to disinvest from unused organs or abilities and invest more in those being more frequently needed for survival. Natural selection must have favored such mechanisms of parsimony.

The phenomenon of disuse atrophy also exemplifies phenotypic plasticity. Everything is not in the genes. An organism can learn, get programmed, get trained, acquire skills, or achieve perfection in one's lifetime provided that its genes have given it the ability to learn. This has given rise to a mosaic of nature and nurture in a wide variety of organisms. Although the naïve debate about whether human behavior is decided more by nature or by nurture went on fruitlessly for decades, the mainstream biological thinking now is that human nature is a complex and inseparable interplay of the two. Biologists are now trying to get insights into the way by which this happens. This is one of the greatest future challenges of biology for perhaps the whole of a century or more of research. What we understand now is that it involves every system of the body in a complex and highly interactive way. The new understanding will come as a slow but massive tornado in which not only naïve philosophies like genetic determinism as well as cultural determinism will be washed off completely but even long-standing and currently held paradigms such as gene-encoded information being the only means of biological inheritance would also lose ground. The limitations of the assumption that the sequence of bases in the DNA is the only heritable information are being increasingly felt and the more complex nature of inheritance being revealed. There are heritable mechanisms that are more flexible than the DNA sequence. There are epigenetic mechanisms that do not change the DNA sequence but shape the way in which genes are expressed, and some of the epigenetic switches can affect the characteristics of offspring, although they are not inherited the way genetic characters are [19]. We also have examples which show effects of maternal and paternal diet or hormone levels on the physiology or gene expression of the offspring later in life [20, 21]. The nature and mechanisms behind such transgenerational effects will be clearer in the decades to come, but it is already apparent that biological inheritance is much more than genes alone. One might see some shades of Lamarckian inheritance in this. It could be called Lamarckian in a very restricted sense. One needs to think of the conditions under which such quasi-Lamarckian mechanisms could have evolved. A pursuit of this question is most likely to tell us that the quasi-Lamarckian mechanisms evolved by Darwinian selection.

Having attempted the difficult task of giving a simple introduction to fundamental evolutionary principles, I want to highlight a few concepts in evolutionary biology on which the central arguments in this book are based.

1. Life history strategies: Different species have different life histories. Some classes of animals have distinct life cycle stages that metamorphose into amazingly different forms. The egg, caterpillar, pupa, and adult in butterflies are a good example. Mammals, on the other hand, do not have morphologically distinct stages, but there are stages such as infant, adolescent, reproductively mature adult, and senescent. Certain principles are common across the widely differing life histories. Growth, maintenance (resulting into longevity), and reproduction are the three basic concerns of life histories. Since an individual has a finite set of resources at any given stage, they need to be partitioned between the three activities. If, for example, more resources are diverted to reproduction, less will be available for maintenance which may reduce life span. Life histories are decided by the dynamics of such compromises and trade-offs in such a way that under the given resource constraints and environmental risks, the lifetime reproductive success is maximized. For example, if reproduction begins at a very early stage, growth may be stunted which would affect future reproductive capacity. So, it may pay better if reproduction is arrested until some threshold growth is achieved. Producing fewer offspring at a time and reserving resources for future reproduction may be a good strategy to avoid competition among the offspring. However, if the risk of predation is high and the chance of surviving for the next reproductive season is low, it is better to make maximum number of offspring and die. A life history evolves through several generations of such compromises and trade-offs. The optimum is dependent on the ecology of the organism such as presence or absence of predators, nutrient availability, competitors, and effects of environmental stochasticity. Grossly applicable for evolution of different life histories in different types of animals, at a subtle level, the theory also applies to fine-tuning of strategies within the lifetime of an individual based on environmental constraints and individual history. For example, a female born in a crowded environment may have a different reproductive optimum than one born in an uncrowded environment. Behavioral ecologists have demonstrated that such a within lifetime flexibility exists although mechanisms of this plasticity are little known. The concept of life history strategies is extremely important for us to

understand diabetes since insulin appears to affect all the three basic concerns of life histories, namely, growth, reproduction, and longevity. As I will argue more elaborately later, it is impossible to understand insulin unless one knows the subtleties of life history tradeoffs. Unfortunately a student of medicine or physiology learns about insulin without having even heard about life history strategies. Needless to say, the theory of life history strategies is going to be an important pillar of my depiction of insulin function in this book.

2. Parental investment, mating systems, and sex ratios of offspring: Parents need to invest in their offspring to ensure its survival and early growth. The amount of time and efforts invested are highly variable across species, humans being on one extreme owing to extended period of dependence on parents. Large variation exists across species between the relative investment by the mother and the father. The difference in parental investment is decided by the ecology of the species, and differential parental investment shapes sexual behavior, mating systems, and sexual dimorphism as well as sex ratios of offspring and the investment in male and female fetuses. The evolutionary logic behind the complex interplay between ecology, parental investment, and reproductive behavior is complex, and we will not get into all of the arguments. There is one that is of direct relevance to us in the context of diabetes which is as follows.

   Inherently the investment per egg is much greater than the investment per sperm. Since both males and females have an upper limit on the total resources available for gamete production, the number of eggs a female would lay in her lifetime is far outnumbered by the number of sperm a male can produce. As a result, a female's total reproductive success is limited by her own capacity to lay eggs or carry fetuses in the womb, whereas a male's upper limit is decided by the number females he can successfully sire. However, since the number of females available is limiting, if one or a few males sire a large number of females, some other males would hardly get any. In effect, a female's reproductive success is decided by her own physical fitness and reproductive capacity, whereas a male's success depends upon his strength and competitiveness relative to other males. It is a simple calculation to show that if the sex ratio is close to 1:1, then the average success of males and females is necessarily the same, but the variance in reproductive success is larger in males as compared to that in females. Stronger males would get a disproportionately larger success than weaker males. This gender difference results in a pattern where among the stronger individuals, males enjoy a much higher reproductive success than females, but among the weaker ones, females do better than males. Recognizing this, Trivers and Willard [22] hypothesized that whenever a female is in a position to invest more in the offspring, she should prefer a male offspring, and whenever resource constraints limit the investment, she should prefer a female-biased probability of offspring sex. The Trivers–Willard hypothesis has attracted a lot of criticism as well as serious efforts to demonstrate the phenomenon with real-life data. In a number of species, if, at the time of conception, the mother's blood sugar is high, signifying a good nutritional status, the offspring sex ratio is male biased [23, 24]. This indicates that blood sugar is a tool to finely manipulate the mother's investment in the fetus. This and similar evolutionary concepts are needed to gain good insight into regulation of the blood sugar level.

   The effects of differential parental investments are more pronounced in species with a promiscuous mating system. Humans have predominantly (but not exclusively) monogamous mating system with biparental care. In such a species the difference between male and female variance is smaller. Also since males invest time and efforts in parental care comparable to a female, the difference in investment is also smaller, and therefore, the effect of TW hypothesis, if any, would also be marginal. Still the TW hypothesis has some relevance to diabetes in humans since it views blood sugar as a mechanism of fine-tuning reproductive strategies. Some of my arguments about regulation of blood sugar will be based on the role of blood sugar in differential investment.

3. Polymorphism and conditions for stable polymorphism: When a given gene locus has more than one allele in a population, the locus is said to be polymorphic. Depending upon dominant–recessive status of the alleles, the allelic frequency decides the distribution of phenotypes in a population. If different phenotypes in relation to a genetically determined character are present in a population, polymorphism would most probably exist at the genetic level too. But why should polymorphism exist in a population? We can think of two alternative scenarios for polymorphism. One is a stable polymorphism where the relative frequencies of the two or more alleles remain constant at equilibrium. This can happen only after a sufficiently long evolutionary history. The other is a transient polymorphism which is not at equilibrium and therefore will change in subsequent generations. This can happen when a new allele has a selective advantage and is spreading in the population or when two populations with different alleles start crossbreeding. Stable polymorphism is of much interest to evolutionary geneticists. It can exist under a restricted set of conditions. If there is negative frequency dependence, i.e., the rarer allele always enjoys an advantage over the commoner allele, a stable equilibrium proportion is reached. Stable equilibrium between two alleles is also reached if a heterozygote has a better fitness than both the homozygotes. Selectively neutral alleles can also coexist if the population size is large. For example, brown-eyed and blue-eyed people can coexist in a large human population since eye color does not play a role in human survival or reproduction. In a small population, however, neutral alleles may be lost by chance alone, a process called genetic drift. Obesity and diabetes do not affect everyone in a population, although they may share diet and lifestyle, and often the reason is assumed to be genetic. If it is genetic, we need to explain why a polymorphism exists in the population and whether it is stable or transient.
4. Evolutionary game theory: In the early part of the twentieth century, mathematicians developed a theory of games which evolutionary biologists led by John Maynard-Smith adopted for handling problems in evolution. Game theory is about optimizing one's strategies, but it differs from other optimization problems. Optimizing to adapt to a given environment involves finding and achieving a unique stable optimum, for example, achieving the optimum rate of perspiration to keep desired body temperature when the outside temperature is say 40°C. Here the actor is optimizing his behavior or metabolism against a fixed environment. The behavior of the environment does not depend upon the strategy adopted by the individual. Optimization in game theory is of a different nature. In game theory there are two or more players, all of whom are free to take decisions to optimize their own benefits. The returns that I get on my decision do not only depend upon what decision I take but also depend upon what decision the other player(s) take(s). The same is true for the other player(s). The nature of the game is situation dependent, and the net payoffs that the players get decide their success. Originally used to handle problems in economics and human behavior, the theory was adopted to evolutionary biology to handle problems in animal behavior. In the human models there may be an assumption of the players being rational. For animals this assumption is replaced by natural selection. The implicit assumption in game theory is that the strategies are decided by genes. Payoffs are measured in terms of propagation of genes so that a strategy having higher payoff invades and replaces other strategies. A number of conflicting situations are analyzed using such games, and the games are variously called as "prisoner's dilemma," "hawk and dove game," "snowdrift game," etc. Game theory has been used not only to analyze behavior of "intelligent" animals but also to handle problems in the biology of bacteria and viruses. For example, yeast growing on sucrose has been

demonstrated to "play" what is called a snowdrift game. The snowdrift game came to be known so from a hypothetical dilemma where two drivers find their road blocked due to a snowdrift. The solution is simply to get down and shovel the snow away. This can be easily done but with an energy cost. There is also a possible social cost of being called a "chicken." If one does the job, both share the benefit, so the net benefit of the idle driver is greater. But if neither does it, both get stuck and are at a considerable loss.

Yeast growing on sucrose face a similar situation. They need to produce an extracellular enzyme invertase that splits sucrose into glucose and fructose. The sucrose dimer cannot be taken in by the cell. Therefore, this enzyme action needs to be extracellular, the breakdown products of which can be taken in and utilized. In a liquid medium if a cell secretes some extracellular enzyme, the cell has no control over the enzymatic reaction and its products. Most of the product drifts away randomly, and only a small fraction enters the cell producing the enzyme. If all cells in a population produce the extracellular enzyme, all are benefitted. But here lies the problem. If some cell decides not to make the enzyme, it will still get benefited from the breakdown products as long as others are making the enzyme, a case similar to that of the nectarless flowers seen earlier. Since a nonenzyme producer does not pay the cost of making the enzyme but gets the benefit from the enzymatic reaction, it will have a greater net benefit and thereby will grow faster than others. So such a cheater cell should replace all cooperating cells eventually. The snowdrift game tells us that it is not quite what actually happens.

The difference in the two examples is that drivers can think while yeast cannot. This is where evolutionary game theory differs from game theory applied to economics and human behavior. Evolutionary game theory does not depend upon thinking and rational decision making by the players. It goes by a different logic. We will assume for the time being that drivers also cannot think, and their behavior is hardwired by their genes. Suppose there is a population of genetically cooperative drivers. All will always cooperate on any snowdrift situation and will share the benefits and costs equally. Now if there is a mutant driver that defects by never being ready to clear the snow, this driver will always enjoy greater net benefit than others since he never pays the cost and always shares the benefit equally. In evolutionary biology benefit is measured in terms of reproductive success. Therefore this hypothetical genetically defector driver will reproduce more than others and in subsequent generations will have a greater population.

We would also examine an upside down scenario. If the population consists of all genetic defectors, they will always get stuck in a snowdrift situation, whereas a cooperator will always clear the snow and progress albeit at a higher cost. This means that in a cheater population, a cooperator gets greater net payoff than the cheaters. This amounts to negative frequency dependence. At some equilibrium frequencies, the net payoffs of both become exactly equal, and these frequencies remain stable in the population. In a slightly technical jargon used by evolutionary game theory, neither pure cooperator nor pure defector is an evolutionary stable strategy or ESS. ESS exists when a population of a given strategy cannot be invaded by any other strategy. In the above case both the driver strategies are not ESS, and ultimately, the population will have a mixture of both types of drivers. This is a clear example of negative frequency dependence giving rise to a stable coexistence at equilibrium proportions. But if we assume the drivers to be rational who can sense which strategy pays more and adopt an appropriate behavior, there would exist a behavioral polymorphism without an underlying genetic polymorphism. Game theory can thus explain coexistence of different behavioral strategies without genetic polymorphism. This may be unlikely to happen in simple organisms like yeast. In real life, the enzymes producing and not producing yeast have been actually shown to coexist at a stable proportion demonstrating that game theory works even in microorganisms [25]. This is more likely to be genetic. But in intelligent and culturally rich species like humans, multiple behavioral phenotypes may exist even without genetic polymorphism.

There are many basic models of game theory. The prisoners' dilemma is a model that is used extensively to handle questions about the evolution and stability of cooperation. The hawk and dove game, which is mathematically very similar to the snowdrift game, is used to address the issues of aggression. The hawk and dove game is the one that we will use extensively in this book, and we will get introduced to it later when we will actually use it.

5. Evolution and homeostasis: We will spend some time in elaborating on an important angle in biology that is not commonly handled by evolutionary biologists but which is central to the theme of this book. This is about the evolution of metabolic regulation. Although there is extensive research on metabolic regulation, it is mostly handled by biochemists and physiologists, rarely by evolutionary biologists. As stated earlier biochemists and physiologists focus mainly on proximate causes or the "how" question more than the "why" question. As a result, a number of possibilities remain untapped, and there remain large voids in the thinking. But before exposing them, we need to introduce some basics briefly.

A central concept in physiology is homeostasis, which refers to the mechanisms of keeping a normal and balanced physiological state and bringing it back if shifted by any perturbation. This is achieved mainly by simple and complex control loops which work at a variety of levels ranging from modulating gene expression to modulating behavior and thinking of an individual. Although all the loops are a part of a single complex network of interconnected pathways, they consist of simple motifs that can be understood in a modular fashion. A simple negative feedback loop can be represented as in Fig. 1.3. Here, in a hypothetical biochemical pathway, the end product *C* reduces the rate at which its precursor *B* is being formed. A negative feedback can be a simple and efficient way of regulating a pathway or the standing concentration of a metabolite to a desired level. Positive feedbacks, on the other hand, can cause a quick escalation. If the effect of *C* on the formation of *B* was enhancing rather than inhibiting, it would lead to a vicious cycle. The greater the concentration of *C*, the greater will be rate of synthesis of *B* which will further increase the concentration of *C* and so on. This can potentially be dangerous for a homeostatic system. And yet, there are many positive feedback loops in the regulatory networks of the body.

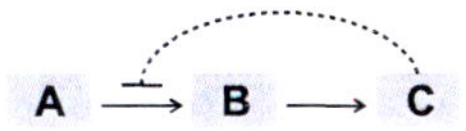

Fig. 1.3 A typical negative feedback loop. *Dotted lines with blunted ends* indicate downregulation, a convention followed throughout the book

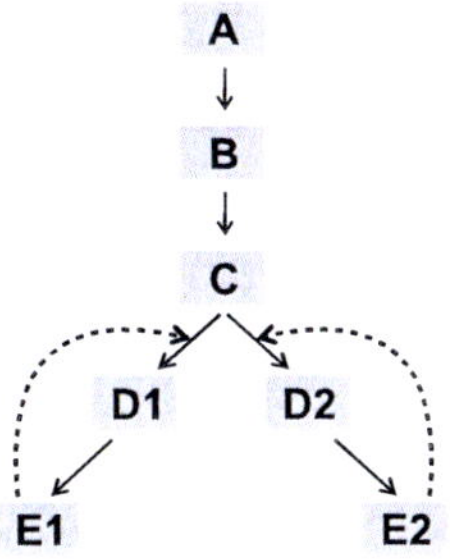

Fig. 1.4 A biochemical switch involving a bifurcated pathway and two positive feedback loops. E_1 facilitates the conversion of C to D_1, and E_2 facilitates the reaction C to D_2. Positive feedbacks are essential for smooth operation of a switching mechanism and the stability of a "switched-on" state

Why and where do we need positive feedback loops then? They are typically needed to operate a metabolic switch (Fig. 1.4). Here, the hypothetical pathway bifurcates into two. The end product of E_1 facilitates the synthesis of its precursor D_1. Similarly, E_2 facilitates D_2. In such a system, if initially the concentration of E_1 is greater, more of it will be formed and formation of E_2 will be suppressed, i.e., pathway 1 is switched on and pathway 2 is switched off. There are many applications of the switching on and off mechanism.

What do you do when you are hungry? The obvious answer is to look for something to eat. And when thirsty, you seek water. In a modern kitchen or restaurant, both are available at a single place. But that is not what our bodies have evolved for. In the wilderness there could be quite a distance between the place you find food and a water hole or river to drink from. In such a case you need to take a decision as to whether you will eat or drink first. What could be the algorithm of deciding this? A simple solution could be to go for food if hunger signal is stronger and for water if thirst signal is stronger. This logic is fine and will work well at this level. However if the algorithm is only to follow whatever is stronger, it can lead to a tricky problem. Let us assume that the thirst signal was stronger, and therefore, you go for water. As you start drinking, the thirst will be gradually quenched and at one stage will be first equal to and then slightly weaker than the hunger signal. If you are following a simple algorithm of going by the stronger thrust, you will have to leave drinking and go for food. But the moment you eat one mouthful, the hunger signal reduces slightly, and thirst becomes stronger. You will run and drink some water, and within moments hunger becomes stronger again. To avoid such a plight, a positive feedback is necessary. The moment you have one sip of water, the thirst signal is strengthened instead of weakening, and you switch off hunger for the time being. If the act of drinking water strengthens thirst, it is a positive feedback. By eating a little bit, hunger becomes stronger, and thirst is temporarily switched off. This makes life a lot more convenient. Wherever there is a choice between two or more options, one of which needs to be chosen while switching off the others, a positive feedback must operate. Without positive feedbacks, clear-cut choices cannot be made. This is particularly important for decisions like whether to fight back, freeze and camouflage, or run away from an enemy. Any of the clear choices could be lifesaving, whereas hesitation, ambiguity, or inability to completely switch off others would be fatal. For example, if an animal decides to freeze and thereby try to escape predator attention, even a slight movement is undesirable, and therefore, fight-or-flight options need to be completely switched off. For the right use of such behavioral options, positive feedbacks are extremely crucial.

The concepts of alternative behavioral strategies, plasticity, and conditional behaviors that we learned above have important implications for homeostasis. The classical physiological concept of homeostasis assumes that there exists a single optimum and ideal level for any parameter such as blood sugar. All mechanisms of homeostasis are designed to restore the normal value by all means. If the metabolite level has something to do with behavior and alternative behavioral strategies exist, the concept of an all-time optimum may not be correct after all. Different behavioral strategies may demand different optimum levels of some crucial metabolites or hormones, which means that there can be more than one optima, and we need mechanisms to change the optimum and achieve the new optimum by altering the homeostatic mechanisms when a different behavioral strategy is adopted. Therefore the concept of homeostasis should be replaced by the more refined concept of allostasis [26] where there is an option of shifting the steady state in a continuous or discrete manner. There are certain obvious examples of allostasis. Animals which hibernate have two different states of metabolism and accordingly two different optima. There is an optimum heart rate in active life and a different lower optimum heart rate during hibernation. Similarly some of the physiological parameters have to change at different altitudes. Life at low altitude has one set of optima and that at high altitude has different. Both are "normal" by all means. Not all examples are as contrasting as the hibernation example. Many behaviors demand only small changes. But it is important to realize that the classical concept of "normal range" in physiology and medicine needs to be reexamined. The normal range is certainly a complex concept, and a change in the level of a metabolite can either be a shift in the normal range itself or a pathological departure from the normal range. A simplistic concept of normal range and homeostasis does not take into account the possibility that an apparent deviation from the "normal" state may be because of a shift in normal range and not a departure from normal. We need to clearly distinguish between the two at least theoretically in order to understand the often thin borderline between altered normal physiology and pathophysiology.

Most of the above concepts are intimately linked to medicine, but unfortunately medicine developed as a science far away from natural history, behavioral ecology, and evolution. Therefore evolutionary concepts are seldom used in medicine. The ones that have been used often lack the rigor that modern evolutionary biology has. This picture has started changing very recently, and Darwinian medicine is increasingly being recognized as a useful aid to medicine. However, as a science, it is still in its infancy, and although potentially it can have many important clinical implications, currently its impact on practicing medicine is little. Only when Darwinian thinking has radical implications for practicing medicine, it will get recognition as a significant contributor to the field. This is likely to happen soon which I hope the reader would appreciate toward the end of this book.

References

1. Birks JB (1962/1963) Rutherford at Manchester. Heywood/Benjamin, London/New York
2. Pepper FS (1987) The Wit and wisdom of the 20th century: a dictionary of quotations. P. Bedrick Books, New York
3. Ferreras P et al (2004) Proximate and ultimate causes of dispersal in the Iberian lynx *Lynx pardinus*. Behav Ecol 15:31–40
4. Drickamer LC, Gillie LL (1998) Integrating proximate and ultimate causation in the study of vertebrate behavior: methods considerations. Am Zool 38:43–58
5. Drickamer LC (1998) Vertebrate behavior: integration of proximate and ultimate causation. Am Zool 38:39–42
6. Thierry B (2004) Integrating proximate and ultimate causation: just one more go! Curr Sci 89:73–79
7. Mayr E (1988) Toward a new philosophy of biology, observations of an evolutionist. Belknap Press of Harvard University Press, Cambridge, MA
8. Dawkins R (1999) The extended phenotype: the long reach of the gene. University Press, Oxford
9. Ridley M (1994) The Red Queen: sex and the evolution of human nature. Macmillan Publishing Company, New York
10. Dawkins R (1989) The selfish gene. Oxford University Press, Oxford
11. Futuyma DJ (2009) Evolution. Sinauer Associates, Sunderland, MA
12. Strickberger MW (2005) Evolution. Jones & Bartlett Learning, Sudbury
13. Thakar JD, Kunte K, Chauhan AK, Watve AV, Watve MG (2003) Nectarless flowers: ecological correlates and evolutionary stability. Oecologia 136:565–570
14. Anand C et al (2007) Presence of two types of flowers with respect to nectar sugar in two gregariously flowering species. J Biosci 32:769–774
15. Belsare PV, Sriram B, Watve MG (2009) The co-optimization of floral display and nectar reward. J Biosci 34:963–967
16. Gould SJ, Vrba ES (1982) Exaptation – a missing term in the science of form. Paleobiology 8:4–15
17. Tinbergen N (1953) The herring gull's world: a study of the social behavior of birds. Praeger, Oxford
18. King MC, Wilson AC (1975) Evolution at two levels in humans and chimpanzees. Science 188:107–116
19. Tinbergen N (1965) Animal behavior. Time-Life Books, New York
20. Wells JCK (2010) Maternal capital and the metabolic ghetto: An evolutionary perspective on the transgenerational basis of health inequalities. Am J Hum Biol 22:1–17
21. Ng SF et al (2010) Chronic high-fat diet in fathers programs β-cell dysfunction in female rat offspring. Nature 467:963–966
22. Trivers RL, Willard DE (1973) Natural selection of parental ability to vary the sex ratio of offspring. Science 179:90–92
23. Cameron EZ, Lemons PR, Bateman PW, Bennett NC (2008) Experimental alteration of litter sex ratios in a mammal. Proc Biol Sci 275:323–327
24. Cameron EZ (2004) Facultative adjustment of mammalian sex ratios in support of the Trivers-Willard hypothesis: evidence for a mechanism. Proc Biol Sci 271:1723–1728
25. Gore J, Youk H, van Oudenaarden A (2009) Snowdrift game dynamics and facultative cheating in yeast. Nature 459:253–256
26. Korte SM, Koolhaas JM, Wingfield JC, McEwen BS (2005) The Darwinian concept of stress: benefits of allostasis and costs of allostatic load and the trade-offs in health and disease. Neurosci Biobehav Rev 29:3–38

2 Diabetes in a Textbook

Before we start thinking about the evolutionary origins of diabetes and related disorders, I need to briefly sketch what is currently known and well accepted about diabetes. This chapter tries to compile a textbook picture of diabetes [1–3] only to serve as a background. Readers who have studied physiology or medicine may skip this chapter straightaway since it does not contain any new argument. It may be necessary and useful for readers who need a fair amount of background information about diabetes before appreciating the paradoxes and puzzles associated with it.

As a student and then as a teacher, I always found textbooks very boring. Since this chapter is an extract of textbooks, it would sound boring too. There is a class of readers that can cope with it. But if I were to be the reader, I would have given up reading the book midway in Chap. 2. To avoid this disaster I would suggest an alternative reading plan. The background textbook information in this chapter is needed in order to understand the questions raised and the stepwise arguments made in further chapters. But it is not necessary to read it right away. So for impatient readers like me, I would suggest skipping this chapter at whichever point it gets boring but returning to it as and when a reference to the basic textbook information is needed. Science becomes lively when one goes beyond textbooks. This can be achieved during lively discussions in a classroom, and that is the approach from the next chapter which I can assure is very different from this chapter.

So here begins the textbook stuff.

Diabetes is among the most widely known and also among the least understood of all disorders. It should also be listed among the earliest known disorders in the history of medicine. We find a description of some of the signs and symptoms of diabetes in ancient Egyptian and Indian literature. Polyuria or frequent and copious urination was among the first symptoms that we find recorded in Egyptian literature over 1500 BC. Indian medicine had noted that the urine of diabetics tasted sweet and would attract ants. By the middle of the eighteenth century, it was shown that the sugar levels in the blood of diabetics were raised substantially. A major contribution in our understanding of diabetes came from Claude Bernard who discovered a number of fundamental principles in physiology such as storage of energy by the liver in the form of glycogen. He demonstrated that damage to medulla oblongata caused severe hyperglycemia and therefore thought that brain was the main center for regulation of blood glucose [2, 4]. However, by the turn of the nineteenth century, it was shown that the removal of pancreas leads to diabetes. Soon the islets of Langerhans were described, and some soluble substance produced from there was suspected to regulate blood sugar. The name "insulin" was first used for this magic molecule in 1909, and by 1921, insulin was purified by Banting, Best, Collip, and Macleod. Almost simultaneously glucagon, the counter signal in glucose regulation, was being discovered. Glucagon was

M. Watve, *Doves, Diplomats, and Diabetes: A Darwinian Interpretation of Type 2 Diabetes and Related Disorders*, DOI 10.1007/978-1-4614-4409-1_2, © Springer Science+Business Media New York 2013

Table 2.1 Factors affecting insulin secretion

Stimulators	Inhibitors
Glucose	Somatostatin
Mannose	2-Deoxyglucose
Amino acids	Mannoheptulose
Intestinal hormones [gastric inhibiting peptide, glucagon-like peptide, gastrin, secretin, cholecystokinin (CCK)]	α-Adrenergic stimulators (epinephrine)
β-Keto-acids	Galanin
Acetylcholine	Diazoxide
Glucagon	K^+ depletion
Cyclic adenosine monophosphate (cAMP)	Alloxan
B-Adrenergic stimulators	Insulin
Theophylline	
Sulfonylureas	

named in 1923. By then Claude Bernard's demonstration of brain damage affecting glucose regulation was almost completely forgotten, and glucose levels were thought to be maintained by the balancing actions of insulin and glucagon.

Insulin, which is believed to be central to diabetes, is a peptide hormone secreted by β cells of pancreas in response to blood glucose levels, amino acids, and nervous stimulation. It is central to regulating carbohydrate and fat metabolism in the body. Normal insulin secretion responses are essential for maintaining normal body functions. Insulin has a wide variety of effects on different tissues of the body. In the adipose tissue it increases glucose uptake, fatty acid synthesis, glycerol phosphate synthesis, triglyceride deposition, activation of lipoprotein lipase, and inhibition of hormone-sensitive lipase. In muscles it increases glucose uptake, glycogen synthesis, K^+ uptake, amino acid uptake, and protein synthesis. Therefore insulin action is important in muscle building and muscle strength. Insulin has an osteoblastic, i.e., bone-forming, function and thereby contributes to bone strength too. In the liver it decreases gluconeogenesis and ketogenesis and facilitates glycogen synthesis, glycolysis, protein synthesis, and lipid synthesis. More generally in the body, insulin is primarily a growth hormone promoting cell growth and protein synthesis. The growth-promoting function of insulin is synergistic with another hormone, the pituitary growth hormone (GH). Insulin has also been shown to be important in the cognitive functions of the brain. Insulin acts as the signaling molecule simultaneously regulating different functions of the body in different ways [3]. Therefore it is logical to expect that when there is some or the other type of problem in insulin function, not only blood sugar is affected but a large number of functions of the body are.

Before understanding the altered patterns of insulin secretion, normal patterns need to be understood. Glucose is a major stimulus for insulin secretion by pancreatic β cells. Insulin production is also stimulated by some other nutrients including amino acids such as arginine. Several other hormones of the body affect basal or glucose-stimulated insulin secretion (Table 2.1). In addition parasympathetic nervous system has a direct control over insulin secretion, but the significance of it in normal or pathological conditions is poorly understood. Under normal conditions out of total insulin secretion in 24 h, ~50% is secreted under basal conditions while the rest is secreted in response to meals. The amount of insulin secreted for different types and timings of food intake in a day such as breakfast, lunch, and dinner does not differ significantly. Insulin secretion is not continuous but is oscillatory in nature. Three types of oscillations have been observed, namely, rapid small amplitude oscillations, ultradian oscillations consisting of large amplitude, slow oscillations, and circadian oscillations in the form of different insulin responses at different times of the day. The exact physiological significance of the oscillations is not clear, but exogenous insulin is more effective in reducing plasma glucose levels when administered in oscillatory doses than at a constant rate.

The insulin response to glucose or a meal has two distinct phases. The first is an acute phase where a brief peak of insulin secretion arises immediately after feeding, even before absorption of food begins. The acute phase insulin response (AIR) is driven by neuronal mechanisms and has sometimes been called cephalic phase insulin response. The second phase appears to be predominated by peripheral mechanisms involving islet stimulation by glucose. The acute phase response is important for the normal glucose tolerance curve. Absence of acute phase response increases the area under the curve for both glucose and insulin. AIR is impaired in diabetes from an early stage.

Diabetes and Its Types

Although insulin has a large number of functions, the definition and diagnosis of diabetes are based on only one of its functions which is regulation of peripheral blood glucose levels. Insulin is a signal molecule, and a signal can function well when it is given out as well as received normally. Endocrine signals are like radio broadcasting where the signal is given globally, but it can be received only where the specific receiving mechanisms exist. Such a signal will fail if either the mechanisms of relaying the signal or the mechanisms of receiving it are impaired. In type 1 diabetes the former is impaired and in type 2 the latter. Defective secretion of insulin causes type 1 diabetes, whereas defective mechanisms of insulin signal reception and downstream action appear to be the root cause of type 2 diabetes. Type 1 diabetes (T1D) most commonly results from autoimmune destruction of insulin-producing β cells of the pancreas. The subsequent lack of insulin leads to increased blood glucose and urine glucose. Although blood and urine glucose are the presenting signs of T1D, most of the pathology of untreated T1D results from deficiency of insulin rather than raised blood sugar.

Type 2 diabetes mellitus (T2D) is classically characterized by high blood glucose which is believed to be a result of a combination of insulin resistance and relative insulin insufficiency. We are going to refer to this condition frequently in this book and will call it as IR-RII throughout. The reduced response of tissues to insulin signaling is called insulin resistance. When insulin resistance sets in, higher amounts of insulin are needed to bring about the same function, just as shouting in louder voice is needed while talking to a hearing impaired. This is what is believed to happen in the body. Insulin resistance is a measurable character and is measured by different techniques tabulated in Table 2.2. The different indices capture somewhat different elements of insulin resistance but generally are intercorrelated sufficiently well. Clinically the most frequently used index of insulin resistance is called HOMA-IR which is proportional to the product of fasting insulin and fasting glucose levels. HOMA-IR reflects muscle as well liver insulin resistance. The gold standard of measuring insulin sensitivity is achieved by using a technique called euglycemic hyperinsulinemic clamp. In this technique insulin is infused continuously to achieve a high and stable plasma level. Because of high insulin level, plasma glucose drops down. The normal levels of glucose can be brought back by infusing glucose. At the stable high insulin level, the amount of glucose infusion needed to retain normal glucose levels is a measure of insulin sensitivity. This measure mainly reflects glucose uptake by muscle by the insulin-dependent pathways. The clamp technique is too complex to be used in clinical practice and is restricted to physiological research.

Insulin resistance is almost always accompanied by higher levels of insulin production such that the resultant function remains normal. This is believed to be because of a compensatory rise in insulin production in response to insulin resistance. High insulin resistance accompanied by raised level of insulin such that blood sugar remains normal is a state that is called insulin-resistant state which generally precedes T2D by several years. We will call the hyperinsulinemic

Table 2.2 Selected measures of insulin sensitivity or resistance (see [5] for details)

Clinically useful methods		Formula	Comments
Methods based on fasting measurements	Fasting insulin	I_0	Simple and feasible methods in a clinical setup. More variability. Do not differentiate between liver and peripheral insulin resistance. Not useful in advanced diabetics
	HOMA	$I_0 \times G_0$/constant	
	QUICKI	$1/[\log(I_0)+\log(G_0)]$	
	G/I	G_0/I_0	
Methods based on OGTT	Cederholm and Wibell	$M/G \times \log I$	More elaborate and reliable but more difficult to perform. Correlated well with clamp methods in normal prediabetics and early diabetics but weakly in advanced diabetics
	Matsuda et al. index	$10{,}000,\ ({}_{G}\,0 \times I0) \times (G \times I)$	
	Insulinogenic index	$(I_{30}-I_0)/(G_{30}-G_0)$	
	Many others		
Methods for research			
Clamp methods	Hyperinsulinemic euglycemic clamp	The quantity of exogenous glucose needed to maintain normal glucose under an induced stable hyperinsulinemic condition	The gold standard of insulin resistance measurement. Highly elaborate and tricky to perform but more reliable and reproducible. Differentially and specifically represents insulin-induced glucose disposal. Physiologically unrealistic conditions
FSIVGTT	Frequently sampled intravenous GTT	Glucose clearance rate per unit change in plasma insulin concentration	Needs over 30 samples in a GTT. Estimates based on a computational model. Physiologically more realistic than clamp

I insulin, *G* glucose, subscript $_0$ indicates fasting, 30 indicates minutes, without subscript mean concentration. *M* glucose load

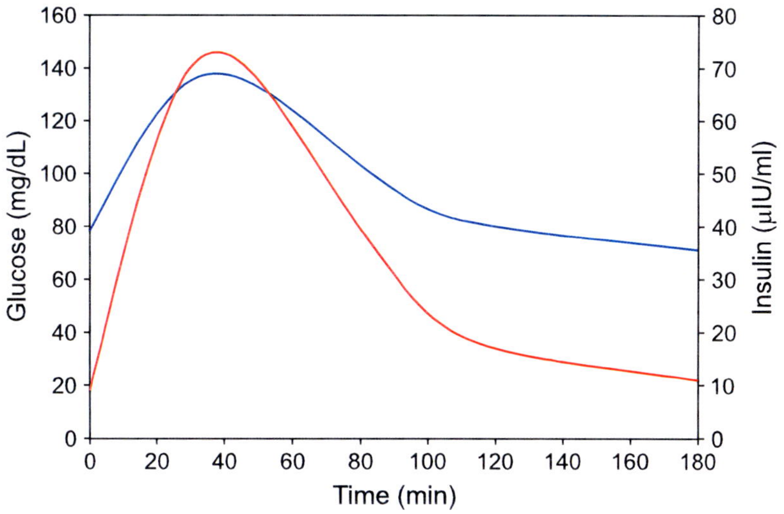

Fig. 2.1 A typical time course of plasma glucose and insulin levels after oral administration of 75 g glucose (oral glucose tolerance test OGTT). The nature of the curve reflects in multiple ways the state of glucose homeostasis in the body (Glucose is shown by *blue line* on primary *Y* axis, insulin in *red* on secondary *Y* axis)

insulin-resistant state as HIIR in which glucose levels are normal. Not all HIIR individuals develop T2D, but many do. This is believed to happen when the β cells are unable to compensate insulin resistance resulting into a relative insulin deficiency. This is said to be because of fatigue or exhaustion of the insulin-producing β cells of the pancreatic islets. At this stage the total insulin production of a diabetic may be higher than that of a nondiabetic and non-insulin-resistant individual, but blood sugars are raised owing to even higher levels of insulin resistance. Thus it is a combined effect of insulin resistance and relative insulin insufficiency (IR-RII) that is believed to lead to diabetes. T2D is the predominant form of diabetes worldwide, constituting about 90% of cases globally. The diabetes epidemic is increasing in both developed as well as developing countries. Globally the number of people with diabetes is expected to rise to above 300 million in 2015 [6], but this is an underestimate because, for each diagnosed case, there is thought to be one undiagnosed case in First World countries and eight in the Third World. Type 2 diabetes has become the world's most important public health problem [6]. The focus of this book is on T2D, although while talking about pathophysiology of diabetic complications, where there is much overlap between T1D and T2D, we will include T1D in the discussions. In some of the older literature, T2D has been often referred to as "mild" diabetes. In the past 40 years, the status of T2D has changed from being considered as a mild disorder of the elderly to one of the major causes of morbidity and mortality increasingly affecting the middle aged and even youth. It is important to note that the rise in prevalence is seen in all six inhabited continents of the globe [7, 8].

The diagnosis of diabetes rests on the measurement of plasma glucose levels. Because plasma glucose concentrations range as a continuum, the criteria are based on estimates of the threshold for the complications of diabetes. Reproducibility of the plasma glucose concentration is an important issue for interpreting the results of diagnostic tests for diabetes. There is significant variation in the results of repeated tests in the adults after 2–6 weeks interval. Thus it is essential that any test showing above normal levels be confirmed by repeated tests. Before overt diabetes appears, alterations in glucose homeostasis are detectable if a curve of changing glucose levels after meals or glucose intake is monitored (Fig. 2.1). This can be done using what is called an oral glucose tolerance test. Although oral glucose tolerance test (OGTT) is an invaluable tool

Table 2.3 Diagnostic criteria for stages of type 2 diabetes

Test	Normoglycemia (mg/dL)	IFG (mg/dL)	IGT (mg/dL)	Diabetes
FPG	<100	100–125		≥126 mg/dL
2-h PG	<140		140–199	≥200 mg/dL
Random plasma glucose concentration				≥200 mg/dL plus symptoms of diabetes

FPG fasting plasma glucose, *PG* plasma glucose, *IFG* impaired fasting glucose, *IGT* impaired glucose tolerance

in research, it is not recommended for routine use in diagnosing diabetes. Levels of glycated hemoglobin A_{1c} (HbA_{1c}) are generally not used for diagnosing diabetes but can be valuable for follow-up. Hemoglobin along with many other proteins get glycated very slowly by spontaneously combining with glucose, and this glycation is directly proportional to the levels of glucose in blood. That makes HbA_{1c} a cumulative estimate of glucose, averaging out the daily fluctuations in glucose levels. Individuals who deviate from the normal glucose tolerance curve but are not overtly diabetic are placed in an intermediate category called impaired glucose tolerance (IGT). The thresholds for the definitions are in Table 2.3.

Type 2 diabetes is traditionally thought to have a strong genetic component. However the increasing incidence of the disorder cannot be of genetic origin since gene(s) that give rise to a disorder will not increase in frequency in a species globally. Even if it does, no allele can increase in frequency dramatically within only a couple of generations. So genes alone cannot be blamed for diabetes. The current thinking is that it is an interaction between genes and environment that is responsible for the current epidemic. Genetically predisposed persons who are exposed to a series of diabetogenic environmental influences develop clinical disease [9–11]. Although T2D is believed to have a strong genetic component, in exceptionally few cases, some specific genes are demonstrably involved. The common forms of diabetes appear to be polygenic in nature and are believed to be due to combination of genes involved in predisposing to obesity, insulin resistance, and abnormal insulin secretion.

The environmental and lifestyle factors predisposing to diabetes are sedentary lifestyle, altered diet leading to a positive energy balance, stressful life, excessive tobacco, and alcohol. In addition a strange pattern was discovered in the early 1990s that individuals born small for gestational age are highly predisposed to diabetes. This trend is consistently observed across the globe. Animal experiments have confirmed that intrauterine growth is crucial in determining the predisposition to diabetes. Intrauterine growth retardation (IUGR) leads to altered metabolic programming which makes a person prone to obesity and insulin resistance in later life.

Obesity and Muscle Insulin Resistance

The primary effect of insulin on glucose is to stimulate translocation of glucose transporters to cell membrane through which glucose enters into the cells and is then utilized for energy production. This happens through an elaborate cell signaling and downstream pathway. It appears that the number of glucose transporters in skeletal muscles of an insulin-resistant person is not different, but the ability of insulin to facilitate this translocation is disrupted. The term insulin resistance indicates an impaired biological response to either endogenously secreted or exogenously administered insulin. There is reduced insulin-mediated glucose transport and metabolism in skeletal muscles and adipocytes and also impaired suppression of hepatic glucose output. Insulin resistance is present in persons predisposed to type 2 diabetes before the onset of hyperglycemia, and so it has been concluded that insulin resistance is the primary abnormality that is responsible for the development of type 2 diabetes. Insulin resistance is present years before the onset of the disease, and this is a consistent finding in type 2 diabetes [12, 13]. A substantial amount of data indicates that insulin resistance plays a major role in the development of glucose intolerance and diabetes. Insulin resistance is associated with the progression to IGT and eventually to type 2 diabetes [14]. Prospective studies show that insulin resistance predicts the onset of disease. Insulin resistance is present in first-degree relatives of type 2 diabetics even when they are not obese. This is in support of having a strong genetic component in the etiology of T2D [12, 14, 15]. There is also a strong influence of environmental factors on the genetic predisposition to insulin resistance [16, 17]. The generally agreed pathway to T2D is that obesity leads to insulin resistance in genetically predisposed individuals which in the long run leads to T2D.

A number of studies show that the risk of insulin resistance and development of T2D increases with increase in body weight and particularly body fat content. It is not clear whether it is the free fatty acids in circulation or the adipose stores of the body that are of primary importance in inducing insulin resistance, but the general role of fat seems to be agreed upon. A close association between obesity and insulin resistance is seen in all ethnic groups and is found across the full range of body weights, across all ages, and in both sexes [18–20]. Absolute amount of body fat has an effect on insulin sensitivity across a broad range [21–23]. However central or visceral (intra-abdominal) adiposity is more strongly linked to insulin resistance, and accumulation of abdominal fat has effects on glucose tolerance independent of total adiposity [24–28]. The reason why intra-abdominal fat is more strongly related to insulin resistance is not clearly known at the proximate and ultimate level. There are a number of possible explanations for this strong association. It is speculated that due to presence of more number of adrenergic receptors as compared to subcutaneous fat, intra-abdominal fat is lipolytically more active [27, 28] generating more free fatty acids (FFAs) which induce insulin resistance. Another reason stated is that the intra-abdominal fat is more resistant to the antilipolytic effects of insulin [29]. One more hypothesis relies on the presence of high concentrations of 11β-hydroxysteroid dehydrogenase type 1 (11βHSD1) that enhances conversion of inactive cortisone to active cortisol, leading to increased local cortisol production, which might be responsible for increased lipolysis. All these hypotheses appear to count on increased FFA production as the proximate cause of insulin resistance. A possible internal contradiction here is that if abdominal fat undergoes rapid lipolysis, it is difficult to account for greater accumulation of fat in the abdomen. Also obese persons have actually been shown to have lower rates of fat oxidation. Increased FFA level is related to insulin resistance in skeletal muscle. The mechanism by which FFAs reduce glucose transport has remained elusive. But it is shown that elevated free fatty acids (FFAs) predict the progression from IGT to diabetes [30, 31] implying some role for FFAs.

In contrast very low levels of lipids in the body or absence of adipose tissue, a condition called lipoatrophy, has a strong association with insulin resistance. So although fat is said to be responsible for inducing insulin resistance, the absence of fat also induces insulin resistance.

Alternatively intramuscular triglycerides (IMTG) rather than FFAs are implicated in insulin resistance. A strong correlation between IMTG concentration and insulin resistance has been demonstrated by evaluating IMTG with biopsy [32]. IMTG may accumulate if the rate of FFA uptake into the muscles and rate of their oxidation mismatch. Again although insulin-stimulated glucose uptake and the amount of IMTG are shown to be inversely related in nonexercising individuals, the mechanism by which IMTGs induce insulin resistance is not clearly known. Furthermore increased IMTG content is not invariably linked to insulin resistance because long-term exercise training is linked with increased IMTG [33], and chronic exercise increases insulin sensitivity as well as IMTG and the capacity for fatty acid oxidation [34–37].

A yet another possibility is that of a central role of malonyl CoA levels. Glucose uptake even in insulin-resistant muscle is higher at the elevated levels of blood glucose as seen in T2D [38, 39]. The resulting high glycolytic activity can generate acetyl CoA which is converted to malonyl CoA. The accumulation of malonyl CoA inhibits the carnitine palmitoyltransferase (CPT I) which resides on outer mitochondrial membrane, inhibiting uptake of acyl CoA [40]. The resulting buildup of acyl CoA and diacylglycerols is proposed to activate one or more protein kinases, resulting in insulin resistance [40]. Insulin-sensitizing effect of exercise supports this hypothesis since exercise lowers intracellular long-chain acyl CoA levels [41]. However, this pathway has a built-in negative feedback since there will be a decreased glucose flux after developing insulin resistance which will reduce the pool of acetyl CoA, eventually reducing malonyl CoA. Therefore this pathway may not induce sustained insulin resistance.

Higher nutrient status is known to activate the mammalian target of rapamycin (mTOR) pathway which through the S6 kinase 1 pathway ultimately decreases the levels of available insulin receptor substrate IRS1 and thereby induces insulin resistance. However insulin signaling itself is needed for expression of mTOR, and therefore, mTOR and insulin signaling appear to be in a negative feedback loop. Negative feedback loops are generally self-regulated and therefore do not lead to escalation. Therefore mTOR activation alone may not be sufficient to explain sustained insulin resistance.

The upcoming alternative explanation for the connection between obesity and insulin resistance is that the adipocytes secrete a number of signal molecules called adipokines which induce insulin resistance. In this line of thinking, the focus is shifted from FFAs to the adipose tissue. A wide variety of compounds that constitute adipokines are known, and their association with insulin resistance has been demonstrated. TNF-α appears to be the most important adipokine in this connection, and the pathway leading to insulin resistance appears to involve inflammatory signals, and yet the precise pathway of induction of muscle insulin resistance remains to be completely elucidated. Therefore although there is a strong correlative evidence for the role of free fatty acids, triglycerides or adipose tissue and its secretions in insulin resistance, and a number of possible pathways are visualized, the exact molecular events which lead to development of sustained insulin resistance in obesity are still not well understood.

There are a number of other possible mechanisms leading to insulin resistance which are not directly dependent on lipid metabolism. The brain monoamines such as serotonin and dopamine influence insulin sensitivity in a complex way, but their actual role in clinical insulin resistance is not clearly known. Insulin resistance is associated with decreased mitochondrial activity. Decreased mitochondrial activity may be responsible for decreased oxidative metabolism and eventually insulin resistance. But a reverse causation is equally possible. Insulin itself is important in upregulating mitochondrial biogenesis, so in an insulin-resistant state mitochondrial activity might be reduced. High-fat diet, although known to induce insulin resistance, was actually shown to increase mitochondrial mass in rats. Therefore, fat-induced insulin resistance is unlikely to work through decrease in mitochondrial activity [42]. The association of decreased mitochondria and insulin resistance is either because of reverse causation or some other causative factor. The central nervous system also plays a role in regulating hyperinsulinemia and insulin resistance. There is evidence for complex interactions between central and peripheral signals modulating insulin sensitivity and secretion [43–46], but the nature of these interactions is not yet understood sufficiently well.

In brief, although the concept of insulin resistance is central to type 2 diabetes, the causes and pathways leading to insulin resistance are not yet clearly known.

Liver Insulin Resistance

Insulin's predominant action on liver is to regulate the hepatic glucose output. Liver has the important job of synthesizing glucose and releasing it into blood in fasting conditions. This is achieved by mobilizing the glycogen stores in the liver (glycogenolysis) or by making glucose from fatty acids or amino acids (gluconeogenesis). Liver glucose production is important during fasting since the brain is crucially dependent on glucose. The fat stores of the body are mobilized under starvation to release fatty acids. Most other tissues can directly use fatty acids as fuel but the brain cannot. Glucose production is therefore most essential for brain function in fasting conditions. However, the glucose production activity needs to be under regulation, and insulin works as a negative regulator of this activity. Impaired insulin signaling in the liver leads to defective regulation of liver glucose production. Although studies have suggested that kidneys can contribute to some endogenous glucose production, the defective glucose production in T2D is primarily in the liver. Role of nervous system and glucose autoregulation of hepatic glucose production is also known but is currently considered to be less important. Insulin decreases endogenous glucose production by direct and indirect mechanisms. In its direct action, portal insulin suppresses glucose production by inhibiting glycogenolysis. Insulin decreases gluconeogenesis simultaneously by directly inhibiting the enzymes required for gluconeogenesis and also through indirect routes by reducing the availability of free fatty acids required for gluconeogenesis and by inhibiting the secretion of glucagon that stimulates glucose production [47–51]. In diabetes insulin is not able to exert an impact on liver, so there is uncontrolled hepatic glucose production [52–54] which leads to raised fasting blood sugar. This is called liver insulin resistance. When the fasting glucose level is marginally high, the contribution is mainly from decline in glucose clearance by muscle and other tissues. However, higher levels of fasting glucose are mainly contributed by hepatic glucose production.

Insulin Resistance: A Broader Picture

There is an increasing realization now that insulin resistance is not restricted to loss of insulin action. A large number of other alterations in hormonal, metabolic, and neuronal pathways accompany insulin resistance. Therefore the expanse of the meaning of the term "insulin resistance syndrome" is increasing day by day. Insulin resistance syndrome is also called "metabolic syndrome" or "syndrome X." Metabolic syndrome is a clinically defined condition with increasing prevalence in the modern world. To clinically identify patients with the metabolic syndrome, the National Cholesterol Education Program Expert Panel on Detection, Evaluation, and Treatment of High Blood Cholesterol in Adults (Adult Treatment Panel III, ATP III) suggested that individuals having three or more of the following criteria are defined as having the metabolic syndrome:

1. Abdominal obesity: waist circumference greater than 40 inches in men and 35 inches in women
2. Hypertriglyceridemia: >150 mg/dL (1.69 mmol/L)
3. Low high-density lipoprotein (HDL) cholesterol: <40 mg/dL (1.04 mmol/L) in men and <50 mg/dL (1.29 mmol/L) in women
4. High blood pressure: >130/85 mmHg
5. High fasting plasma glucose: >110 mg/dL (>6.1 mmol/L)

Although insulin resistance is not included in the clinical definition of metabolic syndrome, it is believed to be the main underlying abnormality. Apart from metabolic syndrome, a series of other disorders are associated closely with T2D and insulin resistance syndrome. They include polycystic ovary syndrome (PCOS), some other sexual dysfunctions, osteopenia, osteoporosis, and certain types of cancers. T2D leads to a variety of complications, some of which are specific to diabetes and others can occur in the absence of diabetes, but the incidence is much higher in diabetes (Table 2.4). It is likely that all the cluster of diseases have a common cause or a cluster of causes. It is believed that the common cause is insulin resistance.

Table 2.4 Diabetic complications

Type	Signs, symptoms, and pathology	Epidemiology	Other comments
Microvascular			
Ocular complications			
Retinopathy	Hemorrhages, microaneurysms, cotton wool spots, hard exudates, venous caliber abnormality, intraretinal hemorrhage, retinal detachment, loss of vision	Seen in more than 60% of diabetic patients	Most common cause of new-onset blindness
Nonproliferative diabetic retinopathy (NPDR)	Progressive pathological defects of mild to severe degree		
Proliferative diabetic retinopathy (PDR)	Vasoproliferation of retina includes new vessels on the optic disc (NVD), new vessels elsewhere on retina (NVE), preretinal hemorrhage (PRH), vitreous hemorrhage, and fibrous tissue proliferation (FP)		
Macular edema	Swelling of the macula, increased permeability, and leakage of fluid. Distortion of macula and blurring or distortion of vision		It may be focal or diffuse macular edema
Other ocular complications include mononeuropathies of third, fourth, and sixth nerve; optic disc pallor; neovascular glaucoma; open-angle glaucoma; corneal damage; and lens changes			
Diabetic nephropathy			
Renal artery stenosis	Atherosclerosis increases risk of renal artery stenosis. Hemodynamic changes, hypertension, pulmonary edema, and progressive impairment in renal function		ACE inhibitors lead to rapid restoration of renal functions
Renal papillary necrosis	Ischemia and medulla and papilla. Flank pain, hematuria, and fever		Ureter obstruction can occur
Diabetic neuropathy			
Subclinical neuropathy	No clinical manifestation. Abnormal electrodiagnostic tests		
Diffuse clinical neuropathy (2 types)			
Proximal motor neuropathies	Affects elderly, begins with pain in thighs, hips, or buttocks. Significant weakness of proximal muscles of lower limb	Incidence is greater in diabetic patients	Demyelination of nerves is seen

Distal symmetric polyneuropathy	Onset is insidious but occasionally acute. Pain and hyperalgesia in lower limbs	Most common and widely recognized form of diabetic neuropathy	
Focal neuropathy	Occur in older population. Acute onset, associated with pain. Course is self-limiting. Results from vascular obstruction		They differ from entrapment syndrome
Autonomic neuropathies	Can cause dysfunction of every part of the body. Complex and confusing symptoms. Often-involved organs are pupil, sweat glands, and genitourinary system. Gastrointestinal system, exercise intolerance, supine or nocturnal hypertension, edema, intolerance to heat. Silent myocardial infarction, respiratory failure		Strict glycemic control and stepwise progressive management of hyperglycemia, lipids, and BP is helpful
Pain in diabetic neuropathies (2 types)		In about 10%	
Acute painful neuropathy	Lasts <6 months. Also called as insulin neuritis		Associated with profound weight loss and severe depression
Chronic painful neuropathy	Lasts for 6 months to a year. There could be loss of sensory quality and spontaneous pain		Can develop tolerance to analgesic and narcotics
Macrovascular			
Cardiovascular disease			
Coronary heart disease	Blockage of coronaries due to atherosclerosis. Myocardial infarction, angina, and sudden death	Two- to threefold increase in the risk of atherosclerosis and cardiovascular disease	Other risk factors obesity, hypertension, dyslipidemia co-occur with T2D
The diabetic foot			
Diabetic foot ulcer	All the three components of neuropathy—sensory, motor, and autonomic—can contribute to ulceration in the foot. Chronic sensorimotor neuropathy is common. Combination of peripheral vascular disease with minor trauma can lead to ulceration	Approximately 5–10% of diabetic patients have had past or present foot ulceration	Patients with retinopathy and renal dysfunction are at increased risk for foot ulceration

Compensatory Hyperinsulinemia

Insulin resistance is accompanied by high levels of insulin production by the insulin-producing β cells of the islets of Langerhans of the pancreas. Insulin sensitivity of tissues and insulin secretion are related in such a way that their multiplication product is a constant. If insulin sensitivity goes down, insulin secretion increases in such a way that the product remains constant. As a result, even if insulin resistance increases, the levels of plasma glucose remain well controlled. The molecular mechanisms by which such a precise compensation takes place are not known. One possibility is that when insulin sensitivity decreases, the blood sugar increases, and increased blood sugar stimulates higher levels of insulin production. The dynamics of such a system can be easily worked out mathematically, and steady state solutions show that even if the insulin response is efficient, the steady state sugar level will be somewhat raised or remain oscillating. However, on the contrary, in early stages of insulin resistance, there is slight hypoglycemia instead of hyperglycemia. It is also known that in insulin-resistant state, the insulin responsivity is increased substantially such that normal levels of sugar elicit a substantially higher insulin response. It appears therefore that some other unknown mechanism increases the glucose responsiveness of β cells to mediate compensatory hyperinsulinemia. For any such mechanism to work, there needs to be some way of sensing the degree of insulin resistance and computing the requirement of insulin. No mechanism by which the pancreas can estimate the level of insulin resistance in muscle and other tissues is known. Yet there appears to be an almost universal agreement that there is precise compensation of insulin resistance by increased insulin secretion. This is perhaps the most mystic part of our current understanding of insulin resistance.

The increased insulin secretion in insulin resistance appears to be mediated by increased β cell mass [55] as well as increased expression of hexokinase which leads to increased secretion of insulin [56]. Beta cell insulin sensitivity varies substantially in individuals with normal glucose tolerance but is correlated to obesity [57, 58].

However, at a later stage in the natural history of insulin resistance, there is progressive degeneration of β cell mass and function. Beta cell failure in type 2 diabetes is a progressive process, which, regardless of the nature of the initial defect, gradually worsens over time. Chronically elevated blood glucose levels adversely affect insulin secretion. There is a well-documented loss of pancreatic β cell mass in type 2 diabetes that almost certainly contributes to the degree of hyperglycemia.

Gordon and Susan have described five stages of β cell failure [59]. (1) Stage 1 is compensation: Insulin secretion increases to maintain normoglycemia in the face of insulin resistance. This stage is characterized by maintenance of differentiated function with intact acute glucose-stimulated insulin secretion (GSIS). (2) Stage 2 occurs when glucose levels start to rise, reaching 5.0–6.5 mmol/l; this is a stable state of β cell adaptation with loss of β cell mass and disruption of function as evidenced by diminished GSIS and β cell dedifferentiation. (3) Stage 3 is a transient unstable period of early decompensation in which glucose levels rise relatively rapidly to the frank diabetes of stage 4. (4) Stage 4 is stable decompensation with more severe β cell dedifferentiation. (5) Finally, stage 5 is characterized by severe decompensation representing a profound reduction in β cell mass with progression to ketosis.

Beta cell dysfunction is necessary for the development of the disease, but the nature of the primary β cell defect is still elusive. Why and how β cells degenerate and become dysfunctional is not precisely known, but a number of possibilities exist. One of the earliest and classical interpretations was that the β cells get exhausted after giving a compensatory hyperinsulinemic response over a long time. The best piece of evidence for this hypothesis has been that exogenous insulin slightly improves β cell performance. This was interpreted as a demonstration of giving "rest" to β cells improving their function. This however fails to work in the long run, and patients chronically under insulin treatment do not appear to improve β cell function anymore. Therefore β cell exhaustion by itself does not appear to explain β cell degeneration. Nevertheless, in a restricted sense, β cell exhaustion is a useful concept and refers to depletion of the readily releasable pool of intracellular insulin following prolonged exposure to a secretagogue [60, 61]. This is certainly a reversible condition.

High level of glucose is implicated in β cell dysfunction, an effect commonly called glucotoxicity. The term glucotoxicity describes the slow and progressively irreversible effects of chronic hyperglycemia on pancreatic β cell function, which occur after prolonged exposure to elevated glucose. Once diabetes is established, chronic hyperglycemia and hyperlipidemia can exert deleterious effects on β cell function, respectively referred to as glucotoxicity and lipotoxicity [62]. Considerable evidence has been reported suggesting that chronic hyperglycemia impairs glucose-induced insulin secretion and insulin gene expression [63]. Glucotoxicity is certainly time course dependent. In the short run higher levels of glucose stimulate β cell proliferation and increase in mass. Only chronic hyperglycemia can result into deleterious effects on β cells. The fact that these associated β cell defects are reversible up until a certain point in time and become irreversible thereafter suggests a continuum between β cell exhaustion and glucotoxicity, the latter becoming predominant after prolonged exposure [64, 65]. In addition to inducing functional changes, chronic hyperglycemia can also decrease β cell mass by inducing apoptosis [55, 66].

Similar to the paradoxically deleterious effects of chronic hyperglycemia, fatty acids (FA), which are essential β cell fuels in the normal state, become toxic when chronically present in excessive levels. Prolonged exposure of pancreatic β cells to FA increases basal insulin release but inhibits glucose-induced insulin secretion [67]. At the molecular level, FAs inhibit insulin gene expression in the presence of elevated glucose levels. Finally, excessive FAs induce β cell death by apoptosis both in vitro and in ZDF rat islets. Fatty acids are demonstrated to cause β cell apoptosis. Elevated levels of circulating FFA and lipoproteins in obese ZDF rats are transported to islets and through a pathway of lipotoxicity leading to diabetes [55, 68]. This is associated with an increase in inducible nitric oxide synthase (iNOS), enzyme expression, and nitric oxide (NO) production, which cause apoptosis. Chronically elevated FA levels do not harm the β cell as long as blood glucose levels are normal but profoundly affect β cell function in the presence of concomitant hyperglycemia [55, 68]. These results clearly support the hypothesis that hyperglycemia is required for lipotoxicity to occur. They are consistent with the clinical observation that the majority of hyperlipidemic individuals are not diabetic. That β cell function is usually normal in patients with disorders of lipid metabolism suggests that obesity or dyslipidemia is not sufficient to cause β cell dysfunction [62].

There is a hen and egg problem in the glucotoxicity–lipotoxicity argument. As long as β cells are healthy and can compensate for insulin resistance, hyperglycemia is unlikely, and if glucotoxicity is primarily responsible for β cell dysfunction, then β cells are unlikely to degenerate. However if β cell degeneration begins due to some other cause or glucose levels rise chronically due to some other cause, then glucotoxicity and lipotoxicity can exert a snowballing effect. That "some other cause" could be crucial to the beginning of diabetes.

Beta cells have been shown to be highly susceptible to oxidative damage [69, 70], and this is likely to be a promising hypothesis for progressive β cell degeneration. Beta cell amyloid deposition has been demonstrated to be associated with obesity and hyperglycemia and appears to be another promising candidate in explaining islet degeneration. Amyloid deposition is not naturally observed in mice. Transgenic mice expressing human amylin show amyloid deposits under certain conditions. These conditions are described by Verchere et al. [71] and other groups [72, 73]. A high-fat diet was the major precipitating factor in the early Seattle studies [71]. High-fat feeding led to significant obesity (doubling in weight), and the transgenic males showed pancreatic islet amyloid deposits and an increased incidence of hyperglycemia when compared with controls. An interesting finding was that a significant number of nonhyperglycemic transgenic males also demonstrated amyloid deposits. So the association of amyloid deposits with hyperglycemia was not very strong.

Patients with type 2 diabetes also demonstrate disproportionate levels of circulating proinsulin, the precursor of insulin and its processing intermediate des-31, 32 proinsulin. These levels are proportional to the degree of hyperglycemia and inversely related to the functional measure [74]. Based on the findings of amyloid and disproportionate hyperproinsulinemia in these two patient populations, it has been postulated that disproportionate hyperproinsulinemia is a marker for the presence of amyloid or amyloid fibrils [75]. If impaired early phase insulin response and greater proportion of proinsulin are taken as markers of β cell dysfunction, then the dysfunction appears to begin very early in the development of T2D, and therefore, hyperglycemia is most unlikely to be responsible for β cell dysfunction. In a nutshell although progressive β cell dysfunction and degeneration are essential for the development of T2D, the precise cause of β cell dysfunction is not yet clearly known.

Beyond Hyperglycemia

Although diabetes is mainly defined and diagnosed by defective regulation of blood sugar, it is not the only defect in the body mechanisms. Diabetes affects many and perhaps every system of the body. Classical thinking is that once the control on glucose homeostasis is lost, the raised levels of glucose along with defective insulin signaling bring about all other pathological changes in different systems of the body. The main common system level effects appear to be:

1. *A low-grade chronic systemic inflammation*. Obesity leads to an overall upregulation of the basal levels of many inflammatory markers and proinflammatory cytokines which alter the migration of phagocytes important in innate immunity. The relationship between obesity, insulin resistance, and inflammation is complex, and it is not clear whether insulin resistance is a cause or effect of inflammation. Nevertheless the association is very strong.
2. *Dysfunction of the vascular endothelium*. Normal functions of endothelial cells, which form the inner lining of blood vessels, include mediation of coagulation, platelet adhesion, immune function, and control of volume and electrolyte content of the intravascular and extravascular spaces. They control the vasodilatation and vasoconstriction responses of blood vessels. The vasodilatation function is mediated through the nitric oxide pathway. In diabetes the delicate balance between vasodilatation and vasoconstriction regulation mechanisms is disturbed.
3. *A generalized oxidative stress*. Reactive oxygen species are molecules such as peroxides that have a highly reactive oxygen moiety. ROS are formed normally during oxidative metabolism and have some signaling functions in normal physiology. High levels of intracellular glucose can increase ROS levels which can cause oxidative damage to proteins and other molecules. This has a logical implication. Since insulin resistance restricts glucose entry in cells, it can protect from oxidative stress. Insulin resistance has been claimed to be a mechanism of protection from free radicals [76]. However the body also has insulin-independent tissues, and these tissues could be mainly susceptible to oxidative damages.

The major detrimental effects of diabetes are brought about by a combination of the three and possibly more pathophysiological processes. Interestingly although the origin of type 1 and type 2 diabetes is quite different, the downstream pathophysiological processes are highly overlapping. The complications accompanying T1D and T2D are largely overlapping despite some differences in the proportion of incidence of some of the complications. The fatal effects of diabetes are through one or more of these complications. Since the common grounds on which type 1 and type 2 diabetes converge is hyperglycemia, it is thought that hyperglycemia is the driver of all pathophysiological processes in both types of diabetes. The precise pathways by which raised blood glucose leads to the complications are not clearly elucidated, but there are many alternative possibilities which are not mutually exclusive:

1. *Increased AGE formation.* Advanced glycation end products (AGE) are glycated proteins which are formed at a slow rate nonenzymatically either by reaction of proteins with glucose or with auto-oxidation products of glucose such as glyoxal or methylglyoxal. The AGEs appear to initiate a number of pathological processes such as production of reactive oxygen species (ROS), activating the nuclear factor kappa B (NFκB), altering gene expression, and initiating inflammatory changes. The effects of AGEs are mediated through specific receptors called RAGE. AGEs decrease elasticity in blood vessels. In the kidneys AGE formation on glomerular basement membrane increases its permeability to albumin initiating nephropathic changes. Blockade of RAGE inhibits many of the diabetic complications and enhances wound healing.
2. *Increased polyol pathway flux.* Aldose reductase, a low-affinity enzyme, diverts glucose to the polyol pathway. Under normal glucose concentrations, negligible quantities of glucose flux are through this pathway, but under increased intracellular glucose concentrations, there is an increased flux through the polyol pathway. This is thought to result into sorbitol-induced osmotic stress and other metabolic alterations. Aldose reductase is inactivated by nitric oxide, and decreased nitric oxide in diabetes is likely to be responsible for activating the polyol pathway flux.
3. *Activation of protein kinase C.* Triggered by metabolic intermediates in the glucose oxidation pathway, certain isoforms of protein kinases are abnormally activated. These are implicated mainly in decreased production of NO.
4. *Increased hexosamine pathway flux.* When intracellular glucose load increases, a significant part of it is shunted through the hexosamine pathway resulting into an increase in Sp1 O-linked *N*-acetylglucosamine (GlcNAc). Since *O*-GlcNAcylation is important in RNA polymerase, transcription of a number of genes can be affected. Of particular relevance is the inhibition of endothelial nitric oxide synthase (eNOS), which ultimately results into decreased NO activity.

Although a large number of different pathways are triggered by hyperglycemia, they seem to converge at two points, namely, decrease in NO production and increase in ROS production. One perplexing fact however is that all the effects of hyperglycemia outlined above are reversible and normalize quickly on normalization of glucose levels. In spite of this normalization, the clinical effects of diabetes remain long lasting. In a somewhat mystic phenomenon called hyperglycemic memory, pathological effects of hyperglycemia are observed even after normalization of blood sugar. What causes these effects is not clearly known, but this phenomenon suggests that there may be some other pathological processes in diabetes which are not yet discovered or appreciated.

Currently the treatment of T2D aims at normalizing blood sugar level as far as possible. Appropriate diet and lifestyle changes are advocated to all patients with T2D. In addition the pharmaceutical treatment takes one of the three lines or a combination of these, the first being use of insulin-sensitizing drugs, second being efforts to increase pancreatic insulin secretion, and the last being infusion of exogenous insulin. None of the treatments appear to "cure" diabetes. Once diabetes is confirmed, the patient remains diabetic for life. The treatment helps keep the sugar levels in "control" for some time. A general pattern with any antidiabetic treatment is that initially it gives dramatic results in that the blood sugar reduces rapidly and approaches normal. However with continued treatment in a variable length of time, the control is gradually lost, and the treatment becomes ineffective in the long run, necessitating change in drug or dose. Insulin is generally the last resort. In long-standing diabetics often high doses of insulin too are unable to normalize blood sugar. Therefore not only the treatment is unable to cure diabetes but in a substantial proportion of patients even unable to keep sugar levels under control.

There are two steps involved in being effective in curing a disorder. One is to understand the basic biology behind it, and the other is to develop the right technology to attack the root cause. Where does the problem lie in the case of diabetes? Has science failed to detect the root cause of diabetes or has technology failed to attack it or is the condition irreversible by nature? If it is irreversible exactly what part of it is irreversible? If not why have we failed to reverse it so far? These are the questions that we need to elaborate sufficiently now.

References

1. Kronenberg H, Williams RH (2008) Williams textbook of endocrinology. Saunders/Elsevier, Philadelphia, PA
2. Pickup JC, Williams G (2003) Textbook of diabetes. Blackwell Science, Oxford
3. Guyton AC, Hall JE (2006) Textbook of medical physiology. Saunders, Philadelphia, PA
4. *Leçons de physiologie opératoire: Bernard, Claude, 1813–1878: Free Download & Streaming: Internet Archive*. http://www.archive.org/details/leonsdephysio 100bernard
5. Monzillo LU, Hamdy O (2003) Evaluation of insulin sensitivity in clinical practice and in research settings. Nutr Rev 61:397–412
6. Zimmet P, Alberti KGMM, Shaw J (2001) Global and societal implications of the diabetes epidemic. Nature 414:782–787
7. Goran MI, Ball GDC, Cruz ML (2003) Obesity and risk of type 2 diabetes and cardiovascular disease in children and adolescents. J Clin Endocrinol Metab 88:1417–1427
8. Diamond J (2003) The double puzzle of diabetes. Nature 423:599–602
9. Cowie CC et al (2006) Prevalence of diabetes and impaired fasting glucose in adults in the U.S. population. Diabetes Care 29:1263–1268
10. Mokdad AH et al (2001) The continuing epidemics of obesity and diabetes in the United States. J Am Med Assoc 286:1195–1200
11. Harris MI, Eastman RC, Cowie CC, Flegal KM, Eberhardt MS (1997) Comparison of diabetes diagnostic categories in the U.S. population according to 1997 American Diabetes Association and 1980–1985 World Health Organization diagnostic criteria. Diabetes Care 20:1859–1862
12. Warram JH, Martin BC, Krolewski AS, Soeldner JS, Kahn CR (1990) Slow glucose removal rate and hyperinsulinemia precede the development of type II diabetes in the offspring of diabetic parents. Ann Intern Med 113:909–915
13. Haffner SM et al (1990) Diminished insulin sensitivity and increased insulin response in nonobese, nondiabetic Mexican Americans. Metab Clin Exp 39:842–847
14. Groop L (2000) Genetics of the metabolic syndrome. Br J Nutr 83(Suppl 1):S39–S48
15. Lehtovirta M et al (2000) Insulin sensitivity and insulin secretion in monozygotic and dizygotic twins. Diabetologia 43:285–293
16. Mayer EJ et al (1996) Genetic and environmental influences on insulin levels and the insulin resistance syndrome: an analysis of women twins. Am J Epidemiol 143:323–332
17. Hong Y, Pedersen NL, Brismar K, de Faire U (1997) Genetic and environmental architecture of the features of the insulin-resistance syndrome. Am J Hum Genet 60:143–152
18. Fujioka S, Matsuzawa Y, Tokunaga K, Tarui S (1987) Contribution of intra-abdominal fat accumulation to the impairment of glucose and lipid metabolism in human obesity. Metab Clin Exp 36:54–59
19. Brambilla P et al (1994) Peripheral and abdominal adiposity in childhood obesity. Int J Obes Relat Metab Disord 18:795–800
20. Berman DM et al (2001) Racial disparities in metabolism, central obesity, and sex hormone-binding globulin in postmenopausal women. J Clin Endocrinol Metab 86:97–103
21. Hu FB et al (2001) Diet, lifestyle, and the risk of type 2 diabetes mellitus in women. N Engl J Med 345:790–797
22. Tuomilehto J et al (2001) Prevention of type 2 diabetes mellitus by changes in lifestyle among subjects with impaired glucose tolerance. N Engl J Med 344:1343–1350
23. Must A et al (1999) The disease burden associated with overweight and obesity. J Am Med Assoc 282:1523–1529
24. Després JP, Tremblay A, Pérusse L, Leblanc C, Bouchard C (1988) Abdominal adipose tissue and serum HDL-cholesterol: association independent from obesity and serum triglyceride concentration. Int J Obes 12:1–13
25. Larsson B et al (1984) Abdominal adipose tissue distribution, obesity, and risk of cardiovascular disease and death: 13 year follow up of participants in the study of men born in 1913. Br Med J (Clin Res Ed) 288:1401–1404
26. Landin K, Krotkiewski M, Smith U (1989) Importance of obesity for the metabolic abnormalities associated with an abdominal fat distribution. Metab Clin Exp 38:572–576
27. Bronnegard M, Arner P, Hellstorm L, Akner G, Gustafsson J-Å (1990) Glucocorticoid receptor messenger ribonucleic acid in different regions of human adipose tissue. Endocrinology 127:1689–1696
28. Nicklas BJ, Rogus EM, Colman EG, Goldberg AP (1996) Visceral adiposity, increased adipocyte lipolysis, and metabolic dysfunction in obese postmenopausal women. Am J Physiol 270:E72–E78

29. Mittelman SD et al (2000) Longitudinal compensation for fat-induced insulin resistance includes reduced insulin clearance and enhanced β-cell response. Diabetes 49:2116–2125
30. Paolisso G et al (1995) A high concentration of fasting plasma non-esterified fatty acids is a risk factor for the development of NIDDM. Diabetologia 38:1213–1217
31. Charles MA et al (1997) The role of non-esterified fatty acids in the deterioration of glucose tolerance in Caucasian subjects: results of the Paris Prospective Study. Diabetologia 40:1101–1106
32. Pan DA et al (1997) Skeletal muscle triglyceride levels are inversely related to insulin action. Diabetes 46:983–988
33. Carlson LA, Ekelund LG, Fröberg SO (1971) Concentration of triglycerides, phospholipids and glycogen in skeletal muscle and of free fatty acids and β-hydroxybutyric acid in blood in man in response to exercise. Eur J Clin Invest 1:248–254
34. Laws A, Reaven GM (1990) Effect of physical activity on age-related glucose intolerance. Clin Geriatr Med 6:849–863
35. Gollnick PD, Saltin B (1982) Significance of skeletal muscle oxidative enzyme enhancement with endurance training. Clin Physiol 2:1–12
36. Turcotte LP, Richter EA, Kiens B (1992) Increased plasma FFA uptake and oxidation during prolonged exercise in trained vs. untrained humans. Am J Physiol 262:E791–E799
37. Romijn JA, Klein S, Coyle EF, Sidossis LS, Wolfe RR (1993) Strenuous endurance training increases lipolysis and triglyceride-fatty acid cycling at rest. J Appl Physiol 75:108–113
38. Kelley DE, Simoneau JA (1994) Impaired free fatty acid utilization by skeletal muscle in non-insulin-dependent diabetes mellitus. J Clin Invest 94:2349–2356
39. Kelley DE, Mandarino LJ (1990) Hyperglycemia normalizes insulin-stimulated skeletal muscle glucose oxidation and storage in noninsulin-dependent diabetes mellitus. J Clin Invest 86:1999–2007
40. Ruderman NB, Saha AK, Vavvas D, Witters LA (1999) Malonyl-CoA, fuel sensing, and insulin resistance. Am J Physiol 276:E1–E18
41. Dean D et al (2000) Exercise diminishes the activity of acetyl-CoA carboxylase in human muscle. Diabetes 49:1295–1300
42. Hancock CR et al (2008) High-fat diets cause insulin resistance despite an increase in muscle mitochondria. Proc Natl Acad Sci USA 105:7815–7820
43. Knauf C et al (2008) Brain glucagon-like peptide 1 signaling controls the onset of high-fat diet-induced insulin resistance and reduces energy expenditure. Endocrinology 149:4768–4777
44. Knauf C et al (2005) Brain glucagon-like peptide-1 increases insulin secretion and muscle insulin resistance to favor hepatic glycogen storage. J Clin Invest 115:3554–3563
45. Perrin C, Knauf C, Burcelin R (2004) Intracerebroventricular infusion of glucose, insulin, and the adenosine monophosphate-activated kinase activator, 5-aminoimidazole-4-carboxamide-1-β-D-ribofuranoside, controls muscle glycogen synthesis. Endocrinology 145:4025–4033
46. Darleen S (2008) CNS GLP-1 regulation of peripheral glucose homeostasis. Physiol Behav 94:670–674
47. Båvenholm PN, Pigon J, Östenson C-G, Efendic S (2001) Insulin sensitivity of suppression of endogenous glucose production is the single most important determinant of glucose tolerance. Diabetes 50:1449–1454
48. Mitrakou A et al (1992) Role of reduced suppression of glucose production and diminished early insulin release in impaired glucose tolerance. N Engl J Med 326:22–29
49. Rebrin K, Steil GM, Mittelman SD, Bergman RN (1996) Causal linkage between insulin suppression of lipolysis and suppression of liver glucose output in dogs. J Clin Invest 98:741–749
50. Mittelman SD, Fu YY, Rebrin K, Steil G, Bergman RN (1997) Indirect effect of insulin to suppress endogenous glucose production is dominant, even with hyperglucagonemia. J Clin Invest 100:3121–3130
51. McCall RH, Wiesenthal SR, Shi ZQ, Polonsky K, Giacca A (1998) Insulin acutely suppresses glucose production by both peripheral and hepatic effects in normal dogs. Am J Physiol 274:E346–E356
52. Hother-Nielsen O, Beck-Nielsen H (1991) Insulin resistance, but normal basal rates of glucose production in patients with newly diagnosed mild diabetes mellitus. Acta Endocrinol 124:637–645
53. Groop LC, Bonadonna RC, Shank M, Petrides AS, DeFronzo RA (1991) Role of free fatty acids and insulin in determining free fatty acid and lipid oxidation in man. J Clin Invest 87:83–89
54. Hother-Nielsen O, Beck-Nielsen H (1990) On the determination of basal glucose production rate in patients with type 2 (non-insulin-dependent) diabetes mellitus using primed-continuous 3-3 H-glucose infusion. Diabetologia 33:603–610
55. Pick A et al (1998) Role of apoptosis in failure of β-cell mass compensation for insulin resistance and β-cell defects in the male Zucker diabetic fatty rat. Diabetes 47:358–364
56. Cockburn BN et al (1997) Changes in pancreatic islet glucokinase and hexokinase activities with increasing age, obesity, and the onset of diabetes. Diabetes 46:1434–1439
57. Kahn SE et al (1993) Quantification of the relationship between insulin sensitivity and β-cell function in human subjects. Evidence for a hyperbolic function. Diabetes 42:1663–1672
58. Toffolo G, Bergman RN, Finegood DT, Bowden CR, Cobelli C (1980) Quantitative estimation of β cell sensitivity to glucose in the intact organism: a minimal model of insulin kinetics in the dog. Diabetes 29:979–990

59. Weir GC, Bonner-Weir S (2004) Five stages of evolving B-cell dysfunction during progression to diabetes. Diabetes 53:S16–S21
60. Sako Y, Grill VE (1990) Coupling of β-cell desensitization by hyperglycemia to excessive stimulation and circulating insulin in glucose-infused rats. Diabetes 39:1580–1583
61. Leahy JL, Bumbalo LM, Chen C (1994) Diazoxide causes recovery of β-cell glucose responsiveness in 90% pancreatectomized diabetic rats. Diabetes 43:173–179
62. Poitout V, Robertson RP (2002) Minireview: secondary β-cell failure in type 2 diabetes – a convergence of glucotoxicity and lipotoxicity. Endocrinology 143:339–342
63. Robertson RP, Harmon JS, Tanaka Y, Trang PO, Poitout V (2004) Glucose toxicity of the β-cell cellular and molecular mechanisms. In: LeRoith D, Taylor SI, Olefsky J (eds) Diabetes mellitus: a fundamental and clinical text. Lippincott Williams and Wilkins, Philadelphia, PA
64. Gleason CE, Gonzalez M, Harmon JS, Robertson RP (2000) Determinants of glucose toxicity and its reversibility in the pancreatic islet β-cell line, HIT-T15. Am J Physiol Endocrinol Metab 279:E997–E1002
65. Moran A et al (1997) Differentiation of glucose toxicity from β cell exhaustion during the evolution of defective insulin gene expression in the pancreatic islet cell line, HIT-T15. J Clin Invest 99:534–539
66. Donath MY, Gross DJ, Cerasi E, Kaiser N (1999) Hyperglycemia-induced β-cell apoptosis in pancreatic islets of Psammomys obesus during development of diabetes. Diabetes 48:738–744
67. McGarry JD, Dobbins RL (1999) Fatty acids, lipotoxicity and insulin secretion. Diabetologia 42:128–138
68. Shimabukuro M et al (1998) Lipoapoptosis in β-cells of obese prediabetic fa/fa rats. Role of serine palmitoyltransferase overexpression. J Biol Chem 273:32487–32490
69. Robertson RP (2004) Chronic oxidative stress as a central mechanism for glucose toxicity in pancreatic islet B cells in diabetes. J Biol Chem 279: 42351–42354
70. Modak MA, Parab PB, Ghaskadbi SS (2009) Pancreatic islets are very poor in rectifying oxidative DNA damage. Pancreas 38:23–29
71. Verchere CB et al (1996) Islet amyloid formation associated with hyperglycemia in transgenic mice with pancreatic B cell expression of human islet amyloid polypeptide. Proc Natl Acad Sci USA 93:3492–3496
72. Höppener JW et al (1999) Extensive islet amyloid formation is induced by development of Type II diabetes mellitus and contributes to its progression: pathogenesis of diabetes in a mouse model. Diabetologia 42:427–434
73. Soeller WC et al (1998) Islet amyloid-associated diabetes in obese Avy/A mice expressing human islet amyloid polypeptide. Diabetes 47:743–750
74. Rørder ME, Porte D Jr, Schwartz RS, Kahn SE (1998) Disproportionately elevated proinsulin levels reflect the degree of impaired B cell secretory capacity in patients with noninsulin-dependent diabetes mellitus. J Clin Endocrinol Metab 83:604–608
75. Porte D Jr, Kahn SE (1989) Hyperproinsulinemia and amyloid in NIDDM. Clues to etiology of islet β-cell dysfunction? Diabetes 38:1333–1336
76. Hoehn KL et al (2009) Insulin resistance is a cellular antioxidant defense mechanism. Proc Natl Acad Sci USA 106:17787–17792

3 Diabetes in an Undergraduate Class

I am an undergraduate teacher and that is what my first love has been. Over the years, learning from others as well as by my own experience, I could figure out that the best method to teach concepts in science to the young minds is the method of doubt and challenge. Doubting and challenging everything that a textbook has or a class teacher says pay large dividends. This way, students learn inferential logic, nature of evidence, experimental designs, hypothesis testing, alternative interpretations, resolving between alternative interpretations, and the required openness of mind to do so in a much more robust and sound way than the conventional "textbook as bible" approach. In the doubt and challenge approach, students are encouraged to contest prevalent concepts and come up with alternative interpretations of the same experiments and data on which the textbook concepts are based. Further, if any of their alternative interpretations look promising even at a speculative level, they are encouraged to suggest experiments that could distinguish between the conventional and alternative explanations of the phenomenon. In order to satisfy ourselves we then have to go beyond textbooks and explore what is happening in that field currently, see whether the experiments that we visualized have been done or are being done anywhere and, if so, what they imply. Most often, the textbook "wins," but not uncommonly we identify significant gaps in the knowledge, some logical inconsistencies that need to be addressed, many unanswered questions, and more frequently many unquestioned answers! It is just too common to find that although the frontiers of research are expanding rapidly, some simple fundamental questions remain not only unanswered but knowingly or unknowingly trivialized and ignored.

The method of doubt and challenge used in undergraduate class has shaped my thinking permanently in such a way that when I got interested in the evolutionary origins of diabetes and related disorders, I found myself extensively using the method of doubt and challenge while trying to read basic textbooks on physiology and pathology of diabetes. Here I was playing the teacher as well as the student myself. Very soon I found myself surprised and amazed by the number of logical inconsistencies and paradoxes present in this field. For any teacher using the method of doubt, it is not uncommon to find some or the other problem in every textbook chapter. But the problems with diabetes appear to be unmatched both qualitatively and quantitatively. Fortunately or unfortunately, I never formally taught the physiology or pathology of diabetes to any class any time. I will try to do that in this chapter now playing both student and teacher myself. As in the classroom, the basic intention here is to develop an insightful understanding. Critical thinking reinforces concepts that are quite sound, refines the ones that are not so much, and makes us rethink about the ones that are not. Only in rare cases the classroom method of doubt turns something upside down. The teacher in me is generally convinced that the textbook is correct most of the times and therefore defends the textbook, whereas the student in me is skeptical.

M. Watve, *Doves, Diplomats, and Diabetes: A Darwinian Interpretation of Type 2 Diabetes and Related Disorders*, DOI 10.1007/978-1-4614-4409-1_3, © Springer Science+Business Media New York 2013

Let us look at the popular textbook portrait of type 2 diabetes from the point of view of this student–teacher duo.

We can best begin by looking at the five classical postulates of T2D based on the textbook picture. Much research in the last two decades has indicated the necessity to drift away from the classical picture, but in order to understand the necessity to drift, we need to take a critical look at the classical picture first:

1. Obesity is a consequence of positive energy balance, that is, the net energy intake being greater than the net energy expenditure.
2. Obesity is the main cause of insulin resistance.
3. The body tries to compensate insulin resistance by making the pancreatic β cells produce more insulin.
4. There is a progressive degeneration of β cells presumably due to "exhaustion" resulting in relative insulin deficiency, that is, insulin levels being short of compensating for insulin resistance. A combination of insulin resistance and relative insulin insufficiency (IR–RII) results in hyperglycemia.
5. The chronically increased blood sugar levels lead to all other pathophysiological consequences of diabetes.

The sequence appears very logical and convincing at a glance. We have seen this in moderate details in the last chapter. Now, we will start looking at the detailed mechanisms and evidence using the method of doubt and challenge in the hypothetical classroom.

1. Energy balance and obesity: The baseline assumption of obesity research is the energy balance equation. It simply says that if the net energy intake of the body is greater than the net energy consumption, body mass will increase and such a continued imbalance over a sufficiently long time will lead to obesity [1]. The equation is nothing but fundamental thermodynamics and therefore cannot be doubted. What the skeptic student wonders is not whether the energy balance equation is right or wrong, but whether it says anything useful about obesity. It is like answering the question "why did you fail in the exams?" by saying "because I got less than the passing marks." This is only saying the same thing in different words, not reasoning out. So the student is not convinced that the energy balance equation "explains" obesity. The student's dissatisfaction is not new. Inadequacy of the thermodynamic energy balance in explaining obesity is recognized from time to time [2, 3]. The teacher, however, knows that the usefulness of the energy balance equation lies in defining the question in more objective terms, not in answering the question. Therefore it is not good to be satisfied by the energy balance equation but at the same time not prudent to discard it. The balance equation forms the platform on which real questioning can begin. In modern lifestyle, food intake has increased, and physical activity has come down resulting in a positive shift in the energy balance. This sounds temptingly simple and logical to believe. In literature, there appears to be no agreement on whether it is increased food intake or decreased activity that is primarily responsible for obesity [4]. The available evidence is contradictory. Experiments with doubly labeled water that measure the total energy expenditure of the body have demonstrated that modern lifestyle has not changed the mean total energy expenditure sufficiently to account for the obesity epidemic [5, 6]. This implies that our energy intake must have gone up. The increased fat content of food is often blamed to be responsible for obesity. This is the most dominating line of thinking in the antiobesity drive. The drive for low-fat diet has resulted in a rapidly growing market for low-fat foods in America, for example, but the increased consumption of low-fat foods has not resulted in curbing obesity [7]. In stark contrast lies the example of the East African Masai tribe whose normal diet has a high proportion of animal fat (60–70% of total energy intake), but obesity, insulin resistance, and hypercholesterolemia are extremely rare among them [8–10].

Perhaps, the teacher tries to defend; in the example of the Masai tribesmen, the energy expenditure is more important than energy intake. The importance of gluttony and sloth could be different in different populations leading to contradictory results across studies. But whether increased energy consumption or decreased expenditure is responsible for the obesity epidemic is not the most important question. Let us assume that both are happening simultaneously resulting in positive energy balance. The real question is what makes the positive energy balance possible when the body has a wide variety of mechanisms that regulate energy intake. Do we eat more than what our body requires only because more is available? Is our energy intake regulated only by how much is available? There is no doubt that the availability of food has increased in many modern human societies [11], and that is believed to be the sole reason for the obesity epidemic by some researchers [12]. The belief that food availability alone drives obesity contradicts the large body of evidence for a variety of elaborate mechanisms of energy homeostasis regulating food intake. A large number of different mechanisms of the body regulate food intake as well as fat storage [13–16]. Signals for regulation for food intake are generated starting from the taste buds and continuing to further levels including distension of the stomach, chemical signals in the upper and lower parts of the gastrointestinal tract, blood levels of sugars and amino acids, insulin signaling, leptin–adiponectin and other signals by adipose tissue, glucose sensing in the portal vein, and glucose sensing neurons in the brain that lead to neuronal circuits regulating food intake (Table 3.1).

History of experimental investigations on food intake control is quite old and interesting. In some of the early rat experiments where there was an open esophageal fistula that drained out the food taken in so that no food entered the stomach, the animals stopped eating after a certain meal size. This indicates that some food intake regulation mechanisms exist even before the food reaches the stomach [17]. The meal size in such sham feeding trials is somewhat larger and the intermeal interval shorter, indicating that sham feeding is unable to induce the same satiety response as that obtained by a distended stomach [17–19]. Nevertheless some regulation does work at the oropharynx level too and is influenced by and interacts with other food-related hormonal signals such as insulin and cholecystokinin (CCK) as well as neuronal processes [17, 20–23]. The stomach level of control was shown to work independent of taste buds. In these experiments, rats were fed liquid diet through a tube bypassing the oropharynx. The animals themselves controlled food intake by pressing a lever for which they were adequately trained. Although these animals had easy, automated, and unlimited access to food, they did not grow obese [24], demonstrating that easy availability of food was not enough. Experiments that excluded both oral and gastric influences and directly introduced food in the jejunum via a fistula demonstrated that regulation mechanisms exist at the level of the intestine independent of oral and gastric signals [25]. This demonstrates that different levels of food intake regulation are at least partly independent of each other and one can function in the absence of the other. Today a large number of molecular and neuronal mechanisms of regulation of energy intake are known that act at acute or chronic level (Table 3.1). It is difficult to perceive how such a system with a series of backups has no role and only the availability of food decides food intake. Therefore there must be something beyond greater availability of food that is responsible for increased food intake.

In the context of the modern obesity epidemic, one should be able to explain why the regulation mechanisms appear to be failing today. Can it be explained by any defect in the regulation mechanisms? Since the energy homeostasis mechanisms are multileveled and independent, if one system fails, others are there to compensate. Therefore a single defect is most unlikely to cause overeating. This is indeed true since the total number of cases in which obesity could be traced back to a single genetic defect is surprisingly small (about 200) [26] in the entire history of medicine and human genetics. On the one hand, the system is difficult to get perturbed by any single defect, and on the other it is most unlikely that many or most of the regulation mechanisms will develop a mutation or an acquired defect simultaneously in a large proportion of the population. Therefore such a large fraction of the population getting obese in spite of the presence of

Table 3.1 Mechanisms of food intake regulation

	Signal	Site of production	Stimulus	Site of action	Action on appetite
	Gut and pancreatic signals (short-term regulation)				
1	Sensory	Oropharynx	Food taste, swallowing	?	↓
2	Physical	Stomach	Distension of stomach	?	↓
3	Insulin	Pancreas	Plasma glucose	ARC, DMH, PVN	↓
4	Pancreatic polypeptide (PP)	Pancreas	Food intake	Area postrema	↓
5	PYY	L cells of GIT	Food intake, particularly fat	Peripheral, arcuate NPY neurons	↓
6	Ghrelin	Gastric oxyntic cells, other parts of GIT	Circadian, stimulated by food anticipation, suppressed by food intake	Hypothalamic ARC	↑
7	GLP-1	L cells of intestine	Food intake	Pancreas, CNS	↓
8	Oxyntomodulin (OXM)	L cells of intestine	Circadian, food intake	Hypothalamus	↓
9	Cholecystokinin (CCK)	Duodenum and jejunum of GIT, CNS	Food intake, aggression, anxiety	CNS—NTS, DMH, area postrema	↓
10	Bombesin	Gut	Food intake	CNS	↓
	Adipocity signals (long-term regulation)				
11	Leptin	Adipose tissue	Increased adiposity	POMC, CART neurons	↓
12	Adiponectin	Adipose tissue	?	Hypothalamus, skeletal muscle	↓, increased energy expenditure
	Central nervous system				
13	Neuropeptide Y	ARC	Fasting	Hypothalamus	↑
14	Melanocortin–POMC products	Hypothalamus	Food intake, insulin, leptin	Hypothalamus—ARC, VMH, and PVN	↓
15	Melanocortin, Agouti, and AgRP	Peripheral, CNS	Fasting	ARC	↑
16	CART	ARC, LHA, PVN	Leptin, fear–anxiety	Hypothalamus	↓
17	Serotonin	Brain	Food and other satiation responses	ARC	↓
18	Glucose sensing neurons	Hypothalamus and other parts of brain	Glucose levels in the brain	Hypothalamus	↓↑
19	Reward circuits opioid, endocannabinoid, dopamine	Brain	Food and other "rewards"	Hypothalamus NAc	↓↑specifically high-energy food

Note: This is unlikely to be a complete list of molecules and mechanisms involved in the regulation of food intake. The table is only intended to illustrate that a large number of mechanisms exist that regulate food intake. Interestingly, there are more anorectic than appetizing molecules. See [13] for a detailed account. Levels of backup and apparent redundancy in these mechanisms are demonstrated by the fact that knockouts of many of these factors have no phenotype. One needs to wonder therefore how one can evade all these mechanisms in order to become obese.

? Stands for no information.

multilevel food regulation mechanisms that evolved and worked well over millennia is an unresolved riddle. Increased availability and security of food is not a sufficient cause. In case it is, one still needs to explain why the regulation mechanisms fail.

A large number of studies have focused on the molecular mechanisms of energy homeostasis, and many of them hope to develop an effective antiobesity drug [16]. There is a major missing link in the whole of this research. Energy homeostasis is not the only thing that regulates food intake. There are many ecological, social, and behavioral factors that are equally important determinants of eating behavior. For example, after a successful hunt, a mother cheetah may not eat in spite of being hungry if feeding the cubs is a priority. A number of animals store and/or hide food. Such an animal would recover the cached food when hungry. However, it may refrain from doing so if another strong and dominant competitor is watching since there is a risk of the food being snatched away. Foraging for food may increase risk of predation, and an animal may choose to stop foraging and eating at much lower levels of satiety in the presence of a predator than in its absence. Not much research has addressed how these behavioral signals interact with the energy homeostasis signals at the molecular and neuronal level. This appears to be due to a tradition in which biochemists generally do not think of behavior as a prime driver of biochemical regulation. As a result, the mainstream thinking appears to have stuck at the narrow vision of energy homeostasis and energy balance equation. The possible effects of behavioral inputs in metabolic regulation have not been adequately considered. But there is a reason to suspect that obesity may be primarily a result of a change in the behavioral components rather than energy homeostasis mechanisms. In the last few decades in which obesity attained epidemic proportions, human ecology, social structure, and behavior have changed substantially, whereas there is no apparent reason why the biological mechanisms of energy homeostasis that have evolved and worked for thousands of generations may suddenly have changed in such a large fraction of the population in a short time. Therefore there is a need to seriously rethink the energy balance and energy homeostasis paradigm. The roots of obesity may lie in social structure and behavior rather than diet. This needs alternative thinking, and we will reserve it for another chapter ahead. For the time being, let us continue to critically examine the textbook picture.

The current thinking about diet composition is closely linked to the obesity centered view. Therefore it would be appropriate to consider it here itself. A number of studies focus on dietary fat rather than accumulated fat as an inducer of insulin resistance [27–29]. Whether high-fat diet or high-carbohydrate diet is linked to obesity and insulin resistance syndrome is highly debated. There has been a fat school and a carbohydrate school, and both have some data in support and possible mechanisms to explain the contribution to obesity [30–40]. A less known possibility is that even high-protein diet would promote insulin resistance. This is logically very sound because the brain needs glucose as the source of energy and it cannot utilize fatty acids or amino acids catabolically. When dietary carbohydrate intake is negligible, the liver needs to make sufficient glucose for the brain and insulin resistance promotes liver gluconeogenesis. Therefore high-protein, low-carbohydrate diet should also induce insulin resistance. Amino acids stimulate mTOR activity [41], and mTOR appears to induce insulin resistance [42]. Thus a mechanism by which high-protein diet can potentially induce insulin resistance also exists. There is an odd piece of evidence in support of it too. Insulin resistance is demonstrated in dolphins under high-protein diet [43]. In humans, high-protein diets do not seem to be generally associated with insulin resistance, but protein deprivation has been reproducibly shown to increase insulin sensitivity, the effect being independent of total calorie intake [44–46]. Therefore some positive association between protein intake and insulin resistance can be shown in humans too. There has been an interesting suggestion that ethnic groups that historically depended more on protein-rich diet evolved a tendency towards insulin resistance (McMichel 2001, source reference [47]). The point I want to

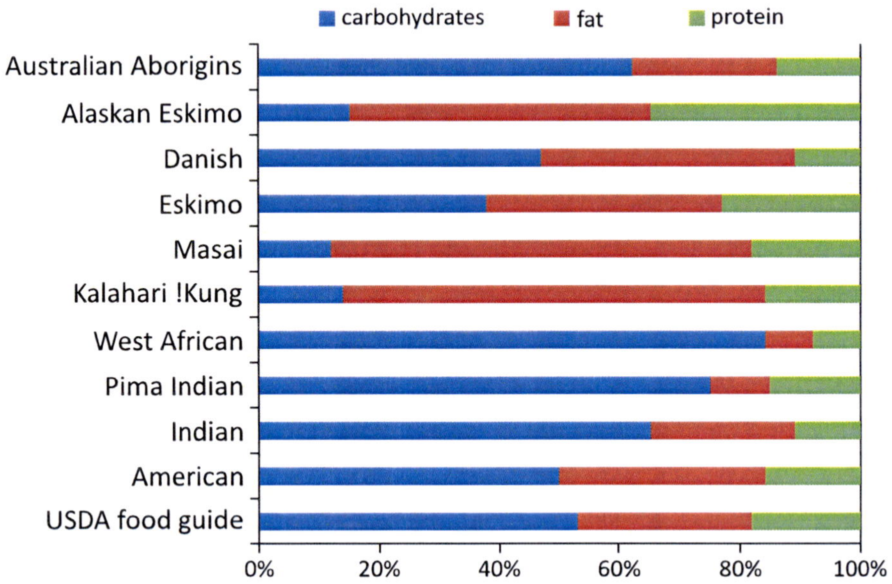

Fig. 3.1 Macronutrient composition of traditional and modern diets around the globe. *Bars* denote the percentage of energy obtained from carbohydrates, fats, and proteins, respectively. It can be noted that average diet composition in America (one of the countries with the highest incidence of obesity) and India (one of the countries with the highest number of type 2 diabetics) is nearest to the prescribed ideal composition. Many other societies with more extreme diets have much lower incidence of obesity, diabetes, and related disorders. Data compiled from multiple sources [32, 50–65]

emphasize here is that there can be arguments and evidence for high-fat, high-carbohydrate, or even high-protein diet causing insulin resistance. Literature on diet and insulin resistance is full of internal contradictions. One would find some evidence for high-fat or high-carbohydrate or high-protein diet causing insulin resistance and also would find sufficient evidence against. Perhaps the effects of dietary composition are non-monotonic and an optimum balance between protein, fat, and carbohydrate could be important, and such recommended ideal composition has been prescribed [48, 49]. Perhaps a balanced diet is the key and loss of balance in any direction could lead to insulin resistance. But while talking about balanced diet, one should not forget what the diversity of people across the globe have been consuming traditionally without developing obesity, T2D, and related disorders on an epidemiological scale.

Even a quick glance at the traditional dietary compositions of different tribes in the world reveals that people have adapted to different extremes of diets under different ecological settings. There are examples of tribes taking over 85% carbohydrates, 70% animal fat, or almost 100% dependence on meat and fish with practically no carbohydrate intake (Fig. 3.1). As compared to the wide variation in diet on which these different societies have survived without significant incidence of the modern lifestyle disorders [50, 66], the dietary changes witnessed by the last few generations in the modern Westernized societies are meager. But there lies the paradox of diet composition. Within the Westernized populations, there is good evidence that high-fat or high-soluble carbohydrate diet causes obesity or insulin resistance [51, 52, 67]. But much more extreme fat- or carbohydrate-consuming tribes do not show obesity and metabolic syndrome [10, 32]. The Masai derive 60–70% energy from animal fat; Kalahari !Kung derive almost the same from plant lipids. West Africans and traditional Pima Indians had a predominantly carbohydrate diet, and Eskimos' traditional diet had hardly any carbohydrates. It may be argued that since these tribes are on extreme diets for several generations, they may have evolved genetic resistance against diet-induced obesity. But this is not true since on adopting modern Western lifestyle, they rapidly develop signs of metabolic syndrome [68–71]. Today the macronutrient composition of the diets in America, where the prevalence of obesity is the highest, or India that has highest number of diabetics is much more "balanced" than these extreme diet communities (Fig. 3.1). But interestingly, the extreme diet communities traditionally had negligibly small prevalence of diabetes, hypertension, or high cholesterol. This does not imply that they have no health problems. The point relevant here is that their health problems are not diabetes, hypertension, and cardiovascular diseases (CVD), as long as they remain in their traditional lifestyle. This suggests that there might be something other than dietary macronutrient composition in the process of acculturation to modern urban lifestyle that is crucial [72]. Normally the human body appears to have the capacity to adapt to widely different dietary compositions.

If we believe that the dietary macronutrient composition drives obesity, the student has another fundamental question which needs an answer. With the exception of essential amino acids and certain fatty acids that our body cannot synthesize, carbohydrates, fats, and proteins are interconvertible in the body, and mechanisms exist to convert carbohydrates into fat or protein and vice versa. These mechanisms are activated according to the metabolic demand, and metabolic demands are largely decided by behavior. A wrestler can build up sufficient muscle protein, even if the diet is predominantly carbohydrates. There are many examples of vegan body builders, wrestlers, and athletes whose total protein intake is substantially less than meat eaters. Masai with high-animal fat diet are lean, insulin sensitive, and having low cholesterol levels [8–10]. Kalahari !Kung that consume high proportion of plant lipids were found to have a marginally impaired glucose tolerance by the Western standards. But this was not due to insulin resistance since their insulin resistance index (HOMA-IR) was invariably low [73].

The real importance of diet composition is in supplying molecules that our body cannot synthesize including vitamins, minerals, essential amino acids, and molecules such as omega 3 fatty acids. Having supplied these according to the requirement, whether energy comes in the form of fat, protein, or carbohydrate should not matter too much since the body can derive energy from all of them. Therefore the relative proportions of carbohydrate, fat, and protein are unlikely to be central to obesity. If negative feedback-based mechanisms exist for regulation of food intake, even the energy density of food should not matter. An energy-dense food may escape short-term regulation mechanisms such as signals given by distension of stomach but is unlikely to escape long-term regulation mechanisms driven by leptin and other adipose tissue signals, provided such mechanisms work. If they do not, the reason for their failure needs to be blamed and not the energy density. The critical factor therefore is not the nature of food but the apparent failure of energy intake regulation mechanisms.

Thus there are problems with all the three lines of mainstream thinking. If we believe that reduced energy expenditure is the main cause, doubly labeled water studies are not in support. If we think that increased energy intake is central to obesity, we have not yet answered why energy intake regulation mechanisms fail. If dietary composition is to be blamed, global anthropological data are not in support. This means that we really do not know the predominant cause of obesity as yet, at least not at a depth sufficient to satisfy students in my undergraduate classroom. Some alternative ways of thinking are coming up recently which we will elaborate subsequently. At this point, I need to keep the discussion on the origins of obesity quite brief for two reasons. One is that sufficient critical literature is already there examining this question (e.g., [33, 47, 74]), and the other is that the origin of obesity is not the prime question of interest here. Two other questions are of greater concern and inadequately addressed so far, namely, whether obesity is central to insulin resistance and whether insulin resistance is central to T2D as generally believed. These two need to be looked at with much more rigor and critical questioning.

2. Obesity as the cause of insulin resistance: There is a strong body of evidence that obesity is

correlated to insulin resistance. However there is much ambiguity about whether obesity causes insulin resistance or insulin resistance leads to obesity. James Neel whose thinking has been central to the prevalent concept of thrifty gene [75] did not think that obesity led to diabetes. He rather stated that a "diabetic tendency" (the concept of insulin resistance was not there when Neel wrote this paper) was responsible for obesity. If the "diabetic tendency" is insulin resistance, then insulin resistance should be the cause of obesity. There is some evidence that insulin resistance could be the cause rather than the effect of obesity. Muscle-specific insulin receptor knockout mice become obese [76–78] demonstrating that induced muscle insulin resistance can lead to fat accumulation provided fat cells remain insulin sensitive. Fat cell-specific insulin resistance, on the other hand, arrests fat accumulation [77, 78]. It is possible therefore that if muscle cells become more insulin resistant than fat cells, then that specific pattern of resistance can lead to obesity. However, whether fat cells remain relatively insulin sensitive in an insulin-resistant individual has not been clearly shown. Other possible mechanisms have also been suggested through which not insulin resistance per se but hyperinsulinemia leads to obesity [79, 80]. This is a part of a different line of thinking that we will see in sufficient depth later.

The current mainstream thinking appears to go in the other direction. Obesity is perceived to be the major clinically important cause of insulin resistance by the current perception. Doubts can be raised whether obesity is the sole factor leading to insulin resistance since a significant proportion of type 2 diabetics or insulin-resistant persons are of normal weight [81–83]. In India, for example, it is just too common to find normal weight or even lean and thin type 2 diabetics. Also obesity is not related to cardiovascular mortality in India as it is in Western countries or Eastern Asia [84]. The teacher now defends the textbook by showing that although the normal weight diabetic individuals may not have a high BMI, they have a higher proportion of body fat and that is what makes them insulin resistant. Here again, the more important question appears to be that if the net energy balance is not positive (as indicated by normal BMI), what leads to higher fat percentage in such individuals? It needs to be appreciated that this is a different question. It is not an energy balance question. If the proportion of fat, not body weight, is what counts, we need to think on a different line. This distinction is often not very clearly made. Body composition is decided by factors other than energy balance, and these factors are not yet clearly known.

The other side of the coin is that not all obese individuals are insulin resistant [83, 85]. This is defended by the teacher saying that it is the individual genetic makeup that decides this. Just as everyone does not get cold when a common cold virus is around, obesity does not lead to insulin resistance in every case. But while saying this, the teacher in me knows that this is a lame argument. We will see in the following chapter all the new evidence (or lack of it?) about the role of genes in obesity and diabetes to realize that after the genome data became available, the genetics argument has almost completely lost ground. But there is another possible argument that might come handy in defense of the obesity school. It is not the body mass index (BMI) or total fat that is so much important. There is evidence that it is central obesity or visceral abdominal fat which induces insulin resistance. Particularly interesting is the case of sumo wrestlers who have a high BMI, a large amount of body fat, and abdominal obesity as well. But as compared to other individuals of equal BMI or equal waist circumference, they are much less insulin resistant as long as they are active wrestlers [86]. So their abdominal obesity does not appear to have the same effect as in other obese individuals. This is defended now by showing that sumo wrestlers have more subcutaneous fat and little visceral fat [87] and therefore they are less insulin resistant. Subcutaneous fat appears to be relatively harmless. There is some evidence in support of this argument [88–90]. The next natural question is exactly what makes the fat stores of sumo wrestlers go subcutaneous and not visceral. Again the energy balance equation is useless to answer this question. Neither total energy intake nor expenditure can explain the distribution of fat in the body.

Therefore there must be something else that decides the distribution of fat, and this something else must be more important than energy intake and expenditure. Currently there is no satisfactory answer to what this "something else" is.

Is the phenotypic association between obesity and insulin resistance reflected at the level of genes as well? A large number of alleles have been shown to have statistically significant associations with obesity and a number of alleles with T2D although the effect size is very small. Out of the recently identified risk alleles for T2D, very few are related to obesity parameters [91]. A locus near IRS1 associated with lower body fat percentage is associated with higher insulin resistance, adverse lipid profile, and low adiponectin in men [92], indicating that the association between obesity and insulin resistance is not obligate at the genetic level.

Nevertheless, let us assume that the association between obesity and insulin resistance is real and causal. The next question is: What is the mechanism by which obesity induces insulin resistance? We have seen in the last chapter that a large number of potential mechanisms have been suggested, but we do not know with sufficient clarity which of them plays a significant role in the current obesity epidemic. Whether FFAs are the main players or adipokines or both is not very clear, and the undergraduate class demands this clarity.

For a long time, circulating lipids and free fatty acids are considered responsible for inducing insulin resistance. However the mechanism by which lipids or FFAs induce muscle insulin resistance is still elusive.A number of hypotheses have been floated, but none has been convincingly demonstrated. For example, one of the earlier hypotheses called Randle hypothesis speculated that fatty acids competed with glucose for substrate oxidation in muscle leading to increased intracellular glucose-6-phosphate [93, 94]. This resulted in accumulation of intracellular glucose and thereby reduced further uptake of glucose by muscle. However later studies in normal individuals with increased fatty acid levels showed that there is actually a fall in the intracellular glucose-6-phosphate and glucose levels. This fails to support the Randle hypothesis. Another hypothesis is based on the observed dysfunction of muscle mitochondria in insulin resistance. Based on correlative evidence, it was proposed that lipid overloading of muscles leads to incomplete fat oxidation by mitochondria. The resultant defective mitochondrial function underlies insulin resistance. However, experiments attempting a direct test of the hypothesis found that lipid accumulation actually increased mitochondrial proteins and their function [95]. If lipids were responsible for reduced mitochondrial function, lowering of lipids should have increased it. However, pharmacological lowering of plasma lipids in insulin-resistant subjects reduced mitochondrial gene expression instead of increasing it [96]. Also demonstration of mitochondrial dysfunction in non-obese first-degree relatives of diabetics [97] suggests that the mitochondrial problem appears to be due to "something else" and is not fat induced.

Another strong line of argument against FFAs being responsible for muscle insulin resistance is based on the measurement of insulin resistance by hyperinsulinemic euglycemic clamp method. In this method, continuous insulin infusion is done to raise and maintain a stable steady high level of insulin in blood. The stable hyperinsulinemia thus achieved rapidly depletes FFAs as has been clearly demonstrated. If standing levels of FFAs were responsible for insulin resistance, then insulin resistance should not be seen in a hyperinsulinemic euglycemic clamp study. But the clamp method is known to measure insulin resistance highly reproducibly, and insulin resistance thus measured matches very well with other methods that work in the physiological range of insulin and thereby not depleting FFAs. If depleting FFAs does not reduce insulin resistance, at least the standing FFA levels cannot be said to be responsible for insulin resistance.

As the mechanisms by which FFAs or circulating lipids induce insulin resistance are problematic, there is a growing body of evidence supporting the role of adipokines in inducing insulin resistance. Adipokines are secreted by adipose tissue, and therefore, the focus shifts from circulating FFAs and lipids to stored lipids.

The truth most inconvenient for this thinking is that if adipocytes secrete proinflammatory and pro-insulin-resistance signals, the same adipocytes also secrete anti-inflammatory and insulin sensitizing signals. Adiponectin is an important signal molecule secreted by adipocytes which has anti-inflammatory, antiobesity, and insulin sensitizing action [98–101] and which is paradoxically under-expressed in obesity although the tissue secreting it is increased in volume. Another adipocyte-secreted protein is sFRP5 which has anti-inflammatory and antidiabetic effects [102]. Normally leptin produced by adipose tissue regulates energy intake and thereby arrests obesity. In healthy life, the adipose tissue signals are bidirectional, and these opposing signals could be balancing out quite well, but in obesity-induced disorders, the proinflammatory and pro-insulin-resistance signals predominate and their counterpart gets suppressed. So it is not the amount of adipose tissue per se, but the shift in the balance of the proinflammatory and anti-inflammatory signals given by the adipose tissue is what appears to be important. A systems biology approach to diabetes in mice has demonstrated that proinflammatory adipokine secretion is detectably greater in diabetes-prone than diabetes-resistant mice at 6 weeks of age when there is no detectable difference in the fat mass of the two. With age and fat feeding, the inflammatory response in diabetes-prone mice increased along with obesity, but that in diabetes-resistant mice did not [103]. This suggests that the shift in balance of proinflammatory and anti-inflammatory actions of adipose tissue might be happening first and obesity arises subsequently. What causes this shift in balance is naturally the question that follows, and there is no textbook answer to it mainly because this question appears to be hardly ever asked before. Again the answer must lie in "something else" and not in obesity per se.

Particularly intriguing is the fact that adiponectin, a molecule that is normally synthesized by adipose tissue in large quantities, is negatively associated with obesity. If it is secreted by adipose tissue, the greater the adipose tissue, the greater should have been the secretion, but in reality it is correlated negatively with obesity. This paradox is well recognized but so far there is no solution. The negative correlation of adiponectin with obesity and the antiobesity action of adiponectin [104, 105] together imply that adiponectin deficiency is more likely to be a cause rather than an effect of obesity. Both logic and evidence support the view that the shift in the balance of adipokines should be arising first and increased fat mass a consequence of it. Since adipocytes produce leptin which regulates food intake and adipogenesis [106, 107], and adiponectin and sFRP5 which arrest fat accumulation, obesity is not possible unless leptin resistance [108] develops and a deficiency of adiponectin and sFRP5 develops. This is not a hen and egg problem. "Obesity first" is just not logical; leptin resistance and adiponectin–sRFP5 deficiency have to arise first, and only then is obesity possible. The shift in balance of adipocyte secretions must clearly precede obesity, and what shifts the balance should be a matter of greater concern than obesity itself. But this question does not have logical solutions so far. This is the major weakness in the prevalent thinking that adipose tissue induces insulin resistance through adipokines. Again we need to bring in that mystic "something else" that changes the adipokines balance. Obesity is made possible by this shift in adipokine balance.

If we assume that this paradox is somehow resolved and satisfy ourselves that the anti-inflammatory adipokines are somehow suppressed and the proinflammatory adipokines have a crucial role in inducing insulin resistance, another contradiction awaits us. Anti-inflammatory steroids such as cortisol should antagonize obesity-induced insulin resistance owing to their anti-inflammatory action. But the reality is diametrically opposite. The anti-inflammatory cortisol has a worsening rather than improving effect on insulin resistance [109]. Therefore assuming inflammatory mediators responsible for the loss of insulin sensitivity is not free of problems.

A different hypothesis linking fat to insulin resistance is adipose tissue expandability hypothesis [85, 110]. It postulates that fat stored in regular adipose depots is not harmful. But when these depots have reached their capacity, there is accumulation of ectopic fat, that is, fat is stored in tissues where it is not normally stored. For example, lipids stored in muscle, called intramuscular triglycerides (IMTG), are said to cause insulin resistance, and this is supported by the observation that insulin-resistant individuals have high IMTG [111–115]. Regarding the role of IMTG, there is a paradox. Athletes are typically rich in intramuscular fat and that is adaptive for them. Regular endurance exercise appears to increase IMTG [116]. In spite of high IMTG, athlete muscles are highly insulin sensitive demonstrating that lipids could not be primarily responsible for muscle insulin resistance [117–119]. The correlation between IMTG and insulin resistance is funny. If only non-exercising individuals are considered, there is a significant positive correlation between IMTG and insulin resistance. If athletes are added to this sample, the correlation gets destroyed since they have high IMTG and still are insulin sensitive [120]. Therefore evidence for IMTG being responsible for insulin resistance is shaky.

An interesting argument has been made to explain why individuals with abnormally low BMI are also frequently glucose intolerant [121]. Lipoatrophy, that is, absence of stored fat in the body is almost invariably associated with insulin resistance [122]. Diabetes associated with undernourishment is also reported [123]. The suggested explanation is related to the adipose tissue expandability hypothesis. The adipose tissue is said to sequester fatty acids and thus has a protective role against lipotoxicity [124]. In lipoatrophic condition, adipose tissue fails to develop, and as a consequence, FFAs keep on circulating in blood and cause insulin resistance. Another piece of evidence that adipose tissue by itself is not responsible for insulin resistance comes from transgenic mice overexpressing phosphoenolpyruvate carboxykinase (PEPCK) in adipose tissue. The enzyme PEPCK facilitates glycerol formation and esterification of FFAs into fat. Adipose tissue-specific overexpression of PEPCK increases fat mass and body weight without changes in circulating glucose, insulin, leptin, or FFAs. This model of obesity is not accompanied by insulin resistance demonstrating that the association of obesity with insulin resistance is not obligate. Rather than adipose tissue itself, something in the process of weight gain and fat accumulation could be causative to insulin resistance. The researchers who reported this finding think it is FFAs that induce insulin resistance, not adipose tissue [125]. Since FFA levels are unaltered in these mice, they remain insulin sensitive in spite of being obese. But also unaltered are leptin and insulin levels, raising the possibility that these might have a central role instead of FFAs.

The student looks too puzzled now. The arguments appear to be maneuvered according to short-term convenience of explanation. It looks as if the conclusion that obesity induces insulin resistance is predecided and details are maneuvered as and when needed in order to keep the conclusion intact. Whenever any evidence is found to contradict the original argument, the argument is maneuvered to explain the inconvenient truth. The new argument also faces some inconvenient truths later and therefore gets further maneuvered. While doing this, the earlier contradiction is often conveniently forgotten. When the mechanism by which FFAs induce insulin resistance is not found, adipose tissue and adipokines are blamed. When the paradox of adipokines is exposed, IMTGs are blamed which face the athletes' paradox anyway. Since people with absence of adipose tissue are found to be insulin resistant, and there can be increased adipose tissue without insulin resistance, the ball is thrown back in the court of FFAs. Such surreptitious skipping between sub-hypotheses makes the main hypothesis unfalsifiable, and the student is certainly reluctant to believe in a theory that is unfalsifiable.

Summarily the relationship between obesity and insulin resistance is not as clean as presumed by the mainstream thinking. Obesity does not appear to meet the "necessary and sufficient" criteria for causation since we do find classes of individuals that are obese but insulin sensitive and also classes that are non-obese but insulin resistant. Also the mechanisms of obesity-mediated insulin resistance are not clearly elucidated so far. There are serious problems with every mechanism suggested so far. Therefore the causal role of obesity in insulin resistance is questionable although a statistical association cannot be denied. In brief, there are a number of gaps, contradictions, and possible flaws in the diet and obesity paradigm. This need not mean that diet and obesity are not important, but it certainly means that this line of thinking is missing something crucial and that "something else" could be more important than obesity itself.

3. Compensatory hyperinsulinemia: We will now examine the next postulate that β cells produce more insulin in order to compensate for the insulin resistance. As seen in the last chapter, in the HIIR state, there is a high level of insulin resistance as well as high levels of insulin in blood. However this is not sufficient to believe that hyperglycemia develops in order to compensate insulin resistance. Association is not a proof of causation. Following a good undergraduate classroom practice, alternative possibilities also need to be examined. The alternative possibilities are that (1) insulin overproduction is primary and since high levels of insulin can lead to hypoglycemia, insulin resistance develops as a compensating mechanism or (2) both are secondary to an unknown primary change that stimulates both the processes simultaneously and in a balanced manner.

If we assume that the primary process begins first and that there is a detectable lag in the response, the first to appear should be the primary cause, and the one to appear later should be the compensatory response. In the normal course of developing insulin resistance, both insulin resistance and hyperinsulinemia are observed simultaneously, and therefore, it is difficult to know what precedes and what follows. Fewer studies have followed the time course of development in sufficient details. In all these experiments, the first detectable change was a slight hypoglycemia [75, 126–129] indicating that hyperinsulinemia was stronger than insulin resistance in the beginning. Neonatal hypoglycemia is extremely common in low birth weight babies [130] that have a high propensity to become insulin resistant later. The early transient hypoglycemic phase in the natural history of T2D is known for a long time. The early hypoglycemic phase was one of the strong supporting pieces of evidence that Neel based his thrifty gene argument on (discussed in details in the following chapter). Neel's arguments are actually based on the assumption that hyperinsulinemia is the starting point of diabetes [75]. It was somewhat later that the followers of thrifty gene theory painted the picture upside down.

But early hypoglycemia by itself may not be considered a conclusive evidence of hyperinsulinemia first. Alternative explanations are possible. For example, it can be argued that the compensatory reaction from pancreas comes instantaneously without any detectable lag and initially the reaction is stronger than needed and finds the right level of response iteratively during which transient hypoglycemia may be observed. A better way to resolve between the alternatives would be to selectively induce one of the processes and see the response of the other. Such situations do exist naturally or experimenters have achieved such feats. In certain types of pancreatic tumors, insulin is overproduced. This overproduction is due to the tumor and therefore can be safely assumed to be primary. An insulin producing tumor is always accompanied by insulin resistance, and if this tumor is removed, insulin sensitivity increases [131–136]. Therefore insulin resistance appears to be the compensatory response to hyperinsulinemia in this case implying that mechanisms do exist by which insulin resistance can develop to compensate hyperinsulinemia.

A reverse experiment in which insulin resistance was artificially induced and therefore we certainly know it was primary is also available. The muscle tissue is responsible for majority of insulin-dependent glucose uptake in the body. If insulin receptors in the muscle are specifically knocked out (such mice, called MIRKO for muscle insulin receptor knockout, are useful models of muscle insulin resistance), making the muscle tissue completely insulin resistant, insulin levels remain surprisingly normal [137]. It appears therefore that at least muscle insulin resistance is not compensated by hyperinsulinemia. A whole body insulin receptor knockout is nevertheless hyperinsulinemic [78]. So there is something else other than muscle insulin resistance that increases the insulin response. One can think of a possible reason that could bring about this change. If insulin exerts a negative feedback on its own secretion and β cell insulin receptors are needed for this action, insulin resistance at the β cell receptor level could result in hyperinsulinemia. But contrary to our expectation, β cell-specific insulin receptor knockouts are deficient in insulin production [138]. It can be concluded therefore that β cell insulin resistance cannot result in hyperinsulinemia. On the other hand, liver-specific insulin receptor knockouts do result in hyperinsulinemia, but this is accompanied by hyperglycemia, and therefore, here insulin production could be simply triggered by hyperglycemia [77]. It appears from these studies that compensatory hyperinsulinemia does not appear to follow insulin resistance in muscle and other tissues unless there is hyperglycemia. Therefore a normoglycemic HIIR state cannot be explained by the compensatory hyperinsulinemia hypothesis.

In contrast with insulin receptor knockouts, knockouts for the insulin receptor substrate (IRS-1 and IRS-2) become hyperinsulinemic [77]. IRS-1 knockouts are hyperinsulinemic in the absence of hyperglycemia. This is an example where induced insulin resistance is accompanied by compensatory hyperinsulinemia. But there is an alternative interpretation. IRS-1 and IRS-2 are also involved in insulin-like growth factor-1-(IGF-1) signaling, and IGF-1 deficiency has been shown to lead to hyperinsulinemia and insulin resistance [139]. So in this case, IGF-1 resistance rather than insulin resistance could be responsible for hyperinsulinemia. As yet, there is no demonstration of induced impairment of insulin signaling that increases insulin response without involving IGF-1. Therefore compensatory hyperglycemia in response to specific impairment of insulin signaling in the absence of hyperglycemia is yet to be clearly demonstrated.

Intrauterine growth retardation is known to lead to insulin resistance and hyperinsulinemia in early life. In some of the models of IUGR, the development of hyperinsulinemia and insulin resistance has been studied in great detail. The FASDEL is such a model in which there is heterozygous deletion of fatty acid synthase. The FASDEL females give rise to small-sized babies owing to retarded intrauterine growth. In the FASDEL model of IUGR mice, at 3 months of age, there is larger islet area and β cell mass accompanied by hyperinsulinemia but normal insulin sensitivity leading to hypoglycemia. However at 12 months of age the picture is reversed. Older mice are glucose intolerant, hyperglycemic, and relatively insulin deficient indicating that hyperinsulinemia appears first in the sequence [140]. The early hyperinsulinemia in this model is certainly not induced by insulin resistance because insulin resistance is not demonstrable till a much later stage. A number of other studies also show a primary occurrence of hyperinsulinemia independent of insulin resistance [128, 141–142].

Let us also look at the mechanisms available that bring about the respective compensatory responses in either case. If pancreas has to compensate insulin resistance, they need to know the level of insulin resistance of the body and accordingly increase the insulin production level. The predominant mechanism that according to the classical theory stimulates insulin production is raised blood glucose. But ironically, the blood sugar is not raised in the early insulin-resistant state. In fact, in the early hyperinsulinemic phase, there could be slight hypoglycemia. Therefore insulin production does not appear to be stimulated by glucose. No other mechanism has been suggested even speculatively that can measure the level of insulin resistance and instruct the β cells to produce appropriate amounts of insulin.

On the other hand, many pathways exist by which insulin resistance can be induced when β cells overproduce insulin. (1) Insulin itself downregulates insulin receptors and desensitizes post-receptor pathways [143]. (2) A negative feedback loop by which insulin action can be regulated appears to be built in the system in the form of the mTOR/S6K1 pathway [42]. Insulin signaling activates mTOR, and mTOR with its downstream S6K1 signaling reduces the insulin receptor substrate response. As a result, if insulin signaling increases unduly, its effect gets downregulated automatically. (3) Another known pathway operates through amylin. Amylin is coproduced by the β cells along with insulin. Whenever the β cells are hyperactive and secrete excessive insulin, they also secrete excessive amylin. Amylin in high concentration has been well demonstrated to induce insulin resistance [144–150]. So the mechanism to avoid the risk of acute hypoglycemia if and when insulin is overproduced is built in the pancreas itself. (4) Alternatively, insulin induces the release of soluble *Klotho* protein from the membrane-bound form [151], and *Klotho* induces insulin resistance [152, 153]. (5) Raised levels of insulin stimulate serotonin secretion [154] in the brain, and chronically increased serotonin signaling in the hypothalamus is also known to induce peripheral insulin resistance [155]. (6) Insulin is known to stimulate lipogenesis, and both plasma triglycerides and adipose fat are implicated causatively in insulin resistance. Suppressing insulin production by somatostatin and its analogues is known to reduce obesity [156] and thereby insulin resistance. Therefore if insulin levels are raised for sufficiently long duration, insulin resistance should automatically follow. Out of the above mechanisms, (1)–(4) are relatively quick-acting mechanisms, and (5) and (6) have chronic actions. At this stage, we do not clearly know the relative importance of the alternative mechanisms. The fact of relevance here is that many mechanisms for compensatory insulin resistance exist, whereas those for compensatory hyperinsulinemia have been elusive.

All the evidence together raises doubts about whether hyperinsulinemia is an attempt to compensate insulin resistance or vice versa. Evidence is clearly in favor of the latter, that is, hyperinsulinemia is a primary response and insulin resistance arises as a compensatory reaction to hyperinsulinemia. If hyperinsulinemia comes first, the logical links between postulates 1–2 and 2–3 are in trouble and postulate 4 just cannot be true. If hyperinsulinemia comes first, β cell degeneration cannot be said to be the cause of failure of compensation, and if that is so, what makes the blood sugar go up? This could be a fatal blow to the classical thinking. Not surprisingly, there is extreme reluctance to accepting or even considering the idea that hyperinsulinemia could be coming first. This is not being said for the first time. In recent years, a number of authors have suggested this possibility giving evidence [75, 157, 158]. But it does not seem to have made a major impact on the mainstream thinking. The teacher in me is more conservative. Although hyperinsulinemia first is sufficient to make the entire theory of diabetes collapse, he convinces the student that it is prudent to ignore this for the time being and continue examine the rest.

If we accept that hyperinsulinemia comes first, we need to worry about what causes hyperinsulinemia. The orthodox theory does not offer any suggestion on this since there has been a mind block on "insulin resistance first." Recently there have been suggestions as to why primary hyperinsulinemia develops in diabetes-prone individuals, and we will discuss them in subsequent chapters. For postulates 1 and 2, we only raised doubts, but for postulate 3, it appears to be a clear case of the textbook being wrong. But if postulate 3 is wrong, there are implications for 1, 2, and 4 as well.

There are a series of horse and cart paradoxes in the current understanding of obesity and diabetes in which the cause–effect relationships are terribly messed up. This is the first of the series. There are many more to come below. We will have to resolve all these paradoxes ultimately. I will call the paradox of whether insulin resistance first or hyperinsulinemia first as the horse and cart paradox I (one) following Shanik et al. [157].

4. Beta cell exhaustion/defect/failure resulting in relative insulin insufficiency leading to hyperglycemia: The next postulate is that of a relative deficiency of insulin causing hyperglycemia. In an early insulin-resistant individual, there is high level of insulin resistance, but glucose level remains normal since the higher levels of insulin are supposed to compensate it. Now since we are examining the current paradigm, let us forget for the time being that hyperinsulinemia comes first. We will continue to follow and examine the classical line of thought. It is believed that by having to produce large amounts of insulin over a long time, the β cells get exhausted and some kind of defects in β cells start accumulating. Histologically progressive islet degeneration can be evidently seen in T2D and is remarkable in the later stages

but could be beginning very early in the history of T2D. Therefore there is no doubt that β cell degeneration does happen. The debate is whether exhaustion is the cause of this degeneration and whether the degeneration is a necessary and sufficient cause of hyperglycemia. Doubts about the β cell exhaustion theory have been raised from the beginning. Neel himself was quite unhappy about the concept of organ exhaustion and has refuted the claim [75].

The idea of chronic fatigue of any organ in the body is quite strange. As a teacher, I like to give analogies to illustrate a point. The problem in this case is that I can find no other example of chronic fatigue. Acute fatigue is quite well known but there is no good example of chronic fatigue. For example, if you run for 5 km at a stretch, your muscles will certainly experience fatigue. This is acute fatigue. But if you start running 5 km every day, then the muscle actually becomes stronger and stops getting exhausted. Except where physical wear and tear is involved such as teeth, this is the typical response of any organs to chronic overuse. Chronic overuse strengthens rather than weakens an organ. Even in the pancreas, there is an increase in the β cell mass in the early HIIR state. Apart from the belief associated with pancreatic islets, there is no other example of any endocrine organ failing or degenerating due to overuse. The adrenal cortex is hyperactive in chronic stress, but there is no demonstration of exhaustion. Thyroid hypertrophy during pregnancy or goiter is well known, but this is not known to be followed by degeneration of the endocrine gland. Still islets are believed to degenerate by chronic overwork.

A knockout experiment helps us to resolve this question once again. IRS-1 deficiency is known to induce insulin resistance and hyperinsulinemia simultaneously but does not directly affect β cell function. In IRS-1 knockouts, diabetes never results in spite of long-standing hypersecretion of insulin. Under this condition, β cells can keep on hypersecreting without getting exhausted. On the other hand, IRS-2 deficiency, in addition to inducing insulin resistance, directly affects β cell development and survival by interfering in the IGF-1 action on β cells [77]. Needless to say, these mice become diabetic. This is a demonstration that hyperinsulinemia by itself does not lead to β cell degeneration and cause diabetes. There needs to be another independent cause of β cell degeneration to develop hyperglycemia. Again there is this mystic "something else."

The organ exhaustion theory is rather old, and now we know many other possible reasons for islet degeneration [159]. They include oxidative stress, glucotoxicity, lipotoxicity, and amyloid deposition, some of which are discussed below. Whether due to exhaustion or any other reason, there is no doubt that both insulin resistance and β cell dysfunction characterize type 2 diabetes. But the most relevant question that we need to pose is whether a combination of these two is both necessary and sufficient to explain fasting and postprandial hyperglycemia and further other complications of diabetes. This question is more important than what causes β cell degeneration. Therefore let us try to handle it first and then return to the causes of β cell degeneration. Overt diabetes is believed to appear when islet degeneration begins. Some kind of defect develops in β cells which are unable to give a compensatory response to insulin resistance.

How can we test this postulate? A simple way to test would be to mimic the hypothetical process and see whether the result is the same. For example, let us take a normoglycemic HIIR individual. Now, if insulin production in this individual is artificially and specifically suppressed, sugar levels should go up. If we can demonstrate this, we will be able to demonstrate that insulin resistance and insulin insufficiency are "sufficient" to explain hyperglycemia.

Such experiments have actually been done repeatedly. The big surprise of these experiments is that when insulin production is suppressed, the insulin sensitivity increases rapidly such that the sugar level remains more or less unaffected. Diazoxide is known to suppress insulin production. In genetically obese Zucker rats, the suppression of insulin production by diazoxide increased insulin sensitivity so much that the fasting sugar levels reduced instead of increasing. Further there was weight loss, increased fat oxidation, and improved lipid profile as well [160–162].

The effect of insulin suppression in humans was of a similar nature. Diazoxide treatment resulted in insulin suppression and simultaneously improved insulin sensitivity [158]. In another study, insulin suppression by diazoxide increased blood glucose levels in non-obese subjects but increased insulin sensitivity without altering glucose in obese subjects [163]. Alternatively insulin can be suppressed by protein deprivation. Using this dietary means to suppress insulin, not only insulin levels decreased but even glucose levels decreased [46], indicating that insulin suppression immediately increased insulin sensitivity. Another insulin suppressing agent octreotide had similar effects in human trials [164–166]. A combination of insulin siRNA and insulin degrading enzyme was able to produce diabetes-like condition in transgenic mice at high levels of suppression of insulin, but my analysis of their data showed that insulin sensitivity actually increased in the insulin-suppressed mice. About 75% of suppression of insulin production was needed to produce a detectable impairment of glucose tolerance [167]. Consistent demonstration of insulin sensitizing effects of insulin suppression by three different pharmacologically active agents as well as by dietary modification makes it highly unlikely that insulin sensitivity was an inadvertent effect of the agent itself. Suppression of insulin in an insulin-resistant state not being able to produce hyperglycemia is what I would like to call horse and cart paradox II. We made the horse move in the direction of diabetes by suppressing insulin but the cart did not. Instead of raising blood sugar, suppression of insulin appears to increase insulin sensitivity for an individual in an insulin-resistant state. The most logical interpretation of these "inconvenient" results is that when insulin production is suppressed, insulin resistance actually seems to adjust itself to keep the sugar levels normal. This compensatory adjustment in insulin sensitivity can fail only when the insulin levels become too low. If we have accepted that hyperinsulinemia is primary and insulin resistance is a compensatory response, then this result is not a surprise at all. If insulin levels are high, insulin resistance is required by the body; otherwise, the sugar levels will drop down to dangerously low levels. But if insulin levels are brought down artificially, then insulin resistance is no more needed, and it comes down to such a level that sugar remains more or less normal.

In these experiments, insulin-resistant state already existed and we suppressed insulin production, so following the classical paradigm, we would have expected the blood sugar level to rise. But the blood sugar level did not go up, indicating that insulin insufficiency on the background of insulin resistance is not "sufficient" to cause hyperglycemia.

Another experiment that raises doubt about IR–RII being sufficient cause of hyperglycemia is again that of MIRKO mice. In these mice, insulin signaling in muscle is reduced by over 95%. This is an extreme possible case of muscle insulin resistance. By classical thinking, we would expect insulin levels to increase in compensation, but they do not as seen earlier. Now if there is extreme muscle insulin resistance and no compensatory insulin production, it must be an ideal situation for blood glucose to go up. But most surprisingly, that does not happen. Sugar levels remain normal in spite of extreme insulin resistance and lack of compensatory rise in insulin [77]. This means that extreme insulin resistance and lack of compensatory insulin response were not "sufficient" to increase blood sugar in MIRKO mice. How did it happen? In the previous experiment, there was an increase in insulin sensitivity when insulin production was suppressed. Here since the insulin receptor itself is knocked out, this possibility is ruled out. There are two other compensatory mechanisms which evidently work to compensate effectively for both insulin resistance and lack of compensatory increase. One is a compensatory glucose uptake by tissues that have intact insulin receptors. Evidence for this comes from the observation that fat cells of MIRKO mice take up excess glucose and increase their fat stores. The other mechanism is that muscles start making use of insulin-independent pathways for picking up glucose. Muscle glucose uptake during exercise is independent of insulin, so insulin-independent pathways do exist. This is evident by the observation that there is no wasting of muscle in MIRKO mice which is expected if insulin action in muscle is completely stopped and not compensated by any other mechanism. The other evidence is that muscle-specific glut-4 knockouts do become hyperglycemic in contrast with MIRKO. This is because ablation of glut-4 blocks the insulin-independent pathways as well [168]. The important insight obtained from these knockout experiments is that mechanisms exist in muscle tissue that can compensate for insulin resistance, but for some strange reason, they do not seem to work in diabetes. Which again points to "something else" being responsible for diabetic hyperglycemia in addition to IR–RII.

Another example of insulin-independent mechanisms compensating for insulin resistance comes from liver-specific insulin receptor knockout (LIRKO) mice. LIRKO mice were observed to be severely insulin resistant as well as hyperglycemic and hyperinsulinemic. We have seen in the last chapter that fasting hyperglycemia mainly reflects liver insulin resistance. As expected, young LIRKO mice had fasting hyperglycemia. However, by 6 months of age, their fasting sugar levels returned to normal without gaining liver insulin receptors back or increasing insulin secretion further. At 2 months, the mean fasting glucose of LIRKO was 132 mg/dl which reduced to 70 mg/dl by 6 months, but the fasting insulin levels remained the same [169]. Since there was no possibility of increasing insulin sensitivity of the liver, we need to infer that some other compensatory mechanism restored the control on liver glucose metabolism resulting in normalization of fasting blood glucose. This example is very dramatic. Completely knocking out the insulin receptor means not only that there is insulin resistance but also that increasing insulin levels cannot compensate for the loss. But in spite of this, fasting sugar levels returned to normal. This means that some compensatory mechanisms do exist which are independent of insulin. Liver glucose production is known to be directly under the control of hypothalamic regulation through the vagus nerve. When insulin mechanism fails to work, this mechanism appears to take over the control of the situation and normalize metabolism. So if insulin signaling in the liver fails, there are compensatory mechanisms to take care. But surprisingly, these compensatory mechanisms appear to keep mum during diabetes. This is one more demonstration that something other than IR–RII is happening in diabetes.

The unanswered question now is what exactly happens in diabetes. By the current belief, in an insulin-resistant state, whenever compensatory insulin production is not adequate, the blood sugar goes up. This is not different from the above experiments in which there were both insulin resistance and lack of compensation by insulin. But what happened in the experiments does not match with what happens in diabetes. The mechanism of adjusting insulin sensitivity to the current insulin production or that of compensating insulin resistance by insulin-independent pathway fails when diabetes sets in. Since insulin resistance and insulin suppression were unable to raise blood sugar, we suspect that some factor other than these two must be responsible for the diabetic hyperglycemia. This something else must be more important and central to T2D than the combination of insulin resistance and insulin decompensation. Possible answers to this question will come later in this book. Here it is sufficient to make a statement that IR–RII is not "sufficient" to cause a rise in blood sugar.

We can start experiments from the other end now to ask whether IR–RII is "necessary" for hyperglycemia. One approach would be to look for natural or experimentally induced cases where hyperglycemia or impaired glucose tolerance is seen without insulin resistance or without β cell degeneration. Two such natural situations exist. One is a condition called anorexia nervosa, generally affecting women and characterized by obsessive fear of becoming fat which adversely affects food intake. As a result, patients with this condition are underweight and scarcely have fat. Although they are highly insulin sensitive, they show impaired glucose tolerance very frequently. Since there is no evidence of β cell degeneration in anorexia nervosa [170], it appears that there can be impaired glucose tolerance in the absence of both insulin resistance and β cell degeneration [171]. Interestingly, anorexia nervosa patients also have high cholesterol despite low calorie intake, and their cholesterol can be normalized by increasing caloric intake and body weight [172–174]. The other example comes from a small study on the Kalahari !Kung which showed that the tribesmen were insulin sensitive but still had impaired glucose tolerance [73] by the Western standards.

I can perceive another experiment where we should take individuals with mild fasting hyperglycemia which is presumably caused by IR–RII. If the popular picture that β cells are trying to keep the glucose levels normal by producing as much insulin as they can but they have reached their limit and are unable to compensate sufficiently for the level of insulin resistance is correct, then we would expect that after bringing down insulin resistance by exercise, the compensation should work and glucose levels should come down towards normal. Insulin levels may reduce following normalization of blood glucose. It is well known that exercise reduces insulin resistance. This has been shown reproducibly by a large number of studies using different measures of insulin resistance. All of them invariably show that exercises increase insulin sensitivity [175–181]. We also have effective insulin sensitizing drugs available. The question of my interest is whether insulin levels come down first or glucose levels come down first. If sugar levels reduce first and in response insulin levels drop, this is compatible with the current understanding. But if the reverse happens, it indicates that something different is happening. Published literature does not give a clear idea as to what comes down first. This is because the time course of events is either not observed or not reported with sufficient resolution. I happened to see some unpublished data in which in a substantial proportion of individuals participating in a trial, exercises reduced insulin resistance and insulin levels dramatically in a few weeks, but fasting sugar levels and HbA_{1c} either did not reduce significantly or did so only after 10–12 weeks.

The question whether insulin comes down first or glucose does not seem to be addressed directly anytime. Nevertheless I will cite a few published reports which do not address this question but we can draw some inferences indirectly from their data. In a study, diabetic mice having fasting plasma glucose close to 9 mmol/l were treated with insulin sensitizing drugs, namely, metformin and rosiglitazone. After a year-long treatment, their glucose-stimulated insulin secretion decreased substantially, but glucose levels did not [182]. Another study compares the glucose tolerance curves of young versus sedentary old and exercising old individuals. As compared to the young age group, the sedentary old individuals have higher glucose levels as well as higher insulin levels in OGTT indicating insulin resistance. On exercise training, the insulin curve of older individuals reduced and became indistinguishable from those of the young. But glucose curves did not change significantly indicating that when insulin sensitivity increased, it was insulin that came down, not glucose [176]. In another study on individuals with obese postmenopausal diabetic women with basal average glycated hemoglobin of 6.9 (the normal range being 5.7–6.4), exercises increased the glucose disposal rates indicating increased insulin sensitivity, but HbA_{1c} did not show any change [183]. Readers may wonder why I suspect that insulin may come down first and why I am specifically interested in this question. This will be answered in a later chapter. The point of relevance here is that if sugar levels remain unchanged even after reducing insulin sensitivity, it might indicate that there is something else that is determining sugar levels apart from IR–RII. Therefore this issue needs to be addressed with experiments specifically designed to answer this question. If we find that insulin levels respond more rapidly to insulin sensitizing treatment than sugar levels, that will make our horse and cart paradox III where we stopped the horse but the cart kept on moving. We reduced insulin resistance but hyperglycemia continued.

An inevitable conclusion of the multiple experiments above is that although insulin resistance and insulin insufficiency due to β cell degeneration are both undoubtedly involved in glucose homeostasis, the two are neither necessary not sufficient to cause hyperglycemia. Since the "necessary and sufficient" criteria are not met, it would be more appropriate to call insulin resistance and β cell degeneration as risk factors for hyperglycemia rather than causes of hyperglycemia. There is something missing, something perhaps more complex or more elusive that appears to determine the blood sugar levels, and both insulin resistance and insulin production appear to adjust themselves to maintain that mystic level. Two more sets of interesting experiments demonstrate that apart from insulin resistance and insulin insufficiency, there is something more involved in deciding blood sugar levels.

Do you think high plasma glucose levels can be reduced by infusing more glucose in the body? This sounds weird but it does happen provided glucose is infused in specific sites in the body. One such site is the portal vein. If glucose is infused directly in the portal vein which carries blood from the gut to the liver, plasma glucose levels drop rapidly [184]. Researchers performing these experiments claim that there is a portal glucose sensor that measures the glucose levels in the portal blood and the rate of change in these levels. Signals from these sensors go to the brain through sensory neurons. Rising levels of portal glucose quickly suppress gluconeogenesis by the liver and stimulate glucose utilization by muscle, both the processes happening apparently independent of insulin. This may result into hypoglycemia. Moreover, if the amount of glucose infused in the portal vein is calculated to match the amount of glucose that liver would have produced otherwise, liver glucose production completely stops [184]. This is as if the liver was assigned a quota for glucose production by someone, and since the liver was then fooled to think that the required amount of sugar was already coming from the gut, it completely stopped producing glucose.

The other example is experimental infusion of glucose in the intracerebroventricular space in the brain which increases CSF glucose without affecting plasma glucose [185]. Such an attempt to increase CSF glucose resulted in substantial reduction in plasma glucose without affecting plasma insulin levels. It is well known that the hypothalamus has a role in glucose regulation, but it is thought that it is activated only under hypoglycemia. Normally as a response to hypoglycemia, the hypothalamus initiates a counterregulatory mechanism that triggers liver glucose production to normalize glucose levels [186]. But if there is adequate fuel for the hypothalamus in the form of glucose or lactate, the counterregulatory response is blocked [187, 188]. However since infusion of glucose in the brain under peripheral normoglycemia lowers plasma glucose, the central mechanisms of glucose regulation appear to be active not only during hypoglycemia but during normoglycemia as well. The role of direct central regulation of glucose is highlighted by experiments in which denervating liver or pancreas changes the glucose tolerance curve [189]. These experiments suggest that there are alternative mechanisms of glucose homeostasis in the body that are partially or completely independent of insulin. This is not new since it is long known that although the muscle uptake of glucose is insulin dependent during resting, during exercise muscle glucose uptake is not completely insulin dependent. It is also known that the brain has glucose sensing neurons and hypothalamus exerts neuronal control over liver glucose production and pancreatic hormone production [189–191]. Apart from glucose sensing by the β cells to produce insulin, there are other sensors and control mechanisms that appear to regulate plasma glucose levels. Glucose homeostasis appears to be complex and multidimensional, and apart from the insulin–glucagon system, there are other important components to it. But in spite of experimental demonstrations of these components, the classical insulin-centered thinking has not attempted to integrate these components in the theory and eventually in clinical practice. This is mainly because there are some missing links in our understanding of central regulation of glucose. As a result so far, there is no integrative understanding of the alternative mechanisms of glucose regulation. Such an integrative thinking is likely to fill the gaps in our current understanding and also help interventional regulation of blood sugar. Unfortunately current medicine has ignored other mechanisms and revolved only around insulin and insulin resistance. This is not to cast any doubt that both of these are important in T2D, but there is something more to glucose homeostasis which the orthodox theory has failed to capture.

Let us for the time being return to the textbook picture and assume that insulin resistance and β cell failure are the only mechanism behind hyperglycemia. The associated question that needs to be handled is what causes the progressive β cell failure and whether this is reversible. The exact nature of β cell degeneration or defect or failure is another elusive concept. Normally like many other endocrine organs, the capacity of the pancreatic islets to produce hormones is in far excess than what is most commonly needed. As a result, in a healthy individual, unless over 85% of the pancreatic mass is removed, hyperglycemia does not result [192]. Postmortem studies on type 2 diabetics have shown that in them reduction in islet mass is often <50% and still hyperglycemia results [193, 194]. This may be argued to be a result of relative rather than absolute β cell deficiency. But what does β cell deficiency actually mean? Let us first take a simple model of β cell deficiency, that is, the number of β cells and their working capacity are insufficient to produce the required insulin to keep the glucose levels in control in spite of insulin resistance. This assumption means that insulin production becomes insufficient only when the demand exceeds the maximum capacity of the islets in their current state. If a person has a fasting hyperglycemia, presumably the capacity of insulin production by β cells had reached their upper limit, and therefore he is unable to normalize glucose levels. If this is true, then such cells should be unable to produce more insulin if stimulated by providing more glucose. But that is not the case. Typically in type 2 diabetics, except in very advanced stages, the insulin-stimulated glucose response is delayed but not dampened. In fact, the total insulin produced by early type 2 diabetics during an OGTT is likely to be substantially more than nondiabetics. This means that β cells had the capacity to produce this much amount of insulin. If this is so, the higher than normal levels of fasting glucose cannot be said to be the result of the inability of β cells to secrete sufficient insulin. They actually demonstrated their ability to produce more insulin after glucose stimulation. Then why did they not produce it at fasting state when sugar levels were above normal? The cells apparently did not use their full capacity of insulin production. Although the standing glucose level was above normal and the β cells had the capacity to make more insulin, they did not, as if "something else" was holding them back or something other than insulin was deciding the glucose levels. So this simple model of quantitative insufficiency in insulin producing capacity does not seem to explain fasting hyperglycemia although it might explain postmeal hyperglycemia.

Perhaps then we need to consider other possible ways in which abnormalities appear in the β cell behavior. The meaning of β cell defect and the time course of the development of β cell defect is highly debated. An alternative interpretation of β cell defect has been in terms of reduced glucose responsiveness rather than reduced insulin making capacity. But data on glucose responsiveness of β cells are contradictory. In vitro studies have shown reduced responsiveness of β cells in terms of higher glucose threshold needed for stimulation [195]. But another set of experiments has different conclusions. Insulin response to glucose follows a saturation curve similar to the Michaelis–Menten kinetics of enzymes. This type of saturation curve has two parameters: one that we will call I_{max}, denotes the maximum saturating capacity of insulin production, and the other is the concentration of glucose at which half the maximum capacity is reached. This parameter indicates the sensitivity to glucose. A comparison of normal and T2D subjects showed that in diabetics, I_{max} is substantially reduced, but the half-saturating concentration is not different from normal subjects. This study claims that it is the capacity not responsiveness that is affected [196]. Consistent with this conclusion is the observation of islet degeneration which appears to be quantitative rather than qualitative since reduction in β cell mass is evident in T2D. When human islets were transplanted into diabetic mice, 1,000 islet equivalents (IEq) were unable to restore euglycemia, but 2,000 IEqs did so for more than 100 days [195]. This implies that it might be only the quantity of β cells that matters. But another experiment contradicts this inference. In this experiment, the β cells of mice were destroyed up to 60%. After 2 weeks, these mice were infused glucose in response to which their β cell mass increased comparable to normal controls. But in spite of the β cell mass normalizing, their insulin response did not increase. This means that something other than cell mass was determining insulin secretion [197]. The picture thus does not go smoothly with either qualitative or quantitative defects. We still do not know with sufficient clarity whether what is affected in diabetes is the maximum capacity of insulin production of β cells; their glucose responsiveness, both simultaneously or both differentially at different stages of T2D; or "something else."

A delay in insulin response to glucose is evident from an early stage of T2D history, but it is debatable whether the cause of the delayed insulin response is a β cell "defect." In a few minutes of food ingestion, there appears a sharp insulin peak which is called the acute-phase insulin response (AIR). In diabetics, the acute-phase response is highly suppressed, and this is believed to be mainly responsible for loss of glycemic control [193, 198–200]. However later the insulin response may be greater than normal. How would a "defect" in β cell explain this delay is not very clear because the same β cell does secrete substantial quantities of insulin in the later phase. There is an alternative explanation for the acute-phase insulin response that involves neuronal stimulation of the pancreas. This is demonstrated by a series of experiments. Two of the most illustrative experiments are as follows. In one set of experiments, the sweet taste of saccharine is shown to induce AIR although there is no glucose ingestion which could stimulate the β cells [201]. In another set of experiments, rats were implanted with gastric drainage fistulas which removed ingested food before any absorption could take place. With this sham feeding too, the acute insulin response was seen [202]. In all these examples, only a feeling of feeding was there with no actual feeding followed by digestion–assimilation. Therefore the AIR could not be one for glucose or any other nutrient levels. It could only be neuronal, stimulated by sensory neuronal inputs from the oropharynx. This was well recognized in the 1980 and has been called a cephalic-phase insulin response [201–204].

A possible weakness in this argument, however, is that intravenous (i.v.) glucose also gives an AIR, albeit substantially smaller (only 30–40%) than oral glucose [205, 206]. The argument against the neuronal inputs is that i.v. glucose evades all sensory inputs from buccal and upper GIT and still gives AIR; therefore it must be independent of any neuronal mechanism. This argument is not foolproof since the possibility of glucose diffusing from plasma to the brain eliciting a neuronal response is not eliminated. There is evidence that even the i.v. glucose AIR could be partially dependent on neuronal mechanisms since denervation of islets further suppresses, although does not abolish, AIR (Fig. 3.2). Therefore suppression of AIR is more likely to be altered neuronal response than β cell defect. The relative contribution of the two is demonstrated from an experiment in dogs in which the pancreas was autotransplanted during which their enervation was lost. With a lost enervation, the ivGTT was not different from normal, but OGTT was significantly altered with prominent hyperglycemia persisting for over 3 h. The total insulin secretion was not different, but the time course of insulin activity was different [207]. This means that altered neuronal component can bring about an impaired GTT similar to pre or early diabetics. This necessitates a reevaluation of the relative contributions of β cell defects and neuronal effects in the impairment of GTT.

There is an alternative mechanistic explanation for AIR. It is presumed that when stimulated by glucose, the β cells release the preformed insulin instantaneously which forms the acute-phase response. In the second phase, newly synthesized insulin is released. If this is true, we should expect a higher AIR for ivGTT when glucose concentration rises suddenly as opposed to OGTT where glucose absorption is expected to cause some delay in stimulating β cells. However the reality is diametrically opposite. AIR is higher in OGTT as compared to ivGTT implying that this explanation is not realistic.

One more possible interpretation of β cell defect is a defect in converting proinsulin to insulin. Proinsulin is the precursor protein of insulin produced in the β cells which undergoes specific

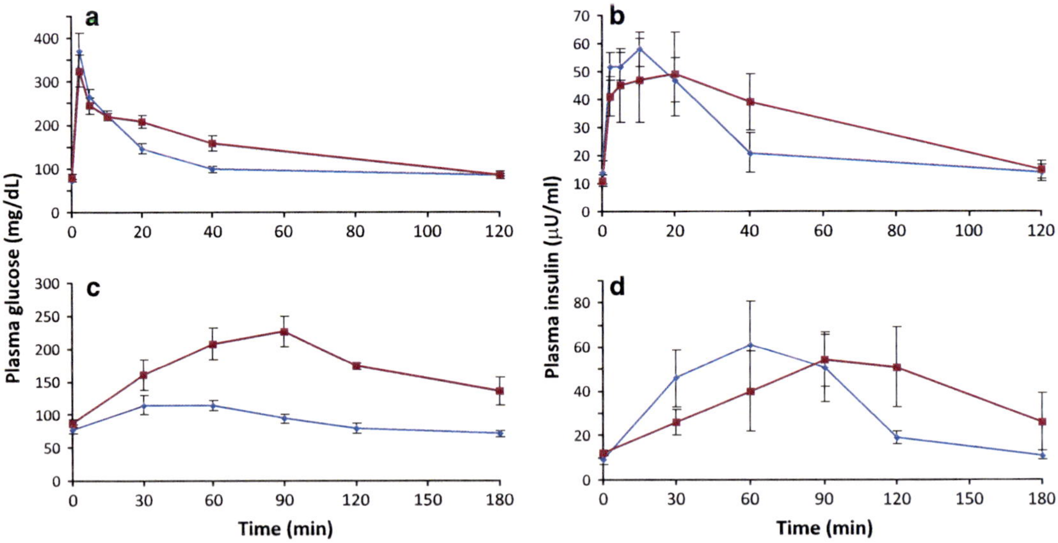

Fig. 3.2 Altered time course curves of plasma glucose and insulin in experimental dogs before and after autotransplantation of pancreas. (**a**) Glucose curve in ivGTT, (**b**) insulin curve in ivGTT, (**c**) glucose curve in OGTT, and (**d**) insulin curve in OGTT. *Blue circles* denote the curves before the operation; *red squares* denote the curves 10–12 weeks after the autotransplantation. Owing to autotransplantation, the nerve connections of the pancreas were lost, but there was no other histological damage. The nearly normal ivGTT demonstrates that the endocrine pancreas was capable of normal insulin secretion. However OGTT was substantially altered demonstrating a significant role played by neuronal component in the normal OGTT curve. (Data from Calhoun et al. [207])

enzymatic cleavage to produce insulin. During this process, another peptide called C-peptide is released. What is surprising is that even in the normal state, between 15% and 20% of circulating insulin is in the form of proinsulin. In diabetes, the proportion of proinsulin in the plasma increases two- to threefold [193]. Going beyond the orthodox view of treating proinsulin only as a precursor of insulin, one could suspect some function for proinsulin itself such that when more proinsulin is required, its conversion to insulin is downregulated. Proinsulin indeed has a function and that is in embryonic neuronal development [208–210]. So far, meager research inputs have gone into the possible functions of proinsulin, so there is very little data. However, doubts can certainly be raised whether larger proportions of proinsulin is a β cell "defect" or it is because greater quantities of proinsulin are needed for some other yet unknown function in the body. The well-known role of insulin in glucose metabolism created a large mental block which slowed the progress of research on other functions of insulin. We know now that insulin is involved in diverse functions of the body including memory and cognitive functions of the brain, sex, and reproduction [211–213]. It will not be a surprise if proinsulin has any such function even in adult life. The C-peptide, which was thought to be only a by-product of insulin synthesis, was later found to have multiple physiological effects [214]. We need to be open for such a possibility before concluding that proinsulin which is secreted so regularly by β cells in normal as well as diabetic state is a β cell "defect."

T2D is believed to be incurable. If we knew the root cause of the problem, we could have targeted it, and removing the root cause should be sufficient to cure the disorder. If obesity and insulin resistance following it are at the roots of the disorder, weight loss would have been sufficient to eliminate diabetes. Although there are demonstrable advantages of weight loss and it certainly increases insulin sensitivity, it is not sufficient to "cure" the disorder. A popular argument is that once the pancreatic islets are damaged, the disorder cannot be reversed. There is a demonstrable regression of islets, and the β cell mass is reduced in advanced T2D. There is hardly any demonstration of current diabetes treatment normalizing the β cells. Based on the belief that the β cell loss is permanent, there have been serious research inputs in stem cell research for generating new β cells. However, on the one hand, islet transplants and stem cell research have met with a very limited and temporary success in normalizing islet function [215, 216]; on the other hand, there is demonstration that the β cells are actually not very difficult to regenerate indigenously. Beta cells are capable of replication as well as can be newly generated from the acinar–ductal stem cells of the pancreas [194, 217–221]. Even in type 1 diabetes where there is extreme β cell loss, there is evidence for continued regeneration of β cells [222]. This means that the belief of β cell loss being the cause of irreversibility does not appear to be true. We are unable to identify the conditions under which β cell regeneration can be effectively brought about and create such conditions. In the absence of these conditions, even implanted stem cells will be unable to sustain for a long time.

In short, there are too many gaps in our understanding of what exactly causes hyperglycemia, and the naïve picture of insulin resistance plus β cell defect being responsible for it appears to be highly inadequate and laden with inconsistencies and paradoxes. I will certainly not be able to convince my undergraduate class that IR–RII is the cause of diabetic hyperglycemia. There are likely to be one or more other processes happening simultaneously, and elucidating them is necessary to increase our understanding of diabetic hyperglycemia. I will try to do it eventually in this book, but we can leave this point here with the conclusion that much of the story is still hidden.

5. Hyperglycemia is the main cause of the pathological consequences of diabetes: I found this point to be the most difficult to convince an undergraduate student in my hypothetical classroom. It is surprising that compelling evidence showing raised blood sugar as the sole or at least the main mechanism triggering all pathological complications is conspicuously absent in literature, but still this has been the most prevalent and largely unquestioned belief. The skepticism of the student in me goes along the following lines:

1. One of the main lines of evidence in support of the glycemic control hypothesis is correlational. A number of studies show that plasma glucose or HbA_{1c} levels are strongly correlated with one or more of the complications of diabetes [223–226]. Correlational evidence is always tricky since correlation does not say much about causation. Before taking these correlations as a support to glycemic control hypothesis, it is necessary to consider and eliminate possible alternative hypotheses that can explain such correlations. It can be argued that cases in which glucose control is difficult to achieve even after treatment are more prone to complications. This can explain the correlation without involving hyperglycemia in a causal role behind the complications. This possibility is supported by the difference in the findings of two different approaches. Epidemiological studies have invariably found a strong association from which it has been estimated that for each 1% increase in HbA_{1c}, the coronary heart disease risk increases by about 10% [226]. However clinical trials comparing intensive versus conventional treatment (see below) fail to show the same effect size [226], that is, reduction in HbA_{1c} by intensive treatment does not show the expected level of improvement. The substantial difference between the inferences of association studies and intervention trials [227] strengthens the possibility that there is a common pathological process that simultaneously drives the loss of glycemic control and development of complications rather than loss of glycemic control causing complications.

In most of the published studies, it is difficult to ascertain whether the higher glucose levels were a result of lack of patients' compliance with the treatment or lack of response to the treatment. Fewer studies have a design where the former possibility can be eliminated, for example, data from

ICU where drugs are administered by the hospital staff and therefore patients' compliance is not a variable. These studies clearly show that those who fail to reduce the glucose levels in response to treatment have a higher mortality rate [228]. This implies that some yet unknown factors independently determine both (1) difficulty of controlling sugar levels in spite of treatment and (2) frequency/severity of complications or mortality. As a result, a strong association is seen, but it does not necessarily mean that sugar levels cause the complications or by controlling sugar levels complications can be prevented. There are other possible interpretations too.

2. We can visualize a hypothetical correlation to illustrate the implication of correlation for causation. Among diabetics under treatment, it would be interesting to look at a correlation between the lifetime intake of all kinds of antidiabetic drugs and the frequency or severity of complications. By all probability, such correlations will be highly significant and positive. Patients having consumed larger amounts of drugs in a lifetime would show greater incidence of complications. This need not be because antidiabetic drugs cause these complications. It could be because most complications arise late in the history of diabetes, and longtime diabetics have consumed more medicines by then. Also, more unresponsive cases need higher doses, and they may be more susceptible to complications. Do we conclude from this correlation that antidiabetic drugs cause diabetic complications? If this conclusion from an observed correlation sounds ridiculous, why do we believe in glucose levels being responsible for the complications from correlational evidence?
3. An ideal experiment to test the glycemic control hypothesis will be to take two randomized groups of patients with comparable baseline glycemic states and treat one with drugs that have specifically glucose lowering action only, leave the other without treatment, and do a long-term follow-up. Since most complications of diabetes only appear after long-standing diabetes, the trial will have to run for several years. Such an experiment is most unlikely to be cleared by any ethics committee since it means denying treatment to the control group over a long time. However, an approach that goes close to this design has been taken by a number of large clinical trials in the last two decades [227, 229–231]. This approach is to compare treatment aiming towards modest glucose control versus aggressive treatment aiming towards completely "normalizing" glucose levels. Results of such trials are highly contradictory. Some of them show a significant reduction in the frequency or severity of specific classes of complications in the intensively treated group as compared to the moderately treated group, but others fail to see so. Generally the macrovascular complications do not seem to respond to tight glucose control, and in some of the trials, there is actually an increase in the mortality due to cardiovascular complications following intensive treatment [230–236] which necessitates a rethinking of the treatment goals [237]. Microvascular complications, on the other hand, appear to respond better to tighter glycemic control, but the relative risk is typically of the order of 0.9 or 0.8 [230, 231, 234] which means that the improvement is about 10–20% only. In no trial tight glycemic control successfully eliminated the effects of diabetes and brought down the risk level at par with nondiabetics. On the other hand, intensive treatment for regulating blood pressure brought about a similar effect size without a change in HbA_{1c} [238] demonstrating that factors other than hyperglycemia are likely to be involved in driving the pathology.

A rather strange concept to explain the failure of glycemic control to reduce the risk of complications is hyperglycemic memory. It is said that even if good glycemic control is achieved at present, a history of hyperglycemia may be sufficient to cause complications. Hyperglycemia in the past somehow keeps a memory somewhere in the body which is sufficient to drive the processes leading to complications. Because of this memory, aggressive glucose lowering therapy is unable to reduce the risk of complications substantially. This is an interesting hypothesis, and the way it is used is even more interesting. Whenever interventional trials failed to give evidence in support of hyperglycemia-driven pathology and even correlational evidence failed, hyperglycemic memory came in handy so that there was no need to think beyond glucose as a driver of pathological processes. All the evidence for hyperglycemic memory is negative in nature, that is, one reaches this conclusion only by ruling out others. If current glucose levels do not explain the complications, then it must be past glucose levels. This means of drawing an inference is not an invalid idea and is sometimes inevitable in science. But in order to use it, one has to think of and rule out all other plausible explanations. It is quite likely that there are other pathophysiological mechanisms in diabetes which are independent of glucose levels and they operate even if glucose control is achieved. This may lead to an apparent dissociation between glycemic control and diabetic complications. But the glucocentric mind block has never allowed this line of thinking to develop. Another problem with the hyperglycemic memory hypothesis is that it can often be shifted back to prediagnostic levels. Diabetes is often diagnosed after being there for quite some time. So there is an unestimated period of prediagnosis diabetes. In addition, prediabetic stages linger around for years or decades. Most often there are no data on it. Often intrauterine conditions can also be said to be the determinant. As a result, the unknown can be easily blamed. This facility to blame the unknown makes the hyperglycemia memory hypothesis unfalsifiable. None of the currently known mechanisms by which hyperglycemia drives pathological processes is irreversible [239], and therefore, there is no reason to believe in hyperglycemic memory until all possible alternatives are ruled out. We will see the alternative ways of thinking in later chapters; in fact, that is the main purpose of this book.

One of the widely agreed conclusions is that from no control to moderate control, there is a positive and impressive effect of glycemic control but from moderate control to tight control, there is little additional benefit. This inference is drawn by a meta-analysis and a Markov decision model [240]. Although the model looks sound, the data are heterogeneous. The former conclusion is mainly based on correlational data and the latter mainly on interventional trials. Therefore the validity of this conclusion still has a question mark.

The only robust inference that can be derived from all the trials and correlations which no one can deny is that keeping sugar level to the normal limits does not bring the diabetic at par with a nondiabetic person in relation to the risk of complications. This means that certainly there are additional pathophysiological mechanisms of diabetes that are independent of sugar levels.

4. Even if we assume that tight glycemic control achieved by drugs results in complete elimination of complications, does it clearly show that raised glucose is responsible for the pathophysiology and that normalizing glucose eliminates all pathological effects? It is possible that medication intended to reduce blood sugar has other independent effects on the real pathophysiological mechanisms responsible for the complications. There is some evidence in this direction too. For example, metformin has been shown to improve vascular function [241, 242], and rosiglitazone is shown to improve endothelial function independent of glucose control [243]. Similarly troglitazone is shown to improve *Klotho* expression [244]. Because such effects are possible and since they have not been thoroughly investigated,

it is difficult to conclude that sugar control prevents complications even if one could show that in intensively glucose-controlled group, there is total elimination of the risk of complications.

5. Not all pathological effects of diabetes can be prevented or reverted by normalizing blood sugar. For example, diabetes in males is frequently associated with erectile dysfunction and is associated with poor glycemic control [245]. Studies have shown that erectile dysfunction does not improve upon achieving good glycemic control [246] but long-term treatment of erectile dysfunction is insulin sensitizing [247].
6. There is a horse and cart problem once again. Hyperglycemia is said to be responsible for the pathophysiology, but the first signs of at least some of the pathophysiological processes start appearing much before overt hyperglycemia appears. Vascular endothelial dysfunction appears prior to the onset of overt diabetes [248–252] although hyperglycemia-induced oxidative stress has been blamed as a cause of it. Diabetes is believed to lead to autonomic neuropathy, but early signs of changes in autonomic function appear much earlier than diabetic hyperglycemia [253, 254]. Some have suspected autonomic dysfunction to be causal to T2D [254–257]. Evidence for causal role of neuropathy in insulin resistance is quite strong since hepatic denervation or cholinergic interruption has been shown to result in insulin resistance [258, 259]. If the classical interpretation that IR–RII causes hyperglycemia is correct, β cell function should deteriorate before the onset of hyperglycemia. But glucotoxicity is believed to be the main cause of β cell degeneration. In all these examples, the putative effects of hyperglycemia appear to precede it in time, and therefore, there is a question mark on causality. This is the horse and cart paradox IV where the cart appears to start running before the horse moves.
7. If some components of the pathophysiological processes begin prior to diabetic hyperglycemia, one can go back a step and ask whether insulin resistance or hyperinsulinemia itself triggers the pathological processes. There have been many claims that hyperinsulinemia itself or in combination with insulin resistance is responsible for hypertension and endothelial dysfunction among other pathological changes. In order to test this hypothesis, we can make use of insulinomas. An insulinoma results in hyperinsulinemia and accompanying insulin resistance, but the etiology is different from that of T2D. Studies that have looked at early signs of diabetic complications in insulinoma patients have failed to detect any [132, 133, 135]. This makes us suspect further that in the etiology of T2D, there is something other than insulin resistance, insulin levels, and hyperglycemia which is behind triggering the cascade of pathological changes. If so, that "something else" is the true culprit. In presence of many such evidences for the hand of "something else," why everyone treats insulin resistance, hyperinsulinemia, and hyperglycemia to be the only factors responsible for all the pathology of T2D and related disorders is very difficult to understand.

The teacher has no answer to the over half a dozen questions from the skeptic student. The teacher himself has taught the students what constitutes scientific evidence of causality. When things do not fall into the required level of scientific evidence, he can no more defend hyperglycemia as being central to diabetic pathology. He can at the most tell them the history of the origin of this concept. The thrust on normalizing blood sugar levels is a burden of history. In the history of discovery of diabetes, the sweetness of urine was the first symptom ever recorded, and this was more than a 1,000 years ago. Many of the ancient schools of medicine seem to have based the diagnosis of diabetes by the ant attracting property of urine or even the taste of urine. In modern medicine, detecting urine sugar or estimating blood sugar was some of the earliest known laboratory investigations. Because of the sequence of these discoveries and the associated clinical practices, all thinking in diabetes is centered on glucose control. Therefore, by the current belief, by bringing back blood sugar levels to normal, all the pathological outcomes could be prevented. A couple of decades ago, when there were data on plasma glucose alone, it was natural to think that the pathological changes are due to hyperglycemia alone. Today we know that a large number of changes take place simultaneously. A large number of hormones and other molecules are involved, and almost every system of the body is affected. This has made it difficult to discern cause–effect relationships clearly from clinical data. Certainly the picture is not as naïve as popularly believed.

But the age-old belief has several effects on the scientific community. It is common that people start seeing what they believe. Also they try to explain what they see only in the confines of what they believe. The mechanisms of hyperglycemia-mediated pathology appear to demonstrate this phenomenon quite well. Let us take the example of "glucotoxicity." The word glucotoxicity has been used in connection with β cell degeneration most frequently, and often it is treated as a mechanism of β cell defect. I will therefore use this example to examine the concept of glucotoxicity. There is a progressive degeneration of pancreatic islets and β cells in particular seen in T2D. The debate is whether raised glucose levels are the main cause of this degeneration and whether the effect of glucose on β cells can be called "toxicity." Glucose is a signaling molecule since a wide variety of cells respond to its presence and concentrations. Particularly important are the glucose-responsive and glucose-sensitive neurons in the brain which are connected to various functions. A signaling molecule is bound to bring about metabolic and hormonal changes in target tissues in a dose-dependent manner. This is what we mean by signaling molecule. The dose and effect of a signaling molecule can be highly nonlinear and even non-monotonic. Signaling molecules can have short-term effects as well as long-term programming effects. Therefore demonstration of some specific reversible or irreversible metabolic change in a tissue cannot be taken as an evidence of "toxicity." However if there is a general breakdown of a vital function such as breakdown of transcription machinery as a whole, it would be called toxicity. All that has been demonstrated under the name of glucotoxicity is decreased insulin production and reduction in insulin transcription levels in response either to transfused glucose or to pathological hyperglycemia [260]. However there is no general breakdown or loss of function of transcriptional machinery. Transcription of other proteins such as Il-1β is functional under the influence of "glucotoxicity" [261]. There are other limitations for the demonstration of glucotoxicity as well. In the natural course of hyperglycemia, since a large number of endocrine and metabolic changes take place simultaneously in the body, it is difficult to show a specific effect of glucose. If glucose levels are artificially increased by transfusing glucose from outside, we can assume that the observed changes are due to altered glucose levels. This cannot be done over a long time. Results of such experiments over moderate time are not consistent across studies, some having reported an increase and others a decrease in insulin production after exposure to high glucose concentrations [193, 262, 263]. Glucose increases insulin production in some studies and decreases in others. Glucose facilitates β cell growth in some studies and degeneration in others. If this is conditional, it is not yet clearly known what the conditions are under which glucose is toxic to β cells. This has created a lot of ambiguity in the concept of glucotoxicity, and I would say if there is true glucotoxicity, it is yet to be clearly demonstrated.

Even if we assume that chronic exposure to high levels of glucose reduces the insulin response of β cells, it is difficult to label it as glucotoxicity. Desensitization of any functional response on chronic overexposure to a normal stimulus is not unique to glucose and insulin production. If you work in a chemical factory which has a peculiar odor, in a few days you stop noticing the odor as there is a specific desensitization of the olfactory receptors. Sitting in a room where a clock ticks continuously, you stop hearing the clock tick after a short time. This is not called "toxicity," it is only habituation. There are similar examples within the endocrine system as well. Patients with choriocarcinoma who have very high levels of human chorionic gonadotropin (hCG) may show no sign of excess gonadotropin and are unresponsive to injected hCG. After removing the tumor, the response may return [157]. Cortisol has an immunosuppressive effect, but chronically elevated cortisol levels may not show the immunosuppressive effects. The phenomenon of homologous desensitization is common in physiology and certainly not sufficient for the definition of "toxicity." Also since signs of β cell defect are said to appear years before diabetic hyperglycemia may lead to glucotoxicity, the causality is in question. Therefore whether there is anything that can be justifiably called glucotoxicity is questionable. But because of the burden of history, people had already decided that glucose has to be blamed for everything bad that happens in diabetes. This prejudice could be the father of the concept of glucotoxicity.

One more paradox associated with the pathophysiology of diabetes is that of diametrically opposite effects in different parts of the body. Diabetes is associated with impaired wound healing on the one hand and retinopathy and nephropathy on the other. In both, angiogenesis dysfunction is an important component, but the directions are diametrically opposite. Impaired angiogenesis is one of the causes of wound healing problems in diabetes [264, 265]. Inefficient angiogenesis mechanisms appear to hamper normal wound healing, and in contrast, diabetic retinopathy and nephropathy are marked by excessive angiogenesis [266–268]. If hyperglycemia were the driving pathogenic factor influencing angiogenesis, why should the effects be diametrically opposite when the same blood, containing the same sugar levels, is being circulated everywhere? This is the horse and cart paradox V. When the horse runs, half of the cart appears to move forward and half backward. This riddle also remains unresolved with the current paradigm and is hardly ever discussed or even acknowledged. We will try to resolve this in a later chapter.

This discussion is not intended to deny any role of hyperglycemia in the pathology of diabetes but certainly to point out that as yet there is no clear demonstration that raised glucose levels are exclusively or even mainly responsible for all the pathological consequences of diabetes. Consequently the assumption that improving glycemic control is the only and sufficient target of diabetes treatment is seriously questionable. Many diabetes researchers are aware of it, and a representative quote by Geoffrey Gill is "Absolute proof that good (glycemic) control can retard or prevent the development of complications has not yet been obtained, but the assumption is accepted by most diabetologists and indeed is almost an article of faith in the current approach to achieve the best possible diabetic control" [269]. Gill wrote this in the context of T1D in 1991. It is equally or perhaps more applicable to T2D as well, and in the last 20 years, situation has not changed. If anything, there are more data to cast doubts on the glucocentric view of pathology.

My arguments go by the logic of a science teacher who is more concerned with the logical structure of science and the nature of evidence. I am aware that clinicians have a somewhat different viewpoint. Their main aim is to treat a patient effectively. If they are able to do so, they may not be worried too much if there are some lapses in the experimental evidence or inferential logic part of the story. So let us examine whether the picture is good enough to keep the clinicians happy. For this, let us assume as the clinicians do that glycemic control is the key to the treatment of diabetes and/or prevention of complications. Everything would be set right on achieving glycemic control. Are the clinicians successful in achieving this goal? Ironically glycemic control itself is progressively lost with increasing attempts to achieve it [228, 270, 271]. It is quite well known that in a large fraction of diabetics, glycemic control goes on becoming progressively more difficult to achieve. Higher and higher doses of drugs are required to achieve the same level of control, and this deterioration cannot be stopped by keeping sugar levels normal [193]. This presumably originates from our inability to stop the progressive degeneration of islets in spite of glucose lowering. This puts a big question mark on the currently perceived cause–effect relationship. Almost every antidiabetic drug shows dramatic effects on first administration. But after the honeymoon period is over, it starts losing its effect progressively such that eventually even higher doses are unable to achieve desired glycemic control. For example, in the United Kingdom Prospective Diabetes Study (UKPDS), after 9 years, only 25% of the subjects achieved HbA_{1c} less than 7%. The treatment goal could be attained in 8% of people on dietary therapy, 13% on metformin, 24% sulphonylurea, and 42% on insulin. Thus any treatment measure currently does not have even 50% success rate in the long run. Why this happens is not clearly understood but can be certainly taken as an indication of failure of clinical practice.

This implies that the picture painted by the orthodox theory and summarized by the five postulates above is certainly not satisfactory. In fact, not even a single of the above postulates has been conclusively demonstrated although the theory of insulin resistance exists for almost half a century now, and the amount of research efforts that have gone in are certainly not meager. It is amazing that so much of research has not been able to give conclusive evidence for even a single component of the theory. It is even more amazing that in spite of the failures to experimentally support any of the postulates, the theory has remained almost unchallenged for so long. This reminds me of an interesting quote. Richard Feynman said, "I think I can safely say that no one understands quantum mechanics." A response to this was that "Yes, we do not understand quantum mechanics but we are well accustomed to it!" This can be even more appropriate to diabetes. There are so many flaws, paradoxes, illogical inferences, and beliefs that would simply amaze any fresh and unbiased reader. But since we keep on hearing about it without bothering to ask critical questions, we all accept it in whatever form. Once we get "accustomed" to it, questions stop bothering us. Our classroom was able to raise so many questions only because we made a fresh beginning and started viewing things critically without having been accustomed to anything.

A class of researchers has certainly realized this, and not only have they pointed out flaws in the mainstream thinking, there are also attempts to develop alternative ways of thinking. For example, it is recognized and argued by many authors that hyperinsulinemia appears first and may lead to both obesity and insulin resistance [2, 33, 80, 157, 165, 166]. But this is only a part of the picture. This work has not substantially affected mainstream thinking and clinical practice since a complete and coherent picture, in which these pieces fit, is yet to emerge.

Some theories are die-hard, and in spite of strong evidence against them, they will continue to survive and dominate [272]. But this is not so much of a surprise. Theories will die under the burden of evidence against them only if alternative theories are emerging. Any promising alternative ways of thinking were not on the horizon so far, and therefore although a large number of researchers and practitioners are aware of its limitations, they still need to stick to it for the want of an alternative platform. Obviously our next target should be to look for such a platform. The alternative platform should allow putting together all the pieces to make a coherent picture and resolve all the paradoxes associated with the classical interpretation including the five important horse and cart paradoxes above and should not create new paradoxes and riddles that are worse than these.

References

1. Spiegelman BM, Flier JS (2001) Obesity and the regulation of energy balance. Cell 104:531–543
2. Lustig RH (2006) Childhood obesity: behavioral aberration or biochemical drive? Reinterpreting the First Law of Thermodynamics. Nat Clin Pract Endocrinal Metab 2:447–458
3. Wells JCK, Siervo M (2011) Obesity and energy balance: is the tail wagging the dog[quest]. Eur J Clin Nutr 65:1173–1189
4. Gary T (2008) The great diet scandal. *The New Scientist*, vol 197:p. 17
5. Speakman JR (1998) The history and theory of the doubly labeled water technique. Am J Clin Nutr 68:932S–938S

6. Westerterp KR, Speakman JR (2008) Physical activity energy expenditure has not declined since the 1980s and matches energy expenditures of wild mammals. Int J Obes (Lond) 32:1256–1263
7. Heini A, Weinsier R (1997) Divergent trends in obesity and fat intake patterns: the American paradox. Am J Med 102:259–264
8. Gibney MJ, Burstyn PG (1980) Milk, serum cholesterol, and the Maasai: a hypothesis. Atherosclerosis 35:339–343
9. Brown GW (1993) Maasai diet. Lancet 341:377
10. Mann GV, Spoerry A (1974) Studies of a surfactant and cholesteremia in the Maasai. Am J Clin Nutr 27:464–469
11. Hall KD, Guo J, Dore M, Chow CC (2009) The progressive increase of food waste in America and its environmental impact. PLoS ONE 4:e7940
12. Chow CC (2010) Summary of SIAM talk. http://sciencehouse.wordpress.com/2010/07/23/summary-of-siam-talk/
13. Stanley S, Wynne K, McGowan B, Bloom S (2005) Hormonal regulation of food intake. Physiol Rev 85:1131–1158
14. Morton GJ, Cummings DE, Baskin DG, Barsh GS, Schwartz MW (2006) Central nervous system control of food intake and body weight. Nature 443:289–295
15. Cummings DE, Overduin J (2007) Gastrointestinal regulation of food intake. J Clin Invest 117:13–23
16. Friedman JM (2000) Obesity in the new millennium. Nature 404:632–634
17. Kraly FS, Carty WJ, Resnick S, Smith GP (1978) Effect of cholecystokinin on meal size and intermeal interval in the sham-feeding rat. J Comp Physiol Psychol 92:697–707
18. Young RC et al (1974) Absence of satiety during sham feeding in the rat. J Comp Physiol Psychol 87:795–800
19. Rayner DV (1992) Gastrointestinal satiety in animals other than man. Proc Nutr Soc 51:1–6
20. Vanderweele DA, Deems RO, Kanarek RB (1990) Insulin modifies flavor aversions and preferences in real- and sham-feeding rats. Am J Physiol 259: 823–828
21. Oetting RL, Vanderweele DA (1985) Insulin suppresses intake without inducing illness in sham feeding rats. Physiol Behav 34:557–562
22. Martin CF, Gibbs J (1980) Bombesin elicits satiety in sham feeding rats. Peptides 1:131–134
23. McGinty D, Epstein AN, Teitelbaum P (1965) The contribution of oropharyngeal sensations to hypothalamic hyperphagia. Anim Behav 13:413–418
24. Epstein AN, Teitelbaum P (1962) Regulation of food intake in the absence of taste, smell, and other oropharyngeal sensations. J Comp Physiol Psychol 55:753–759
25. Hill RG, Ison EC, Jones WW, Archdeacon JW (1952) The small intestine as a factor in regulation of eating. Am J Physiol 170:201–205
26. Mutch DM, Clément K (2006) Unraveling the genetics of human obesity. PLoS Genet 2:e188
27. Kraegen EW et al (1991) Development of muscle insulin resistance after liver insulin resistance in high-fat-fed rats. Diabetes 40:1397–1403
28. Storlien LH et al (1991) Influence of dietary fat composition on development of insulin resistance in rats. Relationship to muscle triglyceride and omega-3 fatty acids in muscle phospholipid. Diabetes 40:280–289
29. Guerre-Millo M et al (2001) PPAR-α-null mice are protected from high-fat diet-induced insulin resistance. Diabetes 50:2809–2814
30. West CE, Sullivian DR, Katan MB, Halferkamps IL, Van Der Torre HW (1990) Boys from populations with high carbohydrate intake have higher fasting triglyceride levels than boys from populations with high fat intake. Am J Epidemiol 131:271–282
31. Atkins Facts – American Heart Association (2011) http://www.atkinsexposed.org/atkins/100/American_Heart_Association.htm
32. Teuscher T, Rosman JB, Baillod P, Teuscher A (1987) Absence of diabetes in a rural West African population with a high carbohydrate/cassava diet. Lancet 329:765–768
33. Taubes G (2004) The diet delusion: challenging the conventional wisdom on diet, weight loss and disease. Vermillion, London
34. Grande F, Anderson JT, Keys A (1970) Comparison of effects of palmitic and stearic acids in the diet on serum cholesterol in man. Am J Clin Nutr 23:1184–1193
35. Keys A (1957) Diet and the epidemiology of coronary heart disease. J Am Med Assoc 164:1912–1919
36. Keys A, Anderson JT, Grande F (1965) Serum cholesterol response to changes in the diet: IV. Particular saturated fatty acids in the diet. Metabolism 14:776–787
37. Yancy WS, Olsen MK, Guyton JR, Bakst RP, Westman EC (2004) A low-carbohydrate, ketogenic diet versus a low-fat diet to treat obesity and hyperlipidemia. Ann Intern Med 140:769–777
38. Samaha FF et al (2003) A low-carbohydrate as compared with a low-fat diet in severe obesity. N Engl J Med 348:2074–2080
39. Bray GA, Paeratakul S, Popkin BM (2004) Dietary fat and obesity: a review of animal, clinical and epidemiological studies. Physiol Behav 83:549–555
40. Lissner L, Heitmann BL (1995) Dietary fat and obesity: evidence from epidemiology. Eur J Clin Nutr 49:79–90
41. Tremblay F, Marette A (2001) Amino acid and insulin signaling via the mTOR/p70 S6 kinase pathway. J Biol Chem 276:38052–38060
42. Zick Y (2005) Ser/Thr phosphorylation of IRS proteins: a molecular basis for insulin resistance. Sci STKE 2005:pe4
43. Venn-Watson SK, Ridgway SH (2007) Big brains and blood glucose: common ground for diabetes mellitus in humans and healthy dolphins. Comp Med 57:390–395
44. Zoppi CC et al (2010) Insulin release, peripheral insulin resistance and muscle function in protein

malnutrition: a role of tricarboxylic acid cycle anaplerosis. Br J Nutr 103:1237–1250
45. Grace CJ, Swenne I, Kohn PG, Strain AJ, Milner RD (1990) Protein-energy malnutrition induces changes in insulin sensitivity. Diabete Metab 16:484–491
46. Schteingart DE, McKenzie AK, Victoria RS, Tsao HS (1979) Suppression of insulin secretion by protein deprivation in obesity. Adv Exp Med Biol 119:125–135
47. Wells JCK (2009) The evolutionary biology of human body fatness. Cambridge University Press, Cambridge
48. Adequate Nutrients Within Calorie Needs. In: Dietary guidelines for Americans 2005 USDA. http://www.health.gov/dietaryguidelines/dga2005/document/html/chapter2.htm
49. Diet and Weight Loss Products|CaloriesPerHour.com. http://www.caloriesperhour.com/products.php
50. Cordain L, Eaton SB, Miller JB, Mann N, Hill K (2002) The paradoxical nature of hunter-gatherer diets: meat-based, yet non-atherogenic. Eur J Clin Nutr 56(Suppl 1):S42–S52
51. Nelson L (1996) Diet composition related to body fat in a multivariate study of 203 men. J Am Diet Assoc 96:771–777
52. Atkin L-M, Davies PS (2000) Diet composition and body composition in preschool children. Am J Clin Nutr 72:15–21
53. Boyce VL, Swinburn BA (1993) The traditional pima Indian diet. Composition and adaptation for use in a dietary intervention study. Diabetes Care 16:369–371
54. Story M et al (1999) The epidemic of obesity in American Indian communities and the need for childhood obesity-prevention programs. Am J Clin Nutr 69:747S–754S
55. Bang H, Dyerberg J, Sinclair H (1980) The composition of the Eskimo food in north western Greenland. Am J Clin Nutr 33:2657–2661
56. Manners J (1997) Kenya's running tribe. Sports Hist 17:14–27
57. Cordain L et al (2000) Plant-animal subsistence ratios and macronutrient energy estimations in worldwide hunter-gatherer diets. Am J Clin Nutr 71:682–692
58. Masai Tribe Diet | LIVESTRONG.COM. http://www.livestrong.com/article/293306-masai-tribe-diet/
59. Macronutrient Ratios | CaloriesPerHour.com. http://www.caloriesperhour.com/tutorial_ratios.php
60. Milton K (2000) Hunter-gatherer diets – a different perspective. Am J Clin Nutr 71:665–667
61. Beis LY et al (2011) Food and macronutrient intake of elite Ethiopian distance runners. J Int Soc Sports Nutr 8:7
62. Vidal FS (1954) Date culture in the oasis of Al-hasa. Middle East J 8:417–428
63. Cooking Practices and Health of Hunter-Gatherers/!Kung San. http://www.beyondveg.com/tu-j-l/raw-cooked/raw-cooked-3f.shtml
64. Gordon R (1999) Hopper changing food production and quality of diet in India, 1947-98. Popul Dev Rev 25:443–477
65. Atkins Facts – American Heart Association. http://www.atkinsexposed.org/atkins/100/American_Heart_Association.htm
66. Eaton SB, Konner M, Shostak M (1988) Stone agers in the fast lane: chronic degenerative diseases in evolutionary perspective. Am J Med 84:739–749
67. Miller W, Lindeman A, Wallace J, Niederpruem M (1990) Diet composition, energy intake, and exercise in relation to body fat in men and women. Am J Clin Nutr 52:426–430
68. Day J, Carruthers M, Bailey A, Robinson D (1976) Anthropometric, physiological and biochemical differences between urban and rural Maasai. Atherosclerosis 23:357–361
69. Schaefer O (1971) When the Eskimo comes to town: nutrition today. Nutr Today 6:8–16
70. Yu CHY, Zinman B (2007) Type 2 diabetes and impaired glucose tolerance in aboriginal populations: a global perspective. Diabetes Res Clin Pract 78:159–170
71. O'Dea K, Patel M, Kubisch D, Hopper J, Traianedes K (1993) Obesity, diabetes, and hyperlipidemia in a central Australian aboriginal community with a long history of acculturation. Diabetes Care 16:1004–1010
72. Eaton C (1977) Part two: diabetes, culture change, and acculturation: a biocultural analysis 1. Med Anthropol 1:41–63
73. Joffe BI et al (1971) Metabolic responses to oral glucose in the Kalahari Bushmen. Br Med J 4:206–208
74. Pond CM (1998) The fats of life. Cambridge University Press, Cambridge
75. Neel JV (1962) Diabetes mellitus: a "thrifty" genotype rendered detrimental by "progress"? Am J Hum Genet 14:353–362
76. Brüning JC et al (1998) A muscle-specific insulin receptor knockout exhibits features of the metabolic syndrome of NIDDM without altering glucose tolerance. Mol Cell 2:559–569
77. Kadowaki T (2000) Insights into insulin resistance and type 2 diabetes from knockout mouse models. J Clin Invest 106:459–465
78. Terauchi Y, Kadowaki T (2002) Insights into molecular pathogenesis of type 2 diabetes from knockout mouse models. Endocr J 49:247–263
79. Isganaitis E, Lustig RH (2005) Fast food, central nervous system insulin resistance, and obesity. Arterioscler Thromb Vasc Biol 25:2451–2462
80. Lustig RH, Sen S, Soberman JE, Velasquez-Mieyer PA (2004) Obesity, leptin resistance, and the effects of insulin reduction. Int J Obes Relat Metab Disord 28:1344–1348
81. Ruderman N, Chisholm D, Pi-Sunyer X, Schneider S (1998) The metabolically obese, normal-weight individual revisited. Diabetes 47:699–713
82. Conus F et al (2004) Metabolic and behavioral characteristics of metabolically obese but normal-weight women. J Clin Endocrinol Metab 89:5013–5020
83. Succurro E et al (2008) Insulin secretion in metabolically obese, but normal weight, and in metabolically healthy but obese individuals. Obesity 16:1881–1886

84. Zheng W et al (2011) Association between body-mass index and risk of death in more than 1 million Asians. N Engl J Med 364:719–729
85. Virtue S, Vidal-Puig A (2008) It's not how fat you are, It's what you do with it that counts. PLoS Biol 6:e237
86. Nesto RW (2005) Obesity. Tex Heart Inst J 32:387–389
87. Matsuzawa Y, Shimomura I, Nakamura T, Keno Y, Tokunaga K (1994) Pathophysiology and pathogenesis of visceral fat obesity. Diabet Res Clin Pract 24(Suppl):S111–S116
88. Kelley DE, Thaete FL, Troost F, Huwe T, Goodpaster BH (2000) Subdivisions of subcutaneous abdominal adipose tissue and insulin resistance. Am J Physiol Endocrinol Metab 278:E941–E948
89. Cnop M et al (2002) The concurrent accumulation of intra-abdominal and subcutaneous fat explains the association between insulin resistance and plasma leptin concentrations. Diabetes 51:1005–1015
90. Goodpaster BH, Thaete FL, Simoneau JA, Kelley DE (1997) Subcutaneous abdominal fat and thigh muscle composition predict insulin sensitivity independently of visceral fat. Diabetes 46:1579–1585
91. Pecioska S et al (2010) Association between type 2 diabetes loci and measures of fatness. PLoS One 5:e8541
92. Kilpelainen TO et al (2011) Genetic variation near IRS1 associates with reduced adiposity and an impaired metabolic profile. Nat Genet 43:753–760
93. Garland PB, Newsholme EA, Randle PJ (1964) Regulation of glucose uptake by muscle. 9. Effects of fatty acids and ketone bodies, and of alloxan-diabetes and starvation, on pyruvate metabolism and on lactate/pyruvate and l-glycerol 3-phosphate/dihydroxyacetone phosphate concentration ratios in rat heart and rat diaphragm muscles. Biochem J 93:665–678
94. Randle PJ, Priestman DA, Mistry SC, Halsall A (1994) Glucose fatty acid interactions and the regulation of glucose disposal. J Cell Biochem 55(Suppl):1–11
95. Turner N et al (2007) Excess lipid availability increases mitochondrial fatty acid oxidative capacity in muscle. Diabetes 56:2085–2092
96. Bajaj M et al (2007) Paradoxical changes in muscle gene expression in insulin-resistant subjects after sustained reduction in plasma free fatty acid concentration. Diabetes 56:743–752
97. Morino K (2005) Reduced mitochondrial density and increased IRS-1 serine phosphorylation in muscle of insulin-resistant offspring of type 2 diabetic parents. J Clin Invest 115:3587–3593
98. Kadowaki T (2006) Adiponectin and adiponectin receptors in insulin resistance, diabetes, and the metabolic syndrome. J Clin Invest 116:1784–1792
99. Berg AH, Combs TP, Scherer PE (2002) ACRP30/adiponectin: an adipokine regulating glucose and lipid metabolism. Trend Endocrinol Metab 13: 84–89
100. Goldstein BJ, Scalia R (2004) Adiponectin: a novel adipokine linking adipocytes and vascular function. J Clin Endocrinol Metab 89:2563–2568
101. Viengchareun S, Zennaro M-C, Pascual-Le Tallec L, Lombes M (2002) Brown adipocytes are novel sites of expression and regulation of adiponectin and resistin. FEBS Lett 532:345–350
102. Ouchi N et al (2010) Sfrp5 is an anti-inflammatory adipokine that modulates metabolic dysfunction in obesity. Science 329:454–457
103. Mori MA et al (2010) A systems biology approach identifies inflammatory abnormalities between mouse strains prior to development of metabolic disease. Diabetes 59:2960–2971
104. Poirier B et al (2005) The anti obesity effect of rimonabant is associated with an improved serum lipid profile. Diabetes Obes Metab 7:65–72
105. Qi Y et al (2004) Adiponectin acts in the brain to decrease body weight. Nat Med 10:524–529
106. Margetic S, Gazzola C, Pegg GG, Hill RA (2002) Leptin: a review of its peripheral actions and interactions. Int J Obes Relat Metab Disord 26:1407–1433
107. Buettner C et al (2008) Leptin controls adipose tissue lipogenesis via central, STAT3-independent mechanisms. Nat Med 14:667–675
108. Munzberg H, Myers MG (2005) Molecular and anatomical determinants of central leptin resistance. Nat Neurosci 8:566–570
109. Andrews R, Walker BR (1999) Glucocorticoids and insulin resistance: old hormones, new targets. Clin Sci 96:513–523
110. Virtue S, Vidal-Puig A (2010) Adipose tissue expandability, lipotoxicity and the metabolic syndrome – an allostatic perspective. Biochim Biophys Acta 1801:338–349
111. Goodpaster BH, Theriault R, Watkins S, Kelley DE (2000) Intramuscular lipid content is increased in obesity and decreased by weight loss. Metabolism 49:467–472
112. Kelley DE, Goodpaster BH, Storlien LH (2002) Muscle triglyceride and insulin resistance. Annu Rev Nutr 22:325–346
113. He J, Watkins S, Kelley DE (2001) Skeletal muscle lipid content and oxidative enzyme activity in relation to muscle fiber type in type 2 diabetes and obesity. Diabetes 50:817–823
114. He J, Kelley DE (2004) Muscle glycogen content in type 2 diabetes mellitus. Am J Physiol Endocrinol Metab 287:1002–1007
115. Pan DA, Lillioja S, Kriketos AD, Milner MR, Baur LA (1997) Skeletal muscle triglyceride levels are inversely related to insulin action. Diabetes 46:983–988
116. Dubé JJ et al (2008) Exercise-induced alterations in intramyocellular lipids and insulin resistance: the athlete's paradox revisited. Am J Physiol Endocrinol Metab 294:E882–E888
117. Stannard SR, Johnson NA (2004) Insulin resistance and elevated triglyceride in muscle: more important for survival than "thrifty" genes? J Physiol 554: 595–607
118. van Loon LJ et al (2004) Intramyocellular lipid content in type 2 diabetes patients compared with overweight sedentary men and highly trained endurance athletes. Am J Physiol Endocrinol Metab 287:E558–E565

119. Tarnopolsky MA et al (2007) Influence of endurance exercise training and sex on intramyocellular lipid and mitochondrial ultrastructure, substrate use, and mitochondrial enzyme activity. Am J Physiol Regul Integr Comp Physiol 292:R1271–R1278
120. Goodpaster BH, He J, Watkins S, Kelley DE (2001) Skeletal muscle lipid content and insulin resistance: evidence for a paradox in endurance-trained athletes. J Clin Endocrinol Metab 86:5755–5761
121. Jauch-Chara K, Schmoller A, Oltmanns KM (2011) Impaired glucose tolerance in healthy men with low body weight. Nutr J 10:16
122. Pardini VC et al (1998) Leptin levels, {β}-cell function, and insulin sensitivity in families with congenital and acquired generalized lipoatropic diabetes. J Clin Endocrinol Metab 83:503–508
123. Rao RH (1988) Diabetes in the undernourished: coincidence or consequence? Endocr Rev 9:67–87
124. Unger RH, Scherer PE (2010) Gluttony, sloth and the metabolic syndrome: a roadmap to lipotoxicity. Trends Endocrinol Metab 21:345–352
125. Franckhauser S et al (2002) Increased fatty acid re-esterification by PEPCK overexpression in adipose tissue leads to obesity without insulin resistance. Diabetes 51:624–630
126. Dubuc PU (1976) The development of obesity, hyperinsulinemia, and hyperglycemia in ob/ob mice. Metabolism 25:1567–1574
127. Dubuc PU (1981) Non-essential role of dietary factors in the development of diabetes in ob/ob mice. J Nutr 111:1742–1748
128. Le Stunff C, Bougnères P (1994) Early changes in postprandial insulin secretion, not in insulin sensitivity, characterize juvenile obesity. Diabetes 43:696–702
129. Wagner JE et al (2006) Old world nonhuman primate models of type 2 diabetes mellitus. ILAR J 47: 259–271
130. Duvanel CB, Fawer CL, Cotting J, Hohlfeld P, Matthieu JM (1999) Long-term effects of neonatal hypoglycemia on brain growth and psychomotor development in small-for-gestational-age preterm infants. J Pediatr 134:492–498
131. Nankervis A, Proietto J, Aitken P, Alford F (1985) Hyperinsulinaemia and insulin insensitivity: studies in subjects with insulinoma. Diabetologia 28:427–431
132. Sawicki P, Baba T, Berger M, Starke A (1992) Normal blood pressure in patients with insulinoma despite hyperinsulinemia and insulin resistance. J Am Soc Nephrol 3:S64–S68
133. Pontiroli AE, Alberetto M, Pozza G (1992) Patients with insulinoma show insulin resistance in the absence of arterial hypertension. Diabetologia 35: 294–295
134. Liu J et al (2000) The intracellular mechanism of insulin resistance in pancreatic cancer patients. J Clin Endocrinol Metab 85:1232–1238
135. Leonetti F et al (1993) Absence of clinically overt atherosclerotic vascular disease and adverse changes in cardiovascular risk factors in 70 patients with insulinoma. J Endocrinol Invest 16:875–880
136. Del Prato S et al (1993) Mechanisms of fasting hypoglycemia and concomitant insulin resistance in insulinoma patients. Metab Clin Exp 42:24–29
137. Kim JK et al (2000) Redistribution of substrates to adipose tissue promotes obesity in mice with selective insulin resistance in muscle. J Clin Invest 105:1791–1797
138. Kulkarni RN et al (1999) Tissue-specific knockout of the insulin receptor in pancreatic β cells creates an insulin secretory defect similar to that in type 2 diabetes. Cell 96:329–339
139. Yakar S et al (2001) Liver-specific igf-1 gene deletion leads to muscle insulin insensitivity. Diabetes 50:1110–1118
140. Chakravarthy MV et al (2008) Decreased fetal size is associated with β-cell hyperfunction in early life and failure with age. Diabetes 57:2698–2707
141. Weyer C, Hanson RL, Tataranni PA, Bogardus C, Pratley RE (2000) A high fasting plasma insulin concentration predicts type 2 diabetes independent of insulin resistance: evidence for a pathogenic role of relative hyperinsulinemia. Diabetes 49:2094–2101
142. Jetton TL et al (2005) Mechanisms of compensatory β-cell growth in insulin-resistant rats. Diabetes 54:2294–2304
143. Garvey WT, Olefsky JM, Marshall S (1986) Insulin induces progressive insulin resistance in cultured rat adipocytes. Sequential effects at receptor and multiple postreceptor sites. Diabetes 35:258–267
144. Koopmans SJ et al (1991) Amylin-induced in vivo insulin resistance in conscious rats: the liver is more sensitive to amylin than peripheral tissues. Diabetologia 34:218–224
145. Frontoni S, Choi SB, Banduch D, Rossetti L (1991) In vivo insulin resistance induced by amylin primarily through inhibition of insulin-stimulated glycogen synthesis in skeletal muscle. Diabetes 40:568–573
146. Molina JM, Cooper GJ, Leighton B, Olefsky JM (1990) Induction of insulin resistance in vivo by amylin and calcitonin gene-related peptide. Diabetes 39:260–265
147. Marzban L, Park K, Verchere CB (2003) Islet amyloid polypeptide and type 2 diabetes. Exp Gerontol 38:347–351
148. Tabata H et al (1992) Islet amyloid polypeptide (IAPP/amylin) causes insulin resistance in perfused rat hindlimb muscle. Diabetes Res Clin Pract 15:57–61
149. Sowa R et al (1990) Islet amyloid polypeptide amide causes peripheral insulin resistance in vivo in dogs. Diabetologia 33:118–120
150. Ye J-M et al (2001) Evidence that amylin stimulates lipolysis in vivo: a possible mediator of induced insulin resistance. Am J Physiol Endocrinol Metab 280:E562–E569
151. Chen C-D, Podvin S, Gillespie E, Leeman SE, Abraham CR (2007) Insulin stimulates the cleavage and release of the extracellular domain of Klotho by ADAM10 and ADAM17. Proc Natl Acad Sci USA 104:19796–19801
152. Kurosu H et al (2005) Suppression of aging in mice by the hormone Klotho. Science 309:1829–1833

153. Bartke A (2006) Long-lived Klotho mice: new insights into the roles of IGF-1 and insulin in aging. Trends Endocrinol Metab 17:33–35
154. Fernstrom JD, Wurtman RJ (1971) Brain serotonin content: increase following ingestion of carbohydrate diet. Science 174:1023–1025
155. Luo S, Luo J, Cincotta AH (1999) Chronic ventromedial hypothalamic infusion of norepinephrine and serotonin promotes insulin resistance and glucose intolerance. Neuroendocrinology 70:460–465
156. Tzotzas T, Papazisis K, Perros P, Krassas GE (2008) Use of somatostatin analogues in obesity. Drugs 68:1963–1973
157. Shanik MH et al (2008) Insulin resistance and hyperinsulinemia: is hyperinsulinemia the cart or the horse? Diabetes Care 31(Suppl 2):S262–S268
158. Ratzmann KP, Ruhnke R, Kohnert KD (1983) Effect of pharmacological suppression of insulin secretion on tissue sensitivity to insulin in subjects with moderate obesity. Int J Obes 7:453–458
159. Prentki M (2006) Islet cell failure in type 2 diabetes. J Clin Invest 116:1802–1812
160. Alemzadeh R, Holshouser S, Massey P, Koontz J (2002) Chronic suppression of insulin by diazoxide alters the activities of key enzymes regulating hepatic gluconeogenesis in Zucker rats. Eur J Endocrinol 146:871–879
161. Alemzadeh R, Karlstad M, Tushaus K, Buchholz M (2008) Diazoxide enhances basal metabolic rate and fat oxidation in obese Zucker rats. Metabolism 57:1597–1607
162. Alemzadeh R, Tushaus KM (2004) Modulation of adipoinsular axis in prediabetic zucker diabetic fatty rats by diazoxide. Endocrinology 145:5476–5484
163. Schreuder T et al (2005) Diazoxide mediated insulin suppression in obese men: a dose response study. Diabetes Obes Metab 7:239–245
164. Velasquez-Mieyer P et al (2003) Suppression of insulin secretion is associated with weight loss and altered macronutrient intake and preference in a subset of obese adults. Int J Obes Relat Metab Disord 27:219–226
165. Lustig RH et al (2005) A multicenter, randomized, double-blind, placebo-controlled, dose-finding trial of a long-acting formulation of octreotide in promoting weight loss in obese adults with insulin hypersecretion. Int J Obes Relat Metab Disord 30:331–341
166. Lustig RH et al (2003) Octreotide therapy of pediatric hypothalamic obesity: a double-blind, placebo-controlled trial. J Clin Endocrinol Metab 88:2586–2592
167. Hwang DY et al (2007) Significant change in insulin production, glucose tolerance and ER stress signaling in transgenic mice coexpressing insulin-siRNA and human IDE. Int J Mol Med 19:65–73
168. Zisman A et al (2000) Targeted disruption of the glucose transporter 4 selectively in muscle causes insulin resistance and glucose intolerance. Nat Med 6:924–928
169. Michael MD et al (2000) Loss of insulin signaling in hepatocytes leads to severe insulin resistance and progressive hepatic dysfunction. Mol Cell 6:87–97
170. Hermans MP, Lambert MJ (2002) HOMA-modelling of insulin sensitivity and β cell function in anorexia nervosa. Eur Eat Disord Rev 10:41–50
171. Kumai M, Tamai H, Fujii S, Nakagawa T, Aoki T (1988) Glucagon secretion in anorexia nervosa. Am J Clin Nutr 47:239–242
172. Nestel PJ (1974) Cholesterol metabolism in anorexia nervosa and hypercholesterolemia. J Clin Endocrinol Metab 38:325–328
173. Mordasini R, Klose G, Greten H (1978) Secondary type II hyperlipoproteinemia in patients with anorexia nervosa. Metabolism 27:71–79
174. Feillet F et al (2000) Plasma cholesterol and endogenous cholesterol synthesis during refeeding in anorexia nervosa. Clin Chim Acta 294:45–56
175. Rice B, Janssen I, Hudson R, Ross R (1999) Effects of aerobic or resistance exercise and/or diet on glucose tolerance and plasma insulin levels in obese men. Diabetes Care 22:684–691
176. Kirwan JP, Kohrt WM, Wojta DM, Bourey RE, Holloszy JO (1993) Endurance exercise training reduces glucose-stimulated insulin levels in 60- to 70-year-old men and women. J Gerontol 48:M84–M90
177. Ligibel JA et al (2008) Impact of a mixed strength and endurance exercise intervention on insulin levels in breast cancer survivors. J Clin Oncol 26:907–912
178. Ivy JL (1997) Role of exercise training in the prevention and treatment of insulin resistance and non-insulin-dependent diabetes mellitus. Sports Med 24: 321–336
179. Heath GW et al (1983) Effects of exercise and lack of exercise on glucose tolerance and insulin sensitivity. J Appl Physiol 55:512–517
180. Goodyear LJ, Kahn BB (1998) Exercise, glucose transport and insulin sensitivity. Annu Rev Med 49:235–261
181. Kahn SE et al (1990) Effect of exercise on insulin action, glucose tolerance, and insulin secretion in aging. Am J Physiol Endocrinol Metab 258: E937–E943
182. Hull RL et al (2005) Long-term treatment with rosiglitazone and metformin reduces the extent of, but does not prevent, islet amyloid deposition in mice expressing the gene for human islet amyloid polypeptide. Diabetes 54:2235–2244
183. Cuff DJ et al (2003) Effective exercise modality to reduce insulin resistance in women with type 2 diabetes. Diabetes Care 26:2977–2982
184. Burcelin R, Dolci W, Thorens B (2000) Portal glucose infusion in the mouse induces hypoglycemia: evidence that the hepatoportal glucose sensor stimulates glucose utilization. Diabetes 49:1635–1642
185. Ono T, Steffens AB, Sasaki K (1983) Influence of peripheral and intracerebroventricular glucose and insulin infusions on peripheral and cerebrospinal fluid glucose and insulin levels. Physiol Behav 30:301–306
186. Karnani MM, Burdakov D (2010) Multiple hypothalamic circuits sense and regulate glucose levels. Am J Physiol Regul Integr Comp Physiol 300:R47–R55

187. Borg MA, Sherwin RS, Borg WP, Tamborlane WV, Shulman GI (1997) Local ventromedial hypothalamus glucose perfusion blocks counterregulation during systemic hypoglycemia in awake rats. J Clin Invest 99:361–365
188. Borg MA, Tamborlane WV, Shulman GI, Sherwin RS (2003) Local lactate perfusion of the ventromedial hypothalamus suppresses hypoglycemic counterregulation. Diabetes 52:663–666
189. Fabris SE, Thorburn A, Litchfield A, Proietto J (1996) Effect of parasympathetic denervation of liver and pancreas on glucose kinetics in man. Metab Clin Exp 45:987–991
190. Perseghin G et al (1997) Regulation of glucose homeostasis in humans with denervated livers. J Clin Invest 100:931–941
191. Rohner-Jeanrenaud F, Jeanrenaud B (1983) The central nervous system-endocrine pancreas axis. Ann Endocrinol (Paris) 44:217–227
192. Orland MJ, Chyn R, Permutt MA (1985) Modulation of proinsulin messenger RNA after partial pancreatectomy in rats. Relationships to glucose homeostasis. J Clin Invest 75:2047–2055
193. Kahn SE (2001) The importance of β-cell failure in the development and progression of type 2 diabetes. J Clin Endocrinol Metab 86:4047–4058
194. Clark A, Jones L, de Koning E, Hansen BC, Matthews DR (2001) Decreased insulin secretion in type 2 diabetes: a problem of cellular mass or function? Diabetes 50:169S–171S
195. Deng S et al (2004) Structural and functional abnormalities in the islets isolated from type 2 diabetic subjects. Diabetes 53:624–632
196. Ward WK, Bolgiano DC, McKnight B, Halter JB, Porte D (1984) Diminished B cell secretory capacity in patients with noninsulin-dependent diabetes mellitus. J Clin Invest 74:1318–1328
197. Bernard C et al (1998) Pancreatic β-cell regeneration after 48-h glucose infusion in mildly diabetic rats is not correlated with functional improvement. Diabetes 47:1058–1065
198. Del Prato S, Tiengo A (2001) The importance of first-phase insulin secretion: implications for the therapy of type 2 diabetes mellitus. Diabetes Metab Res Rev 17:164–174
199. Bruttomesso D et al (1999) Restoration of early rise in plasma insulin levels improves the glucose tolerance of type 2 diabetic patients. Diabetes 48:99–105
200. Del Prato S, Marchetti P, Bonadonna RC (2002) Phasic insulin release and metabolic regulation in type 2 diabetes. Diabetes 51(Suppl 1):S109–S116
201. Berthoud HR, Bereiter DA, Trimble ER, Siegel EG, Jeanrenaud B (1981) Cephalic phase, reflex insulin secretion. Neuroanatomical and physiological characterization. Diabetologia 20(Suppl):393–401
202. Berthoud HR, Jeanrenaud B (1982) Sham feeding-induced cephalic phase insulin release in the rat. Am J Physiol Endocrinol Metab 242:E280–E285
203. Mattes RD, Engelman K, Mattern J, Teff KL (1993) Cephalic-phase insulin in obese and normal-weight men: relation to postprandial insulin. Metab Clin Exp 42:1600–1608
204. Teff KL (2010) Cephalic phase pancreatic polypeptide responses to liquid and solid stimuli in humans. Physiol Behav 99:317–323
205. Perley MJ, Kipnis DM (1967) Plasma insulin responses to oral and intravenous glucose: studies in normal and diabetic subjects. J Clin Invest 46:1954–1962
206. Ahrén B, Winzell MS, Pacini G (2008) The augmenting effect on insulin secretion by oral versus intravenous glucose is exaggerated by high-fat diet in mice. J Endocrinol 197:181–187
207. Calhoun P et al (1986) Evaluation of insulin secretion after pancreas autotransplantation by oral or intravenous glucose challenge. Ann Surg 204:585–593
208. de Pablo F, de la Rosa EJ (1995) The developing CNS: a scenario for the action of proinsulin, insulin and insulin-like growth factors. Trends Neurosci 18:143–150
209. Diaz B, Serna J, De Pablo F, de la Rosa EJ (2000) In vivo regulation of cell death by embryonic (pro)insulin and the insulin receptor during early retinal neurogenesis. Development 127:1641–1649
210. Hernández-Sánchez C, Mansilla A, Rosa EJ, Pablo F (2006) Proinsulin in development: new roles for an ancient prohormone. Diabetologia 49:1142–1150
211. Kern W et al (2001) Improving influence of insulin on cognitive functions in humans. Neuroendocrinology 74:270–280
212. Nef S et al (2003) Testis determination requires insulin receptor family function in mice. Nature 426:291–295
213. Park CR (2001) Cognitive effects of insulin in the central nervous system. Neurosci Biobehav Rev 25:311–323
214. Bloomgarden ZT (2004) Diabetes complications. Diabetes Care 27:1506–1514
215. Hussain MA, Theise ND (2004) Stem-cell therapy for diabetes mellitus. Lancet 364:203–205
216. Ryan EA et al (2005) Five-year follow-up after clinical islet transplantation. Diabetes 54:2060–2069
217. Bouwens L (2006) Beta cell regeneration. Curr Diabetes Rev 2:3–9
218. Suarez-Pinzon WL, Lakey JRT, Brand SJ, Rabinovitch A (2005) Combination therapy with epidermal growth factor and gastrin induces neogenesis of human islet β-cells from pancreatic duct cells and an increase in functional β-cell mass. J Clin Endocrinol Metab 90:3401–3409
219. Tuch BE, Kannangara K (2008) [β] cell regeneration. Drug Discov Today Ther Strateg 5:215–221
220. Bonner-Weir S, Weir GC (2005) New sources of pancreatic β-cells. Nat Biotechnol 23:857–861
221. Baeyens L et al (2005) In vitro generation of insulin-producing β cells from adult exocrine pancreatic cells. Diabetologia 48:49–57
222. Meier JJ, Bhushan A, Butler AE, Rizza RA, Butler PC (2005) Sustained β cell apoptosis in patients with long-standing type 1 diabetes: indirect evidence for islet regeneration? Diabetologia 48:2221–2228

223. Klein R, Klein BEK, Moss SE (1996) Relation of glycemic control to diabetic microvascular complications in diabetes mellitus. Ann Intern Med 124:90–96
224. Klein R, Klein BE (1998) Relation of glycemic control to diabetic complications and health outcomes. Diabetes Care 21(Suppl 3):C39–C43
225. Klein R, Klein BEK, Moss SE, Cruickshanks KJ (1994) Relationship of hyperglycemia to the long-term incidence and progression of diabetic retinopathy. Arch Intern Med 154:2169–2178
226. Gaster B, Hirsch IB (1998) The effects of improved glycemic control on complications in type 2 diabetes. Arch Intern Med 158:134–140
227. Skyler JS et al (2009) Intensive glycemic control and the prevention of cardiovascular events: implications of the ACCORD, ADVANCE, and VA diabetes trials: a position statement of the American Diabetes Association and a Scientific Statement of the American College of Cardiology Foundation and the American Heart Association. Circulation 119: 351–357
228. Lacherade J-C et al (2007) Failure to achieve glycemic control despite intensive insulin therapy in a medical ICU: incidence and influence on ICU mortality. Intensive Care Med 33:814–821
229. Holman RR, Prospective UK (1988) Intensive blood-glucose control with sulphonylureas or insulin compared with conventional treatment and risk of complications in patients with type 2 diabetes (UKPDS 33). UK Prospective Diabetes Study (UKPDS) Group. Lancet 352:837–853
230. The ADVANCE (2008) Collaborative group intensive blood glucose control and vascular outcomes in patients with type 2 diabetes. N Engl J Med 358: 2560–2572
231. Holman RR, Paul SK, Bethel MA, Matthews DR, Neil HAW (2008) 10-year follow-up of intensive glucose control in type 2 diabetes. N Engl J Med 359:1577–1589
232. Dluhy RG, McMahon GT (2008) Intensive glycemic control in the ACCORD and ADVANCE trials. N Engl J Med 358:2630–2633
233. Cefalu WT, Watson K (2008) Intensive glycemic control and cardiovascular disease observations from the ACCORD study. Diabetes 57:1163–1165
234. Duckworth W et al (2009) Glucose control and vascular complications in veterans with type 2 diabetes. N Engl J Med 360:129–139
235. Turnbull FM et al (2009) Intensive glucose control and macrovascular outcomes in type 2 diabetes. Diabetologia 52:2288–2298
236. Nissen SE (2010) The rise and fall of rosiglitazone. Eur Heart J 31:773–776
237. Montori VM, Fernández-Balsells M (2009) Glycemic control in type 2 diabetes: time for an evidence-based about-face? Ann Intern Med 150:803–808
238. Schrier RW, Estacio RO, Esler A, Mehler P (2002) Effects of aggressive blood pressure control in normotensive type 2 diabetic patients on albuminuria, retinopathy and strokes. Kidney Int 61:1086–1097
239. Brownlee M et al (2011) Complications of diabetes mellitus. In: Kronenberg HM, Melmed S, Polonsky KS, Reed Larsen P (eds) Williams textbook of endocrinology, 12th edn. Saunders/Elsevier, Philadelphia, PA
240. Vijan S, Hofer TP, Hayward RA (1997) Estimated benefits of glycemic control in microvascular complications in type 2 diabetes. Ann Intern Med 127:788–795
241. Mather KJ, Verma S, Anderson TJ (2001) Improved endothelial function with metformin in type 2 diabetes mellitus. J Am Coll Cardiol 37:1344–1350
242. Jager J et al (2005) Effects of short-term treatment with metformin on markers of endothelial function and inflammatory activity in type 2 diabetes mellitus: a randomized, placebo-controlled trial. J Intern Med 257:100–109
243. Pistrosch F et al (2004) In type 2 diabetes, rosiglitazone therapy for insulin resistance ameliorates endothelial dysfunction independent of glucose control. Diabetes Care 27:484–490
244. Yamagishi T et al (2001) Troglitazone improves endothelial function and augments renal klotho mRNA expression in Otsuka Long-Evans Tokushima Fatty (OLETF) rats with multiple atherogenic risk factors. Hypertens Res 24:705–709
245. Romeo JH, Seftel AD, Madhun ZT, Aron DC (2000) Sexual function in men with diabetes type 2: association with glycemic control. J Urol 163:788–791
246. Yaman O, Akand M, Gursoy A, Erdogan MF, Anafarta K (2006) The effect of diabetes mellitus treatment and good glycemic control on the erectile function in Men with diabetes mellitus-induced erectile dysfunction: a pilot study. J Sex Med 3:344–348
247. Ayala JE et al (2007) Chronic treatment with sildenafil improves energy balance and insulin action in high fat–fed conscious mice. Diabetes 56:1025–1033
248. Tooke J, Hannemann M (2000) Adverse endothelial function and the insulin resistance syndrome. J Intern Med 247:425–431
249. Tooke JE, Goh KL (1999) Vascular function in Type 2 diabetes mellitus and pre-diabetes: the case for intrinsic endotheliopathy. Diabet Med 16:710–715
250. Tooke JE, Goh KL (1998) Endotheliopathy precedes type 2 diabetes. Diabetes Care 21:2047–2049
251. Leeson CPM, Kattenhorn M, Morley R, Lucas A, Deanfield JE (2001) Impact of Low birth weight and cardiovascular risk factors on endothelial function in early adult life. Circulation 103:1264–1268
252. Meigs JB, Hu FB, Rifai N, Manson JE (2004) Biomarkers of endothelial dysfunction and risk of type 2 diabetes mellitus. J Am Med Assoc 291: 1978–1986
253. Miller AW, Sims JJ, Canavan A, Hsu T, Ujhelyi MR (1999) Impaired vagal reflex activity in insulin-resistant rats. J Cardiovasc Pharmacol 33:698–702
254. Carnethon MR, Jacobs DR, Sidney S, Liu K (2003) Influence of autonomic nervous system dysfunction on the development of type 2 diabetes. Diabetes Care 26:3035–3041
255. Ribeiro RT, Afonso RA, Guarino MP, Macedo MP (2008) Loss of postprandial insulin sensitization

during aging. J Gerontol A Biol Sci Med Sci 63: 560–565
256. Ribeiro RT, Lautt WW, Legare DJ, Macedo MP (2005) Insulin resistance induced by sucrose feeding in rats is due to an impairment of the hepatic parasympathetic nerves. Diabetologia 48:976–983
257. Lindmark S, Wiklund U, Bjerle P, Eriksson JW (2003) Does the autonomic nervous system play a role in the development of insulin resistance? A study on heart rate variability in first-degree relatives of Type 2 diabetes patients and control subjects. Diabet Med 20:399–405
258. Xie H, Lautt WW (1996) Insulin resistance caused by hepatic cholinergic interruption and reversed by acetylcholine administration. Am J Physiol Endocrinol Metab 271:E587–E592
259. Xie H, Tsybenko VA, Johnson MV, Lautt WW (1993) Insulin resistance of glucose response produced by hepatic denervations. Can J Physiol Pharmacol 71:175–178
260. Leahy JL, Cooper HE, Deal DA, Weir GC (1986) Chronic hyperglycemia is associated with impaired glucose influence on insulin secretion. A study in normal rats using chronic in vivo glucose infusions. J Clin Invest 77:908–915
261. Maedler K et al (2002) Glucose-induced β cell production of IL-1β contributes to glucotoxicity in human pancreatic islets. J Clin Invest 110:851–860
262. Kahn SE, Bergman RN, Schwartz MW, Taborsky GJ, Porte D (1992) Short-term hyperglycemia and hyperinsulinemia improve insulin action but do not alter glucose action in normal humans. Am J Physiol Endocrinol Metab 262:E518–E523
263. Ward WK, Halter JB, Beard JC, Porte D (1984) Adaptation of B and A cell function during prolonged glucose infusion in human subjects. Am J Physiol Endocrinol Metab 246:E405–E411
264. Altavilla D et al (2001) Inhibition of lipid peroxidation restores impaired vascular endothelial growth factor expression and stimulates wound healing and angiogenesis in the genetically diabetic mouse. Diabetes 50:667–674
265. Brem H, Tomic-Canic M (2007) Cellular and molecular basis of wound healing in diabetes. J Clin Invest 117:1219–1222
266. Crawford TN, Alfaro DV III, Kerrison JB, Jablon EP (2009) Diabetic retinopathy and angiogenesis. Curr Diabetes Rev 5:8–13
267. Yamamoto Y et al (2004) Tumstatin peptide, an inhibitor of angiogenesis, prevents glomerular hypertrophy in the early stage of diabetic nephropathy. Diabetes 53:1831–1840
268. Zent R, Pozzi A (2007) Angiogenesis in diabetic nephropathy. Semin Nephrol 27:161–171
269. Gill G (1991) Insulin dependent diabetes mellitus. In: Pickup J, Williams G (eds) Textbook of diabetes, vol 1. Blackwell Scientific, Oxford
270. Brown JB, Nichols GA, Perry A (2004) The burden of treatment failure in type 2 diabetes. Diabetes Care 27:1535
271. Koro CE, Bowlin SJ, Bourgeois N, Fedder DO (2004) Glycemic control from 1988 to 2000 among U.S. adults diagnosed with type 2 diabetes. Diabetes Care 27:17–20
272. Lehrer J (2010) The truth wears off: Is there something wrong with the scientific method? The New Yorker, December 13

4 The Rise and Fall of Thrift

Introduction

If we are set to search for an alternative picture, where do we need to begin our search? There is a simple and universal answer to all such questions in biology. Whenever any seemingly paradoxical and intuitively difficult phenomenon is observed in biology, we need to ask why and how it might have originated in evolution. Evolution is fundamental to biology, and as rightly put by the evolutionary geneticist Dobzinski, "Nothing in biology makes sense except in the light of evolution."

But there is a great danger lingering around. Many people frequently use a trivial version of evolutionary logic to somehow "explain away" things. There is no dearth of "just so..." evolutionary stories popular around including the loss of tail, predicted loss of nails and eyebrows in humans, the evolutionary fate of today's monkeys, the long neck of giraffe, and so on. Therefore one needs to be rigorous while trying to apply an evolutionary explanation to anything. Fortunately mainstream evolutionary biology has matured sufficiently today and has developed methods that can save us from falling into such traps.

If we are going to look for an evolutionary theory for T2D and related disorders, let us first list our expectations from an evolutionary theory. (1) It should address the basic riddle why a disorder that has strong familial tendencies and so has some presumably genetic components has suddenly increased its prevalence in the last couple of generations. (2) It should explain the apparent polymorphism in the population, i.e., individual differences in the tendency to become obese and/or diabetic. (3) It should explain the observed epidemiological associations and patterns, for example, the strong association between birth weight and metabolic syndrome. (4) It should not stop at explaining the origins of obesity but also explain why obesity is associated with insulin resistance and its consequences with both proximate and ultimate components of reasoning; more specifically, there needs to be an ultimate reasoning for why obesity induces insulin resistance (if it does) and other metabolic and endocrine changes. (5) It should give logical and testable solutions to all the paradoxes that we saw in the last chapter. (6) It should help us find out why T2D is found to be largely "incurable," at least with the current therapeutic approach. (7) It should be falsifiable and either be supported by existing evidence and/or make a series of predictions that can be tested sooner or later. (8) Optimistically (but optionally), it should lead to some fundamental breakthrough of clinical importance. If it leads to a long-term solution or "cure" for diabetes, sooner or later, it would be really fascinating. This cannot be the primary objective of an evolutionary theory though. Evolutionary theory is not an application-oriented science. Its main objective is to develop fundamental insights into biological reasoning. However, as our insights increase, it is very likely that we find an application.

M. Watve, *Doves, Diplomats, and Diabetes: A Darwinian Interpretation of Type 2 Diabetes and Related Disorders*,
DOI 10.1007/978-1-4614-4409-1_4, © Springer Science+Business Media New York 2013

Origin of Thriftt

Evolutionary approaches to obesity and T2D are not new. There is a history of evolutionary hypotheses intended to do one or more of the above. The first major change in the thinking about diabetes that had a strong evolutionary flavor was brought about single-handedly by a person called James Neel in the early 1960s [1]. Most people at that time believed that diabetes was a genetic disorder much like other well-known genetic disorders such as sickle cell anemia or hemophilia. The belief was so deep rooted that there are papers claiming Mendelian nature of inheritance of type 2 diabetes [2–4], although we know now that T2D is far from being Mendelian. This amusing conclusion appears to arise out of some facets of human nature to which researchers are no exception. When there is a theory and you expect to see or find something based on it, frequently, you start seeing things although they may not actually be there. Also if you do not expect something or do not have a theory for it, you are very likely to miss seeing it, although it may very much be there ("We do not see what we sense. We see what we think we sense" [5]). I have experienced this frequently with undergraduate students. When they are told that the given sample is most likely to contain say staphylococci, every dust particle starts looking like *Staphylococcus*. Perhaps this is how the Mendelian inheritance of diabetes was "observed" by some researchers. Neel also appears to believe quite strongly that inheritance of diabetes was Mendelian. Still Neel appears to be the first person to be convinced that diabetes could not be a "genetic disorder." Not that he doubted the "genetic" part of it, but he raised doubts about diabetes being a "disorder" of genetic origin. All monogenic (where a defective allele at a single gene locus can be clearly shown to be responsible for the disorder) heritable disorders always exist in any population in exceedingly small frequencies, typically one in several thousand. Diabetes, on the other hand, was much more common and notably increasing in frequency by this time. Realizing that no genetic disorder can be as frequent as diabetes is and moreover cannot increase in a population so rapidly, he thought that diabetes should not be treated as a genetic disorder. If the frequency of the diabetic allele is high, there must be a biological reason for it. A frequency of a gene can increase if it is useful in some way so that natural selection favors individuals having that allele. Therefore Neel thought that the genes responsible for diabetes must be adaptive under at least some set of conditions. Since the association of obesity with diabetes was well accepted, he equated the diabetic tendency to the tendency to accumulate fat and argued that a "thrifty" gene helped storage of fat under conditions of better availability of nutrients and allowed reutilization of it under starvation. Neel's definition of thriftiness is "being exceptionally efficient in intake and/or utilization of food." This gives two distinct meanings to thrift. Either a thrifty individual eats more and stores fat or has a less wasteful metabolism which allows greater storage of energy. Let us call the intake efficiency as type 1 thrift and utilization efficiency, i.e., lower rate of burning calories to store more, as type 2 thrift. This classification is extremely important to understand thrift, but most researchers have failed to clearly differentiate between the two, leading to much ambiguity. The two are mutually compatible, and Neel appears to imply both. But as we will see below, both do not apply equally to all arguments about thrift. Therefore, for every argument, there needs to be some clarity as to whether we are talking about type 1 or 2 thrift or both. In Neel's view the thrifty gene was under positive selection pressure in ancestral life when seasonal and climatic conditions resulted into fluctuating food availability often called "feast and famine." The assumption is that human ancestors frequently faced periods of starvation interspersed with periods of food abundance. A genetic tendency to store fat when food was abundant was extremely adaptive under feast and famine conditions and therefore enjoyed a selective advantage. In modern life food is always abundant, something that can be called a "feast and feast" condition, and under such conditions, the same genetically determined tendency leads to obesity. Both types 1 and 2 thrift are compatible with this argument. According to this theory, obesity is the apparent cause of diabetes, but the diabetic tendency is the real cause of obesity. Neel appears to have believed that diabetic tendency leads to obesity. In his own words, "The overweight individual of 40 or 50 with mild diabetes is not so much diabetic because he is obese, as he is obese because he is of a particular (diabetic) genotype." What Neel refers as mild diabetes is type 2 diabetes in the current nomenclature. The parenthesis for the word "diabetic" is original. Neel's argument was based on a number of patterns well known by then including that diabetic mothers (also diabetic fathers!!) have larger babies, that hyperinsulinemia precedes adult-onset diabetes, that insulin has lipogenic action, and also that frequently there is an early history of transient hypoglycemia much before diabetes sets in. Neel thought that hyperinsulinemia is the thrifty tendency. High levels of insulin induce lipogenesis and fat accumulation since it facilitates entry of glucose into fat cells and its conversion to fat. One of his relied evidence was an early hypoglycemic phase in the history of T2D. This hypoglycemia perhaps represented active lipogenic metabolism under the influence of insulin. He thought that at later stages, the high levels of insulin were compensated by "anti-insulin" activity (which could be insulin resistance in today's terms), and eventually, somehow, anti-insulin activity became stronger than insulin action to cause diabetes.

Although I am going to argue against some of Neel's concepts, I must admit first that this paper is one of the greatest papers ever written on diabetes so far. It is full of prophecy and foresight. Neel does not hesitate to make bold speculations whenever he is confident about the logic behind them. It is doubtful whether in today's science publishing trend, a paper as speculative as this one would have ever got published. Speculations are generally unwelcome in today's science, and speculative papers are most likely to be rejected by the current editorial standards. Perhaps things were different in the early 1960s or Neel was extremely lucky. Neel's speculations had a far-reaching vision. Interestingly when Neel wrote this paper, the concept of insulin resistance was not yet established. But Neel appears to have perceived it speculatively and calls it "anti-insulin" activity. Neel also had a very clean insight into a number of issues with which the followers of the thrifty gene hypothesis later messed up completely. Neel seems to be clear in his mind that hyperinsulinemia comes first and anti-insulin activity appears consequently. He also has rejected the idea of β cell "exhaustion" leading to diabetes.

Neel's concept of thrift soon became almost an axiom, although no such "thrifty" gene or set of genes have been convincingly demonstrated any time. Later the observation that individuals born small for gestational age had a greater probability to become obese and type 2 diabetic in later life led to the concept of fetal programming [6–10]. This hypothesis states that if a fetus faces inadequate nutrition in intrauterine life, the body is programmed to be "thrifty" as an adaptation. Although a little later I am going to contest this logic as well, let us first appreciate the great vision of the researchers who detected this pattern for the first time. It is not easy even to imagine that the roots of a disorder that becomes obvious in one's 40s and 50s could have its origins in intrauterine life. But a group of British visionaries discerned such a strange relationship with insightful analysis of data. The concept originated in data on differential death rates in different parts of Britain. Death rates in the newborn in the early decades of 1900s were highest in some of the northern industrial towns and poorer rural areas in the north and west. This geographical pattern closely resembled that in deaths due to coronary heart disease decades later. This led to the suspicion that intrauterine growth retardation had something to do with heart disease in adults. Epidemiological studies were launched based on the simple strategy of examining men and women in middle and late life whose body measurements at birth were recorded. Sixteen thousand men and women born in Hertfordshire during 1911–1930 were traced, and the effort revealed that death rate due to coronary heart disease in the lowest birth weight class was double than in the uppermost class [8]. Later, similar trends were found across most parts of the world and different ethnic groups, showing the robustness of the relationship between fetal growth conditions and adulthood disorders including obesity, type 2 diabetes, hypertension, and coronary heart disease [11–17]. There are two possible components of the proposed thrifty adaptation in response to reduced fetal growth. One relates to an immediate gain in terms of survival during fetal and early infant life. The other is a predictive adaptive response in anticipation of starvation in later life [18]. This distinction is important in understanding the evolution of fetal programming as we will see soon.

Both the concepts of thrifty gene and thrifty phenotype by fetal programming have recently faced serious criticism on several grounds [19–24]. Some of the critics think that the thriftiness concept is flawed and needs to be abandoned [20, 21]. Others have attempted to refine the concept of thrift so as to resolve some of the flaws and paradoxes pointed out by the critics [25–28]. Let us see the classical concept of thrift and its criticism first before considering the refined versions of thrift. Baig et al. [29] integrated the critical arguments challenging the classical thrifty gene and thrifty fetal programming hypotheses using a mathematical model. The model considered three hypothetical genotypes, namely, a non-thrifty wild type having no mechanism for thriftiness, a thrifty genotype which is genetically programmed for thriftiness, and a genotype with a capacity for fetal programming for thriftiness. Taking a year as a natural time unit of seasonality, a simple dichotomy of years was assumed in the model, those with adequate food supply (feast) and those with inadequate food supply (famine). Famines were assumed to occur randomly with some probability. The fitness of an individual with non-thrifty genotype in feast conditions was assumed to be greater than that of an individual with thrifty genotype because of a cost associated with thrift. Obesity and insulin resistance are known to be associated with reduced fecundity [30–33] justifying the cost of thriftiness in feast conditions. Fitness of an individual with non-thrifty genotype in famine conditions was assumed to be less than that of individuals with thrifty genotype. It is simple to visualize from this set of assumptions that at low probability of famine, the non-thrifty genotype will have an advantage, and at high frequency, the thrifty one will have an advantage. It also follows that at no condition, non-thrifty and thrifty genotype would coexist in a stable polymorphism. This is simple to see intuitively as well as to show with simple mathematics as done by the Baig et al. model [29].

Calculation of the fitness of the thrifty programmer was a little more complex. Assuming no correlation between birth and lifetime conditions, the total fitness was written as the sum of all years with the assumption that in the birth year the phenotype was best suited for given conditions. For the rest of the lifetime, the fitness fluctuated according to randomly fluctuating environmental conditions. Analytical solutions to the model showed that at low probabilities of famine, a non-thrifty gene has a selective advantage, and at high probability, thrifty gene would get selected (Fig. 4.1) leaving a very narrow area of advantage for fetal programming. The area was narrower for long-lived species, whereas for short-lived ones, the birth year advantage was large enough as compared to lifetime, and therefore, fetal programming had a much larger width of advantage (Fig. 4.2). Two interesting conclusions of the model are that a thrifty gene can evolve but cannot have a stable polymorphic state and that fetal programming for thrift is most unlikely to evolve by feast and famine conditions in long-lived species [29]. There are a number of other issues about the concept of thrift that need to be addressed.

1. Feast and famine conditions in human ancestry: Is the assumption realistic?

   The fundamental assumption of the thrifty gene hypothesis is that human ancestors suffered from wide fluctuations in food availability either as an effect of season or of year-to-year climatic fluctuations. Thrift evolved in response to these fluctuations. One of the key questions is when in human history could selection for thriftiness, if any, have operated. There are three possible scenarios:

   (a) Selection during hunting-gathering stage: Contrary to the commonly held belief, paleoarcheological as well as anthropological data suggest that chronic starvation was uncommon during hunter-gatherer

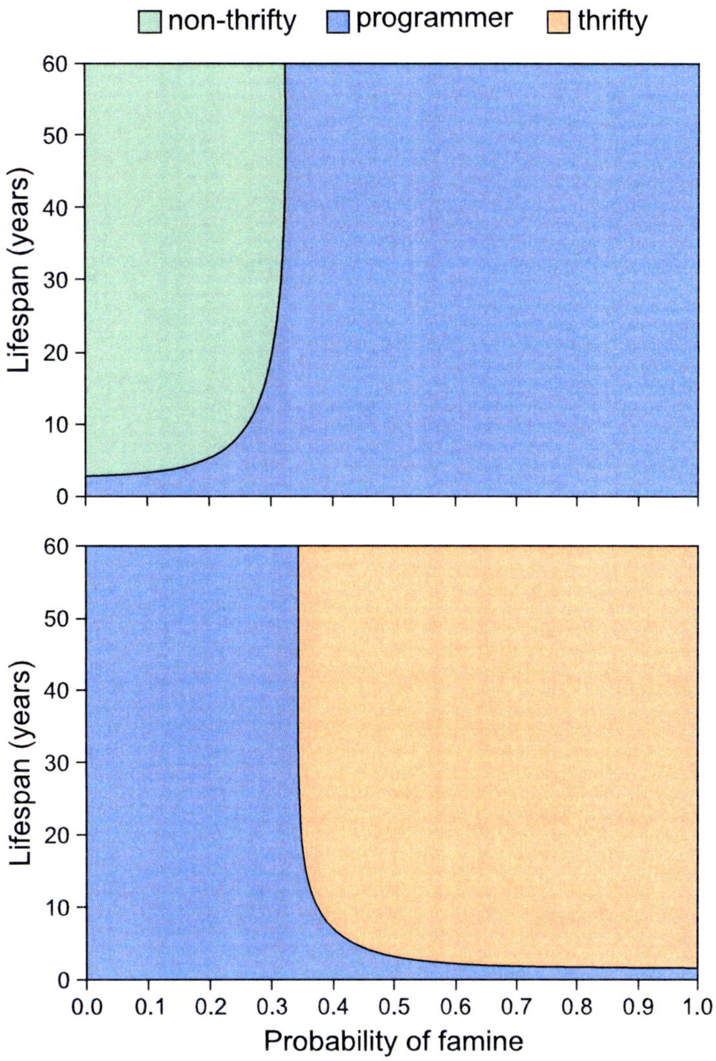

Fig. 4.1 Parameter areas of advantage in the Baig et al. model [29] (*upper*) when only non-thrifty genotype and fetal programmer compete. (*lower*) When only thrifty genotype and fetal programmer compete. For this and all other figures in this chapter, the *colors* denoting advantage areas for the thrifty gene, non-thrifty gene, and fetal programmer remain the same

stage [34, 35]. Today's hunter-gatherer societies do not seem to suffer starvation more frequently or more intensively than agricultural societies [22]. This is in spite of the fact that today, hunter-gatherer societies are pushed to marginal or difficult habitats, and most pristine habitats are occupied by modern man. Therefore the assumption that hunter-gatherer societies suffered frequent starvation is not well supported. But even if we assume hunter-gatherer societies to be prone to feast and famine selection, a number of other questions remain unanswered. Since hominids were hunter-gatherers for the most part of human evolutionary history, selection would have been prolonged, and we would expect alleles to have reached equilibrium frequencies. The Baig et al. model [29] implies that selection cannot result in stable polymorphism of thrifty alleles. In the modern human society, there is considerable variation in the tendency to become obese or diabetic. Therefore polymorphism with respect to genes predisposing to obesity and type 2 diabetes presumably exists. If there is no negative frequency dependence or heterozygote advantage, natural selection will be directional resulting into the fixation of the advantageous

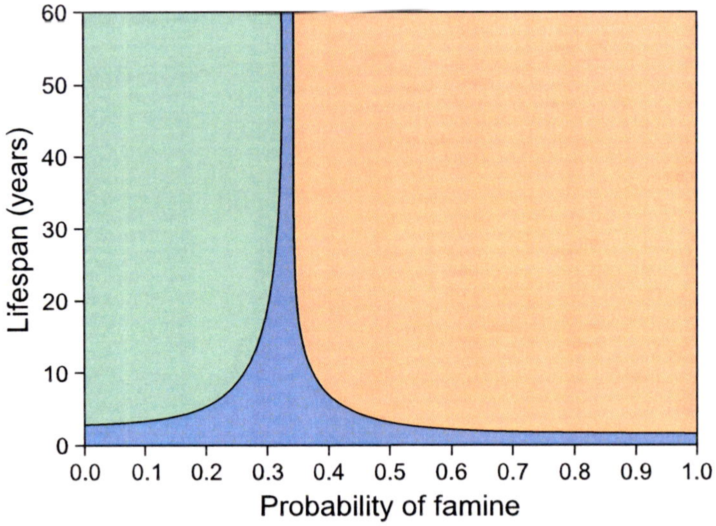

Fig. 4.2 Parameter areas of advantage when non-thrifty genotype, thrifty genotype, and fetal programmer compete simultaneously: Fetal programming can evolve for species with short life span. If the life span is longer, fetal programming is unlikely to offer selective advantage over thrifty or non-thrifty genotypes except for a specific and very narrow range of probability of famine

genotype. At low frequencies of famines, the non-thrifty gene would be the only survivor, and at high frequencies, the thrifty gene would be the only survivor. A critical frequency of famines separates the two, and in no case, the thrifty and non-thrifty genes can coexist stably. Theoretically if a heterozygote of thrifty and non-thrifty alleles gets a dual advantage by expressing the right allele in the right environment, the two alleles can coexist and the population at any given time will consist of thrifty, non-thrifty, and all-time fit heterozygous individuals. Neel assumed that the thrifty gene will coexist in a stable state owing to heterozygote advantage [1]. However, there is no evidence of any heterozygote advantage so far. Stable polymorphism is also possible if there is negative frequency-dependent selection. However if fitness is decided by climatic conditions as assumed by the popular version of thrifty gene hypothesis, frequency dependence is unlikely. Therefore selection during hunter-gatherer stage does not explain the prevalent polymorphism in predisposition to obesity.

(b) Selection after the beginning of agriculture: Chronic starvation due to famines became more serious and common with the beginning of agriculture [36–38]. Signs of chronic starvation such as linear enamel hypoplasia of teeth are more common in early agricultural societies than in hunter-gatherer societies. This is owing to the fact that crops are highly seasonal in nature, and the failure of a single crop leads to long-term food scarcity. Such long-term food shortages are much less probable in hunter-gatherer life, particularly in biodiversity-rich areas. Therefore if selection for thriftiness started acting after the beginning of agriculture, there could be transient polymorphism. A testable prediction of the hypothesis would be that ethnic groups that took to agriculture earlier should show higher tendencies to become obese and diabetic. Data on ethnic groups such as the Australian Aborigines who remained hunter-gatherers until recently is not in support. The recently urbanized individuals of this community developed a surprisingly high prevalence of diabetes and hypertension [39] indicating that selection during agriculture life is unlikely to be responsible for the spread of a thrifty gene. It is difficult to argue therefore that thrifty genes evolved after the advent of agriculture.

(c) Selection in modern times: Intensive agricultural and industrial societies are modern phenomenon just over 200 years old, and it is highly unlikely that this period could have brought about any evolutionary change, although this has been claimed (see below). It can be seen from all the three possible scenarios that natural selection for the hypothetical thrifty gene(s) is unable to explain the apparent polymorphism or high variation in obesity proneness.

2. Does the tendency to become obese protect against a famine?

Whether obese people have a significantly better chance of surviving famines is debatable, but if we assume so, we can use the model to estimate how much advantage is needed for the thrifty gene to evolve. What could be the threshold probability of famine that would permit the evolution of thrifty genotype? Looking at long-term history, Speakman argued that famines with significant mortality occur with a frequency of once in 100–150 years [21]. At this frequency the advantage of thrifty over non-thrifty phenotype in famine conditions should be more than 100 times the relative loss suffered by thrifty gene in feast conditions. Since an obesity-induced reduction in fecundity has been demonstrated, the advantage of thriftiness in famines should be exceedingly large for thrifty genes to evolve. Such a large advantage should be highly evident and easily measurable, but obese people have not been shown to survive famines significantly better than lean individuals [21]. Therefore it is doubtful whether obesity actually offered sufficient advantage during famines to get selected. This is particularly tricky because a major cost associated with obesity is reduced fecundity [30–33, 40, 41]. If there is any advantage of obesity, it should be greater than its reproductive cost, and there have been no attempts to test this quantitatively.

Which particular type of thrift is more relevant here? In principle, both types 1 and 2 thrift may help in a feast and famine or energy limitation problem. But what does evidence show? Which type of thrift is evident in the current obesity epidemic? Although lower metabolic rates have been assumed to lead to obesity, actual measurements of basal metabolic rates have given contradictory results [42–45]. Impaired fat oxidation rather than lower metabolic rate appears to be the main contributor to obesity, and demonstration of impaired fat oxidation in obesity is more consistent across studies [46–50] than basal metabolic rates. Impaired fat oxidation is neither directly implied by type 1 nor by type 2 thrift. If inability to reutilize stored fat is the major cause of obesity, the stored fat is unlikely to help under "famine" conditions, and if this is so, it is a major blow to the thriftiness hypotheses irrespective of whether it is type 1 or 2 thrift. The doubly labeled water studies also suggest that obesity is more a product of hyperphagia than metabolic frugality [51, 52]. Therefore if thrift really exists, it could be type 1. Type 2 thrift stands on very slippery grounds since evidence for lower metabolic rates mainly leading to obesity is laden with contradictions. The known genetic mechanisms of obesity also work by interfering with appetite control rather than through metabolic thrift [53]. Therefore, lower rate of metabolism is unlikely to be the cause of obesity although very widely believed so and the evidence for impaired fat oxidation in obese individuals casts doubt on its usefulness in famine conditions.

3. Are we really adapted to feast and famine?

A number of animals experience periods of extreme feast and famine in their natural environments. There are two major types among these. One consists of animals that take a summer sleep (aestivation) or winter sleep (hibernation). This happens when the winters or summers are so harsh that foraging becomes impossible. The other situation is long-distance migration. Many long-distance migrants including migratory birds travel thousands of kilometers at a stretch which takes several days during which food intake is either absent or marginal but energy expenditure is large. Before either migration or hibernation, animals store large amounts of fat and utilize them during the nonfeeding period with or without active energy expenditure. These animals have fine-tuned their entire metabolism to suit both accumulation of fat as well as utilization of fat. If humans were adapted to periods of food abundance interspersed with periods of starvation, a set of similar metabolic adjustments should be seen in us too.

When migratory birds utilize their fat, they do so with priority, and they can keep on flying until their entire fat is exhausted. Often when migrating birds are found dead, it is due to complete exhaustion of stored energy. When a fat human performs exercises of any intensity, he feels exhausted soon when hardly any portion of the total fat is burnt. If humans starve to death, a substantial amount of fat remains unutilized demonstrating that we have not evolved to be efficient fat utilizers. Most of the deaths during famines are due to infection rather than due to complete energy exhaustion [20, 21]. Humans are equally inefficient in accumulating fat as compared to migrating birds or hibernating animals [54]. When humans starve, the muscle protein starts breaking down much before fat stores are exhausted, whereas in animals adapted to long-time starvation, muscle proteins are conserved until fat depots are exhausted. Also after refeeding muscle strength is regained first in these animals, and fat starts building up later, whereas as in humans, refeeding after fasting starts building fats rapidly keeping muscles weak. While feast and famine animals can migrate or sleep for several weeks or months without feeding with no symptoms or distress, humans feel hungry and restless only after brief fasting, and several symptoms of starvation start appearing much before any significant depletion of fat stores. All these are certainly not signs of being well adapted to feast and famine conditions [55]. If anything we are the poorest performers among all animals facing feast and famine conditions naturally. It is therefore doubtful whether we have evolved any thrift at all. If we have, it does not appear to be in response to seasonal starvation since we do not show any characteristics of animals adapted to seasonal starvation. Interestingly the species most well adapted to overfeeding and starvation do not show adverse effects of obesity as they accumulate fat every year. This suggests that rather than being thrifty, the failure to be thrifty might be the real cause of our obesity problems.

4. Is there a genetic tendency to be obese?

A large body of research has now focused on the genetics of obesity. After the human genome sequence, much deeper insights into the genetics of obesity were expected. An increasing number of loci and mutants associated with obesity are being identified from genome-wide association (GWA) studies. However, there are certain internal paradoxes associated with these data. Studies prior to the genomic era that were based on familial, twin pair, and adoption studies typically predicted a large heritable component in obesity [56]. The GWA studies, on the other hand, have identified a large number of associations, but they together explain a very small fraction of variance in obesity parameters [57–64] leaving a large gap between the pregenomic and emerging genomic picture. The pregenomic studies had estimated between 40 and 80% genetic influence on obesity [56]. In contrast although the genome-wide association studies have revealed a large number of potentially important alleles, they together do not explain more than 2–5% of population variance in body weights. Genes that have a stronger influence on a phenotypic character have a greater probability of being discovered early in such studies. Therefore it is highly unlikely that some gene having a large influence on obesity exists but has not been discovered as yet. But one can expect many more loci with small effects yet to be discovered. How may gene loci have alleles that influence BMI or other parameters of obesity? We do not have a definitive answer to this question as yet. Various estimates have been made based on different considerations and different sources of data which range from a couple of dozen to over 6,000 [65]. Although the exact number is not known today, we can say with confidence that a large number of genes are involved in obesity, each with a small effect size.

This itself means that genes cannot be responsible for variability in the population in the tendency to become obese. The logic is so straightforward and robust that it can be stated almost like a theorem. If obesity were to be decided by a single gene with two alleles such that homozygous for the pro-obesity allele was highly obese, homozygote for the antiobesity allele lean, and the heterozygote intermediate, there would be some mean and variance of population obesity distribution. If we make a simplistic assumption that the frequency of the pro-obesity allele (p) and that of antiobesity allele (q) is equal, by binomial distribution, 25% population will be obese, 50% intermediate, and 25% lean. This is not the reality, but we can take this hypothetical case as a baseline for further mathematical arguments. If we give obesity score of zero to the lean, 0.5 to the heterozygote, and 1 to the homozygous obese, the coefficient of variation works out to be unity. Now if we divide the gene effect to n different gene loci each with one pro-obesity and one antiobesity allele and each locus having a magnitude of effect $1/n^{\text{th}}$ that of the original single gene, we will have a more generalized frequency distribution of the population obesity. Using simple mathematics, it can be shown that the coefficient of variation of this distribution will be ($\sqrt{\ } 2npq/2np$). Going by our original assumption of $p=q$, it simplifies to $1/\sqrt{2n}$. If we go by the most conservative estimate of a couple of dozen gene loci affecting obesity, the coefficient of variation becomes approximately 0.15, indicating that 85% of population variability is lost. If we go by the estimates on the higher side, more than 99% variability is lost. Relaxing the assumption of $p=q$ makes the mathematics more complex, but the conclusion remains the same, i.e., the greater the number of loci, the smaller is the genetic variability in the population.

For readers who do not understand binomial distribution, there is a simple intuitive way to state the same logic. We assume that there are a large number of genes, each with a small effect on obesity and that they segregate independently. Now if I draw a random set of alleles, I will almost invariably get approximately half of them pro-obesity and half antiobesity. It is next to impossible that some individual gets mostly pro-obesity alleles and someone else gets mostly anti-obesity alleles. Therefore all individuals will stand at more or less the same level of obesity. The greater the number of genes involved, the greater will be the similarity between individuals. So it can be stated as a theorem that if obesity is a polygenic trait involving a large number of genes with small individual effects, the observed population variation and apparent heritability cannot be genetic.

In short, based on genomic data, it can be safely concluded that genes have a negligible role in the prevalent epidemic of obesity. We are yet to understand the reasons for the discrepancy between the pregenomic and genomic estimates of the genetic component in obesity. Some of the possible answers are epigenetic mechanisms [66, 67], intrauterine effects, transgenerational effects [25], or familial inheritance of dietary and behavioral traits. The failure to detect genetic influence on obesity is a robust evidence against any hypothesis involving a gene or set of genes for thrift. So with GWA, we can now permanently bury thrifty gene hypothesis. This can be interpreted in favor of the thrifty phenotype or fetal programming hypothesis and its variants.

5. Can lifetime programming for fetal conditions be adaptive?

The thrifty phenotype or fetal programming hypothesis suffers from a different set of problems. Fetal programming can offer two types of potential advantages, short-term survival advantages in fetal and early infant life and long-term predictive adaptive advantages of lifelong duration. If the advantage is of a short duration, it is difficult to explain why a lifetime commitment to a particular metabolic state could have evolved. A number of genes have age-specific expressions. The endocrine and metabolic states of the body change substantially during adolescence, puberty, pregnancy, parenting, menopause, or senescence. This demonstrates the adaptive flexibility of the body with age. Therefore any rigid lifelong programming for short-term advantage is a difficult proposition. If climatic fluctuations were the main selective force, it should evolve metabolic flexibility rather than lifelong rigid programming since climatic conditions may change substantially and unpredictably in one's lifetime. Although we are assuming that there is a short-term advantage of thrift when faced with fetal or infant undernutrition, such an advantage has never been demonstrated. Are the would-be-diabetic individuals really better adapted for early life malnutrition? So far there is no data demonstrating convincingly any such advantage to the thrifty phenotype.

Metabolic programming of a lifelong duration based on intrauterine conditions is unlikely to offer a fitness advantage except in two sets of conditions. As the Baig et al. [29] model suggests, if a species has a very short life span, fetal programming for adapting to the birth year conditions can be beneficial since the birth year itself is a substantial part of the total life span. Assuming 1 year to be the natural unit of seasonal cycles, species with a life span of <3–5 years can be expected to evolve lifelong fetal programming for thriftiness even though the adaptive advantage is of a short duration. For long-lived species, fetal programming is unlikely to evolve unless there is a significant positive correlation between birth conditions and lifelong conditions (Fig. 4.3). Climatic fluctuations from year to year are a complex phenomenon, and since prediction of important climatic features such as rainfall has important implications, there have been serious attempts to detect temporal patterns. However, temporal patterns are of little help in weather prediction since there are no consistent time-lapse correlations in rainfall or other parameters. Since India has the largest population of diabetics and monsoon is the most important determinant of food availability in this region, it would be enlightening to see the patterns in the Indian monsoon. Table 4.1 shows that rainfall in a given year is not correlated with that of the subsequent year, subsequent 10 years or 40 years cumulative [29]. Over the 30 monsoon subdivisions of India, only 6 are statistically significant using individual α level of 0.05, out of which four correlations are negative contrary to the expectation. Using Bonferroni correction for significance level, applicable when a large number of tests are being done together, none of the correlations remain significant. As there is no detectable positive correlation between birth year and lifetime rainfall conditions and no such correlational patterns in any other climatic variables are reported, fetal programming is unlikely to have evolved in anticipation of drought or famine.

6. Does fetal programming account for majority of obese and diabetic people?

   The other problem with fetal programming hypothesis is its limited coverage. Although there is strong statistical evidence that individuals born small for gestational age, presumably due to intrauterine growth retardation, are more likely to become obese, insulin resistant, and develop T2D in adult life, the majority of adult type 2 diabetics are not born with low birth weight [68, 69]. Therefore even if we accept that there is fetal programming for thrift, at the best, it accounts for <50% of type 2 diabetics. The majority of T2D patients are diabetic not due to fetal programming but due to something else. This is a major limitation of all variations of the hypotheses involving fetal programming.

7. Does thrift account for ethnic differences?

   Since different geographic regions of the world differ in the climatic conditions, seasonality, and food availability, ethnic groups

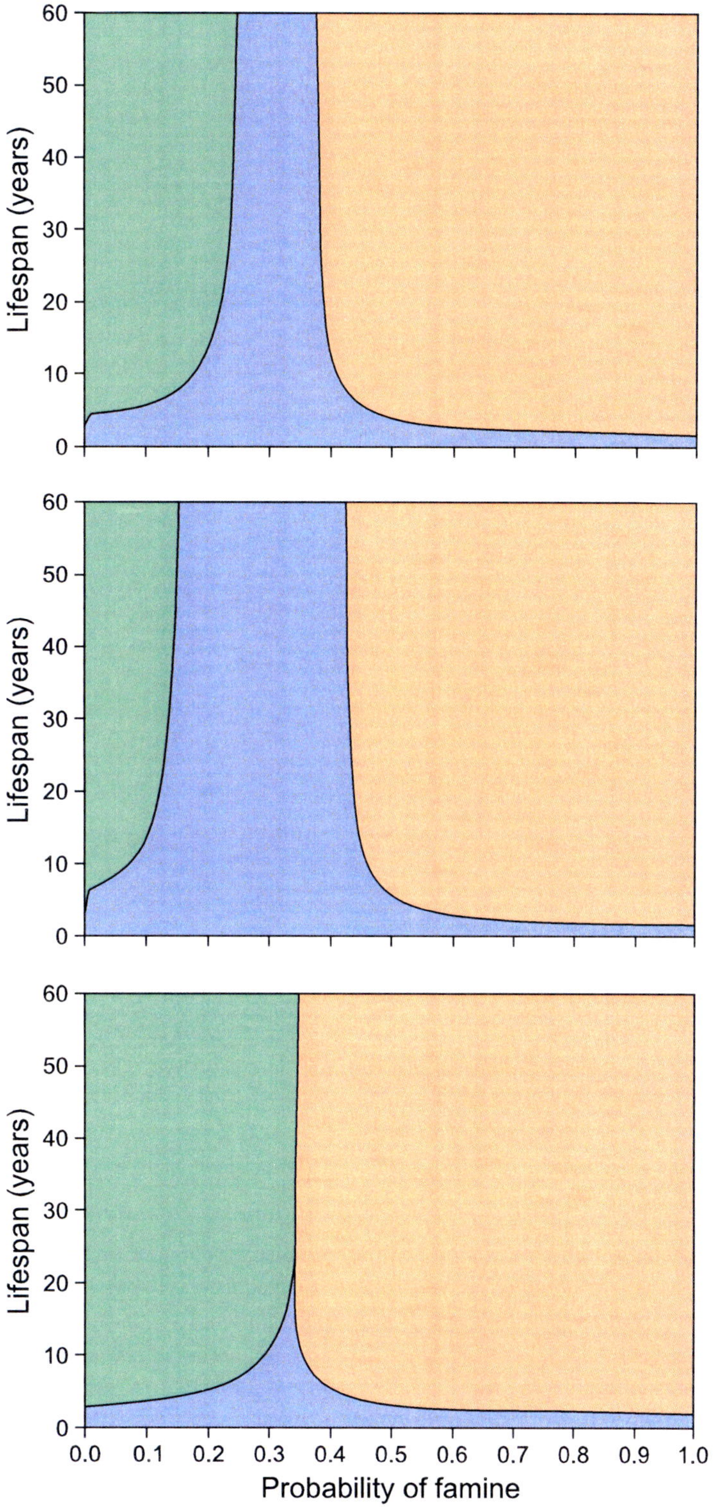

Fig. 4.3 Parameter areas of advantage of the three genotypes when there is a correlation between birth time and lifetime conditions (**a**) $r=0.1$, (**b**) $r=0.2$, and (**c**) $r=-0.05$. With a small positive correlation, the advantage of fetal programmer increases substantially. However, even very weak negative correlations can drive fetal programmer to extinction when life expectancy is high. Selection for fetal programming therefore must be driven by factors that produce significant positive birth time and lifetime correlations. Without such correlations predictive fetal programming is unlikely to evolve

Table 4.1 Correlations of annual rainfall to that of the subsequent year, short-term (10 years) and long-term (40 years) cumulative

S. no.	Subdivision	1 year	10 years cumulative	40 years cumulative
1	Assam Meghalaya	0.047	0.042	−0.029
2	Nagaland, Manipur, Mizoram, and Tripura	0.006	0.010	−0.078
3	Sub-Himalayan West Bengal	−0.061	−0.103	−0.023
4	Gangetic West Bengal	−0.009	0.096	−0.265[a]
5	Orissa	−0.111	0.135	−0.096
6	Jharkhand	−0.054	−0.042	−0.012
7	Bihar	0.070	−0.158	−0.048
8	East Uttar Pradesh	0.096	−0.106	−0.091
9	West Uttar Pradesh plains	−0.036	0.010	−0.066
10	Haryana	−0.010	0.050	0.041
11	Punjab	−0.050	0.099	0.081
12	Rajasthan	0.048	−0.200[a]	−0.094
13	East Rajasthan	0.047	−0.034	−0.017
14	West Madhya Pradesh	0.052	0.114	0.010
15	East Madhya Pradesh	−0.076	0.040	−0.034
16	Gujarat	−0.007	−0.121	−0.110
17	Saurashtra and Kutch	−0.037	−0.118	−0.152
18	Konkan and Goa	0.121	0.154	0.080
19	Madhya Maharashtra	0.227[a]	0.010	−0.163
20	Marathwada	0.171	0.043	−0.101
21	Vidarbha	−0.079	0.026	−0.193[a]
22	Chhattisgarh	0.039	0.278[a]	0.010
23	Coastal Andhra Pradesh	0.082	0.023	−0.053
24	Telangana	0.141	0.110	0.086
25	Rayalaseema	−0.104	−0.008	−0.054
26	Tamil Nadu	0.049	−0.139	−0.169
27	Coastal Karnataka	−0.035	−0.042	0.079
28	North Interior Karnataka	0.132	0.086	−0.020
29	South Interior Karnataka	−0.025	−0.087	−0.202[a]
30	Kerala	0.117	0.083	0.012

[a]Indicate significance of individual correlations at $\alpha=0.05$ level. Since a large number of statistical tests are being performed, a Bonferroni correction to the significance level is applicable. After the Bonferroni correction, none of the correlations are significant

evolved in different areas should show differential predisposition to obesity and related disorders. People evolved in arid or drought-prone areas could have suffered more frequent famines and therefore should have a greater tendency to develop obesity. Also ethnic groups from harsh winter environments were unable to hunt, fish, or farm during the colder months and thus historically could have faced regular feast and famine conditions. Therefore, we may expect higher diabetic tendencies in them. On the other hand, those living in tropical biodiversity-rich areas should be relatively free from the predisposition to obesity or diabetes. Although substantial cross-ethnic differences are observed, the trends are not as expected by the thriftiness hypotheses. Caucasoid are among the ethnic groups most resistant to T2D. Eskimo and some of the Himalayan ethnic groups that have faced harsh winter environments for a large number of generations had a considerably lower frequency of obesity and/or T2D until recently [70–74]. Almost all ethnic groups of warmer tropical habitats have a high incidence on adopting a Western urban lifestyle [39, 75] contrary to the expectation. Cross-ethnic comparisons have a limited value since a large number of confounding factors operate. Also, cross-ethnic differences do not appear to be robust and change with time. For example, Eskimos were once among the people with very low prevalence of diabetes, but the picture is changing very rapidly of late [76]. Interestingly, some proponents of thriftiness hypotheses have tried to use the cross-ethnic differences in support of thrift often twisting history in this attempt. For example, the higher incidence in Indians or Polynesians has been ascribed to the high prevalence of famines or starvation periods these cultures faced [77]. But the question whether these people really faced famines or periodic starvation from other causes for sufficiently long durations to have any evolutionary effects has not been addressed with sufficient rigor. On the contrary, famines and starvation periods were quite common in preindustrial Europe [38, 78] and could surpass those in Asia. Overall, the ethnic difference data are inconclusive and cannot be used as an evidence for or against the thrifty gene hypothesis.

8. Does thrift explain the systems level changes in diabetes?

   Obesity and insulin resistance are associated with a number of changes in the different body systems and their functions as diverse as ovulation, spermatogenesis, sexual behavior, innate immunity, wound healing, angiogenesis, skin architecture, memory, and other cognitive brain functions [23, 79]. The thriftiness hypotheses focus on energy homeostasis alone and offer no explanation as to why these diverse changes are associated with it. Why, for example, acanthosis nigricans accompanies diabetes, why there is a chronic systemic inflammation, why sex and reproduction should be affected in diabetes, why there should be abnormalities in angiogenesis, etc. A good theory should account for all the alterations in body systems at both proximate and ultimate level simultaneously. The thrift hypotheses appear to do neither or at the best give some fragmentary explanations at the proximate level. There are some attempts to explain changes in immunity by saying that a famine could be also accompanied by infectious disease, and therefore, along with thrift, facilitation of the immune system in the form of inflammation may have evolved [80, 81]. However, there is no consistent evidence of increased immunity in "thrifty" individuals. In fact, there appears to be a greater susceptibility to certain classes of infections accompanying T2D as we will see in a later chapter.

   The classical thrift hypotheses appear to stop at the origins of obesity. They assume that obesity inevitably leads to all other pathology, and no evolutionary theory is needed to explain the link of obesity with reproductive, cognitive, immunological, and other changes in the body systems. The simplest and foremost question as to why obesity gives rise to insulin resistance has not been addressed at the ultimate level. Whether insulin resistance is adaptive for obese individuals in some way or the other or whether insulin resistance is mechanistically inevitable following obesity and evolution could not have changed this link remains unaddressed. Owing to the overall failure of the thrift family of hypotheses to account for the system level changes at proximate and ultimate level simultaneously, even if we are reluctant to call the thrifty gene and thrifty phenotype hypotheses "wrong," we need to admit that they are certainly inadequate.

9. What is the mechanism of thrift?

   The original suggestion of Neel [1] was that in individuals prone to obesity and type 2 diabetes (T2D), a "quick insulin trigger" ensures rapid glucose uptake which is then converted into fat. In the early 1960s, it was well known that the levels of insulin are higher in prediabetics, and Neel's original proposal looked sound. However, insulin resistance was discovered soon, and it became clear that the "quick insulin trigger" is unlikely to work as believed by Neel since the insulin trigger would be made ineffective by insulin resistance. Whether insulin resistance is the "diabetic tendency" that Neel presumed to be responsible for obesity is not clear. Evidence for insulin resistance leading to obesity is full of contradictions. Insulin has lipogenic and antilipolytic action. But whether this action remains after development of insulin resistance is questionable. Fat cell-specific insulin receptor knockout (FIRKO) mice are lean and fail to accumulate fat. Out of all the tissue-specific insulin receptor knockouts, only muscle knockout (MIRKO) mice become obese, but they do not become hyperinsulinemic [82]. It appears therefore that if there is muscle insulin resistance without an accompanying fat cell insulin resistance, obesity can result. This could be a possible mechanism of thrift. However, so far, there are no good relative measurements of muscle versus fat cell insulin resistance to test whether this is really the cause of obesity. Therefore whether insulin resistance can be simply equated to thrift is doubtful. As discussed above, whether lower resting metabolic rate accounts for type 2 thrift is questionable [42–45]. If overeating is the meaning of thriftiness, i.e., type 1 thrift, we are yet far from understanding the proximate and ultimate causes of overeating. Since a multitude of feedback mechanisms regulate food intake, what causes failure of these mechanisms is a question that does not have a clean answer as yet. Thus even after half a century of research, we do not have a clear idea as to what exactly is thrift and how it works.

Refining Thrift

Realizing the problems with the classical thriftiness hypotheses, there have been attempts to either refine the concept or suggest alternatives. Possible refinements go along the following lines:

1. Perhaps feast and famine are rather extreme conditions, and thrift may have evolved for subacute energy limitations rather than extreme famines. How do animals adapt to such conditions? An example is the dwarf hamster native to Siberia and Mongolia that do not stop eating and do not hibernate during winters but instead live in underground tunnels. They feed on seeds of grass and other scanty food but keep on foraging under the snow. This is an example of subacute nutrient shortage. As opposed to fat accumulation by animals preparing for hibernation, these animals lose fat as the winter approaches. The adipose tissue decreases from 35% to <5% along with reduction in certain other organs including reproductive organs. This is a strategy to save expenditure of energy which is best achieved by reducing the body mass [55]. In India cattle from drought-prone areas are known to be small and lean rather than large and fat, and they are not obesity prone even if fed well. If subacute energy limitation has driven the evolution of thrift, we should observe a different phenotype evolving, not the obese phenotype of popular perception.
2. Under subacute famine not every individual is undernourished. The socially subordinate ones are more likely to experience greater food insecurity. If this is true, thrift should be adaptive to socially subordinate individuals. Since social subordination is not genetic, rather than thrifty gene(s), conditional thrift linked to social subordination is more likely to evolve. This is an interesting and relevant refinement in the thrift concept and is also supported by theory as well as evidence as we shall see later.
3. Wells [26] suggested that the benefit of fetal programming for thrift goes to the mother rather than the fetus. He argued that there is maternal advantage in fetal programming in the form of optimizing maternal inputs per fetus. Since humans have a long period of dependence on parents even after weaning, the mother has to divide resources between many offspring at different ages simultaneously. Under nutritionally limiting environment, by making a child thrifty, she improves the provision for the next one. This is likely to work for intrauterine investment. But the argument goes beyond that. It proposes that the offspring adapt themselves to the maternal strategy reciprocally. This is beneficial to both the mother and offspring. The offspring thrift assumed in this case is type 2 thrift. This suggestion is not compatible with type 1 thrift. An interesting point of this model is that it expects

thrift to give advantage neither in a predictive manner nor restricted to birth conditions. Thrift is adaptive as long as the offspring has some dependence on the mother. It is possible therefore that since human offspring have long-term dependence on their mother, a long-term thrift might have evolved. However, this long term is still not lifelong. Therefore why there is a lifelong commitment to thrift remains an unanswered question.

4. Apart from genes and fetal programming, there are other possible ways of transgenerational effects on thrift [25, 83]. Therefore, thrift may go beyond the classical thrifty gene and thrifty phenotype models. Thrift need not be only about energy intake and expenditure. As Wells [83] puts it, there is more to fat than thrift and more to thrift than fat. Thrift may take several other forms such as altered growth rate, differential allocation of energy to different organs, changed reproductive strategies, or altered immune system function.
5. Bet hedging: Bet hedging refers to spreading of risk, i.e., not putting all eggs in a single basket. The distribution of genetic information about obesity to several genes, each with a small effect, is thought to be a bet-hedging strategy. If there is variance in the level of thrift among the progeny, at least some of them will have the right level of thrift for the unpredictable feast or famine conditions faced. This is thought to be why genetic elements of obesity are distributed over several genes. Fragmentation of genetic information may also be necessary for phenotypic plasticity [83]. However bet hedging is most unlikely to work by spreading the levels of thrift among the progeny since the variance in obesity decreases with the number of possible gene combinations as illustrated earlier.

It is possible to progressively refine the concept of thrift to accommodate facts including the diverse system level changes accompanying obesity and insulin resistance. Here it appears that the attempts to refine thrift go beyond what is classically implied by the word thrift, and at some stage, the phenomenon would demand some other more appropriate name.

Other Alternative Theories

Realizing the limitations, inadequacies, and flaws of the thrifty gene and fetal programming hypotheses, a number of alternatives have emerged. They have tried to improve upon or replace both the thrifty gene model and fetal programming model:

1. Alternatives to the thrifty gene hypothesis
   (a) Speakman [21] postulated genetic drift rather than selection for obesity-related genes. According to him an upper limit on obesity was set in hunter-gatherer life which was effectively lifted when humans became free of predation. Subsequently the obesity-related genes started spreading by genetic drift.
   (b) Corbett et al. [84] argued that today's obese, diabetic, and PCOS-prone genotype was the ancestral one that had better fertility in famine conditions. In the modern era of food security since 1800 AD, an insulin-sensitive genotype that has better fertility under conditions of food abundance started spreading.
   (c) Moalem et al. [85] hypothesized that high plasma glucose lowers the freezing point of blood which prevents formation of ice crystals in cells through supercooling, and this has been suggested as an adaptation to ice age. This hypothesis was postulated more with reference to type 1 diabetes.
2. Alternative explanations for fetal programming
   (a) It has been argued that the association between low birth weight and insulin resistance may have arisen out of a reverse causation, i.e., babies with insulin resistance genotype are more likely to survive fetal undernourishment [68] or are more likely to be born underweight [86].

Evaluating the Alternatives

We will briefly evaluate the alternative hypotheses now to see whether they offer better explanations, wherever the thriftiness hypotheses are weaker.

1. Polymorphism: Polymorphism in the predisposition to obesity can be potentially explained in three different ways. It is possible that the allelic composition in the population is close to equilibrium, and there is a stable polymorphic equilibrium. Out of all alternative hypotheses postulating a genetic basis, none is able to predict stable polymorphism. An alternative view is that the human population today is not at a stable equilibrium proportion of alleles but is undergoing drift [19, 21] or selection [85]. These alternative hypotheses try to explain polymorphism as a transient state. Speakman [19] tries to quantify the drift dynamics. However his calculation is based on the assumption that human ancestors became free from predators about 1.8 million years ago, and this estimate is debatable [87]. Moreover if the drift is an ongoing process since it is random, it would take a different direction in different populations, and the cross-ethnic differences are expected to be random. Instead, we see that almost all tropical ethnic groups have high tendency to develop obesity and T2D on urbanization, and such a generalized pattern is not expected by the drift hypothesis. Also, as we know now, a large number of genes are associated with obesity, and if drift has to operate independently on all the genes, it is highly unlikely that it will result in a directional change towards increasing obesity. The large number of genes involved makes genetic drift towards obesity highly improbable.

   Corbett et al. [85] presumed that selective pressures changed substantially by 1800 AD, and they claim that 200 years would be sufficient to bring in an evolutionary change. This claim has not been quantitatively tested. Any data to test the dynamics of both the hypotheses using a predictive model are currently lacking.

   The cold adaptation hypothesis of Moalem et al. [86] does not explain stable polymorphism at any of the above levels. If high blood glucose is adaptive in cold environments, then ethnic groups evolved in cold climates should undergo directional selection leading to fixation. Polymorphism in that case can only arise as a transient state or by crossbreeding between ethnic groups coming from warmer and colder environments. It can be argued that since humans originated in the tropics and migrated to colder habitats only recently, there could be a transient polymorphism. In that case ethnic groups that remained in the tropics should not have evolved the tendency to raise blood sugar levels. But the evidence is on the contrary, tropical ethnic groups have greater prevalence of diabetes and hypertension on adopting Western urban lifestyle.

   A different explanation to the apparent polymorphism is that there is no real genetic polymorphism, but individuals are programmed differently depending upon environmental conditions faced in early life. Therefore we find much difference in the tendency to become obese in individuals with similar diet and lifestyle, without true genetic polymorphism.
2. Low birth weight effects: There is globally consistent and strong evidence that a small birth weight is associated with altered metabolic and endocrine states in adult life [11–17]. Different interpretations of the birth weight effects [27, 68, 86] have been offered owing to which the birth weight association itself cannot be considered a convincing evidence of fetal programming. However, rat experiments with induced maternal nutrient deficiencies have also given strong evidence in support of fetal programming [67, 88, 89]. Here the causative role of IUGR is made clear by the experimental design. Therefore some type of fetal programming can be safely inferred. It can nevertheless be debated whether the programming is for thriftiness or for any other adaptation. The Speakman's [19, 21] drifty gene hypothesis, Corbett et al.'s [84] hypothesis of reverse selection in modern times, and Moalem et al.'s [85] cold adaptation hypotheses do not offer any explanation for the birth weight effect. Wells' approach of refining thrift [26] accounts for long-term fetal programming without the need for a predictive adaptation element in it. It attempts to

explain the lifetime commitment in an alternative way. Since the period of dependence for a human child goes much beyond weaning, the mother has to make provisions for children at various ages simultaneously. Under nutrient-limited environments, the mother benefits by making the offspring thrifty beyond the weaning period so that she can provide the younger infant or fetus better. This is an ingenious way of explaining longer time commitment to thrift. However this explanation stops at about prereproductive age after which there appears to be no reason for retaining thrift. Therefore it falls short of explaining lifetime commitment. The other is that it assumes only one type of thrift, that is, type 2 thrift mediated by energy conservation. Type 1 thrift refers to hyperphagia, and this would actually interfere in maternal partitioning of resources. There is increasing evidence that type 1 thrift contributes more than type 2 to obesity as a causal factor [42–45]. Therefore, whether maternal provision could be optimized by offspring thrift is doubtful and remains to be tested. Another problem in bringing long-term dependence of human offspring in the explanation of thrift is the fact that IUGR models in rats and other animals with short infancy are also insulin resistant and show lifetime metabolic programming.

The evolution of lifetime programming based on birth conditions is possible only if there were a positive correlation between birth time and lifetime conditions. The Baig et al. model [29] predicts that for lifetime fetal programming to evolve, a small to moderate positive correlation between intrauterine and lifetime conditions is sufficient (Fig 4.3). The classical thrifty phenotype hypothesis does not suggest what can lead to such correlations. Nevertheless we can think of factors that could potentially cause such correlations. Apart from a climatic "famine," there can be other causes of food deprivation. A high population density can lead to increased competition resulting into undernourishment. Populations of many species oscillate in the wild, and the oscillations typically span over several generations so that an individual born at a high population density is expected to see high population density through most of his life. Therefore periodic food scarcity caused by population oscillations can result into positive correlations between birth year and lifetime nutritional conditions. In social species the social hierarchy might also be related to differential access to food. An individual born smaller and weaker is more likely to face social subordination, and therefore, an anticipatory fetal programming would be adaptive. These social causes can produce positive birth–lifetime correlations, and they are more likely candidates to select for fetal programming of lifelong duration. There exist a broad range of metabolic, endocrinological, behavioral, and cognitive adaptations that accompany social hierarchical positions. Therefore if fetal programming is in response to social hierarchies, we expect that it need not be restricted to diet and energy homeostasis but affects a large number of systems of the body including brain and behavior. This indeed is the case, and type 2 diabetes involves almost all systems of the body [23, 79]. Social subordination is an important predisposing factor for type 2 diabetes [90] too, and this hints towards an alternative interpretation of the phenomenon.

It is possible that on exposure to intrauterine malnourishment, there is fetal programming in anticipation of physical weakness and social subordination or anticipation of high population density rather than anticipation of climatic famine. In primate societies, it is known that social ranks of juveniles and subadults are influenced by the ranks of their mothers [90, 91]. If social rank influences preferential access to food resources, it is sufficient to produce the positive correlation between fetal and lifetime nutritional conditions needed for the evolution of fetal programming. It appears logical therefore that fetal programming could have evolved in anticipation of population and social hierarchy-related factors than anticipation of famines or other climate-related factors. This is a new and promising possibility to which we will return in due course.

3. Genome-wide associations (GWA): Similar to the thrifty gene hypothesis, some of the alternative hypotheses suffer from lack of evidence for substantial genetic influences on obesity. The gap between pregenomic and genomic estimates of the genetic component in obesity is a major paradox, and none of the "gene"-based hypotheses account for it. These hypotheses include Neel's thrifty gene, Speakman's drifty gene, Corbett et al.'s reverse selection, and Moalem et al.'s cold adaptation. With genome-wide association studies yielding poor contribution of genes to the prevalent obesity epidemic, we can safely turn down all these hypotheses right away.
4. Intrauterine effects: On the other hand, fetal programming has a promise for filling the gap between the familial studies and genome-wide association studies. The programming is likely to involve epigenetic mechanisms as well. However, since majority of adult diabetics do not have a history of IUGR, programming by IUGR appears to be highly inadequate as a generalized causative factor. This suggests that apart from fetal nutrition, there could be other stimuli and agents promoting such a programming. It is possible that more than one type of metabolic programming is involved in obesity and insulin resistance which may be induced in various stages of life. There have been no efforts so far to identify alternative programming mechanisms. This appears to be mainly because the thrift paradigm overwhelmed so much that all alternative ways of thinking had been more or less blocked until very recently.
5. The picture across species: Unlike an older belief, insulin resistance and occasionally overt diabetes are not unique to humans. Substantial differences in individual insulin sensitivity and some other markers of metabolic syndrome are observed in primates and other mammals in the wild or semi-wild states [92, 93]. Any evolutionary theory needs to account for these observations too. Any theory based on components of uniquely human evolution such as the origin of agriculture will be inadequate to account for the natural occurrence of insulin resistance in other animals. Some of the existing theories either fail on this ground or have not adequately considered this angle.
6. Beyond obesity: Similar to the thrifty gene and thrifty phenotype hypotheses, most of the alternatives fail to explain the multitude of changes in every system of the body that accompany T2D and related disorders. Obesity and T2D are not only about energy homeostasis but also about changes in innate immunity, sexual and reproductive function, vascular development and function, skin architecture, wound healing and tissue regeneration, memory, cognitive functions, behavior and mechanisms of decision making, social relations, and social signaling. Most of the evolutionary hypotheses so far are too gluco-lipo-centricand therefore fall short of explaining at the proximate and ultimate level the simultaneous involvement of multiple systems. In fact it would be appropriate to say that as yet, there are not hypotheses to explain the evolutionary origins of the insulin resistance syndrome at all. They may at the most explain the origins of obesity. The assumption that obesity inevitably leads to insulin resistance and therefore explaining the origins of obesity is sufficient to explain all the rest is not true. Obesity is linked to insulin resistance through several pathways. Insulin resistance is linked further down to several pathophysiological mechanisms by several pathways. Not all of them can be assumed to be inevitable biochemical effects of FFAs or adipose tissue. Many of the mechanisms involve specific signaling molecules such as TNF-α and other adipokines. Any evolutionary hypothesis needs to explain why adipose tissue secretes these signaling molecules. Certain CNS-mediated mechanisms are also involved in altering muscle glucose uptake or muscle glycogen synthesis or insulin response to glucose under the influence of obesity [94–96]. Recent work is uncovering more complex mechanisms involving synergy between neuronal and peripheral signals that fine-tune insulin

sensitivity [97–99]. Insulin resistance therefore appears to be much more than an inevitable effect of obesity. This means that active mechanisms linking obesity to insulin resistance must have evolved to serve some adaptive purpose. Any evolutionary theory of the origins of diabetes needs to explain why insulin resistance could be adaptive upon developing obesity. Surprisingly hardly any of the existing hypotheses appears to address this question in any significant way. I would expect that a good evolutionary theory should also resolve the existing paradoxes in the pathophysiology of T2D that we saw in the last chapter. This needs to be done simultaneously at the proximate and ultimate levels. So far there has been no theory that works at both proximate and ultimate levels. If any evolutionary theory also suggests some important, practicable, and testable means for preventing, controlling, and at the best curing diabetes, it would be a highly desirable bonus.

I can safely say now that so far, there have been no evolutionary theories for the origin of the insulin resistance syndrome meeting all the criteria that were listed at the beginning of this chapter, although there have been serious and insightful attempts to explain the origins of obesity [55, 83]. I have elaborated sufficiently on how and why the thriftiness paradigm is highly inadequate to account for the evolution, pathogenesis, and epidemiological picture of obesity, T2D, and related disorders. The alternative hypotheses mentioned above are perhaps equally unsatisfactory or only slightly better since they too leave wide gaps in our understanding of the phenomena. The approach of refining the concept of thrift making it more accommodative is an interesting one which can potentially answer a number of questions. But it is likely that by the time it modifies sufficiently to answer most of the questions, it will be no more recognizable as "thrift."

As a teacher I have often experienced that while trying to solve a problem collectively in the class, the thinking often gets stuck at some point and it appears a deadlock. At such time it is often useful to give up that line of thinking, forget everything that has been said so far, and make a fresh beginning right from scratch. This often works better. Therefore although I believe that we can keep on addressing questions one by one and go on modifying existing concepts until things look sufficiently sound, I would prefer to give up this line of thought and make a fresh beginning altogether. We need a real and good alternative theory that fills in the gaps logically, works at both proximate and ultimate levels simultaneously, and is either supported by preexisting evidence or gives rise to predictions that can be tested experimentally and/or epidemiologically. Our efforts should now be focused on finding such an alternative and putting it to test. I would prefer to attempt this by beginning to look at some aspects of life in the wilderness where we evolved.

References

1. Neel JV (1999) Diabetes mellitus: a 'thrifty' genotype rendered detrimental by 'progress'? 1962. Bull World Health Organ 77:694–693
2. Steinberg AG, Wilder RM (1952) A study of the genetics of diabetes mellitus. Am J Hum Genet 4:113–135
3. Steinberg AG (1959) The genetics of diabetes: a review. Ann N Y Acad Sci 82:197–207
4. Allan W (1933) Heredity in diabetes. Ann Intern Med 6:1272–1274
5. Nørretranders T (1998) The user illusion: cutting consciousness down to size. Viking, NY, USA
6. Hales CN et al (1991) Fetal and infant growth and impaired glucose tolerance at age 64. Brit Med J 303:1019–1022
7. Davies DP, Matthes J (1991) Fetal and infant growth and impaired glucose tolerance. Brit Med J 303:1474
8. Barker D (1998) In utero programming of chronic disease. Clin Sci 95:115–128
9. Drake AJ, Walker BR (2004) The intergenerational effects of fetal programming: non-genomic mechanisms for the inheritance of low birth weight and cardiovascular risk. J Endocrinol 180:1–16
10. Hales CN, Barker DJ (1992) Type 2 (non-insulin-dependent) diabetes mellitus: the thrifty phenotype hypothesis. Diabetologia 35:595–601
11. Stein CE et al (1996) Fetal growth and coronary heart disease in south India. Lancet 348:1269–1273
12. Yajnik CS (2001) The insulin resistance epidemic in india: fetal origins, later lifestyle, or both? Nutr Rev 59:1–9
13. Yajnik CS (2004) Early life origins of insulin resistance and Type 2 diabetes in India and other Asian countries. J Nutr 134:205–210

14. Williams MA, Emanuel I, Kimpo C, Leisenring WM, Hale CB (1999) A population-based cohort study of the relation between maternal birthweight and risk of gestational diabetes mellitus in four racial/ethnic groups. Paediatr Perinat Epidemiol 13:452–465
15. McNeely MJ, Fujimoto WY, Leonetti DL, Tsai EC, Boyko EJ (2007) The association between birth weight and visceral fat in middle-age adults. Obesity 15:816–819
16. Pettitt DJ, Jovanovic L (2007) Low birth weight as a risk factor for gestational diabetes, diabetes, and impaired glucose tolerance during pregnancy. Diabetes Care 30:S147–S149
17. Li C, Johnson MS, Goran MI (2001) Effects of low birth weight on insulin resistance syndrome in caucasian and African–American children. Diabetes Care 24:2035–2042
18. Gluckman PD, Hanson MA, Spencer HG (2005) Predictive adaptive responses and human evolution. Trends Ecol Evol (Amst) 20:527–533
19. Speakman JR (2007) A nonadaptive scenario explaining the genetic predisposition to obesity: the 'predation release' hypothesis. Cell Metab 6:5–12
20. Speakman JR (2006) Thrifty genes for obesity and the metabolic syndrome–time to call off the search? Diab Vasc Dis Res 3:7–11
21. Speakman JR (2008) Thrifty genes for obesity, an attractive but flawed idea, and an alternative perspective: the/ drifty gene/' hypothesis. Int J Obes 32:1611–1617
22. Benyshek DC, Watson JT (2006) Exploring the thrifty genotype's food-shortage assumptions: a cross-cultural comparison of ethnographic accounts of food security among foraging and agricultural societies. Am J Phys Anthropol 131:120–126
23. Watve MG, Yajnik CS (2007) Evolutionary origins of insulin resistance: a behavioral switch hypothesis. BMC Evol Biol 7:61
24. Wells JCK (2007) Flaws in the theory of predictive adaptive responses. Trends Endocrinol Metab 18:331–337
25. Wells JCK (2010) Maternal capital and the metabolic ghetto: an evolutionary perspective on the transgenerational basis of health inequalities. Am J Hum Biol 22:1–17
26. Wells JCK (2007) The thrifty phenotype as an adaptive maternal effect. Biol Rev Camb Philos Soc 82:143–172
27. Wells JCK (2009) Thrift a guide to thrifty genes, thrifty phenotypes and thrifty norms. Int J Obes 33:1331–1338
28. Wells JCK (2003) The thrifty phenotype hypothesis: thrifty offspring or thrifty mother? J Theor Biol 221:143–161
29. Baig U, Belsare P, Watve M, Jog M (2011) Can thrifty gene(s) or predictive fetal programming for thriftiness lead to obesity? J Obes 2011:1–11
30. Yilmaz N, Kilic S, Kanat-Pektas M, Gulerman C, Mollamahmutoglu L (2009) The relationship between obesity and fecundity. J Womens Health (Larchmt) 18:633–636
31. Gesink Law DC, Maclehose RF, Longnecker MP (2007) Obesity and time to pregnancy. Hum Reprod 22:414–420
32. Sallmén M, Sandler DP, Hoppin JA, Blair A, Baird DD (2006) Reduced fertility among overweight and obese men. Epidemiology 17:520–523
33. Ramlau-Hansen CH et al (2007) Subfecundity in overweight and obese couples. Hum Reprod 22: 1634–1637
34. Sahlins M. The Original Affluent Society—Marshall Sahlins. http://www.primitivism.com/original-affluent.htm
35. Sahlins M (1974) Stone age economics. Tavistock Publications, London, http://trove.nla.gov.au/work/21605237?selectedversion=NBD347965
36. Cohen MN (1984) Paleopathology at the origins of agriculture. Academic, San Diego, CA
37. Lukacs JR, Walimbe SR (1998) Physiological stress in prehistoric india: new data on localized hypoplasia of primary canines linked to climate and subsistence change. J Archaeol Sci 25:571–585
38. Manning R (2004) Against the grain: how agriculture has hijacked civilization. North Point Press, New York, NY
39. O'Dea K (1991) Westernisation, insulin resistance and diabetes in Australian aborigines. Med J Aust 155:258–264
40. Pasquali R, Gambineri A, Pagotto U (2006) Review article: the impact of obesity on reproduction in women with polycystic ovary syndrome. BJOG: An Int J Obstet Gynaecol 113:1148–1159
41. Norman RJ, Clark AM (1998) Obesity and reproductive disorders: a review. Reprod Fertil Dev 10:55–63
42. Kensara OA et al (2006) Substrate-energy metabolism and metabolic risk factors for cardiovascular disease in relation to fetal growth and adult body composition. Am J Physiol Endocrinol Metab 291: E365–E371
43. Eriksson J, Forsén T, Tuomilehto J, Osmond C, Barker D (2002) Size at birth, fat-free mass and resting metabolic rate in adult life. Horm Metab Res 34:72–76
44. Molnár D, Schutz Y (1997) The effect of obesity, age, puberty and gender on resting metabolic rate in children and adolescents. Eur J Pediat 156:376–381
45. Huang K-C, Kormas N, Steinbeck K, Loughnan G, Caterson ID (2004) Resting metabolic rate in severely obese diabetic and nondiabetic subjects. Obes Res 12:840–845
46. Zunquin G, Theunynck D, Sesboüé B, Arhan P, Bouglé D (2009) Comparison of fat oxidation during exercise in lean and obese pubertal boys: clinical implications. Brit J Sports Med 43:869–870
47. Frisancho AR (2003) Reduced rate of fat oxidation: a metabolic pathway to obesity in the developing nations. Am J Hum Biol 15:522–532
48. Rogge MM (2009) The role of impaired mitochondrial lipid oxidation in obesity. Biol Res Nurs 10:356–373
49. Zurlo F et al (1990) Low ratio of fat to carbohydrate oxidation as predictor of weight gain: study of 24-h RQ. Am J Physiol Endocrinol Metab 259:650–657

50. Sawaya AL, Verreschi I, Tucker KL, Roberts SB, Hoffman DJ (2000) Why are nutritionally stunted children at increased risk of obesity? Studies of metabolic rate and fat oxidation in shantytown children from São Paulo, Brazil. Am J Clin Nutr 72:702–707
51. Speakman JR (1998) The history and theory of the doubly labeled water technique. Am J Clin Nutr 68:932S–938S
52. Westerterp KR, Speakman JR (2008) Physical activity energy expenditure has not declined since the 1980s and matches energy expenditures of wild mammals. Int J Obes (Lond) 32:1256–1263
53. Farooqi IS, O'Rahilly S (2007) Genetic factors in human obesity. Obes Rev 8:37–40
54. Pasquet P et al (1992) Massive overfeeding and energy balance in men: the Guru Walla model. Am J Clin Nutr 56:483–490
55. Pond CM (1998) The fats of life. Cambridge University Press, Cambridge
56. Maes HH, Neale MC, Eaves LJ (1997) Genetic and environmental factors in relative body weight and human adiposity. Behav Genet 27:325–351
57. Rankinen T et al (2006) The human obesity gene map: the 2005 update. Obesity (Silver Spring) 14:529–644
58. Thorleifsson G et al (2009) Genome-wide association yields new sequence variants at seven loci that associate with measures of obesity. Nat Genet 41:18–24
59. Li S et al (2010) Cumulative effects and predictive value of common obesity-susceptibility variants identified by genome-wide association studies. Am J Clin Nutr 91:184–190
60. Mutch DM, Clément K (2006) Unraveling the genetics of human obesity. PLoS Genet 2(12):e188
61. McCarthy MI, Zeggini E (2009) Genome-wide association studies in type 2 diabetes. Curr Diab Rep 9:164–171
62. Sladek R et al (2007) A genome-wide association study identifies novel risk loci for type 2 diabetes. Nature 445:881–885
63. Scott LJ et al (2007) A genome-wide association study of type 2 diabetes in Finns detects multiple susceptibility variants. Science 316:1341–1345
64. Kilpelainen TO et al (2011) Genetic variation near IRS1 associates with reduced adiposity and an impaired metabolic profile. Nat Genet 43:753–760
65. Reed D, Lawer M, Tordoff M (2008) Reduced body weight is a common effect of gene knockout in mice. BMC Genet 9:4
66. Stöger R (2008) The thrifty epigenotype: an acquired and heritable predisposition for obesity and diabetes? Bioessays 30:156–166
67. Tosh DN et al (2010) Epigenetics of programmed obesity: alteration in IUGR rat hepatic IGF1 mRNA expression and histone structure in rapid vs. delayed postnatal catch-up growth. Am J Physiol Gastrointest Liver Physiol 299:G1023–G1029
68. McCance DR et al (1994) Birth weight and non-insulin dependent diabetes: thrifty genotype, thrifty phenotype, or surviving small baby genotype? Brit Med J 308:942–945
69. Boyko EJ (2000) Proportion of type 2 diabetes cases resulting from impaired fetal growth. Diabetes Care 23:1260–1264
70. Swinburn BA (1996) The thrifty genotype hypothesis: how does it look after 30 years? Diabet Med 13:695–699
71. Beck-Nielsen H (1999) General characteristics of the insulin resistance syndrome: prevalence and heritability. European Group for the study of Insulin Resistance (EGIR). Drugs 58(suppl 1):7–10 (discussion 75–82)
72. Vilbergsson S, Sigurdsson G, Sigvaldason H, Hreidarsson ÁB, Sigfusson N (1997) Prevalence and incidence of NIDDM in Iceland: evidence for stable incidence among males and females 1967–1991—the Reykjavik Study. Diabet Med 14:491–498
73. Mouratoff GJ, Scott EM (1973) Diabetes mellitus in Eskimos after a decade. J Am Med Assoc 226:1345–1346
74. Vaz M et al (1999) Body fat topography in Indian and Tibetan males of low and normal body mass index. Indian J Physiol Pharmacol 43:179–185
75. Yudkin JS (1996) Non-insulin-dependent diabetes mellitus (NIDDM) in Asians in the UK. Diabet Med 13:S16–S18
76. Egeland GM, Cao Z, Young TK (2011) Hypertriglyceridemic-waist phenotype and glucose intolerance among Canadian Inuit: the International Polar Year Inuit Health Survey for Adults 2007–2008. Can Med Assoc J 183:E553–E558
77. Diamond J (2003) The double puzzle of diabetes. Nature 423:599–602
78. Zhang DD et al (2011) The causality analysis of climate change and large-scale human crisis. Proc Natl Acad Sci USA 108:17296–17301
79. Koshiyama H, Ogawa Y, Tanaka K, Tanaka I (2008) Diabetes mellitus as dysfunction of interactions among all organs: "ominous orchestra of organs". Clin Med Endocrinol Diabet 1:1–6
80. Fernández-Real JM, Pickup JC (2008) Innate immunity, insulin resistance and type 2 diabetes. Trends Endocrinol Metab 19:10–16
81. Fernández-Real JM, Ricart W (1999) Insulin resistance and inflammation in an evolutionary perspective: the contribution of cytokine genotype/phenotype to thriftiness. Diabetologia 42:1367–1374
82. Terauchi Y, Kadowaki T (2002) Insights into molecular pathogenesis of type 2 diabetes from knockout mouse models. Endocr J 49:247–263
83. Wells JCK (2011) An evolutionary perspective on the trans-generational basic of obesity. Ann Hum Biol 38:400–409
84. Corbett SJ, McMichael AJ, Prentice AM (2009) Type 2 diabetes, cardiovascular disease, and the evolutionary paradox of the polycystic ovary syndrome: a fertility first hypothesis. Am J Hum Biol 21:587–598
85. Moalem S, Storey KB, Percy ME, Peros MC, Perl DP (2005) The sweet thing about Type 1 diabetes: a cryoprotective evolutionary adaptation. Med Hypotheses 65:8–16

86. Hattersley AT, Tooke JE (1999) The fetal insulin hypothesis: an alternative explanation of the association of low birthweight with diabetes and vascular disease. Lancet 353:1789–1792
87. Watve M (1993) Why man has no predator. Curr Sci 65:120–122
88. Simmons RA, Templeton LJ, Gertz SJ (2001) Intrauterine Growth Retardation Leads to the Development of Type 2 Diabetes in the Rat. Diabetes 50:2279–2286
89. Martin JF, Johnston CS, Han CT, Benyshek DC (2000) Nutritional origins of insulin resistance: a rat model for diabetes-prone human populations. J Nutr 130: 741–744
90. Johnson JA (1987) Dominance rank in juvenile olive baboons, Papio anubis: the influence of gender, size, maternal rank and orphaning. Anim Behav 35: 1694–1708
91. Holekamp KE, Smale L (1991) Dominance Acquisition During Mammalian Social Development: The 'Inheritance' of Maternal Rank. Am Zool 31:306–317
92. Wagner JE et al (2006) Old world nonhuman primate models of type 2 diabetes mellitus. ILAR J 47: 259–271
93. Venn-Watson SK, Ridgway SH (2007) Big brains and blood glucose: common ground for diabetes mellitus in humans and healthy dolphins. Comp Med 57:390–395
94. Darleen S (2008) CNS GLP-1 regulation of peripheral glucose homeostasis. Physiol Behav 94:670–674
95. Knauf C et al (2008) Brain glucagon-like peptide 1 signaling controls the onset of high-fat diet-induced insulin resistance and reduces energy expenditure. Endocrinology 149:4768–4777
96. Knauf C et al (2005) Brain glucagon-like peptide-1 increases insulin secretion and muscle insulin resistance to favor hepatic glycogen storage. J Clin Invest 115:3554–3563
97. Chowdhury KK, Legare DJ, Lautt WW (2011) Insulin sensitization by voluntary exercise in aging rats is mediated through hepatic insulin sensitizing substance (HISS). Exp Gerontol 46:73–80
98. Lautt WW (2004) A new paradigm for diabetes and obesity: the hepatic insulin sensitizing substance (HISS) hypothesis. J Pharmacol Sci 95:9–17
99. Lautt WW et al (2001) Hepatic parasympathetic (HISS) control of insulin sensitivity determined by feeding and fasting. Am J Physiol Gastroint Liver Physiol 281:G29–G36

5 Of Hawks and Doves

I will make a fresh beginning now that does not involve any destructive criticism but instead tries to synthesize a new alternative picture that is logically coherent as well as supported by evidence. This necessitates leaving the thread of diabetes for the time being to which we will return shortly. We need to look at certain aspects of the life and behavior of animals in the wild. Since the human race evolved in the same wilderness, it is very likely that any principles learned there may help us give a novel insight in human life, behavior, physiology, and health.

Animals have been used to understand human health and disease for a long time. But we have so far relied almost entirely on captive animals. There are many obvious advantages of using captive animals, the main being the ability to control their environment and thereby minimizing the number of variables. However, in doing so, we are certainly missing an insight into a number of subtle issues in physiology. Wild animals are rarely used to gain insights into physiology for obvious difficulties in controlling the conditions and collecting samples for analysis. Fortunately this picture has started changing of late, at least marginally owing to increased rigor in behavioral ecology research on the one hand and better sampling and assay techniques on the other. Another upcoming trend relevant and useful for us is the insight in human behavior obtained from animal studies. For decades psychology, the science of human behavior, and ethology, the science of animal behavior, have developed parallel to each other without sufficient give and take, often deliberately avoiding any possible points of intersection. For example any reference of animal personalities still evokes strong reactions from many sectors. Personality is about consistent and intrinsic individual differences. There is increasing evidence that individual animals consistently differ in their responses to a given challenge, but still there is some reluctance to call them personality differences. It is only recently that there are some signs of the ice breaking, and there is increasing acceptance to the concept of animal personalities [1–3]. The long-standing reluctance has a historical root in the way Western and Eastern philosophies developed and science originated. In many of the Eastern philosophies, animals are not qualitatively different from humans. The differences perceived are quantitative. In Hinduism, for example, animals are believed to have a soul as much as humans. Not only that, the same soul can take human and animal forms in the natural course of the believed chain of births. The residual effect of this philosophy on the Eastern mind is that animals are viewed at par with humans albeit differences such as expression of thoughts in the form of a spoken language. In contrast in the mainstream traditional Western thought, humans are qualitatively different and are believed to have a soul which animals need not have. Mind and emotions in animals are often denied, or at least doubts are expressed about their existence. This has relevance to science and particularly to medicine. Animals are almost invariably used in research in medicine. Most of our understanding of human physiology has come

M. Watve, *Doves, Diplomats, and Diabetes: A Darwinian Interpretation of Type 2 Diabetes and Related Disorders*,
DOI 10.1007/978-1-4614-4409-1_5, © Springer Science+Business Media New York 2013

from animal experiments. There is no conceptual problem about this. However, that is not the case when it comes to cognition and emotions. Few objections are raised in applying an animal experiment to infer something about human physiology. But this is not acceptable to many in the context of behavior or emotions. Since humans have evolved from animal lineages, it would be natural to assume that the principles of evolution that we learn from plant–animal observations and experiments are applicable to humans as well. But generally this is not the baseline assumption in most branches of humanities. Recently this picture appears to be changing with subjects like evolutionary psychology taking roots as a serious branch of science. But they have not enjoyed a smooth path to acceptance so far.

This debate is relevant to us because I think we can learn much about human health from animal behavior, the link going through human behavior. We need to accept that although the factors and forces shaping human behavior are not identical to those shaping animal behavior, we have evolved following the same evolutionary principles as they have. Many of the fundamental principles of animal behavior were largely applicable to human behavior through much part of our ancestry. People who object to talking about human and animal behavior on the same platform generally do not mind talking about human and animal physiology on the same platform. The parallels or in fact almost identical physiological mechanisms across different species of animals and humans are what the current paradigm accepts very well. Human physiology evolved in an era when we were not too different from animals. Therefore at least in the context of behavior–physiology interface, there should not be any conceptual hurdles about learning some fundamental principles from the animal world.

Although we will use a number of principles from behavioral ecology in understanding human health and disease, we will make a beginning with and focus on multiple behavioral strategies, the ways animals make choices out of the available strategies, and the contextually adaptive or maladaptive effects of these choices. The problem of choosing between alternative behavioral strategies is often addressed using a branch of interesting mathematics called game theory which was introduced briefly in Chap. 1. It is time to make use of that concept now.

The hawk and dove (HD) game is one of the fundamental and earliest models of evolutionary game theory addressing the problem of physical aggression [4]. There are two basic strategies, one called "hawk" which is aggressive and the other called "dove" which avoids physical aggression. A dove may use threatening displays but will retreat invariably if it comes to a physical fight. The assumptions of the game are that there is a resource for which there is some conflict. The nature of the conflict is such that one who wins gets a reward. If the conflict is between a hawk and a dove, dove will always retreat and hawk will get the reward. Therefore we may be tempted to think that it is always good to be a hawk. But this is not true, and we can realize this on looking at what happens when dove meets a dove and hawk meets a hawk. If two doves meet, they may give threat displays and try to force retreat on the other but will never launch an attack. One of the two wins this psychological war eventually and gets the reward. Alternatively, they may negotiate and share the reward. They may have to spend some time and energy in the threatening displays or negotiations, but an actual fight never ensues. On the other hand, when two hawks enter a conflict, a fight is inevitable. One of them wins the fight ultimately and gets the reward. But whenever there is a physical fight, there is a risk of serious injuries which is a great cost. The loser in a hawk–hawk interaction neither gets the reward nor is spared from the risk of injury. The benefit in a dove–dove and hawk–hawk interaction is the same, but the mean cost of injury in a hawk–hawk interaction is large. This is the major disadvantage to hawks.

We can handle this game using the concept of an evolutionarily stable strategy (ESS) now. If the population consists of all doves and if there arises a hawk by mutation or immigration, the hawk will always win and therefore will be at an all-time advantage. Therefore dove is not an ESS. On the other hand, if all are hawks in a population and a dove invades, the dove will never win a battle, but it will also never get injured, whereas hawks will keep on fighting with each other and paying a huge cost of injuries all the time. Therefore a dove might actually do better in an all-hawk population. Therefore hawk is also not an ESS. This can be shown in a payoff matrix (see Appendix III). We are making a simple assumption here that when there is a hawk–hawk encounter, each animal has an equal a priori chance of winning. The same applies to dove–dove encounters.

Since hawk gets an advantage in a dove population and dove does better in a hawk population, there is a negatively frequency-dependent selection. Such a negatively frequency-dependent selection would lead to an optimum proportion of hawks and doves where the mean payoff of hawks and doves is exactly equal. The equilibrium proportion of hawks and doves can be easily calculated by simple algebra (see Appendix III for the mathematical framework).

The algebraic solution of the problem has two possible interpretations. In one, each individual is either a pure hawk or a pure dove, and the population comprises an equilibrium proportion of the two such that the net average payoff of everyone becomes equal. In the other a single individual can behave as a hawk or a dove in an optimum ratio given by the algebra. This is called mixed strategy. A mixed strategy at an optimum proportion can be an ESS provided there is no cost associated with the ability to posses this behavioral flexibility. But if behavioral plasticity has substantial cost, then all the three can coexist, i.e., there can be pure hawks, pure doves, and mixed strategists in the population simultaneously at some equilibrium proportion.

It is very obvious that the model is too naïve as compared to the reality of animal behavior. Realities are more complex than what is assumed in the baseline model. There have been a number of attempts to bring in the complexities of real life into the theoretical model [5, 6]. We will discuss some of the complexities that are relevant to us without going into the depths of theoretical modeling. For example, the simple assumption that each animal has an equal chance of winning in a hawk–hawk encounter is certainly not true. Individuals differ in their size and strength, and a stronger animal will have a higher probability of winning a physical combat. Animals have mechanisms of judging their size and strength vis-a-vis the opponent, and these judgments will influence the behavioral decision. An individual may actually take a decision to behave as a hawk or dove based on this judgment. When faced with a larger and stronger animal, it would prove beneficial to act like a dove, and when interacting with a weaker animal, it will be most appropriate to behave like a hawk. A number of such factors certainly play a role in actual decision making. The model is not intended to apply to behavior of any species in its naïve form. Nevertheless a number of inferences of the model remain robust enough not to be altered by the subtleties and complexities of real life.

One of the important inferences of the solution of HD game is that two or more behavioral strategies can coexist in a population leading either to a genetic polymorphism or to behavioral plasticity. A more complex example of coexistence of alternative strategies is that of lizard males with small territories, those with large territories and those without territories that coexist in a rock–paper–scissor-like game [7]. The rock–paper–scissor (RPS) game is another slightly more complex game in which three strategies can coexist with a negative frequency dependence resulting into stable or oscillating dynamics.

Another important inference of the hawk and dove model is that different alternative behavioral strategies may have their own advantages under different conditions and at different frequencies. The behavior of the dove is of particular interest. Although the dove never escalates aggression and always retreats against a hawk, it does give it a try by giving a threat display. This is a deceptive behavior which works at times. This can be viewed as a kind of social smartness. If an individual is not strong enough to win a physical combat, it can still hope to gain by using tactical deception. Tactical deception is indeed observed in animals, particularly (but not restricted to) primates [8–13], and by the above logic, we expect that physically weaker individuals should display deception more frequently. Owing to the availability of an alternative strategy to gain fitness, subordinate individuals can hope to attain some share of reproductive success. As a result social dominance does not remain synonymous with Darwinian fitness.

There are many real-life examples of hawk- and dove-like strategies in animal populations. In a wide variety of species, males compete for females, and mating behavior necessitates aggression and dominance. However males that are weak, submissive, or subordinate are not complete reproductive failures. In a wide variety of species, subordinate males remain with the harem submissively and sneak mate opportunistically. Sneak mating is a fairly successful reproductive strategy suitable for physically weaker individuals [14–20]. It takes different forms in different species depending on the ecology and sexual behavior. For example, in frogs that have external fertilization, males attract females in the breading season by vocal signals. Larger frogs have large vocal sacs and can therefore attract a large number of females. Smaller frogs that cannot do this linger around larger males, and when the larger males attract females, the satellite males sneak some reproductive success [21]. Researchers claim that in order to do this, they make use of information about neighboring males in an organized fashion [21]. As a result using more than one alternative strategy is common even in these, believed to be less intelligent, species [22]. Because of this kind of smartness, the weaker individuals are not outright losers.

Alternative strategies work not only for mate competition but also apply to food access, social relations, stress response, and other factors. For example while stronger and dominant individuals fight for a patch of food, the submissive ones may wait and watch for opportunities to sneak. This is important to realize because we started with the example of dominant and subordinate males, but it does not imply that the hypothesis being developed applies only to males. We will use the example of alpha male versus subordinate males repeatedly since it is better known to a general reader, has a dramatic appeal, and also much research data are available. But let us take a glance at other examples where being socially smart can partially compensate for not being physically strong. In more intelligent animals higher levels of cognitive abilities are used in tactical deception. A number of interesting observations are depicted by Byrne [23].

The first instance is that of a young male baboon named Paul who used a smart trick to obtain food from a larger and more dominant animal. For example, when an adult female named Mel had just dug up a corm, a prized food item, Paul looked around and seeing nobody looking at them screamed loudly as if someone was harassing him. His mother who was higher ranking than Mel ran into the scene aggressively and chased away Mel, perhaps thinking that Mel was troubling Paul. When his mother was chasing Mel, Paul ate the corm. Paul was observed to use a similar trick more than once but he never used it when in full view of others.

In another instance an adolescent male named Melton played too roughly with an infant and was being chased by the infant's mother since it screamed. He suddenly gave up running and took an alarm posture staring at a rocky hillside as if he had seen a predator or some other threat. Alarmed by this signal the female gave up the chase and started staring in that direction. Observers confirmed that there was no actual threat present in that direction. Melton had used a tactical deception to escape from a difficult situation.

A primatologist friend of mine, Dr. Anindya Sinha, and his team working on bonnet macaques in India have a number of observations on tactical deception in bonnets, and I have borrowed some stories from their diaries for the readers. These stories belong to a yet unpublished study, and I am waiting eagerly for them to get published.

For example AG, a middle-ranking male, was being allogroomed by a middle-ranking female SU on a tree branch. AG observed TE, another male of a close rank, about to copulate with MU, another female in estrus, about 7 m away on the ground. He intently observed them and gave a very loud alarm call that the troop members usually give to a running pack of dogs. There was not a single dog in sight anywhere. Besides, AG did not scan the area at all but gave the call while looking at TE and MU. The pair became very alert, scanned in all directions, and MU moved off. TE again scanned the area and then warning growled and open-mouth threatened AG, who ignored him completely and resumed his grooming session with SU. Clearly, AG had used a misguiding alarm call to successfully interrupt copulation attempt by a competing male.

In another observation, HS, a young adult male, had recently joined the troop of free-ranging macaques within a zoo that held captive primates as well. HS was consorting with RF, an adult female in estrus. On one occasion, as he approached RF, he suddenly saw PK, the alpha male, approaching them. This was a dangerous situation. HS immediately turned around and gave loud warning growls at a golden langur in a neighboring cage. He then intently observed PK and let RF move ahead, intently watching her but not following her anymore. Here HS had successfully avoided aggression by PK by pretending to be engaged in something else.

Not all acts of deception are successful. At times the deceptive acts appear to be identified as

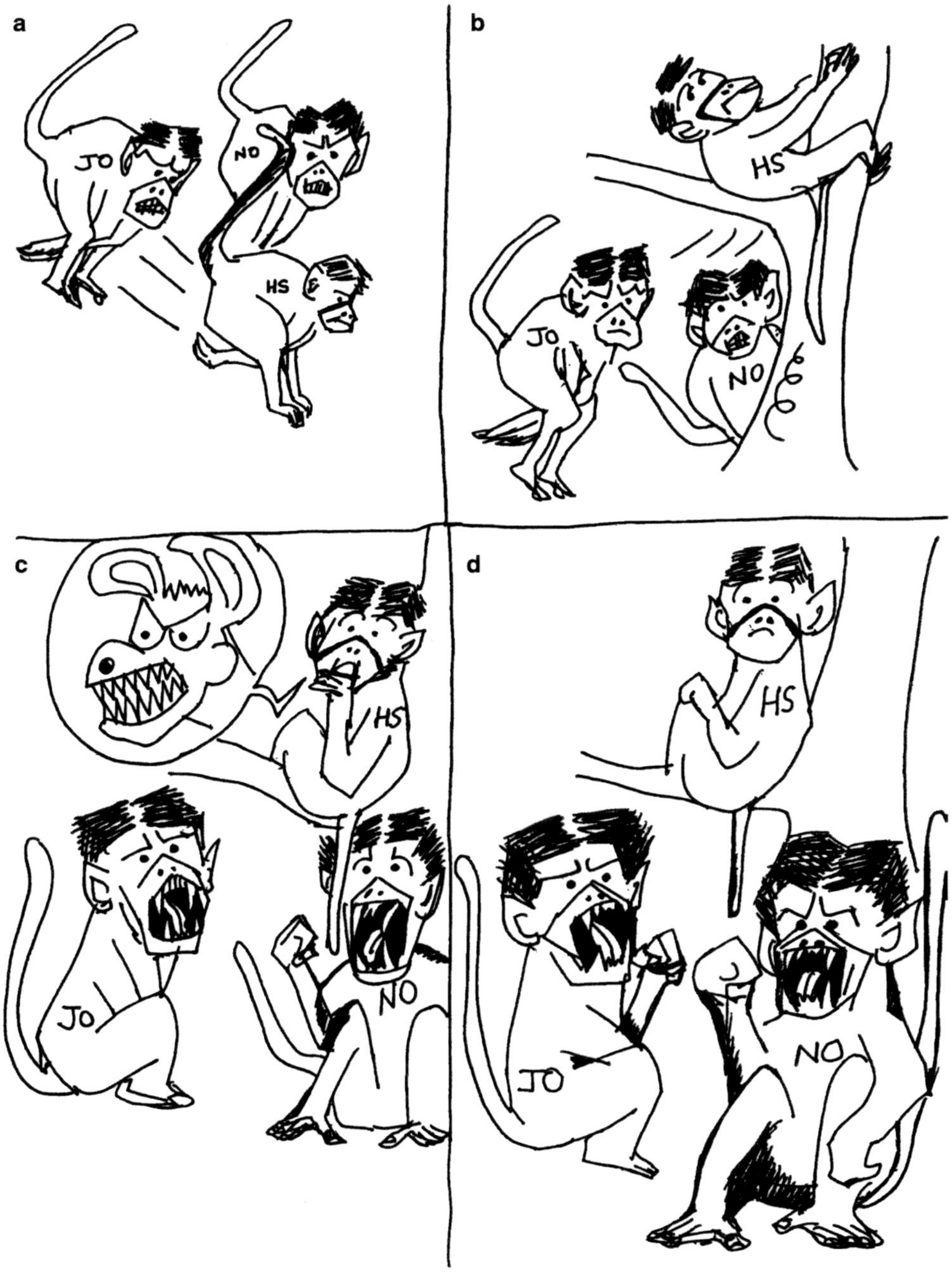

deceptive by the target individuals. This is also a higher-order cognitive task. In another observed story with HS again, he was being chased by JO and NO, two adult females, when he climbed a small tree, looked at a distance, and gave loud alarm calls, normally given to feral dogs. The aggressors, however, ignored the calls completely and continued to direct noncontact aggression towards him. On checking, the observer could not detect any dog or other threat in the vicinity. JO and NO on other occasions responded to threats and alarms normally. But this time they did not as if they knew this was a deceptive alarm call.

It is interesting to note that in majority of the examples of tactical deception, it was the physically weaker individual that initiated the deceptive behavior. This is not to imply that stronger individuals are not intelligent. It rather reflects the greater pressure on the physically weaker individuals to show social smartness in an attempt to compensate for their weakness. Among the less common examples where a high-ranking individual has initiated tactical deception against a weaker individual, most of the situations are such that strength and dominance alone would have failed to work. For example a dominant individual makes a greeting approach towards a subordinate individual who lets him come close. After being close the dominant individual gives a nasty bite. Here the dominant individual would have been unable to punish the subordinate one if the latter kept on running away. Therefore it is unlikely that dominant individuals are unable to show social smartness. Nevertheless subordinates depend more crucially on deception than the dominant ones. Although examples of weaker individuals using tactical deception or showing social smartness abound, ethologists do not appear to have seriously addressed the question whether there is a negative association or trade-off between physical strength and tactical deception. I suspect such a trade-off, and some of my primatologist friends share the impression, although a rigorous quantitative test of the hypothesis is yet to be done.

In real life whether an individual behaves as a hawk or dove will depend on its social surroundings. We may make a fairly safe assumption that every individual has sufficient behavioral flexibility to choose between the two alternatives. However the relatively stronger individuals will have hawk interactions more frequently than weaker individuals. A consistent difference in these frequencies can make "personalities" or "behavioral syndromes." A hawk behavior is primarily about aggression, but there are a number of inevitable correlates of physical aggression that are decided by the social structure, foraging, and mating systems of the species. For example aggression, boldness, and exploration are correlated behaviors in fish that form a behavioral syndrome which is shown to have a genetic basis [24]. There can be differential availability of food for dominant and subordinate individuals shaping different metabolic states. A dominant male has greater access to females in a promiscuous species, and therefore, dominant individuals may be sexually more active. Subordinate males, on the other hand, would be sexually more restrained since their overt advances towards receptive females may not be tolerated by dominant males. Aggressive individuals are more injury prone, and therefore, their immune system needs to be more active and ready at the subcutaneous level. Doves, on the other hand, avoid aggression and therefore avoid injuries too which may shape their immune systems differently. In most species submissive display is a very effective way of avoiding actual physical attack. Animal conflicts are highly restrained in terms of inflicting injuries to each other [25]. Therefore although much of the aggression in a group might be directed against the relatively low-ranking individuals, they are much less likely to get actually injured. Injury is more likely when aggression is allowed to escalate, and escalating aggression is typically a hawk characteristic. Doves may partially disinvest from the innate immune and wound healing mechanisms as a parsimonious measure that can be called an immunological "thrift." If a subordinate individual finds a patch of rich food, there is a risk that a more dominant individual may snatch food from it. However, food once gulped cannot be snatched, and therefore, subordinate animals need to gulp as much as they can before being noticed by a dominant individual. Thus binge eating would be adaptive for a subordinate individual. This implies a peculiar kind of nutritional "thrift" that the physically weak and socially subordinate individuals need to develop. Binge eating directly implies type 1 thrift, but type 2 thrift may also be adaptive for low-ranking individuals who have lower food security. There is a correlated difference in exploring and risk-taking behavior too. The hawks are more likely to be risk takers not only in relation to physical combats but also in relation to foraging. Since physical strength is crucial to their success, they need to obtain better nutrition even if that exposes them to greater predator or any other risk. As a result they are expected to be generally more bold, curious, and exploring their environment with less inhibitions. For weaker individuals, safety is more important than physical strength since physical strength contributes little to their success, but "survive and wait for future opportunities" is a better bet for them. Thus being risk averse in foraging is better for subordinate individuals. Thus the hawk and dove "personalities" go beyond aggression alone and refer to a cluster of correlated behaviors. The two contrasting behavioral syndromes are tabulated in Table 5.1.

Table 5.1 Behavioral syndromes

Hawk/dominant	Dove/subordinate
Physically stronger	Physically weaker
Proactive aggression	Aggression avoidance, occasionally reactive aggression
Snatching	Begging
Injury prone	Less injury prone
Optimum eating behavior	Binge eating on opportunities
Risk takers	Risk and harm avoiders
Rapid responses to challenge	Relatively slower responses
Sexually more active	Sexually restrained

It would therefore be logical to expect that if an individual behaves like a hawk more frequently, its body physiology should be tuned to support a hawk life, and the corollary should be true for dove life too. For example a hawk needs to build and maintain muscle strength. This would need more protein synthesis in muscle for which muscle needs to pick up greater quantities of amino acids as well as glucose. This requires insulin action, and therefore, hawks need to be more insulin sensitive. Physical combat not only requires strength but equally needs agility. Quick reflexes and good nerve muscle coordination are crucial in a fight. Injuries are anticipated in a fight, and the body needs to be prepared for injuries by all possible means. The macrophages should be ready in the subcutaneous tissues to take care of bacteria invading from an open wound. Also the wound healing mechanisms should be more active. This includes the mechanisms of angiogenesis, tissue regeneration, and epidermal growth. For the anticipated loss of blood from injuries, the red cell synthesizing system also needs to be geared up. The social ranks are dynamic, and an alpha position is under continued threat of challenge by some other individual sooner or later. This has two important implications. One is that the stronger individuals need to be more of risk takers to gain and retain an alpha position since the possible rewards from risky behaviors are very high. The other implication is that the hawks need to be sexually more active since they need to reproduce as much as they can before their dominant status is threatened. Not surprisingly sex and aggression mechanisms have a substantial mechanistic overlap. Some of the sex hormones such as testosterone in males have proaggression effects, whereas the effects of estradiol on female aggression are complex and mixed [26–29]. The set of neurons in the hypothalamus involved in attack and in mating also overlap [30]. The link between aggression and reproductive capacities is perhaps much wider. It is recorded in colonial wasps that a new queen cannot have sufficient fecundity unless it displays substantial dominance aggression [31].

The physiological requirements of the dove or subordinate animals are diametrically opposite. Since they are weaker, aggression is counterproductive for them, and they need to completely suppress proactive aggression and as far as possible avoid circumstances necessitating reactive aggression. As a result they become less dependent on strength and more on social manipulation. They can afford to disinvest from muscle building. Amino acid and glucose uptake of muscle can be reduced as a parsimonious or thrifty strategy. But simultaneously cognitive functions of the brain need to be enhanced. Since unlike most other tissues that can use fatty acids for energy, the brain is more specifically dependent on glucose; it is more critical to assure glucose supply to the brain, whereas glucose supply to the muscle may be compromised. Doves need to be more socially supported, and they indulge in activities such as allogrooming that help making alliances and ensuring greater social support. Unlike the dominant individuals, they are expected to be more sexually restrained with lower levels of sex hormones. Since they can effectively avoid fights by increasing submissive displays [25], they are less injury prone and can partially disinvest from peripheral innate immunity, wound healing mechanisms, angiogenesis, and erythropoiesis as compared to a dominant individual. They also need to be more risk avert. Social manipulation is a relatively slow process as compared to a physical fight where split-second decisions are crucial. As a result the responses to challenges by doves could be more sluggish than hawks.

Do our expectations match with data? If our conjunctures are true, a number of subtle physiological differences should be observable between hawks and doves in animal societies. Ethologists generally do not classify individuals as hawks and doves, but in social species where a higher social rank is mainly achieved by physical strength and aggression (notably this is not true for all species), the dominant individuals can be taken equivalent to hawks and subordinate as doves. Our classification should depend on "effective" aggression. Often an alpha male does not have to behave overtly aggressively with other members of the troop very frequently since his dominance is already accepted by others. An intermediate individual may show more frequent aggression which may or may not be too effective in raising its status. It is therefore better to use social ranking that reflects effectiveness of aggression rather than using the frequency of aggression to identify hawks and doves. A note of caution is that this cannot be generalized across species. In primates, for example, physical aggression has a variable role in attaining and maintaining social dominance. The role of aggression can also differ between sexes in a given species. In olive baboon males, overt aggression is the key to attaining dominance, and threat and intimidation are necessary to maintain it [20, 32]. On the other hand, among females of common marmosets, aggression is rare, and dominance does not appear to be dependent on it [32, 33]. Nevertheless both aggression and dominance have important roles in determining reproductive success, although their relative importance and correlation with each other may vary across genders and species. Since our focus is on effective aggression and we are using social dominance as a marker, the discussion below refers to species in which aggression is the predominant determinant of dominance.

In recent years there have been an increasing number of studies on the endocrine and metabolic states of primates and other animals in the wild or socially reared captive colonies. Animal models of insulin resistance are not new. Even primate models for obesity and insulin resistance are known for a long time. In the earlier models some or the other obesifying interventions were deliberately used to induce obesity and insulin resistance. These include dietary manipulations, restricting physical activity, or specific brain lesions interfering with energy homeostasis. Animals do develop obesity and insulin resistance in response to such interventions. However, it is known now that such interventions are not necessary prerequisites of insulin resistance. A significant fraction of individuals in a social group of primates is shown to become insulin resistant around puberty even in the absence of any such interventions [34]. This was demonstrated in bonnet macaques that had normal diet, physical activity, and social interactions. Insulin resistance that developed among the 15% of individuals by peripubertal age was not contributed by high-calorie intake, physical inactivity, or obesity. Interestingly the proportion of insulin-resistant individuals at the physically active peripubertal age was not different from that in the significantly older and sedentary age class [34]. It is possible therefore that insulin resistance in a proportion of individuals is an inherent part of the complex social life in primates and may not have a pathological origin at all.

Data coming from wild or captive groups of animals are fragmentary. Some studies have data on diet but not social interactions, and some others that have looked at social interactions have not reported morphometry and so on. Among the physiological parameters, some studies have monitored corticosteroid levels but not insulin. Some focus on cholesterol but do not estimate plasma glucose. It is possible, nevertheless, to join the pieces together and try to make a coherent picture. Generalizing over a wide variety of species of vertebrates, we can detect certain common patterns. The dominant and aggressive individuals of both sexes typically have high levels of sex hormones, low serotonin and higher dopamine activity in the brain, lower plasma cholesterol, and corticosteroids as well as low plasma insulin. The subordinate ones have the opposite picture of low sex hormones, high serotonin, low dopamine, and higher plasma levels of corticosteroids and cholesterol, more specifically low HDL and high LDL cholesterol [32, 35–57]. There are exceptions though, such as subordinates having lower corticosteroid levels than dominant individuals, but they come from species in which aggression is less important in determining dominance [32]. Fewer studies have looked at the relationship between dominance hierarchy and insulin. One study on wild chimpanzees in the Kibale National Park showed that subordinate individuals have higher urinary C-peptide [58] which reflects higher insulin production and thereby insulin resistance. This pattern was consistent over 13 of 14 seasons monitored and independent of high or low fruit availability. This appears to have come as a surprise to these researchers. They expected the urinary C-peptide to reflect calorie intake and therefore expected the high-ranking individuals to have higher levels. The expected association of calorie intake to insulin levels was seen across seasons. In seasons of better availability of fruits, all individuals had higher levels of C-peptide. This is consistent with other studies showing higher obesity and insulin levels with increased availability of food [59]. But across the society, the effect of social ranking was stronger than the effect of calorie intake. This might be extremely important and relevant to understanding insulin resistance syndrome in humans as we will discuss in the next chapter. Another study looked at insulin-like growth factor 1 (IGF-1), which plays important roles in protein metabolism, muscle physiology, wound healing, erythropoiesis, and immunity and found that social subordination is associated with suppressed levels of IGF-1. This finding in baboons is also relevant since IGF-1 has an important role in controlling insulin resistance as we will see in a later chapter. This study also asserts that the suppressed IGF-1 response is consequential rather than causal to subordination [49]. It is important to realize that some of these studies are on wild primate groups, indicating that these physiological effects are not an artifact generated by captive conditions. The trend in the physiological gradient along the social hierarchy is similar in captive and wild primates. In birds, individuals with a more timid personality have been shown to have greater oxidative stress [60]. Subordinate animals also have higher resting blood pressure [37, 61, 62]. Thus in comparison with hawks, the endocrine and metabolic picture of the doves goes close to metabolic syndrome in humans. In the wild it is rare for subordinate individuals to become overt diabetic and develop clinical complications typical of diabetes. The magnitude of the endocrine and metabolic changes in animals is generally not large enough to cause the pathophysiology commonly seen in the human metabolic syndrome, but the direction of the change is the same. Interestingly higher levels of cholesterol, corticosteroids, and insulin along with moderate insulin resistance are recorded in subordinate individuals in spite of having lower caloric intake as compared to dominant individuals [37, 58]. This apparently counterintuitive example is important in the context of metabolic syndrome in humans as we will see later. Further our expectation that physiological response to social challenges of a hawk would be faster than a dove is also supported by evidence. The stress responses of dominant individuals are typically rapid and sharp. Those of subordinates are sluggish, although they have higher basal levels of stress hormones [37].

The next question of immediate relevance is whether it is the metabolic state that makes an individual dominant or subordinate or it is one's status and behavior in the social environment that alters the metabolic state? The most likely answer to the question is that it is a two-way phenomenon. Manipulation of serotonin levels is known to affect aggressive behavior pointing towards chemistry dictating behavior. But there is evidence for a reverse direction of causation as well. The neuronal nitric oxide synthase (nNOS) knockout mice were found to be aggressive, and their aggression could be reverted by increasing serotonin levels, but for expression of aggression in these mice, transient social isolation was necessary without which the aggressive response was not observed [63]. Brain monoamines monitored during an aggressive confrontation in rats showed that, in the prefrontal cortex, there was a sustained decrease in serotonin to 80% of baseline levels during and after the confrontation and an increase in dopamine to 120% after the confrontation [46]. In cynomolgus monkeys corticosteroid levels prior to social interaction were not good predictors of social rank, but social rank decided their levels [37, 64]. This indicates that aggression changed the chemistry. Many more studies give evidence in either direction. It appears therefore that the relationship is dual and operates as an autocatalytic or positive feedback cycle. This is quite logical since the two behavioral syndromes are diametrically opposite, and adopting one of them needs switching off the other. We have seen earlier (Chap. 1) that positive feedbacks are important operators of such switches. Therefore the debate on whether chemistry shapes behavior or behavior alters chemistry is meaningless. A switch will not operate unless both of them work. Therefore when we observe that rats made diabetic by streptozotocin lose aggression [65], we also need to suspect and examine whether the reverse could also be true. Whether loss of aggression would predispose or even cause diabetes, we will return to this question in due course.

If the endocrine and metabolic picture in hawks and doves is different, we would expect that the two states may be adaptive to the respective behavioral profiles. It is not difficult to visualize how high testosterone, insulin sensitivity, low cholesterol,, etc., are adaptive for a dominant individual. But since the metabolic syndrome-like picture seen in subordinates is linked to a series of health problems, there is a need to explain how it is adaptive for the subordinate individual. The physiological state typical of metabolic syndrome is actually supportive for a subordinate individual in many ways.

1. Insulin resistance leads to disinvestment from muscle: Insulin promotes uptake of glucose and amino acids in muscle tissue, and this is needed to build as well as maintain muscle strength. In the insulin-deficient state typical of type 1 diabetes, there is rapid and progressive wasting of muscle. Type 2 diabetes generally does not lead to the severe wasting of muscle as in type 1. There is only a partial disinvestment from muscle that is mediated by insulin resistance which reduces uptake of glucose and amino acids by muscle on the one hand and on the other permits some level of protein catabolism making the amino acids available for gluconeogenesis in the liver. Since muscles contribute less to success in doves, it makes sense to disinvest from muscle if the energy thus saved can be utilized more fruitfully elsewhere.
2. More glucose may be available for cognitive functions of the brain: Most tissues of the body are able to utilize fatty acids for energy generation in the absence of glucose or effective glucose transport. The brain is more crucially dependent on glucose for energy. When muscle and many other tissues become insulin resistant, their share of glucose uptake is substantially reduced. Since glucose transport from blood to brain is independent of insulin, it is not affected by either insulin resistance or insulin deficiency. Therefore brain can lift a greater share of glucose.

This is extremely important to realize. There are many different glucose transporter molecules active in different tissues with differential dependence on insulin. The main glucose transporter in muscle and viscera is glut-4 whose appearance on the membrane and therefore normal function is dependent on insulin and an intact insulin receptor and downstream pathway. At the other end glut-1, the glucose transporter at the blood–brain barrier, is completely insulin independent. The differential dependence on insulin, changes in insulin levels, and alterations in the insulin receptor and downstream pathway can finely manipulate differential energy budget allocation to different tissues. If differential budget allocation was not necessary, then there is no evolutionary reason for so many different glucose transporters to evolve. A variety of different glucose transporters with differential kinetic properties and insulin sensitivity is not redundant. This variety appears to have specifically evolved, and tissue-specific expression of glucose transporters is seen across widely different vertebrate taxa [66–69]. If such differential budget allocation was not required, a single glucose transporter would have been adequate for all tissues. Moreover insulin itself would not have been required at all. For unicellular organisms no insulin-like molecule is needed for glucose uptake. Regulation by insulin-like central mechanism is needed in complex organisms in which different tissues compete for the same flux of nutrients, and a central mechanism is needed to monitor that different organs pick up the right amount of nutrients. The right amount is decided by the activity, energy requirement, and dispensability of the tissue function. For this there needs to be sufficient flexibility in the budget allocation mechanism since different environmental challenges and different behavioral strategies bring about subtle changes in the differential use of organs and their energy requirement. Under severe nutrient limitation the more indispensible functions need to be catered with priority. The need-based differential allocation of energy is achieved by having different glucose transporters in different tissues whose levels can be differentially regulated; moreover, their sensitivity to the central signal molecule insulin is different so that changes in insulin levels can fine-tune the budget allocation. Insulin resistance is an added layer of this differential budget allocation system. It reduces the nutrient flux to insulin-dependent organs and increases the availability to insulin-independent organs. The evolved role of insulin in this context is not that of a facilitator of glucose transport, as written in many textbooks, but it is that of differential allocator of energy to different tissues based on the needs.

An excellent experimental demonstration of the differential budget allocation effect of insulin resistance is seen in MIRKO mice. In these muscle-specific insulin receptor knockouts, the postmeal glucose uptake by muscle is reduced, but since fat cells remain insulin sensitive, they appear to pick up the share of glucose not picked up by muscle, and therefore, MIRKO mice quickly accumulate large amounts of fat. Interestingly this happens without any detectable changes in plasma glucose or insulin levels. It is likely that very similar dynamics may be taking place between muscle and cognitive areas of the brain. The energy supply to the brain can potentially be increased without affecting plasma glucose levels since muscles develop insulin resistance and thereby reduce their uptake. This appears logical, but such a budget allocation difference has not been experimentally demonstrated in insulin-resistant animals.

We do not know whether doves need more glucose supply for their brain than hawks. This is a tricky point and largely an information void. Brain activity is also involved in physical fight. Rapid nerve muscle coordination is needed by hawks during a fight. This is typically a burst of activity of a short duration. Social manipulation, on the other hand, might need a more sustained energy supply, and the parts of the brain involved are different. It has been shown in nonhuman primates that individuals with larger social network have more gray matter in the brain [70]. Further, glucose consumption of gray matter is two- to fourfold greater than that of white matter [71, 72]. These two findings put together suggest that the brain of a social manipulator may need more glucose than that of a fighter. Comparing across primate species, the relative size of the neocortex correlates positively with mean group size and the prevalence of tactical deception [23]. A larger brain is expected to consume greater quantities of fuel. But as yet there is no direct demonstration that a hawk to dove transition is associated with increased brain glucose demand.

3. Insulin, cholesterol, corticosteroids, and leptin enhance cognitive functions: Although higher glucose pick up by the cognitive areas of the brain in the insulin-resistant state has not been demonstrated, a counterpart of it has substantial experimental evidence. Insulin-resistant state is accompanied by hyperinsulinemia, although what is primary and what is compensatory are debatable. The current argument is valid in either case. Hyperinsulinemia has been shown to facilitate cognitive brain functions [73]. Rats training on a spatial maze-learning task upregulated insulin signaling in the brain [74] demonstrating that cognitive tasks require insulin signaling in the brain. At least one component of the insulin resistance–hyperinsulinemia duo has direct demonstrable enhancing effects

on cognitive functions, and the other is logical but yet to receive experimental verification. But it can be said with confidence that insulin resistance is adaptive to dove behavior in disinvesting from muscle, and hyperinsulinemia is important in facilitating cognitive functions of the brain. In addition to insulin, studies have indicated that cholesterol and leptin as well as corticosteroids enhance learning and cognitive functions [75–79].

It is important to note that apart from humans, the species in which naturally occurring insulin resistance and hyperinsulinemia are known are some nonhuman primates [80] and dolphins [81]. The rat and mice models used for experiments are either mutants, knockouts, or have been made diabetic by some man-made interventions such as extreme diets. But insulin resistance in primates and dolphins has been observed without any such defect or intervention. What is common to species (humans, nonhuman primates, and dolphin) having some natural form of insulin resistance? Why natural forms of hyperinsulinemia occur in these species? A possible answer is relatively larger brain and more cognitively rich life. The hawk–dove dichotomy may be present in a wide variety of species, but in species with large brains and cognitive capacities, the magnitude of the change needed to support higher cognitive functions could be larger. These species perhaps need higher levels of insulin, cholesterol, leptin, and corticosteroids on adopting a dove strategy.

4. Serotonin, cholesterol, and corticosteroids reduce risk-taking and aggressive behaviors: Doves are averters of aggression and risk. Three of the metabolic syndrome-associated molecules help in achieving this behavioral state. One of prime importance is serotonin. Serotonin has a strong aggression suppression action which has been well known in animal models as well as in humans. Manipulated serotonin elevation in a dominant male has been shown to shift the dominant behavior and status towards subordination in fish and lizards [42, 82, 83]. The other is cholesterol which is also associated with reduced aggression, but unlike serotonin where the mechanism is fairly known, the way in which cholesterol reduces aggression is not adequately known. There are indications that cholesterol upregulates serotonergic activity in the brain, thereby reducing aggression [52]. The third is corticosteroids. Corticosteroids reduce risk-taking behaviors and also enhance learning [84, 85] and thereby are adaptive for a dove personality. Corticosteroids are generally associated with stress response, but we will talk about stress and corticosteroids in an evolutionary perspective in a later chapter.
5. Shift from protein-anabolic to lipid-anabolic metabolism: The muscle fiber is chiefly made of proteins, and the major chemical constituent of the brain is fat. Chemically the brain is 60–70% fat [86]. Therefore a muscle-dependent personality needs more protein synthesis, and brain-dependent personality needs more lipid synthesis. It is expected therefore that in hawks, the metabolism has a protein anabolic bias and in doves a lipid anabolic bias. If we take the fate of total carbon intake in a dove, one should see some reduction in the fraction diverted to protein synthesis and compensatory increase in the fraction that goes to lipid synthesis.
6. Altered importance of stored fat: Most aggressive encounters are explosive actions of short duration. Fat degradation is a relatively slow process, and fat becomes available as a fuel only after prolonged physical activity [86]. The immediate source of energy in a physical fight is utilization of ready glucose, aerobically or anaerobically followed by glycogen breakdown. Therefore an aggressive animal should have more of glycogen buildup. Fat as a stored fuel is of less relevance, and therefore, evolution might have built in active mechanisms to enhance glycogen synthesis and to reduce fat in an aggressive personality. Fat might be needed for a different purpose in aggressive life as discussed later in Chap. 11 but not as fuel reserve. This is likely to change after adopting a dove personality.
7. Disinvestment from less important functions—immunity, wound healing, and angiogenesis: Since submissive and socially manipulative animals can effectively avoid physical

injuries, there can be partial disinvestment from peripheral innate immune mechanisms. Disinvestment from wound healing mechanisms including angiogenesis would not be surprising either.

8. Increased investment in the fetus through the placenta: Subordinate individuals have fewer opportunities for reproduction. In primates subordinate females have a higher rate of anovulatory cycles with suppressed estradiol and progesterone levels. In extreme cases hormone levels may be as low as in ovariectomized animals [35]. Since reproductive opportunities are suppressed, they need to make the best use of available opportunities and therefore invest more resources, energy, and efforts towards each offspring. This needs increased investment in the fetus, and insulin resistance helps this investment by a mechanism similar to the mechanism of investing in the cognitive brain. The placenta, similar to the blood–brain barrier, has insulin-independent glucose transport. So if the mother becomes insulin resistant, the transplacental flow of glucose increases significantly, and this results into producing larger babies [87]. It is quite well known that diabetic mothers have heavier babies, and the predominant mechanism is the altered glucose allocation mediated by insulin resistance. This helps in sufficient development of the fetus even if the mother has less food security being a subordinate animal. Similar to humans, in nonhuman primates, gestational insulin resistance has been shown to give rise to larger infants [80].

Thus it can be seen that the metabolic syndrome-like changes in the physiology of physically weak and subordinate animals are adaptive to the subordinate state in multiple ways so that the individual can achieve the best possible genetic fitness within the constraints of the situation. This physiological state of insulin resistance and accompanying alterations is an adaptation, not a disorder. Recent data show that the subordinate state is also accompanied by some of the basic pathological processes of T2D and related disorders including systemic inflammation [88]. This is also a possible consequence of adaptive changes as we will see in a later chapter.

This view differs from earlier evolutionary hypotheses for the origin of metabolic syndrome [89–92] in a number of ways. The synthesis in this and subsequent chapters is more fundamental and broad based and cuts across several taxa. It does not depend upon whether human ancestors faced some specific environmental conditions or not. Qualitatively it would apply to a wide diversity of vertebrate species, although there would be quantitative differences. It does not view the basic evolved physiological mechanisms to be unique to humans. But we need to explain why metabolic syndrome is so common in modern humans. Therefore we need to be careful while taking the theory from animals to humans.

Human societal structure differs from primate social structure in a number of ways. So the hawk and dove logic may not be applicable to us in its original form. But are there any parallels? Is it applicable to us at least in a modified form? Can we get any additional insights from studying the metabolic syndrome-like state in animals? Will it help to not only understand but better cope, treat, and even cure diabetes and related disorders? These are the obvious questions that we need to address now.

References

1. Gosling SD (2001) From mice to men: what can we learn about personality from animal research? Psychol Bull 127:45–86
2. John OP, Gosling SD (1999) Personality dimensions in nonhuman animals: a cross-species review. Curr Dir Psychol Sci 8:69–75
3. Eriksson AC, Booth DJ, Biro PA (2010) Personality' in two species of temperate damselfish. Mar Ecol Prog Ser 420:273–276
4. Smith JM (1982) Evolution and the theory of games. Cambridge University Press, Cambridge
5. Houston AI, McNamara JM (1991) Evolutionarily stable strategies in the repeated hawk–dove game. Behav Ecol 2:219–227
6. Houston AI, McNamara JM (1988) Fighting for food: a dynamic version of the Hawk–Dove game. Evol Ecol 2:51–64
7. Sinervo B, Lively CM (1996) The rock–paper–scissors game and the evolution of alternative male strategies. Nature 380:240–243
8. Byrne RW, Whiten A (eds) (1988) Tactical deception of familiar individuals in baboons. *Machiavellian intelligence: social expertise and the evolution of intellect in monkeys, apes, and humans*. Clarendon, Oxford

9. Bugnyar T, Kotrschal K (2002) Observational learning and the raiding of food caches in ravens, Corvus corax: is it 'tactical' deception? Anim Behav 64:185–195
10. Byrne RW, Whiten A (1992) Cognitive evolution in primates: evidence from tactical deception. Man 27:609–627
11. Byrne RW, Whiten A (1991) Computation and mindreading in primate tactical deception. In: Whiten A (ed) Natural theories of mind: evolution, development and simulation of everyday mindreading. B Blackwell, Oxford
12. Whiten A, Byrne RW (eds) (1988) The manipulation of attention in primate tactical deception. *Machiavellian intelligence: social expertise and the evolution of intellect in monkeys, apes, and humans*. Clarendon, Oxford
13. Byrne RW, Whiten A (1988) Toward the next generation in data quality: a new survey of primate tactical deception. Behav Brain Sci 11:267–273
14. Ohsawa H, Inoue M, Takenaka O (1993) Mating strategy and reproductive success of male patas monkeys (*Erythrocebus patas*). Primates 34:533–544
15. Pilastro A, Bisazza A (1999) Insemination efficiency of two alternative male mating tactics in the guppy *Poecilia reticulata*. Proc Biol Sci 266:1887
16. Parker GA (1990) Sperm competition games: sneaks and extra-pair copulations. Proc Roy Soc Lond B 242:127–133
17. Reichard M, Le Comber SC, Smith C (2007) Sneaking from a female perspective. Anim Behav 74:679–688
18. Digby LJ (1999) Sexual behavior and extragroup copulations in a wild population of common marmosets (*Callithrix jacchus*). Folia Primatol 70:136–145
19. Cristina L-P (2001) Intergroup interactions in wild common marmosets, *Callithrix jacchus*: territorial defence and assessment of neighbours. Anim Behav 62:11–21
20. Smuts B (1986) Sex and friendship in baboons. Aldine Transactions, Piscataway, NJ
21. Berec M, Bajgar A (2011) Choosy outsiders? Satellite males associate with sexy hosts in the European tree frog *Hyla arborea*. Acta Zool Acad Sci Hungar 57:247–254
22. Smith MJ, Roberts JD (2003) Call structure may affect male mating success in the quacking frog *Crinia georgiana* (Anura: Myobatrachidae). Behav Ecol Sociobiol 53:221–226
23. Byrne R (1995) The thinking ape: evolutionary origins of intelligence. Oxford University Press, Oxford
24. Norton WHJ et al (2011) Modulation of Fgfr1a signaling in zebrafish reveals a genetic basis for the aggression–boldness syndrome. J Neurosc 31: 13796–13807
25. Smith JM, Price GR (1973) The logic of animal conflict. Nature 246:15–18
26. Meisel RL, Sterner MR, Diekman MA (1988) Differential hormonal control of aggression and sexual behavior in female Syrian hamsters. Horm Behav 22:453–466
27. Albert DJ, Jonik RH, Walsh ML (1992) Hormone-dependent aggression in male and female rats: experiential, hormonal, and neural foundations. Neurosci Biobehav Rev 16:177–192
28. Albert DJ, Jonik RH, Walsh ML (1992) Interaction of estradiol, testosterone, and progesterone in the modulation of hormone-dependent aggression in the female rat. Physiol Behav 52:773–779
29. Rubenstein DR, Wikelski M (2005) Steroid hormones and aggression in female Galápagos marine iguanas. Horm Behav 48:329–341
30. Lin D et al (2011) Functional identification of an aggression locus in the mouse hypothalamus. Nature 470:221–226
31. Lamba S et al (2007) A possible novel function of dominance behaviour in queen-less colonies of the primitively eusocial wasp *Ropalidia marginata*. Behav Processes 74:351–356
32. Abbott D et al (2003) Are subordinates always stressed? A comparative analysis of rank differences in cortisol levels among primates. Horm Behav 43:67–82
33. Solomon NG, French JA (1997) Cooperative breeding in mammals. Cambridge University Press, Cambridge
34. Kaufman D et al (2005) Early appearance of the metabolic syndrome in socially reared bonnet macaques. J Clin Endocrinol Metab 90:404–408
35. Shively CA, Clarkson TB (1994) Social status and coronary artery atherosclerosis in female monkeys. Arterioscler Thromb 14:721–726
36. Beehner JC, Bergman TJ, Cheney DL, Seyfarth RM, Whitten PL (2005) Testosterone predicts future dominance rank and mating activity among male chacma baboons. Behav Ecol Sociobiol 59:469–479
37. Sapolsky RM (2004) Social status and health in humans and other animals. Annu Rev Anthropol 33:393–418
38. Muller MN, Wrangham RW (2004) Dominance, aggression and testosterone in wild chimpanzees: a test of the 'challenge hypothesis'. Anim Behav 67: 113–123
39. Adams M, Kaplan J, Clarkson T, Koritnik D (1985) Ovariectomy, social status, and atherosclerosis in cynomolgus monkeys. Arterioscler Thromb Vasc Biol 5:192–200
40. Bender N, Heg-Bachar Z, Oliveira RF, Canario AVM, Taborsky M (2008) Hormonal control of brood care and social status in a cichlid fish with brood care helpers. Physiol Behav 94:349–358
41. Blanchard DC, Sakai RR, McEwen B, Weiss SM, Blanchard RJ (1993) Subordination stress: behavioral, brain, and neuroendocrine correlates. Behav Brain Res 58:113–121
42. Larson ET, Summers CH (2001) Serotonin reverses dominant social status. Behav Brain Res 121:95–102
43. King ST et al (1996) Stability of interindividual differences in serotonin function and its relationship to severe aggression and competent social behavior in rhesus macaque females. Neuropsychopharmacology 14:67–76

44. Higley JD et al (1992) Cerebrospinal fluid monoamine and adrenal correlates of aggression in free-ranging rhesus monkeys. Arch Gen Psychiatry 49:436–441
45. Ferrari PF, van Erp AMM, Tornatzky W, Miczek KA (2003) Accumbal dopamine and serotonin in anticipation of the next aggressive episode in rats. Eur J Neurosci 17:371–378
46. van Erp AMM, Miczek KA (2000) Aggressive behavior, increased accumbal dopamine, and decreased cortical serotonin in rats. J Neurosci 20:9320–9325
47. Seo D, Patrick CJ, Kennealy PJ (2008) Role of serotonin and dopamine system interactions in the neurobiology of impulsive aggression and its comorbidity with other clinical disorders. Aggression Violent Behav 13:383–395
48. Sapolsky RM (2005) The influence of social hierarchy on primate health. Science 308:648–652
49. Sapolsky RM, Spencer EM (1997) Insulin-like growth factor I is suppressed in socially subordinate male baboons. Am J Physiol Regulat Integrat Comp Physiol 273:1346–1351
50. Pentürk S, Yalçin E (2003) Hypocholesterolaemia in dogs with dominance aggression. J VetMed Series A 50:339–342
51. Leshner AI, Politch JA (1979) Hormonal control of submissiveness in mice: Irrelevance of the androgens and relevance of the pituitary–adrenal hormones. Physiol Behav 22:531–534
52. Kaplan J et al (1994) Demonstration of an association among dietary cholesterol, central serotonergic activity, and social behavior in monkeys. Psychosomat Med 56:479–484
53. Kaplan JR, Klein KP, Manuck SB (1997) Cholesterol meets Darwin: public health and evolutionary implications of the cholesterol-serotonin hypothesis. Evol Anthropol Issues News Rev 6:28–37
54. Sridhara S, Somasekhar T, John E, Subramanyam MVV, Sundarabai A (1985) Biochemical correlates of agonistic behaviour in *Bandicota bengalensis*: hepatic cholesterol and ascorbic acid. Proc Ani Sci 94:123–127
55. Fernandez X, Meunier-Salaün MC, Mormede P (1994) Agonistic behavior, plasma stress hormones, and metabolites in response to dyadic encounters in domestic pigs: interrelationships and effect of dominance status. Physiol Behav 56:841–847
56. Sapolsky RM, Mott GE (1987) Social subordinance in wild baboons is associated with suppressed high density lipoprotein-cholesterol concentrations: the possible role of chronic social stress. Endocrinology 121:1605–1610
57. Korte SM, Koolhaas JM, Wingfield JC, McEwen BS (2005) The Darwinian concept of stress: benefits of allostasis and costs of allostatic load and the trade-offs in health and disease. Neurosci Biobehav Rev 29:3–38
58. Emery Thompson M, Muller MN, Wrangham RW, Lwanga JS, Potts KB (2009) Urinary C-peptide tracks seasonal and individual variation in energy balance in wild chimpanzees. Horm Behav 55:299–305
59. Kemnitz JW et al (2002) Effects of food availability on serum insulin and lipid concentrations in free-ranging baboons. Am J Primatol 57:13–19
60. Herborn KA, Coffey J, Larcombe SD, Alexander L, Arnold KE (2011) Oxidative profile varies with personality in European greenfinches. J Exp Biol 214:1732–1739
61. Cherkovich GM, Tatoyan SK (1973) Heart rate (Radiotelemetrical registration) in macaques and baboons according to dominant-submissive rank in a group. Folia Primatol 20:265–273
62. Sapolsky RM, Share LJ (1994) Rank-related differences in cardiovascular function among wild baboons: role of sensitivity to glucocorticoids. Am J Primatol 32:261–275
63. Chiavegatto S, Nelson RJ (2003) Interaction of nitric oxide and serotonin in aggressive behavior. Horm Behav 44:233–241
64. Morgan D et al (2000) Predictors of social status in cynomolgus monkeys (*Macaca fascicularis*) after group formation. Am J Primatol 52:115–131
65. Leedom LJ, Meehan WP (1989) The psychoneuroendocrinology of diabetes mellitus in rodents. Psychoneuroendocrinology 14:275–294
66. Díaz M, Antonescu CN, Capilla E, Klip A, Planas JV (2007) Fish glucose transporter (GLUT)-4 differs from rat GLUT4 in its traffic characteristics but can translocate to the cell surface in response to insulin in skeletal muscle cells. Endocrinology 148: 5248–5257
67. Teerijoki H, Krasnov A, Gorodilov Y, Krishna S, Mölsä H (2001) Rainbow trout glucose transporter (OnmyGLUT1): functional assessment in *Xenopus laevis* oocytes and expression in fish embryos. J Exp Biol 204:2667–2673
68. Planas JV, Capilla E, Gutiérrez J (2000) Molecular identification of a glucose transporter from fish muscle. FEBS Lett 481:266–270
69. Lazzari M, Franceschini V (2006) Glucose transporter (GLUT-1) distribution in the brain vessels of the adult Italian wall lizard *Podarcis sicula*. Acta Histochem 108:385–393
70. Sallet J et al (2011) Social network size affects neural circuits in macaques. Science 334:697–700
71. Burkhalter J, Fiumelli H, Allaman I, Chatton J-Y, Martin J-L (2003) Brain-derived neurotrophic factor stimulates energy metabolism in developing cortical neurons. J Neurosci 23:8212–8220
72. Reivich M et al (1979) The [18F]fluorodeoxyglucose method for the measurement of local cerebral glucose utilization in man. Circulat Res 44:127–137
73. Zhao WQ, Alkon DL (2001) Role of insulin and insulin receptor in learning and memory. Mol Cell Endocrinol 177:125–134
74. Zhao W et al (1999) Brain insulin receptors and spatial memory. J Biol Chem 274:34893–34902
75. Farr SA, Banks WA, Morley JE (2006) Effects of leptin on memory processing. Peptides 27:1420–1425
76. Elias PK, Elias MF, D'Agostino RB, Sullivan LM, Wolf PA (2005) Serum cholesterol and cognitive per-

formance in the Framingham heart study. Psychosom Med 67:24–30
77. Harvey J, Shanley LJ, O'Malley D, Irving AJ (2005) Leptin: a potential cognitive enhancer? Biochem Soc Trans 33:1029–1032
78. Smeets T et al (2009) Stress selectively and lastingly promotes learning of context-related high arousing information. Psychoneuroendocrinology 34:1152–1161
79. McEntee WJ, Crook TH (1991) Serotonin, memory, and the aging brain. Psychopharmacology 103:143–149
80. Wagner JE et al (2006) Old world nonhuman primate models of type 2 diabetes mellitus. ILAR J 47:259–271
81. Venn-Watson SK, Ridgway SH (2007) Big brains and blood glucose: common ground for diabetes mellitus in humans and healthy dolphins. Comp Med 57:390–395
82. Lepage O, Larson ET, Mayer I, Winberg S (2005) Serotonin, but not melatonin, plays a role in shaping dominant-subordinate relationships and aggression in rainbow trout. Horm Behav 48:233–242
83. Belsare PV et al (2010) Metabolic syndrome: aggression control mechanisms gone out of control. Med Hypotheses 74:578–589
84. Thaker M, Lima SL, Hews DK (2009) Alternative antipredator tactics in tree lizard morphs: hormonal and behavioural responses to a predator encounter. Anim Behav 77:395–401
85. Thaker M, Lima SL, Hews DK (2009) Acute corticosterone elevation enhances antipredator behaviors in male tree lizard morphs. Horm Behav 56:51–57
86. Pond CM (1998) The fats of life. Cambridge University Press, Cambridge
87. Watve MG, Yajnik CS (2007) Evolutionary origins of insulin resistance: a behavioral switch hypothesis. BMC Evol Biol 7:61
88. Tung J et al (2012) Social environment is associated with gene regulatory variation in the rhesus macaque immune system. Proc Natl Acad Sci USA 109(17):6490–6495
89. Booth FW, Chakravarthy MV, Gordon SE, Spangenburg EE (2002) Waging war on physical inactivity: using modern molecular ammunition against an ancient enemy. J Appl Physiol 93:3–30
90. Neel JV (1999) Diabetes mellitus: a 'thrifty' genotype rendered detrimental by 'progress'? 1962. Bull World Health Organ 77:694–693
91. Eaton SB, Konner M, Shostak M (1988) Stone agers in the fast lane: chronic degenerative diseases in evolutionary perspective. Am J Med 84:739–749
92. Chakravarthy MV, Booth FW (2004) Eating, exercise, and 'thrifty' genotypes: connecting the dots toward an evolutionary understanding of modern chronic diseases. J Appl Physiol 96:3–10

6 Of Soldiers and Diplomats

Perhaps we cannot apply the hawk and dove model of animal behavior and its implications to endocrinology and metabolism to humans as it is. There have been a few attempts to apply these concepts to humans, but there are some problems with this approach. Sapolsky [1] thinks that socioeconomic status (SES) in humans is comparable to social dominance and subordination in animals, and therefore, lower socioeconomic strata suffer from the same health problems as low-ranking primates. There are many problems with the argument. First of all, Sapolsky himself shows that not in all animal species low social rank is associated with high corticosteroids, which is said to be one of the markers of stress-related health problems. The negative association appears to be certainly true for species in which physical strength and aggression are the most important determinants of social rank but not necessarily so in other species. In orangutans, for example, there is a clear developmental and morphometric distinction between the dominant and subordinate males. The subordinate males are "arrested" in a juvenile-like morph, although they are sexually mature. These subordinate males are aggressive and often coercively mate with females. Here, we see that unlike most other species, subordinate males do not have higher glucocorticoid levels [1–6]. This may be because glucocorticoids have a stronger negative association with aggression than with social status. Orangutans are an ideal example where there is sufficient evidence to show that "dominant" and "subordinate" are actually two distinct mating strategies of males. Here the subordinates show aggressive behavior towards females, and their corticoid levels appear to be determined by aggression rather than subordination. Therefore it appears to be not the dominance hierarchy itself but the ways of attaining it and its behavioral consequences that decide the physiological correlates. In modern human society, physical aggression is not the predominant means of attaining high socioeconomic status (SES), and therefore, SES in humans may not be equivalent to social dominance hierarchy in animals as Sapolsky claims. The other reason to doubt Sapolsky's claim is that human data all over the world does not show the same trends across socioeconomic groups. Sapolsky refers to the picture in America where obesity and metabolic syndrome are claimed to be more prevalent in the lower socioeconomic class. But that is not necessarily the case across the globe and at all times. Even within the USA, studies addressing this question are not equivocal [7, 8]. In South Asia and Africa, T2D and CVD have been disorders of the affluent so far [9–12], although the prevalence might be rapidly increasing in lower strata recently. The third reason to doubt Sapolsky's conclusion is that although there is no doubt that humans evolved from animals, human evolution has taken a somewhat different line with respect to cognitive, emotional, and social aspects of life. It is possible therefore that some of the concepts that we discussed in the last chapter apply to humans but others do not. Much careful reexamination of these concepts is therefore needed

M. Watve, *Doves, Diplomats, and Diabetes: A Darwinian Interpretation of Type 2 Diabetes and Related Disorders*, DOI 10.1007/978-1-4614-4409-1_6, © Springer Science+Business Media New York 2013

before we can make any inferences about behavior and physiology of humans.

One obvious common element is that net reproductive success has played a major role in the evolution of both. The mating systems however are considerably different. A "marriage" system appears to have taken root and became almost universal in the human race. Although the details of the form of marriage are quite variable, long-term commitment and investment in child care by both genders are a common norm in marriages. Marriages do not guarantee faithful monogamy, but still monogamy appears to be a modal picture in the human mating system. This is certainly different from the closely related chimpanzees but not unique to humans. Lifelong monogamous pairing is common in some large birds like eagles or cranes. Human males do not have individual territories as in many animals, but some parallels to territoriality in the form of ownership of land or monopoly over a business niche do exist. Also in the strict sense there are no alpha males but a parallel concept of group leaders does exist.

The comparison most relevant to us is in the nature, form, and role of aggression and dominance hierarchy. Here there are substantial differences. In animals although there is substantial variation across species in the nature and form of aggression, escalation, and risk of injury, aggression is primarily physical. Nonphysical aggression, such as threat displays without physical attack, most commonly involves displaying strength or potentially harmful weapons such as canines. It is difficult to find any example where an animal can successfully establish dominance without being physically capable. This can happen in humans. Some anthropologists have differentiated between "dominance" and "prestige," both independently contributing to Darwinian fitness to a similar extent [13]. They define dominance as a superior ability to inflict costs on others and prestige as an individual's relative ability to confer benefits on others. This means that high status in humans can have more than one alternative meaning.

Unlike most animal species physical strength and aggression are only one of the several paths of achieving social status in humans. Apart from physical aggression, there are two other forms of aggression in humans which are either absent or of relatively minor importance in animals. One is verbal aggression that consists of verbally abusing, insulting, psychologically hurting, or threatening someone. The other is political aggression where the aggressor needs to neither inflict physical injuries nor use abusive words or harsh tone towards the victim. He may even talk with very pleasing words but at the same time manipulate circumstances and other individuals to cause injury to the victim. Apart from aggression, high social status can be achieved in humans by altruism, making alliances, or other types of skillful social manipulations. Obviously since the mechanisms and energetics of the alternative means of gaining hierarchical position are different, the physiological needs are also expected to be very different, and we cannot expect to see the same physiological parameters associating with high socioeconomic status irrespective of the means of achieving it.

We categorized animal behavior into hawk and dove strategies based on the aggressive nonaggressive dichotomy. We will not be able to do so in humans unless we segregate the different types of aggressions. We therefore prefer to substitute the terms hawk versus dove with "soldier" or "warrior" versus "diplomat" when it comes to humans. This distinction is not based on the presence and absence of aggression but rather on the nature of expression of aggression. It is also not based on SES but on the means of achieving SES. Physical aggression is a characteristic of a soldier, whereas verbal and political aggression or other nonaggressive means of gaining social position that of a diplomat. The words "soldier" and "diplomat" certainly do not refer to the profession; they rather refer to personalities, behavior, and physiological makeups of individuals. Our classification of soldier and diplomat is different from the dominance–prestige dichotomy of von Rueden et al. [13]. Soldier–diplomat is in a different dimension, and both dominance and prestige can be attained by either soldier or diplomat means. Since the physiological requirements of a warrior are similar to that of a hawk, we expect that the physiological correlates of a hawk should be seen in a soldier and that of a dove in a diplomat. We would expect therefore that soldier personalities would typically have high sex hormone levels, low serotonergic activity, low levels of cholesterol, corticosteroids and insulin in plasma, and higher insulin sensitivity. A diplomat personality, on the other hand, would be low in sex hormones with higher serotonergic activity and has higher plasma cholesterol, cortisol, insulin, and insulin resistance.

The transition from hawk–dove to soldier–diplomat has important consequences. The availability and security of food are lower for the lower ranks in primates as well as in humans. In an animal society physically weaker individual is most likely to have a lower calorie intake than stronger individuals. If a weaker individual finds a source of rich food, there is a high probability of a dominant individual snatching it. The dominant individual would be less interested in snatching less-rich food. Therefore it would be normal for a weaker individual to have a higher proportion of fiber-rich and calorie-poor foods, and evolution might have fine-tuned a physiological link between subordinate status and fiber-rich diet. The physiology of a subordinate or nonaggressive individual might be optimized for a high-fiber low-calorie food. Physically weak individuals having high access to calorie-rich food with substantial food security is perhaps unique to human societies, that too specifically modern human societies. Perhaps in the hunter-gatherer stage, physical strength and fitness would still have been one of the major factors contributing to social dominance and access to food and other resources. Even in agricultural societies, physical strength and activity would be important in deciding productivity and therefore access to food of an individual. But this relation is lost in modern urban life. With the rapid development of civilizations over a relatively brief period, human social structure changed rapidly, and the relevance of physical strength and aggression to social dominance and food security went on decreasing monotonically. This would doubtlessly have physiological consequences that are unique to the modern human society. The period over which the social change took place has been too small for evolution to bring about major changes in human biology. Therefore it is likely that we retain many of the ancestral responses. For example, a subordinate animal that has lower food security would be benefited by binge eating since food once eaten cannot be snatched by another dominant individual. Hyperphagia after social defeat or subordination has indeed been shown in many experiments [14–18]. Glucocorticoid levels that are typically higher in subordinates increase food intake [19]. This tendency which was once adaptive is perhaps still retained by us in spite of changed social structure. Many studies have indicated that social subordination is one of the important factors associated with eating disorders [20, 21], obesity, and T2D [22, 23] in humans too.

This raises a possibility that is speculative at present but appears logical and therefore needs to be pursued seriously. Since weaker individuals in nature are more likely to have lower calorie intake and higher fiber content in food, we would expect some physiological fine-tuning done by evolution to optimize this kind of a diet in physically weaker individuals. If a weak individual eats high-calorie and low-fiber food, there is a mismatch with the evolved tendency which might show up in the form of some pathophysiological processes. On the other hand individuals who are physically strong and have a rough and tough life may consume high-fat high-calorie food without showing any signs of adverse effects as their physiologies would be programmed for this kind of food. If this hypothesis is correct, diet-induced insulin resistance should be seen only in the diplomat class of people and not in the soldier class. This is perhaps what we see in the Masai tribe that consumes a diet rich in animal fat but remain insulin sensitive and hypocholesterolemic [24–27] or in endurance athletes that have high IMTG and are still insulin sensitive [28]. It is likely therefore that a combination of behavioral strategies and diet is what matters and not diet alone.

A good metaphor for the observed effects and importance of diet in the development of metabolic syndrome is that of sugar intake in diabetes or salt intake in hypertension. A large variation in sugar intake is easily tolerated by a nondiabetic individual without losing control of blood sugar, but for a diabetic person, even a moderate intake of sugar may shoot up plasma sugar levels much beyond the normal range. Similarly high-fat diet causes no detectable problem for a warrior, but for a supernormal diplomat, it can lead to signs of obesity and metabolic syndrome in no time. Here lies a possible answer to the diet paradox described in Chap. 3. Across the globe there are many tribes that traditionally live on extremely fat-rich or carbohydrate-rich diets and still show no signs of obesity, hypercholesterolemia, and insulin resistance, whereas a smaller rise in dietary fat or soluble carbohydrates in the modern society is evidently associated with obesity and related disorders. As long as the behavioral components of a hunter–gatherer–warrior lifestyle are intact, dietary macronutrient composition is likely to be of minor importance. Once the deficiency of these behaviors develops, diet suddenly becomes important. It is important to realize that majority of research on diet-induced obesity has been done in the modern urban or semiurban society. Animal research is typically on caged rats. Imagine a hypothetical researcher living in a society made up only of untreated diabetics. This researcher is studying the effect of dietary sugar intake on plasma sugar. Since this person has never examined normal healthy individuals, he will not realize that there is a narrow range of normal glucose levels which is not affected by consuming more or less sugar on a given day. Similarly, a diet school researcher who has never performed his diet experiments in hunter-gatherer societies (without changing their normal behavior) will never realize that macronutrient composition has little to do with body composition and metabolic alterations typical of metabolic syndrome.

The current status of diet-induced obesity in humans is an effect of drifting from hawk–dove to warrior–diplomat. In nature dove would develop thrift but is unlikely to become obese since it is unlikely to have higher access to calorie-rich food. The diplomat has a physiological makeup of a dove but still has higher access to calorie-rich food which might be a root of the problem. Hawks or warriors, on the other hand, may not worry about calories and are unlikely to develop obesity and insulin resistance in spite of having calorie-rich food. Their endocrine and behavioral makeup will convert the extra energy into muscle protein and become stronger than becoming obese. We are developing a hypothesis now that the warrior–diplomat dichotomy is one of the major drivers of physiological state. The deficiency of soldier behavior and perhaps a supernormal diplomat behavior drives neuroendocrine and metabolic changes leading to increased susceptibility to diet-induced metabolic syndrome. But before developing the hypothesis further, we need to be careful about defining the warrior/soldier versus diplomat dichotomy.

What Characterizes Soldiers and Diplomats

Table 6.1 summarizes the postulated differences between soldier and diplomat behaviors which can be easily compared with Table 5.1. The importance of physical activity in controlling obesity and insulin resistance is well accepted by current thinking. But in the current thinking, the importance of physical activity is for burning calories. As I will argue eventually the difference between soldier and diplomat is far beyond the rate of calorie burning. All other elements of the dichotomy listed in the table are absent in the current thinking, and there are reasons to think that they are more important than calorie burning alone as we will see elaborately one by one.

Since the warrior–diplomat dichotomy is primarily (but not exclusively) based on physical aggression, we should start by defining physical aggression in the context of human behavior–physiology interface. This is extremely important since the word aggression has been used in a variety of contexts and with an equally wide variety of connotations. Even the treatment of diabetes is often called aggressive when it intends to control the sugar very tightly by all possible means. In the social context aggression often has a negative connotation and has been commonly viewed in psychiatry as a pathological behavior.

Table 6.1 Typical characteristics of soldiers and diplomats

Soldier/warrior	Diplomat
Physically active	Relatively sedentary
Muscular and strong	Physically weak
Physically aggressive, often proactive	Avoider of physical aggression (other forms of aggression may be seen)
Swift, agile, active complex NMCs	Slower reflexes
Risk taker, exploratory, and adventurous	Risk and adventure avoider
Injury prone	Less injury prone, active harm avoider
Tolerant to physical injuries and discomfort	Lower tolerance to injuries and discomfort
Higher spatial skills and memory	Higher verbal memory and cognitive skills
Socially simple	Socially smart, manipulative, having higher levels of political skills, tactical deception

I will refer to physical aggression here as all the neuromotor acts involved in Stone Age hunting and fighting. These acts include hitting, throwing with aim, punching, kicking, chasing, grabbing, dodging, escaping, etc. Aggression is not anger, irritability, or hostility. Although anger may provoke aggression, anger is not aggression itself. There can be anger without physical aggression and aggression without anger. A carnivore chases and kills a prey not because it is angry on the prey, but it is an act of aggression natural for a carnivore. The same is true for irritability and hostility which may provoke aggression at times but are not identical with aggression. I will use the term aggression independent of its intentions and values such as good or evil. Aggression need not always be antisocial and therefore need not be viewed as bad all the time. In fact for maintenance of peace, law, and order, some amount of aggression needs to be used which can be viewed as good or prosocial aggression. I will use the term aggression to reflect the neuronal, endocrine, and metabolic mechanisms that are involved in Stone Age physical aggression independent of the social and moral context of it. For convenience and clarity, whenever there is a reference to verbal or political aggression, it will be specifically called verbal or political aggression or a more inclusive term—diplomat aggression. When used simply as aggression in this book, the default meaning would be physical aggression.

Aggression is also not equivalent to violence. Modern violence with remotely controlled bombs, missiles, and machine guns need not involve aggression by our definition. Pulling the trigger of a revolver or pressing a button of a remote control may not be recognized as an aggressive neuromuscular act by our Stone Age bodies. On the other hand digging to plant a tree is not a violent act, but it is physiologically aggressive as it involves muscle activity similar to what our Stone Age ancestors used in hunting. In fact digging itself was a common activity in hunting of burrowing animals. But this does not mean any muscle activity is aggressive. The act of walking on a treadmill, for example, does use muscle and burn some calories but has a small, if any, component of aggression.

It is this aggression that I consider one of the most important criteria to distinguish a soldier from a diplomat. Soldiers use physical aggression whereas diplomats avoid it although they may use nonphysical forms of aggression. Is the soldier–diplomat dichotomy an artifact of modern civilization or did it exist from the Stone Age? The latter appears to be more likely since it is a continued and gradual transformation of hawk–dove dichotomy of animal life. There is some evidence that even in human societies, the contribution of physical aggression to reproductive success is negatively frequency dependent as with the hawk and dove model. A comparative study of aggression in two tribes showed that reproductive success of aggressive individuals was greater than submissive ones in the Yanomamo societies but was lower in the Waorani tribe where the frequency of aggressive encounters was much higher [29].

This suggests a negative frequency dependence of reproductive success contribution of aggression. Negative frequency-dependent selection is known to operate in human social systems [30]. Accordingly similar to hawk and dove, soldier–diplomats are likely to have negative frequency dependence and therefore would coexist in a society. The equilibrium proportion of soldier–diplomats would depend on their payoffs that change with the nature of the societal structure. Today's hunter-gatherer societies almost invariably have more or less specialized shamans or magic men who perhaps do less of hunting-gathering themselves but make living by creating, propagating, and exploiting a belief system. Shamans can be viewed as the ancestral diplomats of human civilization. More diplomat niches grew with the gradual transitions into agricultural, urban, industrial, and information-based societies. Throughout this transition the proportion of diplomats has grown monotonically. Nevertheless even in today's diplomat-dominated societies, soldiers do exist in both pro- and antisocial roles.

Having defined aggression, we also need to worry about the conditions under which physical aggression can be adaptive and conditions under which it is maladaptive. Since both the costs and benefits of physical aggression are potentially very large, very subtle mechanisms must have evolved to fine-tune expression versus suppression of aggression and accordingly mobilizing the neuroendocrine and metabolic machinery in support. Since suppression of aggression is particularly important for insulin resistance, we need to look at the conditions for aggression suppression in sufficient details.

When to Express and When to Suppress Aggression

Food and sex are the two major natural causes of aggression in an evolutionary context. Other causes such as territoriality, social ranking, self defense, mate guarding, and maternal aggression are all related to the two basic needs of food and reproduction. In modern life the nature and purpose of aggression may have changed substantially, but the behavior–physiology links have evolved for the Stone Age, and therefore, we need to think of the Stone Age causes of aggression, expression, and suppression. Effective aggression can result into more access to food or better mating opportunities. However, aggression has an energetic cost as well as increased risk of getting injured. Therefore when there is no need for aggression or aggression is unlikely to be effective, it needs to be controlled to avoid unnecessary risk of injuries. Some commonly occurring conditions when aggression needs to be suppressed are:

1. Full meal: A satiated individual does not need to be aggressive and therefore cues of food satiety such as physically having a stomach full, high plasma glucose, or raised insulin levels signal aggression control. This is achieved centrally by stimulating brain serotonin [31, 32]. The role of serotonin in aggression suppression is well known [33–38]. Therefore it is not a coincidence that serotonin and insulin happen to be food satiety signals and simultaneously aggression suppression signals. It is adaptive to suppress aggression after having food, and serotonin serves as the main bridge linking the two.
2. Stored fat: Energy reserve in the form of fat should also signal aggression control since there is little desperation for food. If energy stores are exhausted, the individual has more desperation for food which increases risk-taking behavior and thereby readiness to engage in an aggressive encounter, particularly one related to competition over a food resource. This appears to happen by a number of mechanisms. Higher levels of FFAs facilitate cholesterol synthesis, and cholesterol suppresses aggression by stimulating serotonergic signaling [39–44]. Alternatively leptin also might have aggression suppression effects through activation of POMC neurons [45]. Obesity is also associated with hyperinsulinemia, and insulin has effects on the brain serotonin level which appears to be antagonizing aggression as stated above. High-fat intake has also been shown to reduce aggression in humans and other primates [44, 46, 47].

3. Loss of sexual desire/function: The effects of sexual satiety should be similar to food satiety since sex is an equally strong motivating factor for aggression. Interestingly, the mechanisms by which sexual satiety reduces aggression are also similar to those of food satiety and involve serotonin signaling [48, 49]. Apart from sexual satiety, castration or loss of sexual motivation for any other reason should have an aggression suppression effect too. On the other hand, loss of aggression, dominance, or physical strength should downregulate sexual desire as well. This is because the opportunities and freedom of sex of an individual is much restrained in the presence of a stronger same-sex rival. Engaging in sexual activity in the presence of a stronger rival is inviting attack and injury, and therefore, both sex and aggression should be corepressed by a perception of being subordinate to someone of the same sex. Interestingly in movies, almost all over the world, sex and violence frequently go hand in hand. This association has a possible biological basis. The endocrine as well as neuronal mechanisms involved in the two have a large overlap.
4. Being weaker than the opponent: It makes no sense to initiate aggression when the rival is stronger. Therefore weaker individuals should shun aggression. For this to work, there should be a mechanism to judge one's own strength vis-a-vis that of the other. This is not an easy task. Since any individual cannot see its own size and that of others from the same perspective, a comparison is difficult. Nevertheless since this judgment could be very crucial for survival and reproductive success, elaborate algorithms are likely to have evolved taking data from a variety of sources. Mock fight is a known mechanism; eavesdropping on others' conflicts can give useful data which may be stored in memory and retrieved in the face of a conflict. However, history is not always reliable since relative strengths of individuals are dynamic. There ought to be mechanisms to sense one's own muscle power and continually update these data. This could be done through the sensory nerve endings in muscle. Muscles are known to be rich in sensory nerves. The function of these nerves could not be restricted to muscular pain. The sensory nerves are shown to fire at every contraction of the muscle, but why these signals are generated and what happens to these signals are not clearly known. I think that these signals are extremely important to continuously sense and update a judgment of one's own muscle power which is crucial in making decisions as to when to initiate and when to suppress aggression. A common experience is that when muscles are temporarily weak owing to sickness of any kind, we "feel" the weakness. This self-appraisal might be extremely important in deciding behavioral strategies. I expect therefore that a "feeling" of strong muscle should have a proaggression effect, and a feeling of weakness or fatigue would be anti-aggression factors.
5. Crowding: Although it may sound counterintuitive, crowding suppresses aggression. This argument is very well supported by theory as well as evidence over a wide variety of species. We will look at this much more elaborately in a separate chapter.
6. Mechanistic checkpoint: Before beginning aggression it is necessary to check that all the machinery needed for initiating and executing aggression and coping with the possible consequences is in good condition. Initiation of an act without being prepared for the requirements for execution and coping with possible consequences is an indicator of a badly designed system. Since aggression is an important determinant in evolutionary fitness, evolution would certainly have built in this checkpoint before beginning aggression. Although the concept of aggression checkpoint is new, it is easily testable. A testable prediction would be that any defect or deficiency in factors that build and strengthen bone and muscle such as vitamin D should have anti-aggression effects. Similarly factors arresting or limiting muscle strength such as myostatin should also arrest aggression. Any defect in vascular contraction–relaxation mechanisms should also suppress aggression

since blood flow control is critical in regulating blood supply to muscle as well as arresting bleeding in case of an injury.

Since we are considering the argument that aggression suppression is the main factor linking behavior and lifestyle to metabolic syndrome, we should expect all the ecological and social factors suppressing aggression to be obesogenic and/or diabetogenic. Overeating, stored fat, sexual dysfunction, physical weakness, social subordination, and exposure to crowding should all have significant associations with metabolic syndrome, and we should be able to assign a causative role to them. Much of this is certainly testable, and there is evidence for many in already-published literature as we will eventually see.

Some of the above conditions of aggression suppression work at acute and others at chronic level. For example a full meal should lead to short-term aggression suppression, but stored fat should show a long-term effect. Particularly interesting case is that of muscle weakness. Transient weakness owing to exhaustion or damage is also a disadvantage in an aggressive encounter, and therefore, aggression should be suppressed. Interestingly such short-term effects on muscle have been shown to induce transient insulin resistance. Although exercise is well known to increase insulin sensitivity, muscle exhaustion after marathon running is accompanied by short-term insulin resistance [50]. Eccentric exercises or any other exercises that cause muscle damage or soreness also increase insulin resistance which is not restricted to the damaged muscle [51–57]. This is difficult to explain by the lipid centric theory. In marathon running or any other intensive exercise, substantial fat is burnt, and therefore, insulin resistance should reduce. But damage or exhaustion increases it. Any major debilitating injury [58], burn [59–61], acute infections or sepsis [62–64], acute pain [65], or surgery [66–68] induce transient insulin resistance. In brief, a body condition that induced weakness and thereby transiently makes aggression impossible or unprofitable is bound to induce insulin resistance. This illustrates the tight association of loss of aggression with insulin resistance. Even short-term loss of aggression leads to short-term insulin resistance. Such short-term insulin resistance is unlikely to have any clinical importance. Nevertheless, for us, it is an important demonstration of the more or less obligate association between loss of aggression and insulin resistance. Chronic muscle weakness, on the other hand, is expected to result into chronic loss of aggression and with accompanying long-term insulin resistance. This is certainly likely to have clinical implications.

Apart from aggression there are other characteristics that differentiate soldiers from diplomats. Following Table 6.1 we can make a number of predictions. (1) The first very obvious requirement is strong bone and muscle without which soldier behavior would be counterproductive. It is indeed true that insulin resistance is associated with weak muscle [69–72]. (2) Soldiers need to have quick reflex reactions and agility which the diplomats may afford to lose. Therefore we may expect that insulin resistance may be associated with loss of quick reflex actions. (3) Soldiers need high-level spatial skills since a chase in a forest involves fast running on an unpredictable trajectory, and then it is equally necessary to find the path back. Those who experienced getting lost in a forest will understand the importance of it. Forest-living tribes have an amazing ability of mental mapping and reorientation which urban dwellers find difficult to achieve even with the help of devices such as a magnetic compass and GPS. Diplomats on the other hand need to have better social manipulation skills, facial recognition, and face reading. They need to be better at verbal memory, although they may compromise on spatial memory. (4) Soldiers also need to have a higher tolerance to physical discomfort such as pain from minor injuries, natural fluctuations in temperature, or variation in feeding intervals. (5) Soldiers are more prone to physical injuries, and their immune system needs to be geared up to take care of wounds. Diplomats can avoid physical injuries to a large extent, but if they are in a more social and perhaps enclosed indoor environment, they may be exposed more to respiratory infections. Therefore the behavior of the immune systems of both could subtly differ. (6) It is well known that people are able to solve mathematical problems more easily if presented in an appropriate context than in an abstract form [73]. My speculation is that the context in which soldiers and diplomats would do better mathematics should be detectably different. A soldier should be able to predict the trajectory of a fast-moving object and act accordingly to catch or dodge it within a fraction of a second without understanding the mathematics behind it at a conscious level. A diplomat on the other hand may solve an equally mathematically intricate problem in a social context without understanding the underlying mathematics. In a most likely case they would both fail in each other's mathematical tasks. It can be clear from this discussion that our hypothesis is testable by means of a large number of testable predictions. Many of them are already substantially supported as we will see later. The important point to be illustrated now is that the facets of soldier–diplomat dichotomy other than physical activity can be tested, and these can really differentiate between the older way of thinking which has already recognized the importance of physical activity and the new thinking that has many more components in addition to physical activity.

A soldier or diplomat behavior is highly contextual. Both personality and the nature of challenge decide whether the response to the challenge will be a soldier or a diplomat behavior. Different individuals can give different responses to a given challenge, and this depends predominantly on their personalities particularly in the context of social challenges. However certain types of challenges do not give much choice and compel the responder to behave like a soldier or like a diplomat. The two are also not completely independent, and the nature of challenges faced in the past influences personality at least partially. Genetics certainly plays some role in deciding personality, although currently we do not know to what extent. The terms soldier–diplomat therefore do not simply reflect genetic predispositions or environmental influences alone but are a complex outcome from genetic, developmental, social, environmental, and stochastic elements.

So far I have introduced the concept of hawk–dove or soldier–diplomat as if they are mutually exclusive and strictly dichotomous. In reality as well as in theory, they need not be. I did so for the convenience of description. They should rather be viewed as two ends of a continuum, but a dichotomy is an easier way of expression while introducing a concept. However if we recall the original hawk and dove model, there are two possible solutions which are mathematically equivalent. Either there could be a population equilibrium consisting of an optimum proportion of pure hawks and pure doves or there can be a mixed strategy ESS. The mixed strategy ESS implies that the same individual can behave as a hawk sometime and a dove at other times. Similarly we can think of a mixed soldier and diplomat personality or of soldier and diplomats as two components of personality in a state of balance. There are different physiological requirements of the two components. However the body has sufficient flexibility to give the appropriate response to appropriate behavior. Such flexibility in the physiological responses is quite well known. Problem may arise if this balance is lost.

I am going to argue and demonstrate henceforth through several chapters that it is the lost balance between warrior–diplomat components of personality in the modern urban human behavior that has given rise to the epidemic of obesity and type 2 diabetes. For a person with a moderate mix of soldier and diplomat elements, both the associated physiological responses are adaptive and flexible. The adaptive responses become pathological only after losing the balance and going to the extremes of behavior, in this case giving up soldier behavior almost completely and adopting supernormal diplomat behavior. Steven Pinker [74] has shown that violence has been decreasing in human history, and I think, as compared to violence, physical aggression has decreased orders of magnitude more dramatically. Most of the violence today is mediated more by technology and less by physical aggression. It is quite likely that this is one of the factors behind the epidemic of obesity and metabolic syndrome. In order to understand the pathological consequences of the supernormal diplomat behavior of the modern human society, we need to understand why the endocrine and physiological correlates of diplomat are adaptive for diplomat behavior first. Then it would be easy to extrapolate and see how they turn pathological if stretched beyond the limit in which they evolved.

Much of the logic of these arguments remains the same as that in the last chapter, but we will see now what modifications may be necessary while applying it to humans. Insulin resistance changes the energy budget allocation on adopting a predominantly diplomat lifestyle. This principle is applicable to animals as well as humans, but the change is of a higher order of magnitude in humans. Human brains are much larger, their energetic requirement is substantial, and the contribution of cognitive activities to success is considerably larger in humans. Therefore, in humans, it makes a great sense to finely regulate the brain glucose supply according to the need and insulin action, and insulin resistance is one of the mechanisms of regulating brain glucose supply. The tendency to develop insulin resistance in humans on adopting a diplomat life is expected to be quantitatively much sharper than other primates although qualitatively in the same direction.

Of particular importance is the direct role of insulin in the human brain. Early studies on glucose metabolism in the brain indicated that the uptake of glucose in the brain is independent of insulin, and therefore for several decades, insulin was not perceived as important to brain activities. However insulin receptors are widely distributed in the brain indicating that insulin has some other function in the brain. The role of insulin and its receptors in the cognitive brain functions such as learning and memory has been demonstrated independently by many research groups in animal as well as human system [75–80]. What is of greater interest but unfortunately with scanty research inputs is the specific effects of insulin on mood, emotions, and behavior. Kern et al. [77] studied the effects of insulin on preattentive sensory processing (i.e., before the stimulus gains access to working memory) versus attentive processing (i.e., relatively delayed processing after the stimulus enters working memory, which involves meaning and significance attached by the subject) and demonstrated that insulin impaired information processing at preattentive level but enhanced the slower attentive processing. This might be of great relevance to the soldier–diplomat behavioral axis since hunting or fighting involves a greater component of preattentive processing, and diplomat activities may need more of attentive processing in the hippocampal and frontocortical areas. Insulin infusion also affected mood. The high-insulin infusion group experienced reduced restlessness and annoyance and had less difficulty in thinking [77]. Unfortunately, no more research is found in published literature addressing this aspect of insulin action. But we can make a more generalized testable prediction here that if insulin is infused while keeping glucose levels constant, one should see impairment of soldier-related characteristics such as quick nerve–muscle coordination actions, physical risk-taking, aggression, and tolerance towards physical pain and discomfort. On the other hand one should see simultaneous enhancement of tasks involving memory, thinking, and problem solving in a social context. Such experiments will be of great help in resolving between the two alternative interpretations of insulin resistance syndrome. Currently only the Kern et al. [77] results appear to be the ones available in literature, and they are in support of our expectation that high levels of insulin suppress certain components of soldier behavior and enhance certain components of diplomat behavior.

In apparent contradiction with our expectation that diplomat behavior should be associated with enhanced memory and cognitive functions, there is substantial cognitive impairment in late stages in obesity and diabetes. Discovery of the contribution of insulin towards enhancement of cognitive functions is relatively new and restricted to a handful of research groups. On the other hand, literature showing cognitive degeneration in old diabetics is overwhelming. Is this a contradiction, evidence against our hypothesis, or a paradox? A careful look reveals that it is none. There are a number of possible solutions. Diabetes-related cognitive impairment is typically limited to old age [81]. This could be viewed as an antagonistic pleiotropy effect where a process that enhances fitness during early and prime reproductive life may turn detrimental in later life. In the natural history of type 2 diabetes, early stages are marked by hyperinsulinemia, whereas later stages may see sharp decline in insulin levels. It is likely that cognitive performance declines when a relative insulin deficiency develops in the brain. We will discuss the pathophysiology of cognitive decline in diabetes in a later chapter, but it needs to be stated here that in the early hyperinsulinemic phase of T2D, which is really the diplomat adaptation, there appears to be no evidence of cognitive decline. This question is not yet seriously addressed, but some studies suggest that the relationship of metabolic syndrome with cognitive function is age dependent. For example higher levels of visceral abdominal fat (VAT) were marked by lower cortical thickness in older age group, but in younger age group, the relationship was reversed, and individuals with higher VAT had higher cortical thickness [82]. In another unpublished study that I am aware of, after correcting for birth weight, children with higher levels of insulin resistance showed higher cognitive performance. In the Women's Health Initiative hormone trials, although there was an overall negative association of BMI with cognitive performance, waist to hip ratio was positively associated with cognitive performance assessed by the modified mini-mental state examination (3MSE) [83]. The association of Alzheimer's disease with diabetes in the old age appears to be associated with insulin deficiency rather than hyperinsulinemia, and insulin infusion seems to enhance memory even in AD patients. In Alzheimer's disease, the cerebrospinal fluid to plasma insulin ratio has been shown to be lower than normal, and insulin infusion can have therapeutic effects on AD and some other disorders [77, 84, 85]. This issue is difficult to resolve from the available literature, but a testable prediction can be made at this stage. In young age insulin resistance is expected to show a positive association with cognitive performance which may reverse in old age.

A third possibility is that it is not the cognitive area of the brain but the coordinating area of the brain that degenerates in old age. This degeneration is a result of the lack of complex physical activity, typical of soldier life that needs higher-order nerve muscle coordination. When such activities are chronically lost, there is disuse atrophy. This in turn affects even other functions that need the coordination functions mainly by the cerebellum. This is backed by data that showed cerebellar degeneration rather than cortical degeneration to correlate strongly with cognitive loss in old age [86]. Related to the above there is one more possibility which is more speculative at this stage. As we argued that deficiency of soldier component can lead to disinvestment from muscle deficiency of cognitive brain activities can give rise to disuse atrophy. It is likely that the modern lifestyle is characterized not only by soldier deficiency but also by diplomat deficiency. There are multiple reasons why there can be a double deficiency. (1) Just as modern lifestyle has brought in physical inactivity, the electronic aids for memory, recall, alarm, and directions have brought in mental inactivity as well. (2) As we will see in a later chapter, although diabetes is characterized by high plasma glucose, there can be problems in glucose supply to the brain. Chronically low or inconsistent glucose supply may result into reduced activity of certain parts of the brain. (3) Obesity is having large amounts of energy stores, and under such a condition, there is no need to take efforts on foraging. Just as there is no need to engage in soldier activities on gaining fat, at some stage of obesity, the body may perceive no need to engage in intensive brain activities either. We may have evolved a "withdrawal from all activities" response on having excess stores. In nature, reduced efforts would automatically result into reduced food intake, and the state will be back to normal. This is where modern life differs. If food availability is independent of both physical and mental activity, there can be simultaneous retraction from both, and this can stay for a long time without affecting further food availability. (4) When soldier strategies do not work, one may adopt diplomat behavior. But what if diplomat strategies also do not work? Just as insulin resistance is disinvestment from soldier abilities, there could be mechanisms for disinvestment from diplomat activities. Loss of cognitive function may result in such a case. (5) A recently raised and much promising possibility is that degenerative nervous system changes in diabetes are not related to insulin per se but are caused by deficiency of growth factors and neurotropins. This is discussed much more elaborately later. The point of relevance here is that hyperinsulinemia and/or insulin resistance is most unlikely to be the primary cause of cognitive decline seen in old diabetics.

Thus there are a large number of possible reasons for later cognitive decline in diabetes. We do not know at this stage the relative importance of them. But none of the possible reasons contradict with insulin's role in cognitive enhancement. So there is no real paradox associated with hyperinsulinemia and cognitive functions. Insulin certainly enhances memory and cognitive functions at all ages, and therefore, hyperinsulinemia is adaptive for diplomat life. The decline in β cell function is a pathological change and not a part of the adaptation. We will see a little later a new interpretation about how and why the adaptive hyperinsulinemic response transforms into pathological deficiency of insulin. The argument of relevance at this stage is that enhancement of cognitive function could be the primary adaptive reason for which insulin may be secreted in larger amounts. However, if the brain needs larger amounts of insulin because of greater demand on cognitive functions, it should simultaneously make a provision for reduced insulin sensitivity of muscle and other peripheral tissues without which hypoglycemia will develop which is detrimental to brain functions. Thus the primary requirement of diplomat life could be hyperinsulinemia. Insulin resistance is a necessary compensatory mechanism without which hyperinsulinemia cannot be sustained.

Another aspect of the soldier diplomat difference is important for the pathophysiology of diabetes. In the modern urban life, physical safety has increased to a supernormal level. We noted earlier that by giving submissive displays, subordinate animals can avoid aggressive encounters with conspecifics very effectively. Nevertheless they are prone to minor injuries from many other sources during foraging including pricks or scratches by thorns, insect bites or stings, brushing the body against rough surfaces, etc. These are not injuries of major concern, but they still have a subtle effect on the dynamics of innate immune cells of the body. With every minor injury, some macrophages and neutrophils leave the bloodstream and migrate towards the skin. This is the normal dynamics of the innate immune cells. In the modern urban lifestyle minor injuries have almost disappeared. At home or in the car, bus, or office, we are always in a highly cushioned environment. Penetrative injuries are extremely rare, but even nonpenetrative small-impact injuries are uncommon. In such a supernormally injury-free lifestyle, the dynamics of macrophages can be heavily perturbed. We will discuss at much more depth in a later chapter that this perturbed dynamics of the innate immune mechanisms could be responsible for what is called a low-grade chronic systemic inflammatory state. This inflammatory state is one of the main pathological processes of metabolic syndrome [87–89]. Deficiency of minor injuries also induces a gradual disinvestment from wound healing and tissue regeneration mechanisms. Angiogenesis is one of the essential processes in tissue regeneration. When there is a progressive degeneration of wound healing and regeneration mechanisms, dysfunction of angiogenesis mechanisms also sets in. A number of diabetic complications originate in angiogenesis dysfunction [90–92]. But perhaps the most common and most frequently fatal process is macrovascular pathology that leads to atherosclerosis and cardiovascular disease. This begins with the altered macrophage dynamics. Since the normal rate of migration from blood vessels towards subcutaneous tissue is impaired owing to lack of stimulation by injuries, cells of the monocyte–macrophage lineage accumulate within the blood vessels. An accumulation of macrophages is the earliest process in the eventual formation of a plaque which may ultimately dislodge and block a vital capillary leading to a variety of problems including a heart attack. The other factor important in aggravating the macrophage accumulation in vessel walls is cholesterol, particularly LDL and VLDL cholesterol. Hypercholesterolemia is an important component of metabolic syndrome which has a strong negative association with physical aggression

[39]. Monkeys fed with low-fat, low-cholesterol diet exhibited more aggression than monkeys fed with high-fat, high-cholesterol diet [44]. As opposed to physical aggression, verbal aggression is either not correlated or positively correlated to cholesterol [39, 93]. This is compatible with our assumption that verbal aggression is a "diplomat" rather than a "soldier" trait. Thus nonaggressive and noninjury-prone lifestyle is sufficient to lead to atherosclerotic changes.

The altered macrophage dynamics leads to another biochemical effect. The phagocytic cells of the body are the largest source of reactive oxygen species (ROS) that are responsible for oxidative state. Why do defense cells produce ROS? It is actually meant for killing bacteria [94] during infections. The most frequent occasions to fight bacteria in soldier life are minor epidermal injuries. But now since we do not face injuries for months or years together, the macrophage dynamics of the body changes, and inevitably the ROS dynamics of the body is also expected to change. This is likely to be one of the main factors leading to an "oxidative stress" which is claimed to be responsible for multitude of pathological processes.

Taken together, in a simple and unsophisticated way, we could see how just the lack of aggression and injury proneness leads to almost the entire pathophysiological network of metabolic syndrome. So far I have not explained some other components of T2D including β cell degeneration and hyperglycemia which will be done in later chapters. The adaptive role of insulin resistance depicted here is mainly muscle insulin resistance. Liver insulin resistance is equally or perhaps more important in the pathophysiology of T2D, and as yet we have not wondered why liver should lose its insulin sensitivity. For this the readers will have to wait till Chap. 12. We will spend the next several chapters to elaborate on the pathways by which all the above-mentioned adaptive changes supporting behavioral change are executed. Much can be synthesized from the available literature itself. There are certain gaps in the picture where experimental investigations are needed. Currently in these gaps, we will remain speculative. But as we will see in sufficient details in the following chapters, even at this stage, the new picture is logically much more sound and evidence based than the classical one.

Let us look at the available evidence and its interpretations by the old versus new hypothesis step by step starting with epidemiological, clinical, and observational picture first. We will go to the levels of physiology and molecular biology gradually in subsequent chapters. To begin with, if we take the meaning of soldier and diplomat literally, we do see that the prevalence of metabolic syndrome in the armed forces of any country is reported to be very small as compared to other contemporary citizens [95]. Whatever small incidence is observed among the forces is mostly among the high-rank officers who generally have lost contact with a soldier life and live more of a diplomat life and whose job profile is also more that of a diplomat [95]. What is more interesting is the attitude and norms followed after diagnosing diabetes in any soldier. In many countries, after detection of diabetes, the person is relieved from services. No services are willing to post a diabetic on the war front, and in countries where diabetics are not outright expelled from military services, they are given minor, low-esteem, and low-risk jobs [96]. The picture seems to be changing though with wars being less physically aggressive and more technology oriented. The incidence of type 2 diabetes in armed forces is on the rise and particularly in countries with more advanced technology of war [97].

As opposed to soldiers the incidence of diabetes in typical diplomat professions such as politicians, businessmen, managers, lawyers, etc., has been among the highest of all professions. But these data are of limited value since they do not resolve between the old and the new hypotheses. Both expect a similar trend across professions that are physically active against the ones that are sedentary. In order to differentiate between the two, we need to look at prevalence of metabolic syndrome components across professions that have similar levels of physical activity but differ in other soldier–diplomat traits. For example, driving a car involves physical activity not significantly greater than an office table job. But drivers of racer cars have other elements of soldier behavior such as risk-taking, spatial skills, quick reflexes, and anticipation of injuries. Therefore I would predict that race drivers would have substantially lower prevalence of insulin resistance as compared to politicians, lawyers, and businessmen with comparable calorie expenditures. I did not find data in published literature that would allow us to test the prediction. But this prediction is certainly testable with appropriately collected data and would throw much light on the origins of insulin resistance.

It is well recognized that physically active lifestyle is associated with insulin sensitivity. This is interpreted differently by the old and new hypotheses. The conventional theory believes obesity to be the root cause of diabetes. The simple argument is that individuals that are physically active are less likely to be obese and therefore remain insulin sensitive. Sedentary professionals, on the other hand, have less physical activity which makes them obese and therefore diabetic. Both the traditional and new hypotheses agree that physical activity is insulin sensitizing but for different reasons. The classical theory argues that sedentary life leads to obesity and obesity leads to insulin resistance. The new hypothesis states that physical activities and particularly intensive and aggressive activities by themselves are insulin sensitizing independent of obesity.

If we want to resolve between the two hypotheses, any example where soldier characteristics and obesity are negatively related will be useless, because both the hypotheses appear to explain it equally logically. If obese people are generally less aggressive, for which there is some evidence, we will be unable to segregate the effects of obesity from that of lack of aggression. We would need examples where obesity and aggression or any other soldier characteristics are simultaneously present. Sumo wrestlers provide such an example. They are overweight and have abdominal obesity as well, but at the same time, they play a moderately aggressive game. It is said that sumo wrestlers are insulin sensitive as long as they are active wrestlers. After retirement from the game, they may become insulin resistant rapidly [98, 99]. If this is true then the effect of aggression appears to prevail over the effect of obesity. Unfortunately there are few studies on sumo wrestlers in comparison with nonwrestlers of comparable BMI. But the sumo wrestlers make the diametrically opposite effects of obesity and aggression testable. At the other end of the picture is the example of patients with anorexia nervosa who are extremely thin and hardly have any body fat. There appears to be no data on levels of aggression in anorexia nervosa. But studies have shown a strong association of anorexia nervosa with subordination and harm avoidance behavior [20, 100, 101] which is an indication of low soldier component in personalities. Anecdotally, some of my psychiatrist friends agree that women with anorexia nervosa are generally nice to others, nonaggressive, and who often like to cook tasty food and feed others, keeping little for themselves. Although such individuals are insulin sensitive, they frequently show impaired glucose tolerance and hypercholesterolemia [102–105]. Here again, the effect of absence of soldier personality appears to at least partially prevail over that of nonobesity and gives rise to some components of metabolic syndrome.

Other way to resolve between the two hypotheses is to look at the time course by which exercises improve insulin resistance. Both the hypotheses agree that exercises are insulin sensitizing. If the classical hypothesis is correct, exercises should reduce body fat first, and only then insulin sensitivity will increase. If exercises increase insulin sensitivity without fat reduction, it is not compatible with the obesity-centered hypothesis. The new hypothesis predicts that the beneficial effects of exercise would be seen independent of weight reduction. Weight reduction may accompany exercises, but the beneficial effects could appear before, independent of, or even without weight reduction. A number of studies report simultaneous weight loss and improvement in insulin sensitivity which is compatible with both the hypotheses. A number of other studies show that insulin sensitivity can increase substantially by exercise even without loss of weight or total fat [34, 106–108] which demonstrates that exercise has direct effects on insulin sensitivity independent of weight loss.

Another interesting series of data come from the First and Second World Wars and all other long-drawn wars such as Serbian or Sarajevan Wars. A consistent finding is that in all long-drawn wars, there is a substantial reduction in type 2 diabetes and hypertension in the population [109–111]. The classical theory has tried to explain this phenomenon by saying that there are diet changes during war. Unfortunately proponents of this argument have not given diet data in support of this explanation. The interpretation of the new hypothesis is more directly related to war. War facilitates a soldier attitude even among civilians. Although everyone may not be actually fighting as a soldier, there are certain war-induced elements such as increased patriotism; hearing, thinking, and imagining about a battle; and being prepared for a quick action when say an air raid warning siren is heard. This is likely to change the soldier–diplomat balance in everyone's personality which may affect the physiological states. In order to resolve between the two interpretations of wartime suppression of diabetes, an experiment can be visualized. If we have data on wartime diet, a group of volunteer diabetics can be kept on this diet during peacetime to see whether the impact of war is replicated. As far as I know no peacetime dietary intervention has been as dramatic in reducing diabetes and hypertension as the war itself, but this is a difficult comparison because the diet trials have not taken the wartime diet as a reference. An experiment that can enable a valid comparison has not been done. Therefore we will have to accept that wartime data cannot really resolve between the two hypotheses but is certainly compatible with both. It can be seen however that although the importance of diet is continuously emphasized during the last several decades, the health care systems anywhere in the world have been unable to bring down the incidence to levels comparable to the wartime levels. It is worth at least suspecting that war has effects greater than what diet alone can explain.

Any theory on the origins of diabetes cannot be complete without explaining the well-known effects of fetal programming. In Chap. 4, we have seen that the evidence for fetal programming is robust, although the current theory of programming for thriftiness in anticipation of famines is unsatisfactory. We need an alternative that satisfactorily explains the fetal programming effect. The soldier–diplomat dichotomy explains the lifelong effects of fetal programming in the following way.

If there is intrauterine growth retardation (IUGR), all organs are not equally affected. The least affected is the development of the brain, whereas muscle mass, bone length, and visceral mass are substantially reduced. This phenomenon of preserving the brain development at the cost of other tissues is quite well known as brain-sparing effect [112, 113]. When an individual with IUGR is born underweight, the brain weight is near normal, whereas other organs and particularly muscle mass are subnormal [114–116]. A constraint in mammalian development is that muscle cells do not replicate in adult life. The number of muscle cells remains the same, although they become stronger by exercise training. This means that an upper limit on muscle strength is laid in early life. This is true for neurons as well. They too do not increase in number in adult life. However, neuronal development is almost normal in spite of IUGR. Such an individual born with normal brain but lower muscle mass has to compete with others in real life. It can be easily seen that owing to the inborn disadvantage in muscle, this individual is unlikely to be a successful soldier in competitive life. However, since brain development was comparable to others, he/she can be a successful diplomat more easily. Therefore it is natural for an IUGR baby to mold into a diplomat lifestyle from an early age itself. There is no wonder therefore that hyperinsulinemia and insulin resistance appear early in the life for low birth weight individuals. This accounts for a lifelong programming very logically. Since the limit to the number of muscle cells is set in early life, there are lifelong constraints on the choice of behavioral strategy. This leads to a birth time and lifetime relation that is essential for a lifetime programming to evolve. The old theory is very weak here in explaining why birth time conditions result into a lifetime commitment to a particular physiological state. We have discussed this limitation at length before (Chap. 4). The new hypothesis is certainly more logical than the old one in explaining the birth weight effect on lifelong metabolic programming.

There is a need to go one step backward and examine what causes intrauterine growth retardation. The most predominant cause is believed to be undernutrition of the mother. This may be true but may not be the only cause of IUGR. An alternative possibility is maternal growth factors. Maternal epidermal growth factor (EGF) levels have been shown to affect early fetal growth [117–119]. As we will see in the following chapter, EGF has a strong behavioral connection, and it is one of the most important molecular links between behavior and metabolism. This raises the possibility that maternal behavior could be a contributor to IUGR in addition to or perhaps independent of maternal nutrition.

This also gives us greater insight into what comes first, hyperinsulinemia or insulin resistance. High insulin levels are adaptive for an IUGR baby for two reasons. One is that a compensatory rapid growth is observed in almost all cases of low birth weight if sufficient nutrition is available during infancy. Insulin is a growth-promoting hormone, and in combination with pituitary growth hormone, it facilitates rapid compensatory growth. Higher level of insulin is therefore a requirement for compensatory growth. The other reason is that for an IUGR baby, investment in development of the brain is more crucial. The brain is a developmental priority, and insulin may have an important role in brain sparing. Also an IUGR baby is probabilistically heading for a diplomat life owing to the lifelong limitation on muscle mass. Insulin as well as proinsulin is important in neuronal development [120–122]. Therefore high levels of insulin and presumably proinsulin as well could be adaptive for IUGR infants. Interestingly, but not coincidently, even C-peptide, which is produced during conversion of proinsulin to insulin, has a role in the cognitive brain function, and deficiency of C-peptide can lead to hippocampal neuronal loss [123, 124]. It appears that every component of the insulin gene has a contribution to the development and maintenance of cognitive function. This process may begin in fetal life itself since IUGR babies are often born with higher β cell mass and insulin production with a higher proportion of proinsulin. The problem with high insulin production is that it may lead to dangerously low levels of glucose, and in order to avoid hypoglycemia, insulin resistance is necessary. Mechanisms have therefore evolved to compensate hyperinsulinemia by insulin resistance. We have seen in Chap. 3 earlier that there is stronger evidence for hyperinsulinemia being primary and insulin resistance compensating it rather than the reverse as believed by the old theory. Since muscle insulin resistance has to accompany hyperinsulinemia and insulin resistance arrests protein buildup in muscle, catch-up growth cannot build strong muscle mass and results into greater proportion of fat instead. Thus catch-up growth and diplomat personality are mechanistically handcuffed, the handcuffing being done by hyperinsulinemia.

A critical question here is whether there would be a trade-off between physical strength and cognitive functions, whether there is any evidence for such a trade-off, and will it result into a negative correlation between physical strength and cognitive performance. This is critical to discuss because there would be a temptation to look for such a negative correlation and use it as evidence for or against the soldier–diplomat hypothesis. We need to emphasize that a trade-off is inevitable when resources are limiting. If there is substantial nutritional provision and supporting environment for fetal and infant development, both physical and brain development can be equally good. If both can be there in the same individual, needless to say, natural selection would favor it. This would actually give rise to a positive correlation between physical strength and cognitive performance. Only after removing this effect, any evidence for a trade-off will be apparent. Therefore it is certainly worth looking for a trade-off after removing the effect of early development which can perhaps be done using birth weight, birth morphometry, or such carefully chosen set of parameters reflecting intrauterine development. I know of a yet unpublished study that found higher cognitive abilities in insulin-resistant children, but this trend was significant only after adjusting for the effect of birth weight.

Although the concept of such a trade-off is almost nonexistent among physiologists, there are some suggestive pieces of evidence coming from experimental psychology. A weird-looking experiment in Netherlands demonstrated that the physical act of stepping back enhances cognitive performance as compared to the act of stepping forward [125]. Perhaps stepping back is more than a metaphor and is interpreted by our brain as a retreat in a combat. According to our theory having to retreat should be a signal to lower the soldier component and rely more on the diplomat component. A cognitive enhancement is adaptive in this situation. This crazy experiment is not a freak. Other similar experiments have tested some other approach-versus-avoidance actions of the body and demonstrated that avoidance actions lead to improved cognitive performance [126, 127]. This makes a very good sense to the hawk–dove or soldier–diplomat dichotomy. If you cannot win by being a hawk or a soldier, it is better to try and achieve success by being a good diplomat. This means being more careful, thoughtful, risk avoider, and judgment taker. Other studies have shown that happy mood or positive emotions enhance flexibility and exploratory behavior, whereas sad mood makes the decision making process in accordance with a set of logical rules [128]. These experiments are indicative of the existence of trade-offs very similar to the one expected by our hypothesis. More research is nevertheless needed to elucidate the conditions for the existence of a soldier–diplomat trade-off and the nature of the trade-off at behavioral as well as physiological level.

Why an Adaptation Turns Pathological

We have been talking about two types of adaptive behavioral strategies called soldier and diplomat. Both are adaptations to two different socioecological conditions. The most important question is if they are adaptations, why they give rise to a series of disorders? Under what conditions can an adaptation turn pathological? Another subtle question of interest here is which part of the soldier to diplomat transition is actually the stimulus for physiological changes. Is it more of a deficiency of soldier components or an adoption of diplomat components that drives the physiological change? We will have to wait for an answer to this question. There is certainly evidence to say that deficiency of soldier components matter since even partial fulfillment of the deficient components in the form of exercises has a positive effect. What we do not know is whether the diplomat components such as ability of social manipulations have any independent effect too. More research is needed to probe into this question.

There are two possible explanations for the pathological effects of an adaptive change:

1. Supernormal stimulus and supernormal response: One possible reason is that we are stretching the adaptation beyond the range for which it evolved. We have seen in Chap. 1 with a relevant example how and why this can happen. It is not difficult to visualize how the modern urban life is a supernormal stimulus, part of which is discussed above in the context of minor injuries. More ways of being supernormal will be discussed in subsequent chapters. Since the stimulus is supernormal, the response is also supernormal. This supernormality of today's soldier deficiency and diplomat emphasis turns an evolved adaptive response pathological.
2. A nonsoldier–nondiplomat lifestyle? What if there is a deficiency of diplomat components too in addition to soldier components? What if someone has given up nurturing physical strength, activity, and aggression, but this is not accompanied by increase in cognitive activities? There can be dual deficiency. The modern lifestyle is leading us here as well. With increasing electronic devices for memory, path finding, or numerical calculations, the demand on our cognitive activities may also be on a decline. Many studies have already started demonstrating the cognitive changes accompanying "googling" [129]. What would be the effects of such a dual deficiency? Is it possible that what is turning pathological is not the soldier deficiency alone but a dual deficiency? If diplomat behavior is an evolved response, the physiological mechanisms could not have evolved to lead to a disease. Perhaps

the disorder may be because we are neither soldiers nor diplomats today. Again we do not have a definitive answer right away, but we will move closer to possible answers as we see more of the proximate mechanisms that change the physiological states. This we will start doing from the next chapter.

References

1. Sapolsky RM (2004) Social status and health in humans and other animals. Annu Rev Anthropol 33:393–418
2. Maggioncalda AN, Czekala NM, Sapolsky RM (2002) Male orangutan subadulthood: a new twist on the relationship between chronic stress and developmental arrest. Am J Phys Anthropol 118:25–32
3. Fox E (2002) Female tactics to reduce sexual harassment in the Sumatran orangutan (*Pongo pygmaeus abelii*). Behav Ecol Sociobiol 52:93–101
4. Utami SS, Goossens B, Bruford MW, de Ruiter JR, van Hooff JARAM (2002) Male bimaturism and reproductive success in Sumatran orangutans. Behav Ecol 13:643–652
5. Maggioncalda AN, Sapolsky RM, Czekala NM (1999) Reproductive hormone profiles in captive male orangutans: implications for understanding developmental arrest. Am J Phys Anthropol 109:19–32
6. Maggioncalda AN, Czekala NM, Sapolsky RM (2000) Growth hormone and thyroid stimulating hormone concentrations in captive male orangutans: implications for understanding developmental arrest. Am J Primatol 50:67–76
7. Lynch J et al (2004) Is income inequality a determinant of population health? Part 1. A systematic review. Milbank Q 82:5–99
8. Lynch J, Smith GD, Harper S, Hillemeier M (2004) Is income inequality a determinant of population health? Part 2. U.S. National and regional trends in income inequality and age- and cause-specific mortality. Milbank Q 82:355–400
9. Tan TT et al (1993) Prevalence of NIDDM and impaired glucose tolerance in aborigines and Malays in Malaysia and their relationship to sociodemographic, health, and nutritional factors. Diabetes Care 16:68–75
10. Mohan V et al (2001) Intra-urban differences in the prevalence of the metabolic syndrome in southern India—the Chennai Urban Population Study (CUPS No. 4). Diabet Med 18:280–287
11. Ramachandran A et al (2001) High prevalence of diabetes and impaired glucose tolerance in India: National Urban Diabetes Survey. Diabetologia 44:1094–1101
12. Ramachandran A, Jali MV, Mohan V, Snehalatha C, Viswanathan M (1988) High prevalence of diabetes in an urban population in south India. Brit Med J 297:587–590
13. von Rueden C, Gurven M, Kaplan H (2011) Why do men seek status? Fitness payoffs to dominance and prestige. Proc Biol Sci 278:2223–2232
14. Bartolomucci A et al (2009) Metabolic consequences and vulnerability to diet-induced obesity in male mice under chronic social stress. PLoS One 4:4331
15. Sanghez V et al (2011) Vulnerability to chronic subordination stress-induced depression-like disorders in adult 129SvEv male mice. Prog Neuropsychopharmacol Biol Psychiatr 35:1461–1471
16. Solomon MB, Foster MT, Bartness TJ, Huhman KL (2007) Social defeat and footshock increase body mass and adiposity in male Syrian hamsters. Am J Physiol Regulat Integrat Comp Physiol 292:283–290
17. Shively CA, Register TC, Clarkson TB (2009) Social stress, visceral obesity, and coronary artery atherosclerosis: product of a primate adaptation. Am J Primatol 71:742–751
18. Moles A et al (2006) Psychosocial stress affects energy balance in mice: modulation by social status. Psychoneuroendocrinology 31:623–633
19. Dallman MF et al (2004) Minireview: glucocorticoids—food intake, abdominal obesity, and wealthy nations in 2004. Endocrinology 145:2633–2638
20. Troop NA, Baker AH (2008) The specificity of social rank in eating disorder versus depressive symptoms. Eat Disord 16:331–341
21. Harrell ZAT, Jackson B (2008) Thinking fat and feeling blue: eating behaviors, ruminative coping, and depressive symptoms in college women. Sex Roles 58:658–665
22. Agardh E et al (2007) Socio-economic position at three points in life in association with type 2 diabetes and impaired glucose tolerance in middle-aged Swedish men and women. Int J Epidemiol 36:84–92
23. Agardh EE et al (2003) Work stress and low sense of coherence is associated with type 2 diabetes in middle-aged swedish women. Diabetes Care 26:7 19–724
24. Mann GV, Spoerry A (1974) Studies of a surfactant and cholesteremia in the Maasai. Am J Clin Nutr 27:464–469
25. Day J, Carruthers M, Bailey A, Robinson D (1976) Anthropometric, physiological and biochemical differences between urban and rural Maasai. Atherosclerosis 23:357–361
26. Brown GW (1993) Maasai diet. Lancet 341:377
27. Gibney MJ, Burstyn PG (1980) Milk, serum cholesterol, and the Maasai: a hypothesis. Atherosclerosis 35:339–343
28. Stannard SR, Johnson NA (2004) Insulin resistance and elevated triglyceride in muscle: more important for survival than 'thrifty' genes? J Physiol 554: 595–607

29. Beckerman S et al (2009) Life histories, blood revenge, and reproductive success among the Waorani of Ecuador. Proc Natl Acad Sci USA 106:8134–8139
30. Deaux K, Snyder M (eds) (2012) *The* Oxford handbook of personality and social psychology. Oxford University Press, Oxford
31. Fernstrom JD, Wurtman RJ (1971) Brain serotonin content: increase following ingestion of carbohydrate diet. Science 174:1023–1025
32. Mackenzie RG, Trulson ME (1978) Effects of insulin and streptozotocin-induced diabetes on brain tryptophan and serotonin metabolism in rats. J Neurochem 30:205–211
33. Cleare AJ, Bond AJ (1997) Does central serotonergic function correlate inversely with aggression? A study using -fenfluramine in healthy subjects. Psychiat Res 69:89–95
34. Belsare PV et al (2010) Metabolic syndrome: aggression control mechanisms gone out of control. Med Hypotheses 74:578–589
35. Clotfelter ED, O'Hare EP, McNitt MM, Carpenter RE, Summers CH (2007) Serotonin decreases aggression via 5-HT1A receptors in the fighting fish *Betta splendens*. Pharmacol Biochem Behav 87:222–231
36. Lepage O, Larson ET, Mayer I, Winberg S (2005) Serotonin, but not melatonin, plays a role in shaping dominant-subordinate relationships and aggression in rainbow trout. Horm Behav 48:233–242
37. Duman EA, Canli T (2010) Social behavior and serotonin. In: Müller CP, Jacobs BL (eds) Handbook of the behavioral neurobiology of serotonin, vol 21. Elsevier, Amsterdam, pp 449–456
38. Winberg S, Øverli Ø, Lepage O (2001) Suppression of aggression in rainbow trout (*Oncorhynchus mykiss*) by dietary l-tryptophan. J Exp Biol 204:3867–3876
39. Hillbrand M, Spitz RT (1999) Cholesterol and aggression. Aggress Violent Behav 4:359–370
40. Kaplan JR, Klein KP, Manuck SB (1997) Cholesterol meets Darwin: public health and evolutionary implications of the cholesterol-serotonin hypothesis. Evol Anthropol Issues News Rev 6:28–37
41. Pentürk S, Yalçin E (2003) Hypocholesterolaemia in dogs with dominance aggression. J Vet Med Series A 50:339–342
42. Buydens-Branchey L, Branchey M, Hudson J, Fergeson P (2000) Low HDL cholesterol, aggression and altered central serotonergic activity. Psychiat Res 93:93–102
43. Erickson MT (1997) Lowered serum cholesterol, famine and aggression: a Darwinian hypothesis. Soc Sci Inform 36:211–222
44. Kaplan JR, Manuck SB, Shively C (1991) The effects of fat and cholesterol on social behavior in monkeys. Psychosom Med 53:634–642
45. Mizuno TM et al (1998) Hypothalamic pro-opiomelanocortin mRNA is reduced by fasting and [corrected] in ob/ob and db/db mice, but is stimulated by leptin. Diabetes 47:294–297
46. Kaplan JR, Fontenot MB, Manuck SB, Muldoon MF (1996) Influence of dietary lipids on agonistic and affiliative behavior in *Macaca fascicularis*. Am J Primatol 38:333–347
47. Olson MB et al (2008) Lipid-lowering medication use and aggression scores in women: a Report from the NHLBI-sponsored WISE study. J Womens Health (Larchmt) 17:187–194
48. Lorrain DS, Riolo JV, Matuszewich L, Hull EM (1999) Lateral hypothalamic serotonin inhibits nucleus accumbens dopamine: implications for sexual satiety. J Neurosci 19:7648–7652
49. Lorrain DS, Matuszewich L, Friedman RD, Hull EM (1997) Extracellular serotonin in the lateral hypothalamic area is increased during the postejaculatory interval and impairs copulation in male rats. J Neurosci 17:9361–9366
50. Tuominen JA et al (1996) Postmarathon paradox: insulin resistance in the face of glycogen depletion. Am J Physiol Endocrinol Metab 270:336–343
51. Kirwan JP et al (1992) Eccentric exercise induces transient insulin resistance in healthy individuals. J Appl Physiol 72:2197–2202
52. Asp S, Daugaard JR, Kristiansen S, Kiens B, Richter EA (1996) Eccentric exercise decreases maximal insulin action in humans: muscle and systemic effects. J Physiol 494:891–898
53. Asp S, Daugaard JR, Richter EA (1995) Eccentric exercise decreases glucose transporter GLUT4 protein in human skeletal muscle. J Physiol 482:705–712
54. Del Aguila LF, Claffey KP, Kirwan JP (1999) TNF-α impairs insulin signaling and insulin stimulation of glucose uptake in C2C12muscle cells. Am J Physiol Endocrinol Metab 276:849–855
55. Del Aguila LF et al (2000) Muscle damage impairs insulin stimulation of IRS-1, PI 3-kinase, and Akt-kinase in human skeletal muscle. Am J Physiol Endocrinol Metab 279:206–212
56. Costill DL et al (1990) Impaired muscle glycogen resynthesis after eccentric exercise. J Appl Physiol 69:46–50
57. Asp S, Richter EA (1996) Decreased insulin action on muscle glucose transport after eccentric contractions in rats. J Appl Physiol 81:1924–1928
58. Black PR, Brooks DC, Bessey PQ, Wolfe RR, Wilmore DW (1982) Mechanisms of insulin resistance following injury. Ann Surg 196:420–435
59. Ikezu T, Okamoto T, Yonezawa K, Tompkins RG, Martyn JAJ (1997) Analysis of thermal injury-induced insulin resistance in rodents. J Biol Chem 272:25289–25295
60. Zhang Q et al (2005) Molecular mechanism(s) of burn-induced insulin resistance in murine skeletal muscle: role of IRS phosphorylation. Life Sci 77:3068–3077
61. Shangraw RE et al (1989) Differentiation between septic and postburn insulin resistance. Metabolism 38:983–989
62. Yki-Jarvinen H, Sammalkorpi K, Koivisto VA, Nikkila EA (1989) Severity, duration, and mechanisms of

insulin resistance during acute infections. J Clin Endocrinol Metab 69:317–323
63. Chawla A, Nguyen KD, Goh YPS (2011) Macrophage-mediated inflammation in metabolic disease. Nat Rev Immunol 11:738–749
64. Lang CH (1992) Sepsis-induced insulin resistance in rats is mediated by a β-adrenergic mechanism. Am J Physiol Endocrinol Metab 263:703–711
65. Greisen J et al (2001) Acute pain induces insulin resistance in humans. Anesthesiology 95:578–584
66. Thorell A, Efendic S, Gutniak M, Häggmark T, Ljungqvist O (1994) Insulin resistance after abdominal surgery. Brit J Surg 81:59–63
67. Soop M, Nygren J, Myrenfors P, Thorell A, Ljungqvist O (2001) Preoperative oral carbohydrate treatment attenuates immediate postoperative insulin resistance. Am J Physiol Endocrinol Metab 280:576–583
68. Thorell A et al (1999) Surgery-induced insulin resistance in human patients: relation to glucose transport and utilization. Am J Physiol Endocrinol Metab 276:754–761
69. Barzilay JI et al (2009) Insulin resistance is associated with decreased quadriceps muscle strength in nondiabetic adults aged ≥70 years. Diabetes Care 32:736–738
70. Abbatecola AM et al (2005) Insulin resistance and muscle strength in older persons. J Gerontol A Biol Sci Med Sci 60:1278–1282
71. Benson AC, Torode ME, Singh MAF (2006) Muscular strength and cardiorespiratory fitness is associated with higher insulin sensitivity in children and adolescents. Int J Pediatr Obes 1:222–231
72. Karelis AD et al (2007) Association of insulin sensitivity and muscle strength in overweight and obese sedentary postmenopausal women. Appl Physiol Nutr Metab 32:297–301
73. Bjorklund DF, Hubertz MJ, Reubens AC (2004) Young children's arithmetic strategies in social context: how parents contribute to children's strategy development while playing games. Int J Behav Dev 28:347–357
74. Pinker S (2011) The better angels of our nature: why violence has declined. Viking Adult, New York, NY
75. Zhao W-Q, Chen H, Quon MJ, Alkon DL (2004) Insulin and the insulin receptor in experimental models of learning and memory. Eur J Pharmacol 490:71–81
76. Gerozissis K (2010) Brain insulin: regulation, mechanisms of action and functions. Cell Mol Neurobiol 23:1–25
77. Kern W, Born J, Fehm HL (2002) Role of insulin in Alzheimer's disease: approaches emerging from basic animal research and neurocognitive studies in humans. Drug Dev Res 56:511–525
78. Kern W et al (2001) Improving influence of insulin oncognitivefunctionsinhumans.Neuroendocrinology 74:270–280
79. Benedict C et al (2004) Intranasal insulin improves memory in humans. Psychoneuroendocrinology 29:1326–1334
80. Benedict C, Kern W, Schultes B, Born J, Hallschmid M (2008) Differential sensitivity of men and women to anorexigenic and memory-improving effects of intranasal insulin. J Clin Endocrinol Metab 93:1339–1344
81. Ryan CM, Geckle M (2000) Why is learning and memory dysfunction in type 2 diabetes limited to older adults? Diabet Metab Res Rev 16:308–315
82. Isaac V, Sim S, Chee M (2011) Adverse associations between visceral adiposity, brain structure, and cognitive performance in healthy elderly. Front Ag Neurosci 3:12
83. Kerwin DR et al (2010) The cross-sectional relationship between body mass index, waist–hip ratio, and cognitive performance in postmenopausal women enrolled in the Women's Health Initiative. J Am Geriatr Soc 58:1427–1432
84. Hallschmid M et al (2008) Towards the therapeutic use of intranasal neuropeptide administration in metabolic and cognitive disorders. Regul Pept 149:79–83
85. Schmidt H, Kern W, Giese R, Hallschmid M, Enders A (2008) Intranasal insulin to improve developmental delay in children with 22q13 deletion syndrome: an exploratory clinical trial. J Med Genet 46:217–222
86. Hogan MJ et al (2011) Cerebellar brain volume accounts for variance in cognitive performance in older adults. Cortex 47:441–450
87. Esposito K, Giugliano D (2004) The metabolic syndrome and inflammation: association or causation? Nutr Metab Cardiovasc Dis 14:228–232
88. Monteiro R, Azevedo I (2010) Chronic inflammation in obesity and the metabolic syndrome. Mediat Inflamm 2010:1–10
89. Sutherland JP, McKinley B, Eckel RH (2004) The metabolic syndrome and inflammation. Metab Syndr Relat Disord 2:82–104
90. Boodhwani M, Sellke FW (2009) Therapeutic angiogenesis in diabetes and hypercholesterolemia: influence of oxidative stress. Antioxid Redox Signal 11:1945–1959
91. Facchiano F et al (2006) Glycated fibroblast growth factor-2 is quickly produced in vitro upon low-millimolar glucose treatment and detected in vivo in diabetic mice. Mol Endocrinol 20:2806–2818
92. Nakagawa T, Kosugi T, Haneda M, Rivard CJ, Long DA (2009) Abnormal angiogenesis in diabetic nephropathy. Diabetes 58:1471–1478
93. Hillbrand M et al (2005) Serum cholesterol concentrations and non-physical aggression in healthy adults. J Behav Med 28:295–299
94. Craig M, Slauch JM (2009) Phagocytic superoxide specifically damages an extracytoplasmic target to inhibit or kill *Salmonella*. PLoS One 4:4975
95. Shetty K (1988) Diabetes in the armed forces. Int J Diabetes Develop Ctries 8:3–11
96. World MJ (2008) Diabetes mellitus and the armed forces. Brit J Diabetes Vasc Dis 8:55–60
97. Paris RM, Bedno SA, Krauss MR, Keep LW, Rubertone MV (2001) Weighing in on type 2 diabetes in the military. Diabetes Care 24:1894–1898

98. Matsuzawa Y, Shimomura I, Nakamura T, Keno Y, Tokunaga K (1994) Pathophysiology and pathogenesis of visceral fat obesity. Diabetes Res Clin Pract 24(Suppl):S111–116
99. Nesto RW (2005) Obesity a major component of the metabolic syndrome. Tex Heart Inst J 32:387–389
100. Bailer UF et al (2004) Altered 5-HT(2A) receptor binding after recovery from bulimia-type anorexia nervosa: relationships to harm avoidance and drive for thinness. Neuropsychopharmacology 29: 1143–1155
101. Bailer U et al (2007) Exaggerated 5-HT1A but normal 5-HT2A receptor activity in individuals ill with anorexia nervosa. Biol Psychiatry 61:1090–1099
102. Nestel PJ (1974) Cholesterol metabolism in anorexia nervosa and hypercholesterolemia. J Clin Endocrinol Metab 38:325–328
103. Feillet F et al (2000) Plasma cholesterol and endogenous cholesterol synthesis during refeeding in anorexia nervosa. Clin Chim Acta 294:45–56
104. Mordasini R, Klose G, Greten H (1978) Secondary type II hyperlipoproteinemia in patients with anorexia nervosa. Metabolism 27:71–79
105. Kumai M, Tamai H, Fujii S, Nakagawa T, Aoki T (1988) Glucagon secretion in anorexia nervosa. Am J Clin Nutr 47:239–242
106. Duncan GE et al (2003) Exercise training, without weight loss, increases insulin sensitivity and postheparin plasma lipase activity in previously sedentary adults. Diabetes Care 26:557–562
107. Ross R (2003) Does exercise without weight loss improve insulin sensitivity? Diabetes Care 26: 944–945
108. van der Heijden GJ, Wang ZJ, Chu Z, Toffolo G, Manesso E, Sauer PJJ, Sunehag AL (2010) Strength exercise improves muscle mass, insulin sensitivity in obese youth: abstract and introduction. Med Sci Sports Exerc 42:1973–1980
109. Westlund K (1966) Incidence of diabetes mellitus in Oslo, Norway 1925 to 1954. Br J Prev Soc Med 20:105–116
110. Kulenovic I, Robertson A, Grujic M, Suljevic E, Smajkic A (1996) The impact of war on Sarajevans with non-insulin-dependent diabetes mellitus. Eur J Public Health 6:252–256
111. Hermanides J, Belqhazi L, Michels RPJ, Hoekstra JBL (2008) Lower incidence of type 2 diabetes mellitus with changes in lifestyle: clues from World War II. Ned Tijdschr Geneeskd 152:2415–2417
112. Lumbers ER, Yu Z, Gibson KJ (2001) The selfish brain and the barker hypothesis. Clin Exp Pharmacol Physiol 28:942–947
113. Watve MG, Yajnik CS (2007) Evolutionary origins of insulin resistance: a behavioral switch hypothesis. BMC Evol Biol 7:61
114. Schain RJ, Watanabe KS (1973) Effects of undernutrition on early brain growth in the rabbit. Exp Neurol 41:366–370
115. Winick M, Brasel JA, Velasco E, Rosso P (1974) Effects of early nutrition on growth of the central nervous system. Birth Defects Orig Artic Ser 10:29–36
116. Winick M (1974) Malnutrition and the developing brain. Res Publ Assoc Res Nerv Ment Dis 53:253–262
117. Lindqvist P, Grennert L, Maršál K (1999) Epidermal growth factor in maternal urine—a predictor of intrauterine growth restriction? Early Hum Dev 56:143–150
118. Kamei Y et al (1999) Maternal epidermal growth factor deficiency causes fetal hypoglycemia and intrauterine growth retardation in mice: possible involvement of placental glucose transporter GLUT3 expression. Endocrinology 140:4236–4243
119. Popliker M et al (1987) Onset of endogenous synthesis of epidermal growth factor in neonatal mice. Dev Biol 119:38–44
120. Diaz B, Serna J, De Pablo F, de la Rosa EJ (2000) In vivo regulation of cell death by embryonic (pro)insulin and the insulin receptor during early retinal neurogenesis. Development 127:1641–1649
121. Hernández-Sánchez C, Mansilla A, Rosa EJ, Pablo F (2006) Proinsulin in development: new roles for an ancient prohormone. Diabetologia 49:1142–1150
122. de Pablo F, de la Rosa EJ (1995) The developing CNS: a scenario for the action of proinsulin, insulin and insulin-like growth factors. Trends Neurosci 18:143–150
123. Sima AAF, Li Z (2005) The effect of C-peptide on cognitive dysfunction and hippocampal apoptosis in type 1 diabetic rats. Diabetes 54:1497–1505
124. Sima AA et al (2001) C-peptide prevents and improves chronic type I diabetic polyneuropathy in the BB/Wor rat. Diabetologia 44:889–897
125. Koch S, Holland RW, Hengstler M, van Knippenberg A (2009) Body locomotion as regulatory process. Psychol Sci 20:549–550
126. Koch S, Holland RW, van Knippenberg A (2008) Regulating cognitive control through approach-avoidance motor actions. Cognition 109:133–142
127. Fayant M-P, Muller D, Nurra C, Alexopoulos T, Palluel-Germain R (2011) Moving forward is not only a metaphor: approach and avoidance lead to self-evaluative assimilation and contrast. J Exp Soc Psychol 47:241–245
128. de Vries M, Holland RW, Corneille O, Rondeel E, Witteman CLM (2010) Mood effects on dominated choices: positive mood induces departures from logical rules. J Behav Dec Making 81:74–81
129. Sparrow B, Liu J, Wegner DM (2011) Google effects on memory: cognitive consequences of having information at our fingertips. Science 333:776–778

7 The Physiology of Aggression

So far I have been arguing at the level of ultimate reasoning to address the question why physical activity, aggression, and other components of the soldier behavior syndrome should have a negative relationship with components of the insulin resistance syndrome. Now we need to look at the proximate mechanisms which execute this relationship. In order to understand the mechanisms and pathways involved in this connection, we need to know the physiology of aggression–dominance syndrome first.

In traditional thinking testosterone has been associated with aggressive behavior. Testosterone is believed to facilitate aggression, and there is sufficient evidence to support this. It is predominantly a male aggression hormone. Since, generally, aggression research is male biased, the importance of testosterone appears to be overplayed. It also needs to be appreciated that the role of testosterone in aggression is contextual. A high-testosterone individual is not always aggressive but rather is more likely to be aggressive in certain contexts. It is much more important to realize that it is not testosterone alone that modulates aggression. Associating testosterone with aggression is a sign of a typical old-fashioned reductionism in biology. Today we have better tools with which we can understand the wide array of molecules involved in aggression. For example, Chiavegatto [1] lists knockout mouse models with altered aggression. This list has 36 genes and is by no means a complete list of genes involved in aggression. The list of knockouts with altered aggression illustrates a number of things that were not appreciated earlier. For example, not only androgen receptor mutants lose aggression, estrogen receptor α mutants and aromatase (the enzyme that converts testosterone to estradiol) mutants also lose aggression indicating an important role for estrogens in mediating aggression as well.

In addition to gene knockouts, another source of data is studies on gene expression in animals selected for aggression versus that in wild types. A pleasant surprise here is that there are some data on maternal aggression in females too [2]. These studies tell us that there are a large number of players with different roles in the physiology of aggression, and the testosterone-centered picture is too narrow and piecemeal. Maternal aggression, for example, is not chiefly testosterone driven but has been demonstrated to have the same beneficial effects against at least some of the components of metabolic syndrome such as hypertension [3]. Nevertheless, since there is a large volume of work on testosterone, we will have to continue referring to it every now and then, but we need to understand that it is not the only aggression hormone and also that aggression is not a male-specific behavior.

The pattern I discovered while scanning the literature on aggression was that majority of the aggression-associated molecules are also associated with metabolic syndrome in some way or the other. For example, out of the 36 genes listed by Chiavegatto [1], 28 are connected with metabolic syndrome in some way or the other. The total number of signal molecules that link aggression

M. Watve, *Doves, Diplomats, and Diabetes: A Darwinian Interpretation of Type 2 Diabetes and Related Disorders*,
DOI 10.1007/978-1-4614-4409-1_7,

to metabolic syndrome is about 70, and the list is growing. Here I include hormones, neuropeptides, growth factors, and transcriptional regulators as well as dietary signals under the generic term "signal molecule." For interested readers a complete list of these signal molecules with their nature of association with aggression and with metabolic syndrome with select references is compiled in Appendix I. The width of this association by itself is astonishing. It is difficult to believe that this is nothing more than a coincidence. It certainly points to a highly evolved link between the aggression pathways and the endobolic (i.e., endocrine and metabolic) pathways related to metabolic syndrome. Unless this association has evolved over the millennia, it is unlikely to involve such a wide array of links. Perhaps it is possible to find the evolutionary origins of this link. So far I have been unable to gather sufficient data to trace the origin, but there are indications that it is very old and conserved. We find this link in *Drosophila*, for example, which generally is a nonaggressive species. Fruit flies do not have any lethal weapons, and therefore, their fights are never fatal. Nevertheless, mild male–male aggression is observed during courtship over access to the females. A gene called *sugarbabe* which downregulates insulin production in the fly is overexpressed during courtship behavior in the fly [4]. Since in these flies there is little aggression apart from courtship, it is difficult to separate aggression from courtship. It is very likely therefore that *sugarbabe* overexpression is actually linked to courtship-related aggression. This raises a possibility that the connection between sex, aggression, and insulin may be very ancient and has been conserved during evolution. This needs to be thoroughly investigated.

Belsare et al. [5] showed that most molecules associated positively with aggression have a negative association with metabolic syndrome, and the ones negatively associated with aggression have a pro-obesity or pro-insulin-resistance action. This reverse association is highly significant. However, although I was a contributor to this paper, I now think that the categorization of positive and negative associations with aggression is too naïve. In complex control loops it is difficult to assign positive or negative signs to associations. For example, if A directly upregulates B and B through a negative feedback loop controls A, do we call the association of A and B as positive or negative? Endorphins or brain-derived neurotrophic factor (BDNF) are related to aggression in a negative feedback loop as we will see a little later. Aggressive encounter upregulates endorphin or BDNF, but both of them control aggression. This relationship is therefore positive in one direction and negative in the other. Until the exact role of a signal molecule is elucidated sufficiently well, it is difficult to infer anything from positive and negative associations. A number of molecules in the aggression regulation circuits have such relationships as we will see soon. Therefore rather than looking at positive and negative associations, it is better to look at the specific roles of each molecule and interpret them in the right context which I will attempt to do in this chapter. Although at this stage we do not know the specific role of every molecule, the fact remains that a large number of molecules appear to link aggression with components of the metabolic syndrome in some way or the other, and therefore, it is not possible to understand metabolic syndrome without understanding aggression.

The Bridge Between Aggression–Dominance and Metabolic Syndrome: Some Important Pillars

A. Serotonin: The endocrinology of territorial aggression and dominance in animals is well known. Perhaps the single most important molecule central to the suppression of aggression is serotonin in the brain as we have seen in the previous chapters [6–14]. Dominant individuals have low serotonin as compared to the subordinate ones among animal societies [15–19], and an intervention to raise brain serotonin activity in a dominant individual changes its behavior to that of a subordinate [15, 20, 21]. The arrow of causation appears to be dual, i.e., change in serotonin levels affects aggression and aggressive behavior modulates serotonin levels [6, 10]. In humans low serotonin activity is associated with impulsive aggression or violent suicide [8, 16, 22].

Serotonin is also central to metabolic syndrome in a number of ways. The relationship of serotonin with obesity is complex. Serotonin works as a satiety signal and thereby controls hunger. Therefore serotonin action should be antiobesity. Knocking out or blocking serotonin receptors leads to obesity as expected [23, 24]. Serotonin reuptake inhibitors have been proposed as potential antiobesity drugs, and some serotonin reuptake inhibitors have been shown to reduce body fat in short-term studies [25–27]. The effect, however, may be reversed in the long run. Many of the serotonin reuptake inhibitors that raise the standing serotonin levels lead to a paradoxical weight gain and insulin resistance [28–30]. It is possible that serotonin has an antiobesity action in the short run, but chronic elevation of serotonin gives rise to a state that can be called serotonin resistance. The concept of serotonin resistance is used in a slightly different context [31] but is similarly applicable here. Serotonin resistance may lead to obesity and insulin resistance. This is supported by data that demonstrate association of low serotonin responsiveness with insulin resistance [32, 33]. Chronically elevated levels of serotonin in the ventromedial hypothalamus have been shown to induce insulin resistance [34]. Serotonin reuptake inhibitors which increase basal serotonin levels are shown to induce insulin resistance in vitro [35]. Chronic signaling by serotonin along with norepinephrine in the ventromedial hypothalamus increases glucose sensitivity of the pancreatic β cells [36] and thereby increases the production of insulin. Reciprocally hyperinsulinemia and insulin resistance result in elevated serotonin levels [37]. Similar to insulin, leptin also increases serotonin turnover in the brain [38]. Thus there appears to be a mutually supportive mechanism. High levels of insulin and leptin which mark insulin resistance upregulate the serotonin levels, and chronically high serotonin induces insulin resistance. This is a typical positive feedback cycle that stabilizes a state. Serotonin reduces bone mass and strength on the one hand [39–42] and increases memory and cognitive functions on the other [43–48] with the exception of spatial working memory [47, 49, 50]. This is compatible with a warrior to diplomat transition. Serotonin also decreases sexual motivation and function [51, 52]. Thus serotonin seems to do almost everything that happens in the early insulin resistance syndrome and related conditions.

I must admit here that the picture painted above is rather oversimplified. Serotonin has a number of receptors having partly overlapping and partly differential functions. Science is just beginning to understand the roles of different serotonin receptors under different sets of conditions. Therefore a comprehensive picture of the network of effects of serotonin signaling is yet to emerge. Nevertheless, a robust message that everyone will agree with is that it is essential to better understand the aggression suppression hormone serotonin in order to gain sufficient insight into the insulin resistance syndrome.

B. Dopamine: The relation of dopamine with aggression appears to be diametrically opposite to that of serotonin. Dopamine activity is positively associated with aggression, and the arrow of causation also appears to be dual [6, 22, 53]. Dopamine in contrast to serotonin facilitates physical strength. The deletion of DAT gene (dopamine transporter) results in osteopenia and deficiencies in skeletal structure and integrity illustrating the role of dopamine in strengthening the bone [54]. This is compatible with dopamine's proaggression role. It is no surprise then that dopamine has antiobesity and antidiabetic action. In mice, the use of dopamine receptor antagonist is associated with weight gain and insulin resistance, whereas agonists of the D1 and D2 dopamine receptor isoforms decrease food intake and improve insulin sensitivity [55, 56]. Dopamine may be extremely important in the chain of pathways leading to diabetes since diabetes is associated with decreased dopamine activity [57]. Dopamine, in contrast with serotonin, is crucial in spatial learning and memory [47, 58–60]. It is not difficult to visualize that spatial learning and memory are of critical importance to a hunter or warrior. Therefore a proaggression signal enhances spatial memory, whereas an aggression suppression signal enhances verbal memory, which is a diplomat requirement. Therefore serotonin which is a general memory enhancer does not boost spatial memory [47, 49, 50].

C. Cholesterol: Perhaps next to serotonin, cholesterol appears to have a major role in the suppression of aggression. The negative association between cholesterol and physical aggression is quite robust [61–64]. In contrast to physical aggression, nonphysical aggression has a positive association with cholesterol [65]. Cholesterol seems to have not only an association but a causal role in aggression control since cholesterol lowering appears to increase aggression [66–69]. But unlike serotonin, the mechanism by which cholesterol affects behavior is largely unknown. Cholesterol apparently does not cross the blood–brain barrier but still affects serotonin activity in the brain [70, 71]. This has been claimed as an adaptive response under famine conditions where a famine-induced lowering of cholesterol would lower serotonin activity allowing aggressive behavior that might be needed under intense competition for food [72]. But food and fat intake are not the only factor determining the blood cholesterol levels. There appears to be a CNS-mediated control on cholesterol too [73, 74]. It is likely therefore that there are behavioral inputs in the regulation of cholesterol, but not much research has gone into this aspect of cholesterol. There is no need to reemphasize the role of cholesterol in metabolic syndrome. Cholesterol strongly links aggression suppression to metabolic syndrome, but much research is needed to elucidate its precise role in both.

D. Corticosteroids: Similar to cholesterol, low cortisol levels are associated with aggression [75–78] raising the possibility that cortisol may serve as an aggression control signal. This interpretation is muddled by the concept of it being a "stress" hormone. We will devote a separate chapter to discuss the concept of stress, and therefore, I will not discuss it here except a passing remark that I think aggression suppression and risk avoidance are the major roles for which the corticosteroid response has evolved. I will discuss evidence for this statement in relation to stress more elaborately later since it deserves much more space. Cortisol is known to induce insulin resistance [79, 80] and therefore is another potential link between aggression suppression and insulin resistance.

E. Sex hormones: Sex and aggression are positively associated in a wide variety of species including vertebrates and invertebrates [81–83], many sex hormones serving as aggression hormones simultaneously. Dominant and physically aggressive individuals have high testosterone/estradiol levels [84–88]. Testosterone's role in male aggression is well recognized. However, endocrinology of female aggression is not very clearly elucidated. Aggression is equally important in female life as well, although the context, nature, and intensity of aggression could be different from that of the males. There are certain female-specific forms of aggression such as maternal aggression. Unfortunately aggression research is largely centered on male aggression. I will have to repeatedly talk about testosterone in this as well as the following chapters, and contextually it would be more applicable to males. This is a bias that I do not like to have, but it is inevitable with the currently available data. I hope to see more research in the female-specific forms of aggression which would help us remove this bias eventually. Interestingly, in a species of mouse that shows biparental care, males show paternal aggression similar to the maternal aggression in females with much overlap in the mechanisms involved [89]. This could be particularly relevant to the human species which is characterized by biparental care. It appears that the importance of testosterone in aggression is overplayed at the cost of a number of other molecules involved in aggression in both genders. For example, similar to androgen receptors, estrogen receptor knockout mice also lose aggression [1]. But the role of estradiol in aggression is not as clearly recognized as that of testosterone. Nevertheless, there is evidence for an important but complex role of estradiol in aggression [90–95]. Apart from testosterone, dehydroepiandrosterone (DHEA) is an adrenal androgen precursor and is important for the expression of animal aggression in nonbreeding season when gonadal testosterone synthesis is low [96].

Testosterone facilitates glucose transport in muscles [97], reduces oxidative stress and accompanying cell damage [98, 99], and stimulates angiogenesis and vascular regrowth [100]. The multiple health beneficial effects of testosterone have given rise to suggestions that testosterone deficiency may be the prime factor behind metabolic syndrome, although the causal role of testosterone is debated [101]. Estradiol is also shown to promote growth and angiogenesis [102]. Estrogens, especially 17 β-estradiol and estriol, themselves are naturally occurring antioxidants [103]. Owing to their multiple actions, both male and female sex hormones have protective effects against metabolic syndrome in the respective sex [87, 104], although they may have different effects in the other sex [104]. DHEA is also protective against obesity, insulin resistance, and some of the mechanisms of diabetic complications [105–111].

F. Insulin: It is likely that hyperinsulinemia and insulin resistance itself are involved in aggression control. Parameters of insulin sensitivity were shown to be positively correlated to violent suicide [112], an association similar to low cholesterol. Since insulin enhances serotonin response, the aggression suppression action of insulin is likely to be mediated through serotonin. Insulin has a complex regulatory role in aggression, impulsivity, restlessness, and other behavioral characteristics [113], which appears to operate through the effects of insulin on the monoamine dynamics in the brain [114–118]. This role of insulin in the brain monoamine dynamics has been realized recently, and a clear picture is yet to emerge. But we know now that insulin substantially modulates monoamine actions which are important behavioral regulators. I suspect that insulin has a direct role in the suppression of anger, aggression, and impulsivity and hope that this is clearly demonstrated in the near future. On the other hand insulin has a direct positive role in cognitive functions of the brain [119–124].

G. IGF-1: A significant positive association has been shown between IGF-1 mRNA expression and the level of aggression in fish [125]. Dominant pudu males had higher IGF-1 levels than their subordinate pen mates [126]. IGF-1 expression is shown to be suppressed in subordinate baboons [127]. This dominance–aggression-associated molecule has a protective effect against obesity and insulin resistance. IGF-1 reduced hyperphagia, obesity, hyperleptinemia, and hyperinsulinemia in rats [128, 129]. Liver IGF-1-deficient mice showed insulin resistance [130]. IGF-1 also has vascular protective effects [131].

H. Cholecystokinin (CCK): Higher expressions of CCK and CCK2R were associated with aggression in transgenic mice overexpressing progastrin [132]. In normal mice brain, CCK levels elevated following aggressive encounters in adults or a rough and tumble play in adolescents [133]. The antiobesity and antidiabetic effects of CCK are well demonstrated [134–136]. CCK promotes growth of endocrine pancreas [137] and enhances wound healing [138].

I. Ghrelin: Ghrelin levels are negatively associated with cholesterol, and violent individuals have low cholesterol and leptin along with elevated ghrelin [139]. Although ghrelin is known to enhance appetite [140, 141] and thereby expected to increase obesity, low plasma ghrelin levels are associated with type 2 diabetes, insulin resistance, and hypertension [142, 143].

J. EGF, NGF and other growth factors: Aggression stimulates salivary epidermal growth factor (EGF) secretion [144–146]. Testosterone induces EGF synthesis but not secretion [147, 148]. EGF release is under autonomic nervous control [148–150]. Therefore a combination of testosterone and sympathetic stimulation, a characteristic of aggressive behavior, is necessary for EGF release. The adaptive logic behind why salivary glands are the major sources of EGF and why aggression stimulates their secretion is that aggression anticipates injuries and EGF is needed in wound healing. Since animals lick wounds, salivary glands are the best places for the secretion of these growth factors. In T2D, there is a deficiency of EGF [151–153] which is likely to be due to deficiency of aggressive behavior. This is one of the reasons for delayed wound healing in diabetes. Apart from wound healing, EGF has an important role in β cell regeneration too [154], and patents have been filed for therapeutic use of EGF in stimulating β cell growth [155]. EGF also plays a role in retinal and neuronal integrity [156, 157]. The aggression-stimulated EGF thus has multiple roles in protecting against diabetes.

Similar to EGF, the secretion of nerve growth factor (NGF) is also stimulated by aggression [158–163] or adventure [164]. Apart from a general role in wound healing and tissue regeneration, NGF has specific roles in maintaining nerve integrity as well as retinal integrity, and deficiency of NGF is one of the suspected causes of diabetic neuropathies [165, 166] and retinopathy [167].

Apart from EGF and NGF, a number of other growth factors play a role in aggression, injuries, and β cell regeneration that have not attracted sufficient research as yet. They include platelet-derived growth factor (PDGF), fibroblast growth factor (FGF), growth hormone (GH), prolactin , placental lactogen, parathyroid hormone-related peptide (PTHrP), gastrin, gastrin-releasing peptide (GRP), glucagon-like peptide, and betacellulin. As we will see more elaborately in a later chapter, many of them are involved in wound healing and are synthesized locally or systemically after even minor injuries. Interestingly most of the growth factors needed for wound healing are also involved in β cell regeneration. This is a factor linking aggression–dominance syndrome to diabetes in a different way. Hawk or warrior individuals are more prone to injuries, and owing to the fitness importance of injuries, we can expect evolution to have built-in upregulation of wound healing mechanism on adopting aggressive behavior. Since a number of growth factors are common, this upregulation could also strengthen endocrine pancreas. A deficiency of aggressive behaviors is likely to lead to a cumulative deficiency of growth factors and thereby affect β cell function. In addition to helping β cell regeneration, some of the growth factors have been shown to exert systemic anti-inflammatory action [168].

K. BDNF: The BDNF bridge between aggression and diabetes appears to be built in a different way. Frequent aggression [169] and particularly repeated defeat in fight appear to upregulate BDNF [170], and BDNF controls aggression. BDNF deficiency leads to aggressive phenotype [171, 172]. There exists a link between EGF, NGF, and BDNF such that EGF and NGF together enhance BDNF activity [173]. The role of BDNF in diabetes is that it is insulin sensitizing [174] and has a protective role against neuropathies [165, 175]. BDNF heterozygous knockout mice are obese and insulin resistant [176]. BDNF levels are reduced in diabetes [177] and Alzheimer's disease [177]. Unlike all the bridges listed above, BDNF is an anti-aggression molecule and still has a protective role against diabetic complications. Does this go against our hypothesis? We will soon see that it does not. But to understand that argument, we need to understand first the socioecophysiology of aggression in a holistic way.

L. Endorphins: The brain opioid peptides including endorphin are aggression control signals [178, 179]. It has two apparently contradicting links with obesity and insulin resistance. On the one hand it facilitates overeating and is positively associated with obesity and insulin resistance [180–183]. On the other hand it is involved in exercise-induced insulin sensitization [184, 185]. This apparent contradiction will also get resolved when we will see the socioecophysiology of aggression.

M. Gamma-aminobutyric acid (GABA): The role of brain GABAergic activity in aggression and impulse control is similar and perhaps synergistic to that of serotonin [186–189]. Although GABA is involved in aggression control, the knockout of GABA-synthesizing enzyme $GAD65^{-/-}$ is paradoxically nonaggressive. This paradox will also get resolved when we understand aggression holistically. GABA is also intricately involved in food intake regulation and thereby obesity [190–195].

N. Growth hormone (GH): GH administration increases isolation-induced aggression [196–198]. GH is essential in building up muscle mass, although its contribution to muscle strength is debated [199–203]. In obese individuals the GH response is blunted [204], and GH deficiency is characterized by central obesity. The growth-promoting action of GH is in synergy with the action of insulin, and therefore, it is expected that GH would upregulate insulin secretion, which may be accompanied by increased resistance. The action of GH does increase insulin resistance

[205, 206]. However, the GH signaling pathway involves IGF-1 which is insulin sensitizing [129, 130, 207], and therefore, the net effect of GH would be enhancing insulin action.

O. Progastrin–gastrin: Transgenic mice overexpressing progastrin showed increased aggression [132]. On the other hand central obesity and insulin resistance were increased in mice lacking gastrin expression [208]. Gastrin in combination with EGF stimulates β cell regeneration and therefore has antidiabetic action [154, 209].

P. Myostatin: Myostatin limits muscle growth, and increasing levels of myostatin are thought to be responsible for age-induced sarcopenia. The aggression hormone testosterone suppresses myostatin [210] and thereby promotes muscle growth. Myostatin sets the limit to muscle growth [211] presumably by downregulating mTOR pathway [212]. Along with downregulation of mTOR, myostatin increases insulin resistance [213], and myostatin inhibitors, known to facilitate muscle growth, prevent the development of obesity and type 2 diabetes [214]. Myostatin thus mediates a negative association between muscle strength and insulin resistance.

Q. Nitric oxides (NO): Nitric oxides and the NO synthases have complex involvement in both aggression and metabolic syndrome. NO signaling is increasingly known to be involved in diverse functions of the body. Reduced bioavailability of NO characterizes obesity and related disorders [193]. Interestingly endothelial NO synthase (eNOS) and neuronal NO synthase (nNOS) have different effects on aggression and accordingly have differing effects on metabolic states too [10, 215–224]. The eNOS is proaggression and insulin sensitizing, whereas nNOS has mixed effects on aggression and also mixed effects on insulin resistance.

R. Phosphoenolpyruvate carboxykinase (PEPCK): PEPCK is a key enzyme in glucose metabolism and has an interesting relationship with insulin resistance and with aggression. Overexpression of this enzyme in liver leads to liver insulin resistance and upregulation of gluconeogenesis [225, 226]. Its overexpression in adipose tissue leads to increased accumulation of fat. This kind of obesity is not accompanied by higher levels of FFAs in blood. Also in spite of obesity, there is no change in insulin levels or insulin sensitivity [227]. This demonstrates that if only fat deposition is specifically stimulated without other accompanying changes, insulin resistance does not follow, implying that some other process accompanying development of obesity could be more important for insulin resistance than obesity itself. More interesting is the case with muscle. If this enzyme is engineered to overexpress in muscle, it increases endurance making the muscle almost immune to fatigue. Interestingly animals with such increased muscle strength increase their aggression levels substantially. These aggressive animals have high levels of intramuscular triglycerides, but they are highly insulin sensitive [228]. PEPCK overexpression in muscle is not a naturally occurring condition, and therefore, it may not be of any clinical relevance. But it gives us an important insight into the physiology of aggression as well as the origin of metabolic syndrome, which we will see shortly.

S. Mammalian target of rapamycin (mTOR): The mTOR is a protein in the downstream pathway of insulin or IGF-1 and IGF-2 and regulates cell growth. Increased mTOR activity reduces insulin action which works like a feedback regulator of insulin action [229–231]. The mTOR signaling pathway is activated by excess nutrients such as FFAs or amino acids and insulin itself. This is likely to be the mechanism of diet-induced insulin resistance. Therefore mTOR is thought to be a crucial link in insulin resistance. Interestingly mTOR is activated by exercise [232–236] and testosterone [237, 238], but both exercise and testosterone are insulin sensitizing. This means that physical activity and testosterone, both important components of physical aggression, increase the mTOR threshold for insulin resistance. There is evidence for interaction between testosterone, insulin, and mTOR pathway [239]. mTOR is involved in protein synthesis and cell growth, which is particularly important in strengthening muscle. The action of myostatin on mTOR appears to be opposite that of testosterone and exercise. Myostatin suppresses mTOR but induces insulin resistance. So fine-tuning of mTOR pathway could be the key link between aggression, muscle strength, and insulin action.

T. Sympathetic activation components: What is perhaps the most important mechanism linking aggression to obesity is the interplay between sympathetic nervous system, adipose tissue, and testosterone. Visceral adipose tissue is innervated by autonomic nerves. Sympathetic activation of the adipose tissue is long known to have a lipolytic action. This action is through β adrenergic receptors, and testosterone enhances the expression of β adrenergic receptors on adipocytes [240, 241]. Therefore when testosterone levels are adequate, with every environmental and social stimulus that leads to sympathetic activation, there is lipolysis. However in testosterone deficiency, typical of nonaggressive personality, lipolysis is arrested. A combination of testosterone and sympathetic actions, a characteristic of aggressive encounter, prevents accumulation of visceral fat [242–246]. The other effects of sympathetic activation including liver glucose production are not affected by low testosterone. In effect, sympathetic activation may release more glucose for lipogenesis. Thus, in the presence of testosterone, sympathetic activation has antiobesity effect, but in its absence, the same sympathetic stimulation may contribute to fat accumulation.

U. Osteocalcin: The sex and aggression signals is an adjective to testosterone and estradiol to facilitate osteocalcin production [247–251]. A marker of bone formation and strength, osteocalcin induces β cells to release insulin and also induces the release of adiponectin [248] which has a strong antiobesity, anti-inflammatory, and insulin-sensitizing action. Osteocalcin also has insulin-sensitizing and antiobesity actions [248–252].

V. Oxytocin: Oxytocin has a somewhat mystic role in the network of links between aggression and metabolic syndrome. The behavioral effects of oxytocin are increasing trust and affection on the one hand [253–261] and increasing aggression on the other [255, 262–266]. As an expected corollary, the effects of oxytocin on obesity, hypertension, and insulin resistance are also mixed: It facilitates lipogenesis but has antiobesity effects, and it facilitates liver gluconeogenesis yet reduces insulin resistance [257, 258, 267–277]. This needs careful interpretation, but as you will see below once we understand the ecophysiology of aggression in a broader perspective and place oxytocin in its natural role, the apparent contradiction gives way to logical coherence.

W. Glucose: Low levels of plasma glucose are long known to stimulate aggression [278–280]. Anthropologists have discussed and debated over the hypoglycemia–aggression hypothesis explaining the differences in aggressive behaviors within and between ethnic groups [281–283]. Close to the physiological range, plasma glucose is negatively related to aggression. Low glucose levels promote aggression, but unusually low glucose levels arrest aggression.

X. Tryptophan and arginine: Tryptophan is one of the fewer clearly known links between diet and aggression. Tryptophan supplementation reduces and depletion increases aggression [13, 20, 284–288]. This is because tryptophan is a precursor of serotonin in the brain and can pass the blood–brain barrier. Arginine is another known amino acid bridge between insulin and aggression. Although a number of amino acids stimulate insulin release, individual amino acids differ in the elicited response. Arginine has a specific insulin secretion stimulatory effect [289–291]. Arginine reduces aggressive response [292] as expected for an insulin-secreting agent.

Y. Vitamin D: Vitamin D is an important essential vitamin that is known to be important for bone strength and integrity. Disruption of the vitamin D receptor gene results in loss of aggression in mice [293] accompanied by an increase in grooming behavior [293, 294]. Compatible to our hypothesis, vitamin D deficiency is identified as a significant contributor to insulin resistance [295], and vitamin D_3 supplementation is shown to be an effective insulin sensitizer [296].

Z. Fat and triglycerides: Plasma triglyceride levels negatively correlated with aggression [297] consistent with the hypothesis. Leptin levels which are positively correlated with fat are negatively associated with aggression [69, 298]. Monkeys fed with a high-fat high-cholesterol diet were observed to be less aggressive than those on low-fat low-cholesterol diet [71, 298, 299]. This is perplexing at a glance since both fatty acids and cholesterol do not cross the blood–brain barrier to reach the brain and exert any behavioral effects. An interesting mechanism is suggested as to how fat diet may reduce aggression [69, 112, 300]. Fatty acids compete with tryptophan for albumin binding. As a result when FFAs are high, more tryptophan remains in the free state. Since only free tryptophan can pass the blood–brain barrier, more of it is available in the brain when plasma FFAs are high. Tryptophan is a precursor of serotonin, and its concentration can be a limiting factor in serotonin synthesis. Therefore high levels of FFAs in plasma can increase serotonin synthesis in the brain without crossing the blood–brain barrier. Although speculative at this stage, the hypothesis is promising and may be a strong link between diet and behavior. FFAs therefore are important molecules bridging behavior and biochemistry. It is likely that many other dietary factors influence behavior, but research in this direction is scanty.

This list, although A–Z coincidentally, is not really the A–Z of the links between aggression and metabolic syndrome. This is only an incomplete list of molecules associated with aggression and injuries that are also associated with metabolic syndrome but includes the ones about which there are sufficient data so that we can appreciate their relevance to metabolic syndrome. For example, the peptide proopiomelanocortin (POMC) and its cleavage products are intimately involved in aggression control as well as in obesity and insulin resistance, but the POMC system by itself is so complex that we may have to wait for a few more years' research to make logical sense out of all the components of POMC and their interactions. Nevertheless, there are clear indications that POMC and component peptides are very important links between aggression and obesity. The total number of links between aggression and metabolic syndrome is much greater (Appendix I) and will probably need another few decades of research to understand them in a coherent and comprehensible way. Here I need to restrict myself to less than half of them. Having looked at the role of some of the individual molecules, we need to join the pieces together to make a coherent picture, although the picture will be far from complete at this stage.

Often when pieces of a complex system are joined together, some emergent properties are seen which are not displayed by any of the components themselves. In fact, such emergent properties are at the core of biological systems. Therefore we need to look at the aggression–dominance syndrome in the behavioral, ecological, and neurophysiological context at the same time. This is particularly important. Aggression has been addressed by behavioral ecologists, psychologists, and neurophysiologists for a long time. But each group has created its own niche, and they rarely come out of their expertise castles to exchange ideas. Therefore a holistic picture of aggression has failed to emerge. Since I am not an expert in any of these fields, I can comfortably try to combine the piecemeal research on aggression and try to construct a coherent picture.

The Complex Physiology of Aggression

Physical aggression is an event that has important lifelong consequences. The reward as well as the risk levels of aggression are very high, and therefore, natural selection must have fine-tuned all the mechanisms involved. The decision to initiate aggression or retaliation is an extremely important decision, and a wrong decision can cost life. Therefore mechanisms must have evolved to check and countercheck the decision before implementing it. Before initiating an aggressive act, it is necessary to check that all mechanisms needed for effective implementation of aggression as well as for coping with possible consequences are functioning well. If one or more of them are likely to fail, it is better to avoid aggression. This I called as the "aggression checkpoint" in the last chapter. Let us see now what the essential requirements for execution of aggression are and accordingly what the essential checkpoints are.

It is obvious that bone and muscle strength is needed for effective physical aggression. This is evident by the loss of aggression and increased submissive behavior associated with vitamin D deficiency which weakens the bone architecture. It is crucial that some mechanism for assessing muscle strength and muscle energy supply should exist, and they must feed this information into the decision-making system for initiating aggression.

Not much research has addressed this question directly, but there are indications that this could be happening. Everyone is familiar with the feeling of fatigue or weakness when muscle energy sources are depleted. This feeling makes aggression less likely. It is possible that the sensory nerve endings in muscle that are known to be stimulated by every contraction continually keep on collecting this information. Myokines that are released with every muscle contraction are also likely signal molecules. This is currently speculative. But a recent piece of experiment demonstrates how muscle energetics influences aggression. When experimenters engineered overexpression of PEPCK in muscle, the muscle energy metabolism received an extra boost and became almost immune to exhaustion. This was demonstrated by the sustained hyperactivity of these animals. The most interesting point to be noted is that these animals became highly aggressive. The experimenters had manipulated only the energy supply to muscle and nothing else. There was no primary neuronal or hormonal change that could have resulted in aggression. Since only increasing muscle energy increased aggression, it suggests that there must be some mechanism that constrains aggression based upon muscle capacity. Muscle capacity must have been an important limiting factor in aggression, and when this limit was lifted, aggression control got lifted too. I feel confident in concluding that the checkpoint of muscle and bone strength/capacity must be an important controller of aggression. If this is true, it implies that any energetic disinvestment from muscle will control aggression. Insulin resistance or insulin deficiency results into energetic disinvestment from muscle, and therefore, it is no surprise that diabetes is accompanied by loss of aggression [301].

Strength is not the only requirement for successful aggression. Aggression has conflicting hemodynamic requirements. On the one hand there is explosive muscle action that needs increased blood supply. Vasodilatation is one of the mechanisms to increase the blood flow. On the other there is increased risk of injury, and in order to minimize bleeding, vasoconstriction is also adaptive. Therefore both vasodilatation and vasoconstriction mechanisms must be boosted for successful aggression. Nitric oxides are crucial in vasodilatation and endothelins in vasoconstriction. The fact that eNOS knockouts [302] as well as ET-1 knockouts [303] lose aggression illustrates that there exists a hemodynamic checkpoint too. The involvement of both NO and ET-1 in an aggressive act is downstream to decision making in the process of aggression. Still impairment in NO or ET-1 dynamics inhibits aggression. Perhaps similar is the case with adrenalectomy which results in loss of aggression [304]. Aggression induces salivary EGF production in anticipation of injuries. However, removal of salivary gland in mice results into a delayed suppression of aggression and enhancement of defensive behavior [305]. Extracts of salivary glands induce aggression [306] demonstrating that signals from salivary glands are involved in initiating aggression. Both salivary and adrenal hormones are downstream in the aggression pathway, but they are essential in starting aggression. Barring checkpoint it is difficult to think of any other explanation for a downstream factor arresting an upstream process.

For the success of an aggressive encounter and from the point of view of lifetime fitness of an individual, a decision to stop an aggressive act at the right time is as important as the decision of initiating it. Therefore before initiating aggression, there must be a checkpoint for mechanisms of stopping it. The decision to stop is mainly driven by impulse control mechanisms involving serotonin and GABA. It is interesting to note that some of the GABA-related knockouts $GAD65^{-/-}$ and serotonin-related knockouts $5\text{-}HT_{1A}^{-/-}$ and $5\text{-}HTT^{-/-}$ lose aggression. This is apparently paradoxical. GABA and serotonin are involved in aggression control, so their elimination should increase aggression, but in reality, it is shown to decrease. This paradox is resolved by the concept of aggression checkpoint. If the mechanism to stop aggression is impaired, starting aggression can be counterproductive, and evolution is likely to have hardwired mechanisms of countercheck. Thus there are multiple evidences for aggression checkpoints, but we do not know as yet how exactly they work. This is mainly because the concept is new, and no research inputs have gone into this question as yet.

The situation at the beginning of an aggressive encounter is likely to change rapidly during the encounter. Accordingly the decision may have to be revised as well as reversed as per the need. These mechanisms also must have evolved since these are life and death decisions. If the fight is prolonged, at some stage muscles will get exhausted and one has to stop. However it is unwise to wait for complete exhaustion of energy. In most species intraspecific aggression is not fatal, it is meant to test the relative strength and endurance of the fighters. It would be adaptive then to fight till the last second possible. However, fight with a given opponent is not the only concern for survival. There are other elements including predators. If a predator is sensed immediately after a fight, there needs to be sufficient energy reserve to run for life. Therefore it is not good to fight till exhaustion of available energy. A stop signal must come much before actual energetic exhaustion. This function is served by the central fatigue signals which get activated and stop the aggressive act much before peripheral energy supplies are burnt out. Interestingly the central fatigue signals are also driven by serotonin and GABA. Since serotonin and GABA are negatively correlated with aggression, in a test of endurance, a less aggressive individual is likely to get exhausted first and withdraw from fight. A more aggressive individual is more likely to win. But here lies a trade-off. Aggressive individuals are more likely to win a given combat but would be left with lower energy reserves. As opposed to that in nonaggressive individuals, the basal serotonin and GABA signaling would be high, and therefore, the central fatigue mechanisms would stop physical activity at an earlier stage. This would burn less energy but increases the chances of a defeat. There is an interesting modern consequence of this mechanism. When nonaggressive individuals, i.e., ones with high levels of the central fatigue signals, try to exercise, they get premature fatigue, and the energy reserves are not burnt by exercise as much as expected. This could be one of the mechanisms by which loss of aggression can contribute to obesity.

Other signals of exhaustion include endorphins and cortisol. Exercise research has shown that physical activity increases endorphin and cortisol, but this increase is not linear. Both endorphin and cortisol are not primarily needed for physical activity or aggression. The mechanisms primarily needed such as catecholamines that ensure energy release increase linearly with exercise [307, 308]. Endorphin and cortisol, on the other hand, start increasing only after prolonged and exhaustive activity [307, 308], and they are actually meant for signaling exhaustion and controlling aggression. Therefore the anti-aggression activities of endorphin and cortisol are adaptive.

The conditions for initiating and arresting aggression need to be weighed against levels of desperation in the context of life history trade-offs. An individual with little energy reserve is more desperate for food and therefore should be more aggressive in competition. It is not surprising therefore that hunger signals such as ghrelin have proaggression effects. An individual with larger energy reserves can afford to wait for better and less competitive opportunities in the future and therefore retract from aggressive encounter earlier. Here again serotonin and endorphin appear to play a role. Leptin upregulates serotonin, and therefore, obese individuals would activate central fatigue signals and withdraw from aggressive encounter faster. FFAs also play a role in upregulating serotonin as seen earlier. Energy-depleted individuals have a reduced chance of long-term survival, and therefore, they should utilize the present more desperately rather than relying on the future. Therefore they are expected to take more risks, fight longer, and exhaust their energy reserves more. Although sounding somewhat paradoxical, it is adaptive for individuals with higher energy reserves to spend less energy and those with lower reserves to spend more in highly competitive situations.

Similar to exhaustion, aggression needs to be controlled after perceived defeat. Defeat increases the central opioid peptides [178, 309, 310]. Brain opioid activity decreases aggression [178]. This negative feedback loop appears to have evolved as an adaptive response. Proactive aggression will be counterproductive after exhaustion or in front of a stronger opponent, and therefore, aggression suppression mechanisms are needed along with suppression of pain. The defeat response suppresses pain as well as aggression. This appears to be mediated by brain opioids. BDNF works in a similar negative feedback loop but on a more chronic scale. Repeated defeats are needed for raising BDNF [170]. Cortisol after defeat will have a similar function of suppressing aggression and risk-taking behavior. Losing is not only about controlling aggression. There is a need to cope with other consequences of losing. There could be injuries, and both immediate and delayed strategies are needed to take care of the wounds. The acute requirement is to prevent bleeding, and vasoconstriction is important for this purpose. After withdrawing from aggression, vasodilatation is no more needed, but vasoconstriction is since bleeding would continue even after the end of the fight. Endothelin is an important player here. So the balance between NO and ET-1 would be tilted towards endothelin on perceiving defeat and shunning aggression. The HIIR state appears to be involved in shifting this balance [311]. Endothelin and insulin are in a positive feedback loop. Insulin signaling upregulates endothelin production [121, 312, 313], and endothelin induces insulin resistance [314–317] as well as insulin secretion [318]. This may be adaptive on the face of a possible defeat where it is necessary to suppress aggression and reduce peripheral blood flow in anticipation of injuries. It is not surprising then that endothelin levels are generally increased in obesity and diabetes [319–322]. It is no surprise therefore that aggression suppression is associated with hypertension [323, 324].

As opposed to the loser the winner has a different metabolic state. In a winner, testosterone and dopamine activities go up rapidly, and cortisol is lowered. Both testosterone and dopamine signal the capability and readiness for aggression in a future encounter. This again is a positive feedback loop in which a winner attains a hormonal and metabolic state that increases aggressiveness, mobilizes mechanisms of strengthening muscle and bone, and thereby increases the probability of winning in future. Loser, on repeated defeats, attains a hormonal and metabolic state that prompts submissive behavior and risk avoidance. This positive feedback has long-term consequences and is likely to give rise to sustained behavioral syndromes that we are referring to as hawk–dove or warrior–diplomat (Fig. 7.1).

The long-term sustenance of hawk or dove strategies is very different from a win–lose outcome in a short-term encounter. Acute and chronic suppression of aggression have different consequences, and it is a crucial point to realize. Aggression suppression due to exhaustion is temporary and can reverse quickly with replenishment of energy. In this context, it is appropriate that the exhaustion signal endorphin mediates hyperphagia in an attempt to restore energy [325, 326] but has insulin-sensitizing action on the background of recent physical activity. Aggression suppression necessitated by being substantially weaker than the opponent cannot and should not be reversed unless there is dramatic change in situation. The anti-aggression activity of exhaustion signals such as endorphin has evolved to stop aggression at the right time, whereas chronic aggression suppressors such as serotonin and cortisol are evolved to arrest initiation of aggression itself. The acute and chronic aggression suppression mechanisms are not two different mutually exclusive compartments but are two ends of a continuum. Some signals are involved in both, and some like BDNF appear to take an intermediate position. BDNF activity increases after repeated defeat [170]; BDNF is an aggression suppressor and has synergy with serotonin activity [327, 328]. However BDNF levels are not sustained on chronic aggression suppression. This is because BDNF is not a part of the positive feedback cycle of insulin resistance syndrome described below. As a result the initial phases of commitment to a dove or diplomat strategy might witness a high BDNF level which will be depleted eventually. Similar is likely to be the case with NGF, which is also not a part of the diplomat positive feedback circuit. Therefore a prolonged period of diplomat behavior and lack of soldier behavior would eventually lead to deficiencies of BDNF and NGF, although

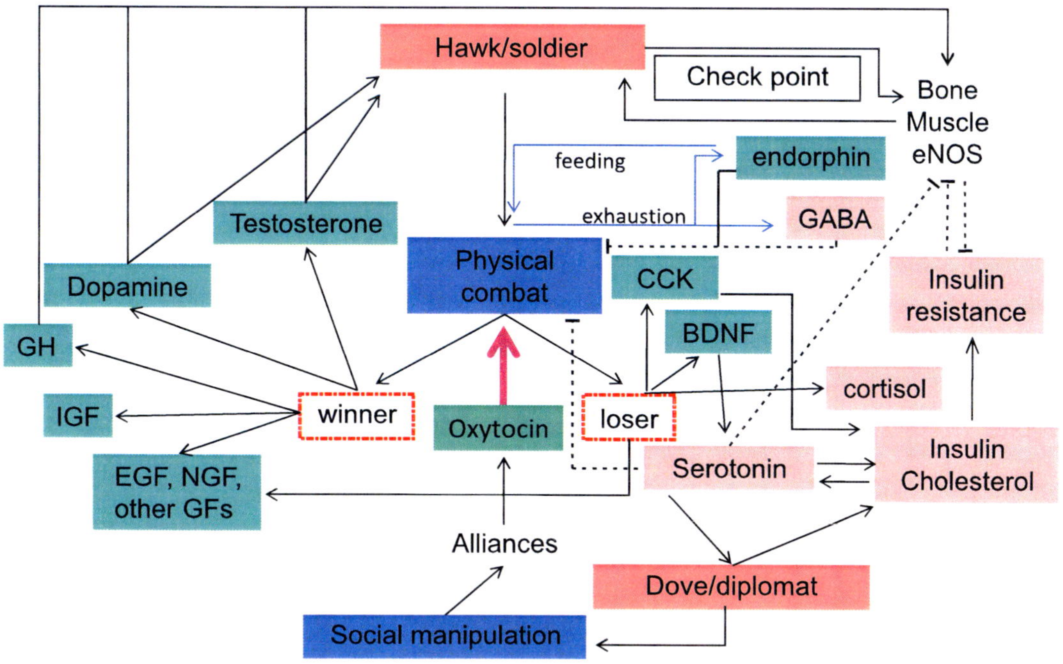

Fig 7.1 The physiology of aggression and its relevance to the insulin resistance syndrome. *Pointed arrows* on *solid lines* denote upregulation, and *blunt arrows* on *dotted lines* denote downregulation. There are two positive feedback loops that give rise to two different basins of attraction which are stable by themselves. Factors in *green* are known to have insulin-sensitizing, antiobesity, or antidiabetic effects. Factors in *purple* are known to be involved in the insulin resistance syndrome

that may not be the case initially. In contrast with BDNF, the raised levels of serotonin and cortisol are sustained in chronic aggression suppression as they are a part of the positive feedback cycle, and they are the true stable makers of a diplomat mode of life. Similarly testosterone and dopamine may be sustained by a winner, and they mark a hawk or warrior strategy. Many other molecules upregulated in a defeat response do not sustain their raised levels chronically. These include EGF, NGF, and other growth factors. The autonomic nervous system also undergoes shift in activity. A sympathetic response is typically needed for aggression, whereas loss of aggression would lead to an inclination towards parasympathetic activity. Parasympathetic activity enhances pancreatic insulin response [329–331], and this may have a crucial role in maintaining the hyperinsulinemic state of a dove or diplomat. A hyperinsulinemic state needs to be necessarily accompanied by two compensatory changes, without which the organism will not survive. One is insulin resistance. Hyperinsulinemia without insulin resistance will lead to fatal hypoglycemia, and insulin resistance must accompany hyperinsulinemia to avoid this. A number of short- and long-term mechanisms exist that induce insulin resistance in response to hyperinsulinemia. The second response is that of disruption of food intake regulation mechanisms. Insulin is a potent regulator of food intake. If substantially raised insulin levels are sustained with normally functioning anorectic reflexes, the individual will almost never eat and die of starvation. In order to avoid this the appetite regulation mechanisms need to be disrupted, and evolution appears to have built-in mechanisms to do this as well. This can lead to type 1 thrift which is adaptive for a subordinate individual for reasons as under.

The hyperinsulinemic diplomat state has adequate provision of positive feedbacks to sustain itself. Hyperinsulinemia induces insulin resistance which results into disinvestment from muscle. The partially energetically deprived muscles give aggression suppression signals and prevent a transition back to aggression. Here lies the difference in the fine-tuning of strategies. If temporary exhaustion is the cause of aggression control which can be reversed after energy replenishment, disinvestment from muscle should not happen. This is ensured by the insulin-sensitizing action of the combination of exercise and endorphin. Thus short-term aggression suppression is prevented from getting into the vicious cycle. But if aggression suppression is chronic, it gets into the positive feedback cycle to stabilize the state. There is a positive feedback loop between insulin and serotonin as seen earlier which is stabilizing. The combined effect of serotonin and parasympathetic stimulation would raise insulin production. High insulin levels are adaptive for diplomat life since they enhance cognitive functions. Similarly leptin [332] and cholesterol are also needed for a similar purpose. But both these molecules have an anti-aggression effect further strengthening and stabilizing the state. High levels of insulin mediate insulin resistance and the chain of metabolic alterations downstream to it.

"Thrift" is particularly adaptive for the dove or subordinate life for multiple reasons. One is that subordinate individuals have less access to food and lower food security. Since dominant individuals can snatch food, the subordinates need to gulp as much as they can whenever it is available. Another reason is that since the chances of gaining immediate reproductive success are lower, they have to wait for a future opportunity. This necessitates investing in the future in the form of energy reserves. For this purpose hyperactivity of central fatigue signals is highly adaptive. Risk avoidance behavior in the present must accompany any strategy of investing in future. This role is mainly played by cortisol. Although energy intake may be increased, insulin resistance and cortisol arrest energy supply to muscle, and energy flow is diverted to lipid synthesis instead. Here lies a potential conflict of mechanisms, and evolution appears to have resolved that as well. Since leptin has important role in the cognitive brain functions, higher levels of leptin are needed. However, high leptin levels will prevent energy intake. This conflicts with the simultaneous requirement of thrift. The possible conflict appears to be resolved by evolution through mechanisms of leptin resistance. Leptin resistance allows hyperphagia and fat accumulation in spite of high leptin levels. There is some evidence that hyperinsulinemia and insulin resistance are sufficient to induce leptin resistance [333, 334]. Once leptin resistance develops, there is a clear path towards accumulating fat. Fat further reduces aggression by making more tryptophan available for serotonin synthesis completing one more turn of the cycle. These multiple positive feedback loops make the insulin-resistant state highly stable.

If both the aggressive hawk and nonaggressive dove/diplomat states are stabilized by positive feedbacks, is there any way to transit from one state to the other? There surely is. The transition from hawk to dove can be induced by repeated defeats, weakness due to any disease or major debilitating injury, or any other condition that makes aggression impossible for sufficiently long duration. Under such conditions it is adaptive to shun aggression for some time and focus on eating and resting to regain strength as much as possible. This marks a transition from hawk to dove state. In social species it is good for a weaker individual to be subordinate, groom other individuals, be more social, and make alliances. This may prepare one for getting back into the aggressive social rank struggle in the future. In social primates it is common for weaker males to make alliances and challenge the dominant alpha male. This strategy often succeeds. If the social strategies are successful, there can be a chance of reversal of the dove state. Mechanisms for this reversal also exist. For example, successful social relationships increase oxytocin which counteracts the effects of corticosteroids. After lowering corticosteroid activity, there can be greater risk tolerance, and slowly the behavior pattern may shift away from dove. Somewhat less known is the effect of oxytocin on protein synthesis [335]. By stimulating protein synthesis oxytocin could shift the metabolic bias from being lipid anabolic to protein anabolic, and this could be important in shifting behavioral strategies too. Getting back into the aggression cycle after making alliances is a tricky process. Stable alliances need trust and affection and utilizing the alliance to challenge a rival group needs aggression. Aggression and affection need to coexist during this transition, and this is appropriately achieved by oxytocin. Oxytocin has complex context-dependent behavioral effects, and it facilitates both affection and aggression [253]. Further oxytocin has complex effects on obesity and insulin resistance as well which are also likely to be context dependent. Our understanding of oxytocin so far is contradictory and confusing since evidence is contradictory. I think this apparent contradiction is an effect of trying to label oxytocin actions out of context. Oxytocin is a player on tricky grounds, so it needs to have evolved complex actions in combination with other hormonal and metabolic signals and under varied social and ecological conditions. Research now needs to focus on studying oxytocin action as a potential pathway for the transition from dove to hawk behavior using social alliances. In short, although both the aggressive and nonaggressive states are robust and self-supporting owing to stabilizing positive feedback loops, transition from one stable state to the other is possible under appropriate conditions, and pathways and mechanisms for both the transitions have evolved.

Here lies the difference between animal and modern human societies. In animals dove is frequently a reversible state, and for gaining higher ranking, one has to get back into physical dominance, although social alliances help in doing so. Since the dove state is reversible eventually, the metabolic state would also reverse substantially. In the modern human society, gaining higher status is possible, in fact easier, by being and remaining a diplomat. Therefore the need to reverse a diplomat state is hardly ever felt. It is interesting to note that the pathological effects of a diplomat metabolic state, i.e., an insulin-resistant state, are not observed for a long time. This is perhaps because we have evolved mechanisms to cope with this state for quite some time, until getting the right circumstances to reverse it. Basically the dove response has evolved as a transient adaptive state to reverse ultimately to the higher fitness-giving hawk state. This transition does not happen now with the modern human social structure, but we are still evolved for reversibility of the state after some interval. It is only after surpassing the time for which a diplomat state would be naturally sustained that we start observing the pathological effects. It is possible therefore that our diplomat state is not only quantitatively supernormal but is also temporally supernormal, and this supernormal temporal response leads to pathological consequences after several years.

The picture constructed above by joining several pieces of a jigsaw puzzle is a generalized common minimum principle. Although I expect it to be qualitatively similar and conserved, there would be subtle quantitative refinements in different species and different ecological setups as well as difference between individuals shaped by genetic, epigenetic, developmental, social, and other stochastic factors. These differences can be appreciated only when a common principle platform exists. Much more research is needed to understand how fine-tuning of these mechanisms takes place under varying conditions. Here I have only attempted to construct such a common platform in a brief comprehensive way. This should not be taken as an invariable necessity.

After having made a comprehensive sketch of the physiology of aggression in a behavioral ecology setting (Fig. 7.1), let us look at how it reflects on the clinical aspects of T2D and related disorders. It can be seen that almost all molecules that are acutely activated during aggressive encounter, irrespective of triumph or defeat, have anti-obesity and antidiabetic action. These include sex hormones, dopamine, endorphins, BDNF, CCK, EGF, NGF, and other growth factors. Also markers of sustained hawk strategy including testosterone and dopamine are insulin sensitizing. Only the chronic dove or diplomat state markers such as serotonin and cortisol have pro-insulin-resistance action. This means that an aggressive encounter, irrespective of the win–lose outcome, will be insulin sensitizing, but chronically shunning aggression will lead to a state of insulin resistance. The story does not stop at inducing insulin resistance. It also suggests a possible cause of β cell problems in the form of chronic deficiency of gastrin, EGF, and other growth factors needed for β cell regeneration. Further it accounts for deficiency of EGF, NGF, and BDNF which are likely to play an important role in diabetic neuropathy and retinopathy as we will see in a later chapter.

The positive feedback mechanisms stabilizing the diplomat state also explain why types 1 and 2 diabetes have very similar pathological consequences in spite of having very different etiologies. Aggression suppression is the cause of type 2 diabetes, whereas aggression suppression is an effect of type 1 diabetes. Insulin deficiency in type 1 diabetes results into disinvestment from muscle. This would induce aggression suppression through checkpoint mechanisms. The two types of diabetes have different entry points into the cycle, but the cycle runs in very similar way, and therefore, the downstream effects of both types of diabetes are highly overlapping if not identical.

After having tried to make a new synthesis, we need to address the question how we can differentially test the new theory in comparison with the old one. The answer is a little tricky but not very difficult to perceive. It is tricky because it is not a clear case of two hypotheses being completely different and nonoverlapping. It is not a simple case of differentiating between whether X causes A or Y causes A since Y causes X and X likely causes Y too. Fat and cholesterol modulate aggressive behavior, and aggression hormones affect fat metabolism. Therefore it is difficult to decide what comes first, what is really important, and what is not. But the subtleties of the suggested pathways and mechanisms help us in differentiating between the two. The aggression suppression hypothesis involves downregulation of testosterone, EGF, NGF, BDNF, and other growth factors in the suggested pathophysiological mechanisms. Therefore this hypothesis predicts altered levels of these markers preceding T2D and related disorders. If positive energy balance, obesity, and obesity-induced insulin resistance are the root cause of T2D, then there is no reason why one should find deficiencies of testosterone, EGF, NGF, BDNF, and other growth factors in diabetes. This therefore becomes a differential testable prediction. There is indeed reproducible evidence for deficiency of testosterone [86, 87, 104, 245, 336–338], EGF [152, 339], NGF [166, 340–344], and BDNF [165, 175, 345] in T2D. In addition there are a few more indications in support of the aggression suppression hypothesis:

1. Anemia and iron metabolism: Aggression has a number of downstream physiological effects that must have evolved in anticipation of injuries. Just as salivary EGF is produced in anticipation

of injuries, there are likely to be other preparatory alterations in anticipation of injuries. Injuries imply blood loss, and therefore, the body should be prepared for making up by increasing the rate of synthesis of hemoglobin and red blood cells. It is no surprise then that aggression stimulates erythropoietin synthesis through the agency of testosterone [346–348]. Erythropoietin is an important signal involved in red blood cell formation. Some diabetics become anemic to a variable extent. This anemia is generally not due to iron, vitamin B_{12}, or any such dietary deficiency. It is erythropoietin deficiency that characterizes diabetes-associated anemia [349–354]. Anemia in diabetes is often associated with excess iron deposits in the liver, a condition called iron overload [355, 356]. It is even been suggested that iron overload may contribute to insulin resistance [355]. This indicates that the raw material for the synthesis of hemoglobin and red blood cells could be there, but the stimulus for synthesis is lacking which is very likely to be due to behavioral deficiency than dietary deficiency. The lipocentric paradigm does not have an adequate explanation for erythropoietin deficiency anemia and iron overload accompanying diabetes.

2. Sexual dysfunction: Similarly the lipocentric view does not adequately account for sexual dysfunction accompanying diabetes. The aggression suppression hypothesis makes a stronger case supported by evidence. The aggression suppression mediator serotonin has been shown to decrease sexual motivation [357, 358]. Dopamine, on the other hand, has facilitative effect on sexual motivation [52]. Sex hormones of both sexes, testosterone and estrogens, are associated with aggression [359–361]. Dominant males that are sexually more active have higher testosterone levels than subordinate ones [362]. Obesity and metabolic syndrome, on the other hand, are associated with lower testosterone levels [86, 87, 104, 245, 336–338]. It is interesting to note that although erectile dysfunction in diabetic males has a positive correlation with plasma glucose levels [363–366], glycemic control does not improve erectile dysfunction [367], but chronic treatment of erectile dysfunction is insulin sensitizing [368]. This reflects on the possible direction of the arrow of causation.
3. The origin of endothelial dysfunction: The three isoforms of NOS, namely, endothelial (eNOS), neuronal (nNOS), and inducible (iNOS), have complex effects on aggressive behavior as well as on metabolic syndrome. Male mice with targeted deletion of eNOS gene display dramatic reduction in aggression [217]. eNOS knockout mice were hypertensive and had fasting hyperinsulinemia, hyperlipidemia, and insulin resistance [218]. eNOS gene polymorphisms have a significant association with type 2 diabetes [219]. eNOS also has a significant role in facilitating wound repair and growth-factor-stimulated angiogenesis [369, 370]. Based on different model systems and different cell types/tissues examined, eNOS has both pro-inflammatory and anti-inflammatory effects [371], but it has a protective role against pancreatitis [302]. The effects of nNOS on aggression are sex specific in mice. In males nNOS knockout mice showed large increase in aggressive behavior and excess sexual behavior [220]. Inhibitors of nNOS also increase aggression in males [216, 221]. Female nNOS knockouts do not show increased aggression and may in fact show a decrease [222]. As opposed to eNOS knockouts which develop insulin resistance in both liver and peripheral tissue, nNOS knockouts develop only the latter [223]. It is apparent therefore that eNOS which has clearly a proaggression action is antidiabetic, and nNOS that has a mixed action on aggression also has mixed effects on insulin resistance. The inducible form iNOS has not been studied from the behavioral point of view, but targeted disruption of iNOS synthase protects against obesity-induced IR [224]. The role of NOS in sexual function is well known, and sildenafil, a popular drug used for the treatment of erectile dysfunction, has insulin-sensitizing effects [372], and interestingly, aggressive behavior has been noted as a “side effect” [373, 374]. Since nitric

oxide signaling is directly involved in aggression, a chronic loss of aggression can affect NO levels which has important vascular effects. The traditional interpretation of diabetes holds raised plasma glucose levels to be responsible for vascular pathology. However signs of vascular pathology appear much before hyperglycemia. Therefore I feel that the traditional view has not adequately explained the origin of vascular pathology in diabetes.

There is one possible argument against the aggression suppression hypothesis. Aggression is characterized by sympathetic activation and suppression of aggression by increased parasympathetic tone. Contrary to our expectation, in T2D, there is increase in the markers of sympathetic tone. This contradiction is obvious but only apparent and stems from the naïve association of sympathetic activity with aggression. Although a physically aggressive act necessarily involves sympathetic activation, the reverse cannot be inferred. Sympathetic activation does not necessarily increase aggression. Evidence for association of sympathetic activation and aggression is contradictory [324, 375–379]. There are two reasons for the association being nonobligate. First is that the effect of sympathetic activation depends upon the standing metabolic, endocrinal, and behavioral state. For example sympathetic stimulation has a lipolytic action. But this action is mediated through β adrenergic receptors in visceral adipose tissue. The expression of β adrenergic receptors is testosterone dependent, and therefore, lipolytic effects of sympathetic action will be seen only if testosterone levels are adequate. Similarly sympathetic action will increase aggression only if all other factors are favorable for aggression. The other reason why sympathetic tone does not necessarily associate with aggression is that the effect of sympathetic activity on aggression is nonmonotonic. Moderate activation of central noradrenergic system increases aggression, whereas strong activation decreases it [376]. This makes sense for optimizing behavioral strategies. If sympathetic stimulation is induced by aggressive competition and the response is proportional to the level of threat, the nonmonotonicity can be easily explained. A weaker opponent poses less of a threat, and being aggressive is a fruitful strategy, but if the opponent is too strong the threat level is high, it is better to avoid aggression.

There are multiple reasons for which sympathetic activity does not have a one-to-one correlation with aggression. Therefore higher level of sympathetic markers in diabetes is not evidence against the aggression suppression hypothesis. There is a different reason for higher nonaggressive sympathetic activation in diabetes. This sympathetic activity is extremely important and central to diabetic hyperglycemia which we will see in Chap. 12. As yet we have not developed sufficient background to understand why sympathetic tone rises in diabetes. To understand this we need to wait a little more. But at this stage I want to make a subtle distinction. There is a fundamental difference in endocrine and neuronal signals in the body. Endocrine signals are like radio broadcasting. The signal reaches every corner more or less simultaneously and with the same intensity. Neuronal signals are specific to the recipient like a one-to-one telephone call. One muscle may be instructed to contract and the very neighboring muscle to relax at the same time. We should not forget that the autonomic system is a neuronal system primarily, although they also alter levels of epinephrine/norepinephrine generally. Owing to the specificity, one autonomic function could be on and while others are kept off. Therefore equating aggression with sympathetic activation and expecting that, since sympathetic tone is high in diabetes, aggression would aggravate diabetes is illogical.

Exercise and Aggression

Both the lipocentric and aggression centric views converge on the importance of physical activity or exercise. However there are important differences. Exercise is viewed in terms of burning calories by the lipocentric view, whereas exercise is a necessary component of physical aggression from the new viewpoint. There are two possible ways in which we can resolve between the two alternative ways in which exercise and its effects can be interpreted. First if exercise is only calorie burning, then all types of exercises with equal calorie expenditure should have equal effects on the endobolic state. The alternative view would expect a difference between the effects of different types of exercises, particularly between more aggressive and less aggressive exercises. By our definition, aggressive exercises comprise neuromotor actions involved in Stone Age hunting or fighting as defined in Chap. 5. So far there is little research investigating the effects of aggressive exercise, but experiments have shown that different types of exercises have different effects. For example, some studies found resistance exercise more insulin sensitizing than aerobic exercise [380–382]. Of particular interest is the finding that extremely short-duration (30 s) high-intensity exercises improve insulin sensitivity and other health parameters surprisingly well in spite of very low total calorie expenditure [383–386]. These exercises are similar to a short-duration intensive chase during hunting. Such exercises have been shown to increase aggression hormones and growth factors including testosterone, GH, and IGF binding protein [387]. Therefore it can be suspected that the beneficial effects of these exercises are through their aggression component rather than their energetic component. A yet unpublished study compared three groups of young males doing no exercise, less aggressive exercises such as swimming and running, and more aggressive exercises such as boxing or kickboxing. There was no difference in the insulin sensitivity and LDL cholesterol of the no exercise and nonaggressive exercise group, but the aggressive exercise group was significantly more insulin sensitive and had lower LDL and higher HDL levels [388]. More experiments specifically designed to test the effects of aggressive exercises in comparison with nonaggressive exercises are needed to resolve between the two alternative possible effects of exercise.

The other possible means of resolving between the two alternative interpretations is to look at the pathway through which exercise brings about its beneficial effects. If the lipid view is correct, the beneficial effects should be routed through energy pathways such as fat oxidation, glycogenolysis, and glycolysis pathways. If the aggression hypothesis is true, the effects should be brought about through pathways involving one or more of the aggression signal pathways described above including EGF, NGF, BDNF, endorphin, and sex hormones. Certain types of exercises upregulate testosterone [387, 389], EGF [390, 391], NGF [392, 393], and BDNF [394, 395] as well as endorphin [307, 308]. If exercise was related only to calorie burning, there is no reason why exercise should stimulate these molecules.

Mitochondrial function is said to be an important determinant of insulin resistance [396, 397]. If altered mitochondrial function has a causal role in insulin resistance, then it would be enlightening to look at the effects of lipids and those of exercise on mitochondrial function. Exercise training is reproducibly shown to improve mitochondrial function [398–401]. In synergy with exercise, the sex and aggression hormones testosterone and estradiol also enhance muscle mitochondrial capacity [336, 402], whereas corticosteroids reduce it [403]. On the other hand evidence for lipids or FFAs lowering mitochondrial function is shaky. In one study lipid accumulation actually increased mitochondrial proteins and their function [404]. If lipids were responsible for reduced mitochondrial function, lowering of lipids should have increased it. However, pharmacological lowering of plasma lipids in insulin-resistant subjects reduced mitochondrial gene expression instead of increasing it [405]. This strengthens the hypothesis that physical aggression and activity have direct insulin-sensitizing effects independent of their effects on lipid metabolism.

For an aggressive and routinely exercising individual, muscle building is of prime importance. Muscle building does not take place during exercise but takes place during postfeeding and resting time. For this to happen, muscle uptake of nutrients needs to be facilitated, and this is possible if muscles are insulin sensitive. Therefore there needs to be a mechanism that makes muscle tissue insulin sensitive upon exercising regularly. A new pathway with a large potential importance has come into light recently but is not yet completely worked out. It involves postfeeding hepatic production of a soluble signal molecule called

HISS (hepatic insulin-sensitizing substance) that increases insulin activity specifically in muscle and thereby facilitates muscle building. HISS activity is shown to increase by exercise [406]. The compound responsible for HISS signaling is yet to be isolated, but available evidence shows that this pathway involves parasympathetic activity, NO, and insulin action in the liver [407–410]. It appears therefore that specialized mechanisms exist to fine-tune insulin sensitivity of muscle, and muscle insulin resistance is not a passive and inevitable process induced by fat. Exercise creates the need to invest energy in muscle, and the need stimulates the HISS pathway and perhaps other mechanisms to increase muscle insulin sensitivity independent of fat metabolism.

In reality the two effects of exercise, i.e., fat reduction and triggering aggression pathways, are not mutually exclusive, and both would be at work simultaneously. Research should then address the question of evaluating the relative importance of calorie burning versus aggression pathways in the beneficial effects of exercise. The increment in the levels of the above factors can be used to quantify the aggression component of an exercise schedule, and studies can be designed to evaluate the role of aggression component of exercise vis-a-vis the energy component. There is no research input in this question so far because the aggression component of exercise was not yet appreciated. This is evident by the fact that the otherwise comprehensive accounts of exercise physiology do not even mention that exercise stimulates EGF, NGF, BDNF, sex hormones, and other factors [411] although there is adequate evidence for it. Exercise physiologists need to consider the possibility that at least part of the exercise effect is mediated by the aggression pathways rather than the energy pathways.

Since evidence as well as pathways exists through which loss of aggression can lead to metabolic syndrome disorders, there is a need to explore the possibility further to test whether loss of aggression is both necessary and sufficient to cause any or all of the pathophysiological processes of this set of disorders. We should not repeat the mistake that was done with the classical theory of T2D where nobody really bothered to test whether obesity, insulin resistance, and insulin insufficiency comply to the necessary and sufficient criteria of causation for type 2 diabetes and all its complications. Since there is sufficient preliminary evidence for aggression deficiency to be the potential cause of insulin resistance or T2D, research in the near future should directly address the question whether it fulfills the necessary and sufficient condition for causation of insulin resistance and/or type 2 diabetes. Such experiments need to be sufficiently long drawn since the effects of behavioral deficiency are necessarily chronic. A note of caution is that the aggression that we address here is a normal and natural form of aggression innate to an individual's physiology. There can be pathological forms of aggression that need to be carefully excluded. A pathological form of aggression is one that evades the physiological controls described in this chapter. For example, if the mechanisms of aggression checkpoint are impaired, a person with weak muscle or impaired hemodynamic response might still behave aggressively. This may or may not give the correlates of aggression that are expected in normal physiology. Such conditions are rare and will not disturb any study with a sufficiently large sample size and randomization. This warning is only for those who may like to come up with some isolated exceptional case and a pretense of having disproved me.

Some of the possible lines of investigations could be (1) to see whether there are strong, muscular, and physically aggressive individuals that are insulin resistant or type 2 diabetic. This needs to be studied in comparison with nonobese type 2 diabetics. We know that a significant proportion of diabetics belong to the latter. But there are no epidemiological studies looking at the former. Anecdotally I have seen many thin T2D patients, but I am yet to see one who is strong, muscular, and physically aggressive. Beyond personal impressions this question needs to be seriously addressed. If there is significant proportion of strong muscular and aggressive diabetics, it can be concluded that loss of aggression syndrome is not necessary for being a type 2 diabetic (just as being obese is not necessary). (2) If aggressive exercises in the absence of weight loss are unable to increase insulin sensitivity and other parameters of exercise benefit, it can be concluded that loss of aggression is not necessary for insulin resistance. Such experiments can be carefully designed by giving an intensive aggressive exercise regime along with sufficient caloric supplementation to ensure no loss of weight or loss of fat mass. (3) In the above experiment levels of testosterone, EGF, NGF, BDNF, and other growth factors can be monitored to test whether their levels increase in the absence of loss of weight or fat mass. (4) On the other hand, if a group of young and aggressive individuals is made to retract from physical activity and aggression for a sufficient duration and their weight is maintained at basal level by providing calculated dietary energy, if they fail to decrease their insulin sensitivity, loss of activity and aggression is not sufficient to induce insulin resistance. Here again the aggression markers can help in quantifying the loss of aggression. These and perhaps a few other ways would exist that can test the necessary and sufficient conditions for a causal relationship between loss of aggression and insulin resistance. It would not be easy to begin such experiments. Much thinking and pilot experiments would be needed to standardize and choose parameters. That would be the appropriate way to handle and develop the new way of thinking. This line of evidence will take a long time to come in hand but will prove to be the right direction of research.

There is a quick alternative way to test necessary and sufficient conditions for a putative causative factor of a phenomenon, that is, to construct an inclusive mechanistic model of the system as far as we know it. In such an artificial system perturbations can be introduced and the effect monitored. In the age of computers a simulated system can be a useful and powerful tool to study the possible effects of any change. This is like shaking a spider web at one place and seeing what moves and how. For this to work, the individual threads of the spider web need to be known based on which web can be constructed in a hypothetical cyberspace and then studied. Individual threads in this condition are single regulation reactions such as glucose stimulates insulin production, glucagon facilitates gluconeogenesis, or EGF stimulates β cell regeneration. With the involvement of some students and colleagues, we constructed a network model of all known signaling and regulatory pathways of the body concerned with T2D. The multiorgan physiological network consists of over 70 different players and more than 250 positive or negative regulatory actions. Only those regulatory actions were included that were clearly demonstrated in literature. We made an attempt to include all pathways that we could find. An attempt was made to exclude redundancy to keep the model to a simplest possible level. Even after that, it makes a very complex network (see Appendix II for details). Since most of the available information on signaling mechanisms is qualitative in nature, we look at the model qualitatively. Each node has only three possible states, normal, upregulated, and downregulated. An upregulated node can alter other nodes to which it is connected by positive or negative links. Through such links, a signal can travel through the entire network and change the state of the network.

Once such a network is constructed, one can introduce environmental perturbations singly or in combination and see the effect on the entire network. The network model can also be used to see the possible system-wide effects of any drug or remedial measure. In a complex network a large number of possibilities exist and analyzing such a network considering all possible perturbations and remedies is a huge amount of work. The analysis is far from being complete as of today. The network itself is perhaps incomplete since all regulatory pathways may not have been discovered so far. But even with an incomplete network, a network approach is certainly more comprehensive than reductionist thinking.

With no claims of foolproof results, the preliminary results of the model that have started emerging are relevant here and therefore necessary to discuss. These results throw light on the possible action of etiological factors including those classically perceived and those proposed here. If we make the network system forcefully eat more food for some time and then leave it on its own, it develops hyperinsulinemia and insulin resistance. This means that the classical model of diet-induced obesity and insulin resistance works well. But we can observe that this state is accompanied by behavioral changes as well, including loss of aggression. To resolve between the effects of hyperphagia from those of behavioral changes, we can freeze aggression at the normal default level in the model. If this is done, hyperphagia is unable to induce and sustain insulin resistance. Hyperphagia is unable to induce obesity in the network model unless there is independent suppression of adiponectin production. Also once we withdraw the induced hyperphagia and let the system feed ad lib according to its built-in regulation mechanisms, it returns back to normal or oscillates around the normal. However if behavioral changes are allowed to accompany higher food intake, then after a threshold time of induced hyperphagia, the system gets locked into the altered state owing to behavioral positive feedback mechanisms described earlier in this chapter. As a result, even if the user-induced hyperphagia is withdrawn, the system remains in the same state. Thus after induced hyperphagia of sufficiently long duration, the personality becomes nonaggressive and metabolic state adopts itself to the nonaggressive lifestyle. This result brings the classically recognized and newly realized etiological factors together in a coherent picture. Hyperphagia and obesity indeed induce insulin resistance but only through behavioral modulation.

If we forcibly increase adipose tissue in the model, insulin resistance is induced, but hyperglycemia fails to follow. On the other hand suppression of aggression alone is sufficient to induce insulin resistance followed by hyperglycemia. This works even if obesity is frozen in the default normal state. Suppression of adiponectin alone is sufficient to cause obesity followed by all downstream changes typical of T2D. (In this book we have not yet wondered about the primary causes of adiponectin suppression which we will do in Chap. 9 when some additional risk factors will be uncovered.) Increase in diplomat-like brain activities is also sufficient to induce a series of changes in the same direction, inducing suppression of aggression on the way. The network model therefore leads us towards inferring that obesity comes along the pathway towards T2D but is unable to cause T2D without involving behavioral mechanisms. On the other hand behavioral changes are both necessary and sufficient for the diabetic state even if obesity is forcibly kept in control.

I will certainly not call this a conclusive statement but a seemingly robust argument supported by a network model based on the state-of-the-art knowledge of regulatory pathways across different organs and tissues of the body. This model needs to be developed further with more empirical data and used to make useful inferences that could take us from correlations to causality.

References

1. Chiavegatto S (2006) Using mouse models to unravel aggressive behavior. In: Canli T (ed) Biology of personality and individual differences. The Guilford Press, New York, NY
2. Gammie SC et al (2007) Altered gene expression in mice selected for high maternal aggression. Genes Brain Behav 6:432–443
3. Hahn-Holbrook J, Holt-Lunstad J, Holbrook C, Coyne SM, Lawson ET (2011) Maternal defense. Psychol Sci 22:1288–1295
4. Ellis LL, Carney GE (2011) Socially-responsive gene expression in male drosophila melanogaster is influenced by the sex of the interacting partner. Genetics 187:157–169
5. Belsare PV et al (2010) Metabolic syndrome: aggression control mechanisms gone out of control. Med Hypotheses 74:578–589
6. Ferrari PF, Van Erp AMM, Tornatzky W, Miczek KA (2003) Accumbal dopamine and serotonin in anticipation of the next aggressive episode in rats. Eur J Neurosci 17:371–378
7. Westergaard GC, Suomi SJ, Higley JD, Mehlman PT (1999) CSF 5-HIAA and aggression in female macaque monkeys: species and interindividual differences. Psychopharmacology (Berl) 146:440–446
8. Cleare AJ, Bond AJ (1997) Does central serotonergic function correlate inversely with aggression? A study using—fenfluramine in healthy subjects. Psychiatry Res 69:89–95
9. Fachinelli C, Sargo S, Bataller R, Rodríguez Echandía EL (1989) Effect of 5-HTP and ketanserin on the aggressive reaction induced by food competition in dominant and submissive pigeons (*Columba livia*). Behav Brain Res 35:265–270
10. Chiavegatto S, Nelson RJ (2003) Interaction of nitric oxide and serotonin in aggressive behavior. Horm Behav 44:233–241

11. Adams CF, Liley NR, Gorzalka BB (1996) PCPA increases aggression in male firemouth cichlids. Pharmacology 53:328–330
12. Clotfelter ED, O'Hare EP, McNitt MM, Carpenter RE, Summers CH (2007) Serotonin decreases aggression via 5-HT1A receptors in the fighting fish *Betta splendens*. Pharmacol Biochem Behav 87:222–231
13. Höglund E, Bakke MJ, Øverli Ø, Winberg S, Nilsson GE (2005) Suppression of aggressive behaviour in juvenile Atlantic cod (*Gadus morhua*) by l-tryptophan supplementation. Aquaculture 249:525–531
14. Peremans K et al (2005) The effect of citalopram hydrobromide on 5-HT2A receptors in the impulsive-aggressive dog, as measured with 123I-5-I-R91150 SPECT. Eur J Nucl Med Mol Imaging 32:708–716
15. Larson ET, Summers CH (2001) Serotonin reverses dominant social status. Behav Brain Res 121:95–102
16. Linnoila VM, Virkkunen M (1992) Aggression, suicidality, and serotonin. J Clin Psychiatry 53:46–51
17. Semsar K, Perreault HAN, Godwin J (2004) Fluoxetine-treated male wrasses exhibit low AVT expression. Brain Res 1029:141–147
18. Higley JD et al (1992) Cerebrospinal fluid monoamine and adrenal correlates of aggression in free-ranging rhesus monkeys. Arch Gen Psychiatry 49:436–441
19. Higley JD et al (1996) Stability of interindividual differences in serotonin function and its relationship to severe aggression and competent social behavior in rhesus macaque females. Neuropsychopharmacology 14:67–76
20. Winberg S, Øverli Ø, Lepage O (2001) Suppression of aggression in rainbow trout (*Oncorhynchus mykiss*) by dietary l-tryptophan. J Exp Biol 204:3867–3876
21. Lepage O, Larson ET, Mayer I, Winberg S (2005) Serotonin, but not melatonin, plays a role in shaping dominant-subordinate relationships and aggression in rainbow trout. Horm Behav 48:233–242
22. Seo D, Patrick CJ, Kennealy PJ (2008) Role of serotonin and dopamine system interactions in the neurobiology of impulsive aggression and its comorbidity with other clinical disorders. Aggress Violent Behav 13:383–395
23. Vickers SP, Dourish CT (2004) Serotonin receptor ligands and the treatment of obesity. Curr Opin Investig Drugs 5:377–388
24. Butler AA, Cone RD (2001) Knockout models resulting in the development of obesity. Trends Genet 17:S50–S54
25. Hefti FF, Hadham M (2000). Use of an NK-1 receptor antagonist and an SSRI for treating obesity. US Patent No. 6,162,805
26. Halford JCG, Harrold JA, Boyland EJ, Lawton CL, Blundell JE (2007) Serotonergic drugs: effects on appetite expression and use for the treatment of obesity. Drugs 67:27–55
27. Brown M et al (2001) Sibutramine reduces feeding, body fat and improves insulin resistance in dietary obese male Wistar rats independently of hypothalamic neuropeptide Y. Brit J Pharmacol 132:1898–1904
28. Raeder MB, Bjelland I, Emil Vollset S, Steen VM (2006) Obesity, dyslipidemia, and diabetes with selective serotonin reuptake inhibitors: the Hordaland Health Study. J Clin Psychiatry 67:1974–1982
29. Schwartz TL, Nihalani N, Jindal S, Virk S, Jones N (2004) Psychiatric medication induced obesity: a review. Obesity Rev 5:115–121
30. Harvey BH, Bouwer CD (2000) Neuropharmacology of paradoxic weight gain with selective serotonin reuptake inhibitors. Clin Neuropharmacol 23:90–97
31. Smolin B, Klein E, Levy Y, Ben-Shachar D (2007) Major depression as a disorder of serotonin resistance: inference from diabetes mellitus type II. Int J Neuropsychopharmacol 10:839–850
32. Muldoon MF et al (2004) Low central nervous system serotonergic responsivity is associated with the metabolic syndrome and physical inactivity. J Clin Endocrinol Metab 89:266–271
33. Muldoon MF et al (2006) The metabolic syndrome is associated with reduced central serotonergic responsivity in healthy community volunteers. J Clin Endocrinol Metab 91:718–721
34. Luo S, Luo J, Cincotta AH (1999) Chronic ventromedial hypothalamic infusion of norepinephrine and serotonin promotes insulin resistance and glucose intolerance. Neuroendocrinology 70:460–465
35. Levkovitz Y et al (2007) Antidepressants induce cellular insulin resistance by activation of IRS-1 kinases. Mol Cell Neurosci 36:305–312
36. Liang Y, Luo S, Cincotta AH (1999) Long-term infusion of norepinephrine plus serotonin into the ventromedial hypothalamus impairs pancreatic islet function. Metabolism 48:1287–1289
37. Fernstrom JD, Wurtman RJ (1971) Brain serotonin content: increase following ingestion of carbohydrate diet. Science 174:1023–1025
38. Calapai G et al (1999) Leptin increases serotonin turnover by inhibition of brain nitric oxide synthesis. J Clin Invest 104:975–982
39. Haney EM et al (2007) Association of low bone mineral density with selective serotonin reuptake inhibitor use by older men. Arch Intern Med 167:1246–1251
40. Warden SJ, Bliziotes MM, Wiren KM, Eshleman AJ, Turner CH (2005) Neural regulation of bone and the skeletal effects of serotonin (5-hydroxytryptamine). Mol Cell Endocrinol 242:1–9
41. Warden SJ, Nelson IR, Fuchs RK, Bliziotes MM, Turner CH (2008) Serotonin (5-hydroxytryptamine) transporter inhibition causes bone loss in adult mice independently of estrogen deficiency. Menopause 15:1176–1183
42. Richards JB et al (2007) Effect of selective serotonin reuptake inhibitors on the risk of fracture. Arch Intern Med 167:188–194
43. McEntee WJ, Crook TH (1991) Serotonin, memory, and the aging brain. Psychopharmacology 103:143–149
44. Clarke HF, Dalley JW, Crofts HS, Robbins TW, Roberts AC (2004) Cognitive inflexibility after prefrontal serotonin depletion. Science 304:878–880

45. Amin Z, Canli T, Epperson CN (2005) Effect of estrogen–serotonin interactions on mood and cognition. Behav Cogn Neurosci Rev 4:43–58
46. Schmitt JAJ, Wingen M, Ramaekers JG, Evers EAT, Riedel WJ (2006) Serotonin and human cognitive performance. Curr Pharm Des 12:2473–2486
47. Luciana M, Collins PF, Depue RA (1998) Opposing roles for dopamine and serotonin in the modulation of human spatial working memory functions. Cereb Cortex 8:218–226
48. Richter-Levin G, Segal M (1996) Serotonin, aging and cognitive functions of the hippocampus. Rev Neurosci 7:103–113
49. Majlessi N, Naghdi N (2002) Impaired spatial learning in the Morris water maze induced by serotonin reuptake inhibitors in rats. Behav Pharmacol 13:237–242
50. Ueda S, Sakakibara S, Yoshimoto K (2005) Effect of long-lasting serotonin depletion on environmental enrichment-induced neurogenesis in adult rat hippocampus and spatial learning. Neuroscience 135:395–402
51. Rosen RC, Lane RM, Menza M (1999) Effects of SSRIs on sexual function: a critical review. J Clin Psychopharmacol 19:67–85
52. Hull EM, Muschamp JW, Sato S (2004) Dopamine and serotonin: influences on male sexual behavior. Physiol Behav 83:291–307
53. van Erp AMM, Miczek KA (2000) Aggressive behavior, increased accumbal dopamine, and decreased cortical serotonin in rats. J Neurosci 20:9320–9325
54. Bliziotes M et al (2000) Bone histomorphometric and biomechanical abnormalities in mice homozygous for deletion of the dopamine transporter gene. Bone 26:15–19
55. Cincotta AH, Tozzo E, Scislowski PWD (1997) Bromocriptine/SKF38393 treatment ameliorates obesity and associated metabolic dysfunctions in obese (ob/ob) mice. Life Sci 61:951–956
56. Bliziotes M, Gunness M, Eshleman A, Wiren K (2002) The role of dopamine and serotonin in regulating bone mass and strength: studies on dopamine and serotonin transporter null mice. J Musculoskelet Neuronal Interact 2:291–295
57. Murzi E et al (1996) Diabetes decreases limbic extracellular dopamine in rats. Neurosci Lett 202:141–144
58. Xing B et al (2010) Dopamine D1 but not D3 receptor is critical for spatial learning and related signaling in the hippocampus. Neuroscience 169:1511–1519
59. El-Ghundi M et al (1999) Spatial learning deficit in dopamine D1 receptor knockout mice. Eur J Pharmacol 383:95–106
60. Mura A, Feldon J (2003) Spatial learning in rats is impaired after degeneration of the nigrostriatal dopaminergic system. Mov Disord 18:860–871
61. Hillbrand M, Spitz RT (1999) Cholesterol and aggression. Aggress Violent Behav 4:359–370
62. Zhang J, Muldoon MF, McKeown RE, Cuffe SP (2005) Association of serum cholesterol and history of school suspension among school-age children and adolescents in the United States. Am J Epidemiol 161:691–699
63. Pentürk S, Yalçin E (2003) Hypocholesterolaemia in dogs with dominance aggression. J Vet Med Series A 50:339–342
64. Mufti RM, Balon R, Arfken CL (1998) Low cholesterol and violence. Psychiatr Serv 49:221–224
65. Hillbrand M et al (2005) Serum cholesterol concentrations and non-physical aggression in healthy adults. J Behav Med 28:295–299
66. Golomb BA, Kane T, Dimsdale JE (2004) Severe irritability associated with statin cholesterol-lowering drugs. Quart J Med 97:229–235
67. Olson MB et al (2008) Lipid-lowering medication use and aggression scores in women: a Report from the NHLBI-sponsored WISE study. J Womens Health (Larchmt) 17:187–194
68. Muldoon MF, Manuck SB, Mendelsohn AB, Kaplan JR, Belle SH (2001) Cholesterol reduction and non-illness mortality: meta-analysis of randomised clinical trials. Brit Med J 322:11–15
69. Golomb BA (1998) Cholesterol and violence: is there a connection? Ann Intern Med 128:478–487
70. Buydens-Branchey L, Branchey M, Hudson J, Fergeson P (2000) Low HDL cholesterol, aggression and altered central serotonergic activity. Psychiatry Res 93:93–102
71. Kaplan J et al (1994) Demonstration of an association among dietary cholesterol, central serotonergic activity, and social behavior in monkeys. Psychosomat Med 56:479–484
72. Erickson MT (1997) Lowered serum cholesterol, famine and aggression: a Darwinian hypothesis. Soc Sci Inform 36:211–222
73. Shanygina KI, Fomina MP, Parfenova NS, Kalashnikova NM (1981) Changes in the cholesterol metabolism of rats following sympathetic and parasympathetic denervation of the liver. Vopr Med Khim 27:505–509
74. Scott LM, Tomkin GH (1985) Cholesterol metabolism: regulatory effects of the vagus in the normal and diabetic animal. Diabetes Res 2:313–317
75. McBurnett K (2000) Low salivary cortisol and persistent aggression in boys referred for disruptive behavior. Arch Gen Psychiatr 57:38–43
76. Sever'ianova LA (1981) Role of ACTH and corticosteroids in the aggressive–defensive behavior of rats. Fiziol Zh SSSR Im I M Sechenova 67:1117–1122
77. Herzog AG, Edelheit PB, Jacobs AR (2001) Low salivary cortisol levels and aggressive behavior. Arch Gen Psychiatry 58:513–515
78. Reinecke DM (2011) Children's cortisol in preschool and aggression one year later in kindergarten. PhD thesis, Auburn University
79. Rizza RA, Mandarino L, Gerich J (1982) Cortisol-induced insulin resistance in man: impaired suppression of glucose production and stimulation of glucose utilization due to a postreceptor defect of insulin action. J Clin Endocrinol Metab 54:131–138

80. Phillips DIW et al (1998) Elevated plasma cortisol concentrations: a link between low birth weight and the insulin resistance syndrome? J Clin Endocrinol Metab 83:757–760
81. Elias M (1981) Serum cortisol, testosterone, and testosterone binding globulin responses to competitive fighting in human males. Aggress Behav 7:215–224
82. Giammanco M, Tabacchi G, Giammanco S, Di Majo D, La Guardia M (2005) Testosterone and aggressiveness. Med Sci Monit 11:RA136–RA145
83. Lamba S et al (2007) A possible novel function of dominance behaviour in queen-less colonies of the primitively eusocial wasp *Ropalidia marginata*. Behav Processes 74:351–356
84. Kumagai S, Holmäng A, Björntorp P (1993) The effects of oestrogen and progesterone on insulin sensitivity in female rats. Acta Physiol Scand 149:91–97
85. Galipeau D, Verma S, McNeill JH (2002) Female rats are protected against fructose-induced changes in metabolism and blood pressure. Am J Physiol Heart Circ Physiol 283:2478–2484
86. Holmäng A, Björntorp P (1992) The effects of testosterone on insulin sensitivity in male rats. Acta Physiol Scand 146:505–510
87. Muller M, Grobbee DE, den Tonkelaar I, Lamberts SWJ, van der Schouw YT (2005) Endogenous sex hormones and metabolic syndrome in aging men. J Clin Endocrinol Metab 90:2618–2623
88. Andersson B et al (1997) Estrogen replacement therapy decreases hyperandrogenicity and improves glucose homeostasis and plasma lipids in postmenopausal women with noninsulin-dependent diabetes mellitus. J Clin Endocrinol Metab 82:638–643
89. Trainor BC, Finy MS, Nelson RJ (2008) Paternal aggression in a biparental mouse: parallels with maternal aggression. Horm Behav 53:200–207
90. Trainor BC, Kyomen HH, Marler CA (2006) Estrogenic encounters: how interactions between aromatase and the environment modulate aggression. Front Neuroendocrinol 27:170–179
91. Trainor BC, Greiwe KM, Nelson RJ (2006) Individual differences in estrogen receptor α in select brain nuclei are associated with individual differences in aggression. Horm Behav 50:338–345
92. Trainor BC, Sima Finy M, Nelson RJ (2008) Rapid effects of estradiol on male aggression depend on photoperiod in reproductively non-responsive mice. Horm Behav 53:192–199
93. Finney HC, Erpino MJ (1976) Synergistic effect of estradiol benzoate and dihydrotestosterone on aggression in mice. Horm Behav 7:391–400
94. Albert DJ, Jonik RH, Walsh ML (1992) Interaction of estradiol, testosterone, and progesterone in the modulation of hormone-dependent aggression in the female rat. Physiol Behav 52:773–779
95. Albert DJ, Jonik RH, Walsh ML (1991) Hormone-dependent aggression in the female rat: testosterone plus estradiol implants prevent the decline in aggression following ovariectomy. Physiol Behav 49:673–677
96. Soma KK, Scotti M-AL, Newman AEM, Charlier TD, Demas GE (2008) Novel mechanisms for neuroendocrine regulation of aggression. Front Neuroendocrinol 29:476–489
97. Tsai LW, Sapolsky RM (1996) Rapid stimulatory effects of testosterone upon myotubule metabolism and sugar transport, as assessed by silicon microphysiometry. Aggress Behav 22:357–364
98. Ahlbom E, Prins GS, Ceccatelli S (2001) Testosterone protects cerebellar granule cells from oxidative stress-induced cell death through a receptor mediated mechanism. Brain Res 892:255–262
99. Túnez I et al (2007) Effect of testosterone on oxidative stress and cell damage induced by 3-nitropropionic acid in striatum of ovariectomized rats. Life Sci 80:1221–1227
100. Franck-Lissbrant I, Häggström S, Damber J-E, Bergh A (1998) Testosterone stimulates angiogenesis and vascular regrowth in the ventral prostate in castrated adult rats. Endocrinology 139:451–456
101. Traish AM, Miner MM, Morgentaler A, Zitzmann M (2011) Testosterone deficiency. Am J Med 124:578–587
102. Dabrosin C, Palmer K, Muller WJ, Gauldie J (2003) Estradiol promotes growth and angiogenesis in polyoma middle T transgenic mouse mammary tumor explants. Breast Cancer Res Treat 78:1–6
103. Mooradian AD (1993) Antioxidant properties of steroids. J Steroid Biochem Mol Biol 45:509–511
104. Ding EL, Song Y, Malik VS, Liu S (2006) Sex differences of endogenous sex hormones and risk of type 2 diabetes. J Am Med Assoc 295:1288–1299
105. Cleary MP, Zisk JF (1986) Anti-obesity effect of two different levels of dehydroepiandrosterone in lean and obese middle-aged female Zucker rats. Int J Obes 10:193–204
106. López-Marure R, Huesca-Gómez C, Ibarra-Sánchez Mde J, Zentella A, Pérez-Méndez O (2007) Dehydroepiandrosterone delays LDL oxidation in vitro and attenuates several oxLDL-induced inflammatory responses in endothelial cells. Inflamm Allergy Drug Targets 6:174–182
107. Liu D et al (2008) Dehydroepiandrosterone stimulates endothelial proliferation and angiogenesis through extracellular signal-regulated kinase 1/2-mediated mechanisms. Endocrinology 149:889–898
108. Hansen PA, Han DH, Nolte LA, Chen M, Holloszy JO (1997) DHEA protects against visceral obesity and muscle insulin resistance in rats fed a high-fat diet. Am J Physiol 273:1704–1708
109. Han DH, Hansen PA, Chen MM, Holloszy JO (1998) DHEA treatment reduces fat accumulation and protects against insulin resistance in male rats. J Gerontol A Biol Sci Med Sci 53:19–24
110. Varet J et al (2004) Dose-dependent effect of dehydroepiandrosterone, but not of its sulphate ester, on angiogenesis. Eur J Pharmacol 502:21–30

111. López-Marure R, Huesca-Gómez C, Ibarra-Sánchez Mde J, Zentella A, Pérez-Méndez O (2007) Dehydroepiandrosterone delays LDL oxidation in vitro and attenuates several oxLDL-induced inflammatory responses in endothelial cells. Inflamm Allergy Drug Targets 6:174–182
112. Golomb BA et al (2002) Insulin sensitivity markers: predictors of accidents and suicides in Helsinki Heart Study screenees. J Clin Epidemiol 55:767–773
113. Kern W et al (2001) Improving influence of insulin oncognitivefunctionsinhumans.Neuroendocrinology 74:270–280
114. Singh RB, Pella D, Mechirova V, Otsuka K (2004) Can brain dysfunction be a predisposing factor for metabolic syndrome? Biomed Pharmacother 58(Suppl 1):56–68
115. Buijs RM, Kreier F (2006) The metabolic syndrome: a brain disease? J Neuroendocrinol 18:715–716
116. Orosco M et al (1992) Striatal dopamine metabolism is differentially affected by insulin according to the genotype in Zucker rats: a microdialysis study. Psychoneuroendocrinology 17:443–452
117. Schoffelmeer ANM et al (2011) Insulin modulates cocaine-sensitive monoamine transporter function and impulsive behavior. J Neurosci 31:1284–1291
118. Figlewicz DP, Szot P, Chavez M, Woods SC, Veith RC (1994) Intraventricular insulin increases dopamine transporter mRNA in rat VTA/substantia nigra. Brain Res 644:331–334
119. Strachan MWJ (2005) Insulin and cognitive function in humans: experimental data and therapeutic considerations. Biochem Soc Trans 33:1037–1040
120. Stockhorst U, de Fries D, Steingrueber H-J, Scherbaum WA (2004) Insulin and the CNS: effects on food intake, memory, and endocrine parameters and the role of intranasal insulin administration in humans. Physiol Behav 83:47–54
121. Hu RM, Levin ER, Pedram A, Frank HJ (1993) Insulin stimulates production and secretion of endothelin from bovine endothelial cells. Diabetes 42:351–358
122. Benedict C et al (2011) Intranasal insulin as a therapeutic option in the treatment of cognitive impairments. Exp Gerontol 46:112–115
123. Benedict C et al (2006) Intranasal insulin improves memory in humans: superiority of insulin aspart. Neuropsychopharmacology 32:239–243
124. Schmidt H, Kern W, Giese R, Hallschmid M, Enders A (2008) Intranasal insulin to improve developmental delay in children with 22q13 deletion syndrome: an exploratory clinical trial. J Med Genet 46:217–222
125. Vera Cruz EM, Brown CL (2007) The influence of social status on the rate of growth, eye color pattern and Insulin-like Growth Factor-I gene expression in Nile tilapia, *Oreochromis niloticus*. Horm Behav 51:611–619
126. Bartos L, Reyes E, Schams D, Bubenik G, Lobos A (1998) Rank dependent seasonal levels of IGF-1, cortisol and reproductive hormones in male pudu (Pudu puda). Comp Biochem Physiol A Mol Integr Physiol 120:373–378
127. Sapolsky RM, Spencer EM (1997) Insulin-like growth factor I is suppressed in socially subordinate male baboons. Am J Physiol Regul Integr Comp Physiol 273:1346–1351
128. Vickers MH, Ikenasio BA, Breier BH (2001) IGF-I treatment reduces hyperphagia, obesity, and hypertension in metabolic disorders induced by fetal programming. Endocrinology 142:3964–3973
129. O'Connell T, Clemmons DR (2002) IGF-I/IGF-binding protein-3 combination improves insulin resistance by GH-dependent and independent mechanisms. J Clin Endocrinol Metab 87:4356–4360
130. Yakar S et al (2001) Liver-specific igf-1 gene deletion leads to muscle insulin insensitivity. Diabetes 50:1110–1118
131. Conti E et al (2004) Insulin-like growth factor-1 as a vascular protective factor. Circulation 110:2260–2265
132. Li Q, Deng X, Singh P (2007) Significant increase in the aggressive behavior of transgenic mice overexpressing peripheral progastrin peptides: associated changes in CCK2 and serotonin receptors in the CNS. Neuropsychopharmacology 32:1813–1821
133. Burgdorf J, Panksepp J, Beinfeld MC, Kroes RA, Moskal JR (2006) Regional brain cholecystokinin changes as a function of rough-and-tumble play behavior in adolescent rats. Peptides 27:172–177
134. Ahrén B, Holst JJ, Efendic S (2000) Antidiabetogenic action of cholecystokinin-8 in type 2 diabetes. J Clin Endocrinol Metab 85:1043–1048
135. Cooper S, Dourish C, Clifton P (1992) CCK antagonists and CCK-monoamine interactions in the control of satiety. Am J Clin Nutr 55:291S–295S
136. Timothy HM (2000) Cholecystokinin and satiety: current perspectives. Nutrition 16:858–865
137. Kuntz E, Pinget M, Damgé P (2004) Cholecystokinin octapeptide: a potential growth factor for pancreatic β cells in diabetic rats. J Pancreas 5:464–475
138. SchmassmannA,ReubiJC(2000)Cholecystokinin-B/gastrin receptors enhance wound healing in the rat gastric mucosa. J Clin Invest 106:1021–1029
139. Atmaca M et al (2006) Serum ghrelin and cholesterolvaluesinsuicideattempters.Neuropsychobiology 54:59–63
140. Tschop M, Smiley DL, Heiman ML (2000) Ghrelin induces adiposity in rodents. Nature 407:908–913
141. Wren AM et al (2001) Ghrelin enhances appetite and increases food intake in humans. J Clin Endocrinol Metab 86:5992
142. Pöykkö SM et al (2003) Low plasma ghrelin is associated with insulin resistance, hypertension, and the prevalence of type 2 diabetes. Diabetes 52:2546–2553
143. Tschöp M et al (2001) Circulating ghrelin levels are decreased in human obesity. Diabetes 50:707–709
144. Nexo E, Hollenberg M, Bing J (1981) Aggressive behavior in mice provokes a marked increase in both plasma epidermal growth factor and renin. Acta Physiol Scand 111:367–71
145. Lakshmanan J (1986) Aggressive behavior in adult male mice elevates serum nerve growth factor levels. Am J Physiol 250:E386–392

146. Nexø E, Olsen PS, Poulsen K (1984) Exocrine and endocrine secretion of renin and epidermal growth factor from the mouse submandibular glands. Regul Pept 8:327–334
147. Roberts ML (1974) Testosterone-induced accumulation of epidermal growth factor in the submandibular salivary glands of mice, assessed by radioimmunoassay. Biochem Pharmacol 23:3305–3308
148. Byyny RL, Orth DN, Cohen S, Doyne ES (1974) Epidermal growth factor: effects of androgens and adrenergic agents. Endocrinology 95:776–782
149. Davison JS (2003) The cervical sympathetic trunk—submandibular gland neuro-endocrine axis: its role in immune regulation. Biomed Res 14:30–37
150. Lamey PJ, Savage AP, Fisher BM, Bloom SR, Frier BM (1990) Secretion of epidermal growth factor in parotid saliva in diabetic patients: role of autonomic innervation. J Oral Pathol Med 19:351–354
151. Kasayama S, Ohba Y, Oka T (1989) Epidermal growth factor deficiency associated with diabetes mellitus. Proc Natl Acad Sci USA 86:7644–7648
152. Oxford GE et al (2000) Salivary EGF levels reduced in diabetic patients. J Diabet Complicat 14:140–145
153. Noguchi S, Ohba Y, Oka T (1990) Involvement of epidermal growth factor deficiency in pathogenesis of oligozoospermia in streptozotocin-induced diabetic mice. Endocrinology 127:2136–2140
154. Brand SJ et al (2002) Pharmacological treatment of chronic diabetes by stimulating pancreatic β-cell regeneration with systemic co-administration of EGF and gastrin. Pharmacol Toxicol 91:414–420
155. Nielsen JH, Svensson C, Douglas Galsgaard E, Moldrup A, Billestrup N (1999) Beta cell proliferation and growth factors. J Mol Med 77:62–66
156. Anchan RM, Reh TA, Angello J, Balliet A, Walker M (1991) EGF and TGF-α stimulate retinal neuroepithelial cell proliferation in vitro. Neuron 6:923–936
157. Peng H et al (1998) Epidermal growth factor protects neuronal cells in vivo and in vitro against transient forebrain ischemia- and free radical-induced injuries. J Cereb Blood Flow Metab 18:349–360
158. Spillantini MG et al (1989) Nerve growth factor mRNA and protein increase in hypothalamus in a mouse model of aggression. Proc Natl Acad Sci USA 86:8555–8559
159. Alleva E, Aloe L, Cirulli F, Della Seta D, Tirassa P (1996) Serum NGF levels increase during lactation and following maternal aggression in mice. Physiol Behav 59:461–466
160. Lakshmanan J (1986) Aggressive behavior in adult male mice elevates serum nerve growth factor levels. Am J Physiol Endocrinol Metabol 250:386–392
161. Maestripieri D, De Simone R, Aloe L, Alleva E (1990) Social status and nerve growth factor serum levels after agonistic encounters in mice. Physiol Behav 47:161–164
162. Lakshmanan J (1986) β-Nerve growth factor measurements in mouse serum. J Neurochem 46:882–891
163. Lakshmanan J (1987) Nerve growth factor levels in mouse serum: variations due to stress. Neurochem Res 12:393–397
164. Aloe L et al (1994) Emotional stress induced by parachute jumping enhances blood nerve growth factor levels and the distribution of nerve growth factor receptors in lymphocytes. Proc Natl Acad Sci USA 91:10440–10444
165. Nitta A et al (2002) Diabetic neuropathies in brain are induced by deficiency of BDNF. Neurotoxicol Teratol 24:695–701
166. Fernyhough P et al (1995) Human recombinant nerve growth factor replaces deficient neurotrophic support in the diabetic rat. Eur J Neurosci 7:1107–1110
167. Hammes HP, Federoff HJ, Brownlee M (1995) Nerve growth factor prevents both neuroretinal programmed cell death and capillary pathology in experimental diabetes. Mol Med 1:527–534
168. Circolo A, Pierce GF, Katz Y, Strunk RC (1990) Antiinflammatory effects of polypeptide growth factors. Platelet-derived growth factor, epidermal growth factor, and fibroblast growth factor inhibit the cytokine-induced expression of the alternative complement pathway activator factor B in human fibroblasts. J Biol Chem 265:5066–5071
169. Lang UE et al (2009) Higher BDNF concentrations in the hippocampus and cortex of an aggressive mouse strain. Behav Brain Res 197:246–249
170. Berton O et al (2006) Essential role of BDNF in the mesolimbic dopamine pathway in social defeat stress. Science 311:864–868
171. Coppola V, Tessarollo L (2004) Control of hyperphagia prevents obesity in BDNF heterozygous mice. Neuroreport 15:2665–2668
172. Lyons WE et al (1999) Brain-derived neurotrophic factor-deficient mice develop aggressiveness and hyperphagia in conjunction with brain serotonergic abnormalities. Proc Natl Acad Sci USA 96:15239–15244
173. Tirassa P, Triaca V, Amendola T, Fiore M, Aloe L (2003) EGF and NGF injected into the brain of old mice enhance BDNF and ChAT in proliferating subventricular zone. J Neurosci Res 72:557–564
174. Nakagawa T et al (2002) Brain-derived neurotrophic factor (BDNF) regulates glucose and energy metabolism in diabetic mice. Diabetes Metab Res Rev 18:185–191
175. Seki M et al (2004) Involvement of brain-derived neurotrophic factor in early retinal neuropathy of streptozotocin-induced diabetes in rats. Diabetes 53:2412–2419
176. Duan W, Guo Z, Jiang H, Ware M, Mattson MP (2003) Reversal of behavioral and metabolic abnormalities, and insulin resistance syndrome, by dietary restriction in mice deficient in brain-derived neurotrophic factor. Endocrinology 144:2446–2453
177. Krabbe KS et al (2006) Brain-derived neurotrophic factor (BDNF) and type 2 diabetes. Diabetologia 50:431–438
178. Tordjman S et al (2003) Aggression and the three opioid families (endorphins, enkephalins, and dynorphins) in mice. Behav Genet 33:529–536
179. Kotegawa T, Abe T, Tsutsui K (1997) Inhibitory role of opioid peptides in the regulation of aggressive and

sexual behaviors in male japanese quails. J Exp Zool 277:146–154
180. Margules D, Moisset B, Lewis M, Shibuya H, Pert C (1978) β-Endorphin is associated with overeating in genetically obese mice (ob/ob) and rats (fa/fa). Science 202:988–991
181. David LM (1979) Beta-endorphin and endoloxone: hormones of the autonomic nervous system for the conservation or expenditure of bodily resources and energy in anticipation of famine or feast. Neurosci Biobehav Rev 3:155–162
182. Giugliano D et al (1987) Hyperglycemia and obesity as determinants of glucose, insulin, and glucagon responses to β-endorphin in human diabetes mellitus. J Clin Endocrinol Metab 64:1122–1128
183. Giugliano D et al (1988) Altered metabolic and hormonal responses to epinephrine and β-endorphin in human obesity. J Clin Endocrinol Metab 67:238–244
184. Su CF et al (2005) Mediation of β-endorphin in exercise-induced improvement in insulin resistance in obese Zucker rats. Diabetes Metab Res Rev 21:175–182
185. Su C-F et al (2004) Infusion of β-endorphin improves insulin resistance in fructose-fed rats. Horm Metab Res 36:571–577
186. Brady KT, Myrick H, McElroy S (1998) The relationship between substance use disorders, impulse control disorders, and pathological aggression. Am J Addict 7:221–230
187. Kaufman KR, Kugler SL, Sachdeo RC (2002) Tiagabine in the management of postencephalitic epilepsy and impulse control disorder. Epilepsy Behav 3:190–194
188. Liu G et al (2007) Reduced aggression in mice lacking GABA transporter subtype 1. J Neurosci Res 85:649–655
189. Lee R, Petty F, Coccaro EF (2009) Cerebrospinal fluid GABA concentration: relationship with impulsivity and history of suicidal behavior, but not aggression, in human subjects. J Psychiatr Res 43:353–359
190. Fisler JS, Shimizu H, Bray GA (1989) Brain 3-hydroxybutyrate, glutamate, and GABA in a rat model of dietary obesity. Physiol Behav 45:571–577
191. Coscina DV, Nobrega JN (1984) Anorectic potency of inhibiting GABA transaminase in brain: studies of hypothalamic, dietary and genetic obesities. Int J Obes 8:191–200
192. Tong Q, Ye C-P, Jones JE, Elmquist JK, Lowell BB (2008) Synaptic release of GABA by AgRP neurons is required for normal regulation of energy balance. Nat Neurosci 11:998–1000
193. Gruber H-J et al (2008) Obesity reduces the bioavailability of nitric oxide in juveniles. Int J Obes 32:826–831
194. Coscina DV, Lloyd KG (1980) Medial hypothalamic obesity: association with impaired hypothalamic GABA synthesis. Brain Res Bull 5:793–796
195. Boutin P et al (2003) GAD2 on chromosome 10p12 is a candidate gene for human obesity. PLoS Biol 1:361–371
196. Jönsson E, Johnsson JI, Björnsson BT (1998) Growth hormone increases aggressive behavior in juvenile rainbow trout. Horm Behav 33:9–15
197. Matte AC (1981) Growth hormone and isolation-induced aggression in wild male mice. Pharmacol Biochem Behav 14:85–87
198. Mondal M, Rajkhowa C, Prakash BS (2006) Relationship between plasma growth hormone concentrations and temperament in mithuns (*Bos frontalis*). Horm Behav 49:190–196
199. Rosén T, Johannsson G, Bengtsson B (1994) Consequences of growth hormone deficiency in adults, and effects of growth hormone replacement therapy. Acta Paediatr 83:21–24
200. Cuneo RC, Salomon F, Wiles CM, Hesp R, Sonksen PH (1991) Growth hormone treatment in growth hormone-deficient adults I. Effects on muscle mass and strength. J Appl Physiol 70:688–694
201. Wallymahmed ME et al (1997) Quality of life, body composition and muscle strength in adult growth hormone deficiency: the influence of growth hormone replacement therapy for up to 3 years. Clin Endocrinol 47:439–446
202. Johannsson G, Grimby G, Sunnerhagen KS, Bengtsson B-Å (1997) Two years of growth hormone (GH) treatment increase isometric and isokinetic muscle strength in GH-deficient adults. J Clin Endocrinol Metab 82:2877–2884
203. Yarasheski KE, Zachwieja JJ, Campbell JA, Bier DM (1995) Effect of growth hormone and resistance exercise on muscle growth and strength in older men. Am J Physiol Endocrinol Metab 268:E268–E276
204. Scacchi M, Pincelli AI, Cavagnini F (1999) Growth hormone in obesity. Int J Obes Relat Metab Disord 23:260–271
205. Rizza RA, Mandarino LJ, Gerich JE (1982) Effects of growth hormone on insulin action in man. Mechanisms of insulin resistance, impaired suppression of glucose production, and impaired stimulation of glucose utilization. Diabetes 31:663–669
206. Krag MB et al (2007) Growth hormone-induced insulin resistance is associated with increased intramyocellular triglyceride content but unaltered VLDL-triglyceride kinetics. Am J Physiol Endocrinol Metab 292:E920–E927
207. Kolaczynski JW, Caro JF (1994) Insulin-like growth factor-1 therapy in diabetes: physiologic basis, clinical benefits, and risks. Ann Int Med 120:47–55
208. Cowey SL et al (2005) Abdominal obesity, insulin resistance, and colon carcinogenesis are increased in mutant mice lacking gastrin gene expression. Cancer 103:2643–2653
209. Suarez-Pinzon WL, Lakey JRT, Brand SJ, Rabinovitch A (2005) Combination therapy with epidermal growth factor and gastrin induces neogenesis of human islet β-cells from pancreatic duct cells and an increase in functional β-cell mass. J Clin Endocrinol Metab 90:3401–3409
210. Kovacheva EL, Sinha Hikim AP, Shen R, Sinha I, Sinha-Hikim I (2010) Testosterone supplementation

reverses sarcopenia in aging through regulation of myostatin, c-Jun NH2-terminal kinase, notch, and AKT signaling pathways. Endocrinology 151:628–638
211. Lee SJ, McPherron AC (2001) Regulation of myostatin activity and muscle growth. Proc Natl Acad Sci USA 98:9306–9311
212. Amirouche A et al (2009) Down-regulation of Akt/mammalian target of rapamycin signaling pathway in response to myostatin overexpression in skeletal muscle. Endocrinology 150:286–294
213. Hittel DS, Berggren JR, Shearer J, Boyle K, Houmard JA (2009) Increased secretion and expression of myostatin in skeletal muscle from extremely obese women. Diabetes 58:30–38
214. McPherron AC, Lee S-J (2002) Suppression of body fat accumulation in myostatin-deficient mice. J Clin Invest 109:595–601
215. Chiavegatto S et al (2001) Brain serotonin dysfunction accounts for aggression in male mice lacking neuronal nitric oxide synthase. Proc Natil Acad Sci 98:1277–1281
216. Demas GE et al (1997) Inhibition of neuronal nitric oxide synthase increases aggressive behavior in mice. Mol Med 3:610–616
217. Demas GE et al (1999) Elimination of aggressive behavior in male mice lacking endothelial nitric oxide synthase. J Neurosci 19:RC30
218. Duplain H et al (2001) Insulin resistance, hyperlipidemia, and hypertension in mice lacking endothelial nitric oxide synthase note added in proof. Circulation 104:342–345
219. Monti LD et al (2003) Endothelial nitric oxide synthase polymorphisms are associated with type 2 diabetes and the insulin resistance syndrome. Diabetes 52:1270–1275
220. Nelson RJ et al (1995) Behavioural abnormalities in male mice lacking neuronal nitric oxide synthase. Nature 378:383–386
221. Kriegsfeld LJ, Dawson TM, Dawson VL, Nelson RJ, Snyder SH (1997) Aggressive behavior in male mice lacking the gene for neuronal nitric oxide synthase requires testosterone. Brain Res 769:66–70
222. Gammie SC, Nelson RJ (1999) Maternal aggression is reduced in neuronal nitric oxide synthase-deficient mice. J Neurosci 19:8027–8035
223. Shankar RR, Wu Y, Shen HQ, Zhu JS, Baron AD (2000) Mice with gene disruption of both endothelial and neuronal nitric oxide synthase exhibit insulin resistance. Diabetes 49:684–687
224. Perreault M, Marette A (2001) Targeted disruption of inducible nitric oxide synthase protects against obesity-linked insulin resistance in muscle. Nat Med 7:1138–1143
225. Valera A, Pujol A, Pelegrin M, Bosch F (1994) Transgenic mice overexpressing phosphoenolpyruvate carboxykinase develop non-insulin-dependent diabetes mellitus. Proc Natl Acad Sci 91:9151–9154
226. Sun Y et al (2002) Phosphoenolpyruvate carboxykinase overexpression selectively attenuates insulin signaling and hepatic insulin sensitivity in transgenic mice. J Biol Chem 277:23301–23307
227. Franckhauser S et al (2002) Increased fatty acid re-esterification by PEPCK overexpression in adipose tissue leads to obesity without insulin resistance. Diabetes 51:624–630
228. Hanson RW, Hakimi P (2008) Born to run; the story of the PEPCK-Cmus mouse. Biochimie 90:838–842
229. Zick Y (2005) Ser/Thr phosphorylation of IRS proteins: a molecular basis for insulin resistance. Sci STKE 2005:pe4
230. Khamzina L, Veilleux A, Bergeron S, Marette A (2005) Increased activation of the mammalian target of rapamycin pathway in liver and skeletal muscle of obese rats: possible involvement in obesity-linked insulin resistance. Endocrinology 146:1473–1481
231. Haar EV, Lee S, Bandhakavi S, Griffin TJ, Kim D-H (2007) Insulin signalling to mTOR mediated by the Akt/PKB substrate PRAS40. Nat Cell Biol 9:316–323
232. Dreyer HC et al (2008) Leucine-enriched essential amino acid and carbohydrate ingestion following resistance exercise enhances mTOR signaling and protein synthesis in human muscle. Am J Physiol Endocrinol Metab 294:E392–E400
233. Bolster DR et al (2003) Immediate response of mammalian target of rapamycin (mTOR)-mediated signalling following acute resistance exercise in rat skeletal muscle. J Physiol 553:213–220
234. Bodine SC (2006) mTOR signaling and the molecular adaptation to resistance exercise. Med Sci Sports Exerc 38:1950–1957
235. Deldicque L, Theisen D, Francaux M (2005) Regulation of mTOR by amino acids and resistance exercise in skeletal muscle. Eur J Appl Physiol 94:1–10
236. Williamson DL, Kubica N, Kimball SR, Jefferson LS (2006) Exercise-induced alterations in extracellular signal-regulated kinase 1/2 and mammalian target of rapamycin (mTOR) signalling to regulatory mechanisms of mRNA translation in mouse muscle. J Physiol 573:497–510
237. Wu Y, Bauman WA, Blitzer RD, Cardozo C (2010) Testosterone-induced hypertrophy of L6 myoblasts is dependent upon Erk and mTOR. Biochem Biophys Res Commun 400:679–683
238. Altamirano F et al (2009) Testosterone induces cardiomyocyte hypertrophy through mammalian target of rapamycin complex 1 pathway. J Endocrinol 202:299–307
239. Allemand MC et al (2009) Effect of testosterone on insulin stimulated IRS1 Ser phosphorylation in primary rat myotubes—a potential model for PCOS-related insulin resistance. PLoS One 4:e4274
240. Xu X, De Pergola G, Björntorp P (1990) The effects of androgens on the regulation of lipolysis in adipose precursor cells. Endocrinology 126:1229–1234
241. Xu XF, De Pergola G, Björntorp P (1991) Testosterone increases lipolysis and the number of β-adrenoceptors in male rat adipocytes. Endocrinology 128:379–382

242. Mårin P, Odén B, Björntorp P (1995) Assimilation and mobilization of triglycerides in subcutaneous abdominal and femoral adipose tissue in vivo in men: effects of androgens. J Clin Endocrinol Metab 80:239–243
243. Khaw K-T, Barrett-Connor E (1992) Lower endogenous androgens predict central adiposity in men. Ann Epidemiol 2:675–682
244. Rebuffé-Scrive M, Mårin P, Björntorp P (1991) Effect of testosterone on abdominal adipose tissue in men. Int J Obes 15:791–795
245. Mårin P et al (1992) The effects of testosterone treatment on body composition and metabolism in middle-aged obese men. Int J Obes Relat Metab Disord 16:991–997
246. Mrin P, Arver S (1998) Androgens and abdominal obesity. *BailliÃ¨re's clinical*. Endocrinol Metab 12:441–451
247. Falahati-Nini A et al (2000) Relative contributions of testosterone and estrogen in regulating bone resorption and formation in normal elderly men. J Clin Invest 106:1553–1560
248. Ferron M, Hinoi E, Karsenty G, Ducy P (2008) Osteocalcin differentially regulates β cell and adipocyte gene expression and affects the development of metabolic diseases in wild-type mice. Proc Natl Acad Sci USA 105:5266–5270
249. Pittas AG, Harris SS, Eliades M, Stark P, Dawson-Hughes B (2009) Association between serum osteocalcin and markers of metabolic phenotype. J Clin Endocrinol Metab 94:827–832
250. Rached M-T et al (2010) FoxO1 expression in osteoblasts regulates glucose homeostasis through regulation of osteocalcin in mice. J Clin Invest 120:357–368
251. Saleem U, Mosley TH, Kullo IJ (2010) Serum osteocalcin is associated with measures of insulin resistance, adipokine levels, and the presence of metabolic syndrome. Arterioscler Thromb Vasc Biol 30:1474–1478
252. Kanazawa I et al (2009) Serum osteocalcin level is associated with glucose metabolism and atherosclerosis parameters in Type 2 diabetes mellitus. J Clin Endocrinol Metab 94:45–49
253. Kosfeld M, Heinrichs M, Zak PJ, Fischbacher U, Fehr E (2005) Oxytocin increases trust in humans. Nature 435:673–676
254. Kirsch P et al (2005) Oxytocin modulates neural circuitry for social cognition and fear in humans. J Neurosci 25:11489–11493
255. Anne C (2008) Attachment, aggression and affiliation: The role of oxytocin in female social behavior. Biol Psychol 77:1–10
256. Gulledge AK, Hill M, Lister Z, Sallion C (2007) Non-erotic physical affection: it's good for you. In: L'Abate L (ed) Low-cost approaches to promote physical and mental health. Springer, New York
257. Light KC, Grewen KM, Amico JA (2005) More frequent partner hugs and higher oxytocin levels are linked to lower blood pressure and heart rate in premenopausal women. Biol Psychol 69:5–21
258. Holt-Lunstad J, Birmingham WA, Light KC (2008) Influence of a 'warm touch' support enhancement intervention among married couples on ambulatory blood pressure, oxytocin, α amylase, and cortisol. Psychosomat Med 70:976–985
259. Gordon I, Zagoory-Sharon O, Leckman JF, Feldman R (2010) Oxytocin and the development of parenting in humans. Biol Psychiatr 68:377–382
260. Kortesluoma S, Karlsson H (2011) Oxytocin, a neuropeptide regulating affection and social behavior. Duodecim 127:911–918
261. Naber F, van IJzendoorn MH, Deschamps P, van Engeland H, Bakermans-Kranenburg MJ (2010) Intranasal oxytocin increases fathers' observed responsiveness during play with their children: a double-blind within-subject experiment. Psychoneuroendocrinology 35:1583–1586
262. Bosch OJ, Meddle SL, Beiderbeck DI, Douglas AJ, Neumann ID (2005) Brain oxytocin correlates with maternal aggression: link to anxiety. J Neurosci 25:6807–6815
263. Ferris CF et al (1992) Oxytocin in the amygdala facilitates maternal aggression. Ann N Y Acad Sci 652:456–457
264. DeVries AC, Young WS III, Nelson RJ (1997) Reduced aggressive behaviour in mice with targeted disruption of the oxytocin gene. J Neuroendocrinology 9:363–368
265. Bales KL, Carter CS (2003) Sex differences and developmental effects of oxytocin on aggression and social behavior in prairie voles (*Microtus ochrogaster*). Horm Behav 44:178–184
266. Ferris CF (2005) Vasopressin/oxytocin and aggression. Novartis Found Symp 268:190–198 (discussion 198–200, 242–253)
267. Stock S, Granström L, Backman L, Matthiesen AS, Uvnäs-Moberg K (1989) Elevated plasma levels of oxytocin in obese subjects before and after gastric banding. Int J Obes 13:213–222
268. Kublaoui BM, Gemelli T, Tolson KP, Wang Y, Zinn AR (2008) Oxytocin deficiency mediates hyperphagic obesity of Sim1 haploinsufficient mice. Mol Endocrinol 22:1723–1734
269. Takayanagi Y et al (2008) Oxytocin receptor-deficient mice developed late-onset obesity. Neuroreport 19:951–955
270. Tolson KP et al (2010) Postnatal Sim1 deficiency causes hyperphagic obesity and reduced Mc4r and oxytocin expression. J Neurosci 30:3803–3812
271. Floyd K, Mikkelson AC, Hesse C, Pauley PM (2007) Affectionate writing reduces total cholesterol: two randomized, controlled trials. Hum Commun Res 33:119–142
272. Mirsky I, Perisutti G (1961) The insulin-like action of oxytocin on adipose tissue. Biochim Biophys Acta 50:603–604
273. Braun T, Hechter O, Rudinger J (1969) 'Insulin-like' action of oxytocin: evidence for separate oxytocin-sensitive and insulin-sensitive systems in fat cells. Endocrinology 85:1092–1096

274. Augert G, Exton JH (1988) Insulin and oxytocin effects on phosphoinositide metabolism in adipocytes. J Biol Chem 263:3600–3609
275. Whitton PD, Rodrigues LM, Hems DA (1978) Stimulation by vasopressin, angiotensin and oxytocin of gluconeogenesis in hepatocyte suspensions. Biochem J 176:893–898
276. Camerino C (2009) Low sympathetic tone and obese phenotype in oxytocin-deficient mice. Obesity 17:980–984
277. Altszuler N, Hampshire J (1981) Oxytocin infusion increases plasma insulin and glucagon levels and glucose production and uptake in the normal dog. Diabetes 30:112–114
278. Andrade ML, Benton D, Brain PF, Ramirez JM, Walmsley SV (1988) A reexamination of the hypoglycemia-aggression hypothesis in laboratory mice. Int J Neurosci 41:179–186
279. Benton D (1988) Hypoglycemia and aggression: a review. Int J Neurosci 41:163–168
280. Bolton R (1976) Hostility in fantasy: a further test of the hypoglycemia-aggression hypothesis. Aggr Behav 2:257–274
281. Lewellen TC et al (1981) Aggression and hypoglycemia in the andes: another look at the evidence [and comments and replies]. Curr Anthropol 22:347–361
282. Bolton R et al (1984) The hypoglycemia-aggression hypothesis: debate versus research [and comments and reply]. Curr Anthropol 25:1–53
283. Bolton R (1973) Aggression and hypoglycemia among the Qolla: a study in psychobiological anthropology. Ethnology 12:227–257
284. DeNapoli JS, Dodman NH, Shuster L, Rand WM, Gross KL (2000) Effect of dietary protein content and tryptophan supplementation on dominance aggression, territorial aggression, and hyperactivity in dogs. J Am Vet Med Assoc 217:504–508
285. Chamberlain B, Ervin FR, Pihl RO, Young SN (1987) The effect of raising or lowering tryptophan levels on aggression in vervet monkeys. Pharmacol Biochem Behav 28:503–510
286. Cleare AJ, Bond AJ (1995) The effect of tryptophan depletion and enhancement on subjective and behavioural aggression in normal male subjects. Psychopharmacology 118:72–81
287. Bjork JM, Dougherty DM, Moeller FG, Cherek DR, Swann AC (1999) The effects of tryptophan depletion and loading on laboratory aggression in men: time course and a food-restricted control. Psychopharmacology 142:24–30
288. LeMarquand DG et al (1998) Tryptophan depletion, executive functions, and disinhibition in aggressive, adolescent males. Neuropsychopharmacology 19:333–341
289. Schmidt HH, Warner TD (1992) Insulin secretion from pancreatic beta cells caused by L-arginine-derived nitrogen oxides. Science 255:721–723
290. Mulloy A, Kari F, Visek W (2008) Dietary arginine, insulin secretion, glucose tolerance and liver lipids during repletion of protein-depleted rats. Horm Metab Res 14:471–475
291. Floyd JC, Fajans SS, Conn JW, Knopf RF, Rull J (1966) Stimulation of insulin secretion by amino acids. J Clin Invest 45:1487–1502
292. Weber M (2008) Influence of dietary arginine on behavior. The Ohio State University, Department of Psychology Honors Theses
293. Kalueff AV et al (2006) Behavioural anomalies in mice evoked by 'Tokyo' disruption of the Vitamin D receptor gene. Neurosci Res 54:254–260
294. Kalueff AV, Lou Y-R, Laaksi I, Tuohimaa P (2004) Increased grooming behavior in mice lacking vitamin D receptors. Physiol Behav 82:405–409
295. Holick MF (2007) Vitamin D deficiency. N Engl J Med 357:266–281
296. Borissova AM, Tankova T, Kirilov G, Dakovska L, Kovacheva R (2003) The effect of vitamin D3 on insulin secretion and peripheral insulin sensitivity in type 2 diabetic patients. Int J Clin Pract 57:258–261
297. Scheffel A (1996) Serum cholesterol, triglycerides, HDL and LDL in aggressive elderly patients with dementia. Psychiatr Pol 30:159–170
298. Kaplan JR, Manuck SB, Shively C (1991) The effects of fat and cholesterol on social behavior in monkeys. Psychosom Med 53:634–642
299. Kaplan JR, Fontenot MB, Manuck SB, Muldoon MF (1996) Influence of dietary lipids on agonistic and affiliative behavior in *Macaca fascicularis*. Am J Primatol 38:333–347
300. Golomb B (2011) Low cholesterol and violence—further findings, and evidence serum cholesterol needn't be a surrogate for brain cholesterol. Response to Sheehan et al aberrant cholesterol and lipoprotein levels in aggressive male adolescents. WebmedCentral Psychitry 2(10):WMC002346
301. Leedom LJ, Meehan WP (1989) The psychoneuroendocrinology of diabetes mellitus in rodents. Psychoneuroendocrinology 14:275–294
302. DiMagno MJ, Williams JA, Hao Y, Ernst SA, Owyang C (2004) Endothelial nitric oxide synthase is protective in the initiation of caerulein-induced acute pancreatitis in mice. Am J Physiol Gastrointest Liver Physiol 287:G80–87
303. Kurihara Y et al (2000) Role of endothelin-1 in stress response in the central nervous system. Am J Physiol Regul Integr Comp Physiol 279: R515–R521
304. Brain PF, Nowell NW, Wouters A (1971) Some relationships between adrenal function and the effectiveness of a period of isolation in inducing intermale aggression in albino mice. Physiol Behav 6:27–29
305. Bigi S, Huber C, De Acetis L, Alleva E, Dixon AK (1994) Removal of the submaxillary salivary glands first increases and then abolishes the agonistic response of male mice in repeated social encounters. Physiol Behav 55:13–19
306. Taha M, McMillon R, Napier A, Wekesa KS (2009) Extracts from salivary glands stimulate aggression and inositol-1,4,5-triphosphate (IP3) production in the vomeronasal organ of mice. Physiol Behav 98:147–155

307. Farrell PA, Kjaer M, Bach FW, Galbo H (1987) B endorphin and adrenocorticotropin response to supramaximal treadmill exercise in trained and untrained males. Acta Physiol Scand 130:619–625
308. Schwarz L, Kindermann W (2008) β-Endorphin, catecholamines, and cortisol during exhaustive endurance exercise*. Int J Sports Med 10:324–328
309. Miner LL, Elmer GI, Pieper JO, Marley RJ (1993) Aggression modulates genetic influences on morphine analgesia as assessed using a classical Mendelian cross analysis. Psychopharmacology 111:17–22
310. Miczek KA, Thompson ML, Shuster L (1986) Analgesia following defeat in an aggressive encounter: development of tolerance and changes in *Opioid Receptors*. Ann New York Acad Sci 467:14–29
311. Kim J, Montagnani M, Koh KK, Quon MJ (2006) Reciprocal relationships between insulin resistance and endothelial dysfunction. Circulation 113:1888–1904
312. Ferri C et al (1995) Insulin stimulates endothelin-1 secretion from human endothelial cells and modulates its circulating levels in vivo. J Clin Endocrinol Metab 80:829–835
313. Wolpert HA, Steen SN, Istfan NW, Simonson DC (1993) Insulin modulates circulating endothelin-1 levels in humans. Metab Clin Exp 42:1027–1030
314. Wilkes JJ, Hevener A, Olefsky J (2003) Chronic endothelin-1 treatment leads to insulin resistance in vivo. Diabetes 52:1904–1909
315. Ottosson seeberger A, Lundberg JM, Alvestrand A, Ahlborg G (1997) Exogenous endothelin 1 causes peripheral insulin resistance in healthy humans. Acta Physiol Scand 161:211–220
316. Juan CC et al (1996) Endothelin-1 induces insulin resistance in conscious rats. Biochem Biophys Res Commun 227:694–699
317. Piatti PM et al (2000) Relationship between endothelin-1 concentration and metabolic alterations typical of the insulin resistance syndrome. Metab Clin Exp 49:748–752
318. Gregersen S, Thomsen JL, Brock B, Hermansen K (1996) Endothelin-1 stimulates insulin secretion by direct action on the islets of Langerhans in mice. Diabetologia 39:1030–1035
319. Ferri C et al (1995) Plasma endothelin-1 levels in obese hypertensive and normotensive men. Diabetes 44:431–436
320. Takahashi K, Ghatei MA, Lam H-C, O'Halloran DJ, Bloom SR (1990) Elevated plasma endothelin in patients with diabetes mellitus. Diabetologia 33:306–310
321. Seligman BG, Biolo A, Polanczyk CA, Gross JL, Clausell N (2000) Increased plasma levels of endothelin 1 and von Willebrand factor in patients with type 2 diabetes and dyslipidemia. Diabetes Care 23:1395–1400
322. Schneider JG et al (2002) Elevated plasma endothelin-1 levels in diabetes mellitus. Am J Hypertens 15:967–972
323. Perini C, Müller FB, Bühler FR (1991) Suppressed aggression accelerates early development of essential hypertension. J Hypertens 9:499–503
324. Perini C, Müller FB, Rauchfleisch U, Battegay R, Bühler FR (1986) Hyperadrenergic borderline hypertension is characterized by suppressed aggression. J Cardiovasc Pharmacol 8(Suppl 5):S53–S56
325. Cooper SJ, Jackson A, Kirkham TC (1985) Endorphins and food intake: kappa opioid receptor agonists and hyperphagia. Pharmacol Biochem Behav 23:889–901
326. David McKay L, Kenney NJ, Edens NK, Williams RH, Woods SC (1981) Intracerebroventricular β-endorphin increases food intake of rats. Life Sci 29:1429–1434
327. Martinowich K, Lu B (2007) Interaction between BDNF and serotonin: role in mood disorders. Neuropsychopharmacology 33:73–83
328. Mattson MP, Maudsley S, Martin B (2004) BDNF and 5-HT: a dynamic duo in age-related neuronal plasticity and neurodegenerative disorders. Trends Neurosci 27:589–594
329. Xie H, Lautt WW (1996) Insulin resistance of skeletal muscle produced by hepatic parasympathetic interruption. Am J Physiol Endocrinol Metab 270:858–863
330. Strubbe JH (1992) Parasympathetic involvement in rapid meal-associated conditioned insulin secretion in the rat. Am J Physiol Regul Integr Comp Physiol 263:R615–R618
331. D'Alessio DA, Kieffer TJ, Taborsky GJ, Havel PJ (2001) Activation of the parasympathetic nervous system is necessary for normal meal-induced insulin secretion in rhesus macaques. J Clin Endocrinol Metab 86:1253–1259
332. Farr SA, Banks WA, Morley JE (2006) Effects of leptin on memory processing. Peptides 27:1420–1425
333. Lustig RH, Sen S, Soberman JE, Velasquez-Mieyer PA (2004) Obesity, leptin resistance, and the effects of insulin reduction. Int J Obes Relat Metab Disord 28:1344–1348
334. Isganaitis E, Lustig RH (2005) Fast food, central nervous system insulin resistance, and obesity. Arterioscler Thromb Vasc Biol 25:2451–2462
335. Krahl ME (1964) Specificity of insulin or oxytocin stimulation of protein synthesis in adipose tissue. Am J Physiol Legacy Content 207:1169–1172
336. Pitteloud N et al (2005) Relationship between testosterone levels, insulin sensitivity, and mitochondrial function in men. Diabetes Care 28:1636–1642
337. Traish AM, Saad F, Guay A (2009) The dark side of testosterone deficiency: II. Type 2 diabetes and insulin resistance. J Androl 30:23–32
338. Kaplan SA, Meehan AG, Shah A (2006) The age related decrease in testosterone is significantly exacerbated in obese men with the metabolic syndrome. What are the implications for the relatively high incidence of erectile dysfunction observed in these men? J Urol 176:1524–1528

339. Kasayama S, Ohba Y, Oka T (1989) Epidermal growth factor deficiency associated with diabetes mellitus. Proc Natl Acad Sci USA 86:7644–7648
340. Hellweg R, Hartung HD (1990) Endogenous levels of nerve growth factor (NGF) are altered in experimental diabetes mellitus: A possible role for NGF in the pathogenesis of diabetic neuropathy. J Neurosci Res 26:258–267
341. Hellweg R, Raivich G, Hartung H-D, Hock C, Kreutzberg GW (1994) Axonal transport of endogenous nerve growth factor (NGF) and NGF receptor in experimental diabetic neuropathy. Exp Neurol 130:24–30
342. Apfel SC, Arezzo JC, Brownlee M, Federoff H, Kessler JA (1994) Nerve growth factor administration protects against experimental diabetic sensory neuropathy. Brain Res 634:7–12
343. Anand P et al (1996) The role of endogenous nerve growth factor in human diabetic neuropathy. Nat Med 2:703–707
344. Apfel SC et al (1998) Recombinant human nerve growth factor in the treatment of diabetic polyneuropathy. Neurology 51:695–702
345. Gray J et al (2006) Hyperphagia, severe obesity, impaired cognitive function, and hyperactivity associated with functional loss of one copy of the brain-derived neurotrophic factor (BDNF) gene. Diabetes 55:3366–3371
346. Malgor L, Fisher J (1970) Effects of testosterone on erythropoietin production in isolated perfused kidneys. Am J Physiol Legacy Content 218:1732–1736
347. Rishpon-Meyerstein N, Kilbridge T, Simone J, Fried W (1968) The effect of testosterone on erythropoietin levels in anemic patients. Blood 31:453–460
348. Moriyama Y, Fisher J (1975) Effects of testosterone and erythropoietin on erythroid colony formation in human bone marrow cultures. Blood 45:665–670
349. McGill JB, Bell DSH (2006) Anemia and the role of erythropoietin in diabetes. J Diabetes Complicat 20:262–272
350. Pavkovic P, Mrzljak V, Profozic V, Metelko Z (2004) Erythropoietin in the treatment of patients with type 2 diabetes mellitus and anemia. Diabetologia Croat 33:13–16
351. Bosman DR, Winkler AS, Marsden JT, Macdougall IC, Watkins PJ (2001) Anemia with erythropoietin deficiency occurs early in diabetic nephropathy. Diabetes Care 24:495–499
352. Thomas MC, Cooper ME, Tsalamandris C, MacIsaac R, Jerums G (2005) Anemia with impaired erythropoietin response in diabetic patients. Arch Intern Med 165:466–469
353. Dikow R, Schwenger V, Schomig M, Ritz E (2002) How should we manage anaemia in patients with diabetes? Nephrol Dial Transplant 17:67–72
354. Thomas M (2006) The high prevalence of anemia in diabetes is linked to functional erythropoietin deficiency. Semin Nephrol 26:275–282
355. Fernández-Real JM, López-Bermejo A, Ricart W (2002) Cross-talk between iron metabolism and diabetes. Diabetes 51:2348–2354
356. Jiang R et al (2004) Body iron stores in relation to risk of type 2 diabetes in apparently healthy women. J Am Med Assoc 291:711–717
357. Hull EM et al (1999) Hormone–neurotransmitter interactions in the control of sexual behavior. Behav Brain Res 105:105–116
358. Lorrain DS, Riolo JV, Matuszewich L, Hull EM (1999) Lateral hypothalamic serotonin inhibits nucleus accumbens dopamine: implications for sexual satiety. J Neurosci 19:7648–7652
359. van Bokhoven I et al (2006) Salivary testosterone and aggression, delinquency, and social dominance in a population-based longitudinal study of adolescent males. Horm Behav 50:118–125
360. Michael RP, Zumpe D (1993) A review of hormonal factors influencing the sexual and aggressive behavior of macaques. Am J Primatol 30:213–241
361. Albert DJ, Jonik RH, Walsh ML (1992) Hormone-dependent aggression in male and female rats: experiential, hormonal, and neural foundations. Neurosci Biobehav Rev 16:177–192
362. Blanchard DC, Sakai RR, McEwen B, Weiss SM, Blanchard RJ (1993) Subordination stress: behavioral, brain, and neuroendocrine correlates. Behav Brain Res 58:113–121
363. Lu C-C et al (2009) Association of glycemic control with risk of erectile dysfunction in men with type 2 diabetes. J Sex Med 6:1719–1728
364. Awad H, Salem A, Gadalla A, El Wafa NA, Mohamed OA (2010) Erectile function in men with diabetes type 2: correlation with glycemic control. Int J Impot Res 22:36–39
365. Hidalgo-Tamola J, Chitaley K (2009) Review type 2 diabetes mellitus and erectile dysfunction. J Sex Med 6:916–926
366. Romeo JH, Seftel AD, Madhun ZT, Aron DC (2000) Sexual function in men with diabetes type 2: association with glycemic control. J Urol 163:788–791
367. Yaman O, Akand M, Gursoy A, Erdogan MF, Anafarta K (2006) The effect of diabetes mellitus treatment and good glycemic control on the erectile function in men with diabetes mellitus-induced erectile dysfunction: a pilot study. J Sex Med 3:344–348
368. Festa A et al (2000) Chronic subclinical inflammation as part of the insulin resistance syndrome: the insulin resistance atherosclerosis study (IRAS). Circulation 102:42–47
369. Lee PC et al (1999) Impaired wound healing and angiogenesis in eNOS-deficient mice. Am J Physiol Heart Circ Physiol 277:H1600–H1608
370. Fukumura D et al (2001) Predominant role of endothelial nitric oxide synthase in vascular endothelial growth factor-induced angiogenesis and vascular permeability. Proc Natl Acad Sci USA 98:2604–2609

371. Ying L, Hofseth LJ (2007) An emerging role for endothelial nitric oxide synthase in chronic inflammation and cancer. Cancer Res 67:1407–1410
372. Ayala JE et al (2007) Chronic treatment with sildenafil improves energy balance and insulin action in high fat-fed conscious mice. Diabetes 56:1025–1033
373. Milman HA, Arnold SB (2002) Neurologic, psychological, and aggressive disturbances with sildenafil. Ann Pharmacother 36:1129–1134
374. Hotchkiss AK et al (2005) Aggressive behavior increases after termination of chronic sildenafil treatment in mice. Physiol Behav 83:683–688
375. Sgoifo A, De Boer SF, Haller J, Koolhaas JM (1996) Individual differences in plasma catecholamine and corticosterone stress responses of wild-type rats: relationship with aggression. Physiol Behav 60:1403–1407
376. Haller J, Makara G, Kruk M (1997) Catecholaminergic involvement in the control of aggression: hormones, the peripheral sympathetic, and central noradrenergic systems. Neurosci Biobehav Rev 22:85–97
377. Korzan WJ, Summers TR, Summers CH (2000) Monoaminergic activities of limbic regions are elevated during aggression: influence of sympathetic social signaling. Brain Res 870:170–178
378. Korzan WJ, Summers TR, Ronan PJ, Summers CH (2000) Visible sympathetic activity as a social signal in anolis carolinensis: changes in aggression and plasma catecholamines. Horm Behav 38:193–199
379. Schaan BD et al (2005) Sympathetic modulation of the renal glucose transporter GLUT2 in diabetic rats. Auton Neurosci 117:54–61
380. Eriksson J et al (2007) Aerobic endurance exercise or circuit-type resistance training for individuals with impaired glucose tolerance? Horm Metab Res 30:37–41
381. Eriksson J et al (2007) Resistance training in the treatment of non-insulin-dependent diabetes mellitus. Int J Sports Med 18:242–246
382. Cuff DJ et al (2003) Effective exercise modality to reduce insulin resistance in women with type 2 diabetes. Diabetes Care 26:2977–2982
383. Babraj JA et al (2009) Extremely short duration high intensity interval training substantially improves insulin action in young healthy males. BMC Endocr Disord 9:3
384. Nybo L et al (2010) High-intensity training versus traditional exercise interventions for promoting health. Med Sci Sports Exerc 42:1951–1958
385. Richards JC et al (2010) Short–term sprint interval training increases insulin sensitivity in healthy adults but does not affect the thermogenic response to β adrenergic stimulation. J Physiol 588:2961–2972
386. Hawley JA, Gibala MJ (2009) Exercise intensity and insulin sensitivity: how low can you go? Diabetologia 52:1709–1713
387. Meckel Y et al (2009) The effect of a brief sprint interval exercise on growth factors and inflammatory mediators. J Strength Cond Res 23:225–230
388. Al Rashid et al (2010) Effects of aggressive and non-aggressive exercise on insulin sensitivity in young men: a cross sectional pilot study. MRes thesis, University of Glasgow
389. Volek JS, Kraemer WJ, Bush JA, Incledon T, Boetes M (1997) Testosterone and cortisol in relationship to dietary nutrients and resistance exercise. J Appl Physiol 82:49–54
390. Nexo E, Hansen MR, Konradsen L (1988) Human salivary epidermal growth factor, haptocorrin and amylase before and after prolonged exercise. Scand J Clin Lab Invest Informa Healthcare 48:269–273
391. Lukaszyk A et al (1998) The role of epidermal growth factor in platelet-endothelium interactions. J Physiol Pharmacol 49:229–239
392. Chae C-H, Kim H-T (2009) Forced, moderate-intensity treadmill exercise suppresses apoptosis by increasing the level of NGF and stimulating phosphatidylinositol 3-kinase signaling in the hippocampus of induced aging rats. Neurochem Int 55:208–213
393. Ang ET, Gomez-Pinilla F (2007) Potential therapeutic effects of exercise to the brain. Curr Med Chem 14:2564–2571
394. Widenfalk J, Olson L, Thorén P (1999) Deprived of habitual running, rats downregulate BDNF and TrkB messages in the brain. Neurosci Res 34:125–132
395. Stranahan AM et al (2009) Voluntary exercise and caloric restriction enhance hippocampal dendritic spine density and BDNF levels in diabetic mice. Hippocampus 19:951–961
396. Ritz P, Berrut G (2005) Mitochondrial function, energy expenditure, aging and insulin resistance. Diabet Metab 31:5S67–5S73
397. Stark R, Roden M (2007) Mitochondrial function and endocrine diseases. Eur J Clin Invest 37:236–248
398. Farrar RP, Martin TP, Ardies CM (1981) The interaction of aging and endurance exercise upon the mitochondrial function of skeletal muscle. J Gerontol 36:642–647
399. Tonkonogi M, Sahlin K (2002) Physical exercise and mitochondrial function in human skeletal muscle. Exerc Sport Sci Rev 30:129–137
400. Menshikova EV et al (2005) Effects of weight loss and physical activity on skeletal muscle mitochondrial function in obesity. Am J Physiol Endocrinol Metab 288:E818–E825
401. Menshikova EV et al (2006) Effects of exercise on mitochondrial content and function in aging human skeletal muscle. J Gerontol A Biol Sci Med Sci 61:534–540
402. Breda E et al (1992) Modulation of fatty-acid-binding protein content of rat heart and skeletal muscle by endurance training and testosterone treatment. Pflugers Archiv Eur J Physiol 421:274–279

403. Kimberg DV, Loud AV, Wiener J (1968) Cortisone-induced alterations in mitochodrial function and structure. J Cell Biol 37:63–80
404. Turner N et al (2007) Excess lipid availability increases mitochondrial fatty acid oxidative capacity in muscle. Diabetes 56:2085–2092
405. Bajaj M et al (2007) Paradoxical changes in muscle gene expression in insulin-resistant subjects after sustained reduction in plasma free fatty acid concentration. Diabetes 56:743–752
406. Chowdhury KK, Legare DJ, Lautt WW (2011) Insulin sensitization by voluntary exercise in aging rats is mediated through hepatic insulin sensitizing substance (HISS). Exp Gerontol 46:73–80
407. Schafer J, Legare DJ, Lautt WW (2010) Acetylcholinesterase antagonist potentiated insulin action in fed but not fasted state. J Pharmacol Exp Ther 333:621–628
408. Lautt WW (2004) A new paradigm for diabetes and obesity: the hepatic insulin sensitizing substance (HISS) hypothesis. J Pharmacol Sci 95:9–17
409. Correia NC, Guarino MP, Raposo J, Macedo MP (2002) Hepatic guanylyl cyclase inhibition induces HISS-dependent insulin resistance. Proc West Pharmacol Soc 45:57–58
410. Lautt WW et al (2001) Hepatic parasympathetic (HISS) control of insulin sensitivity determined by feeding and fasting. Am J Physiol Gastrointest Liver Physiol 281:G29–G36
411. McArdle WD, Katch FI, Katch VL (2009) Exercise physiology: nutrition, energy, and human performance. Lippincott Williams & Wilkins, Philadelphia, PA

8 Deploying the Immunological Garrison

Many of the comorbidities of the metabolic syndrome are now known to be due to inflammatory changes [1–4]. There is said to be a low-grade chronic systemic inflammation, and serum levels of various inflammatory cytokines, chemokines, and inflammatory markers including TNF-α; interleukins IL-1β, IL-8, and IL-10; monocyte chemoattractant protein-1 (MCP-1); macrophage inflammatory protein-1α (MIP-1α); growth-regulating oncogene-α (GRO-α); inducible protein-10 (IP-10); and C-reactive protein (CRP) are increased. Adipocytes are active secretors of various inflammatory and chemotactic cytokines, and a substantial portion, if not all, of the raised serum levels are contributed by the adipose tissue [5–7]. Obesity, systemic inflammation, and insulin resistance are linked in such a way that it is difficult to decide what comes first. Perhaps the three are in an autocatalytic loop.

At the level of ultimate cause, there is no satisfactory explanation as yet for why adipose tissues secrete chemokines. If this property has been consistently shown, it must have evolved under some kind of selective pressure. Chemokine secretion by adipocytes must be adaptive under at least certain set of conditions, although it is pathological in the context of metabolic syndrome. There have been attempts to explain ultimate causes behind the obesity-induced hypercytokinemia under the thriftiness paradigm. It has been interpreted as raised level of innate immunity [8], and its evolution is said to be a response to increased chances of infection under starvation conditions in which the thrifty phenotype evolved [9]. If starvation and infection challenges co-occurred during hunter-gatherer life, thrifty genotype and infection-resistant genotype may have coevolved. This is in support to the notion that infections accompany famines, and most of the deaths during famines are due to infections than to starvation [10, 11]. However, an inherent weakness of this explanation is that in obesity or insulin resistance, there is no evidence of increased resistance to infections. The raised levels of inflammatory cytokines have not been demonstrated to confer increased overall resistance to infections. On the contrary many specific infections, including infections of the skin and urinary tract, are more common in diabetic patients [12]. Other infections occur with increased severity and are associated with an increased risk of complications in patients with obesity and diabetes [13].

The behavioral origin hypothesis attempts to explain the inflammatory changes in a different way. In a soldier life the risk of getting wounds and injuries is substantially greater as compared to a diplomat life. Therefore the immune cells should move peripheral, i.e., towards subcutaneous tissue. The phagocytic cells of the body, namely, neutrophils and macrophages, are produced centrally in the bone marrow. They circulate in blood, macrophages in blood being in the form of monocytes. Any injury causes the extravasation, i.e., movement of cells across vessel walls and migration of these cells from local blood vessels towards the site of injury. This is the normal dynamics of these cells. In the natural ancestral environment, minor skin injuries would

M. Watve, *Doves, Diplomats, and Diabetes: A Darwinian Interpretation of Type 2 Diabetes and Related Disorders*,
DOI 10.1007/978-1-4614-4409-1_8, © Springer Science+Business Media New York 2013

be common, and stimulated by such injuries, these cells will continually migrate towards the skin. There is hardly any backflow in this dynamics. But the normal dynamics would keep a steady-state density of cells in blood vessels and a certain density in subcutaneous tissue as well. Whenever there is a shift to a diplomat state, epidermal injuries are less probable and the normal dynamics would be altered. There will be reduced migration from blood vessels to skin. As a result the density of macrophages in subcutaneous tissue would be less than normal, and their density in flowing blood and presumably attached to vessel walls would increase. In other words the immune system could be retracting from the periphery and as a result would be centrally hyperactive. This results into delayed wound healing on the one hand and increased inflammatory tendency of vascular tissue and some other central organs on the other.

Can we test the hypothesis? First of all, does a negative association or a trade-off between cutaneous and central immunity exist? If yes, does it change in obesity and metabolic syndrome? If yes, what are the molecular mechanisms involved? There is some evidence for trade-off or shift of balance among the innate immune mechanisms in obesity and insulin resistance. There is a negative association between atopic allergy and obesity [14]. Cutaneous delayed-type hypersensitivity is reduced in both type 1 and type 2 diabetes [15, 16], suggesting that at least some of the components of the immune system are partially retracted from the skin. The peripheral density of macrophages [17] and mast cells [18] is reported to be reduced in diabetes. Along with macrophages, mast cells also have important functions in wound healing [19–22]. On the other hand macrophages appear to accumulate in large numbers in adipose tissue [23–26] and so do mast cells [27]. Macrophage accumulation in blood vessels is an essential and very early phase in macrovascular complications in diabetes [28]. In diabetic nephropathy, macrophage accumulation is an important contributor and arresting macrophage accumulation can prevent diabetic nephropathy. There is no overall increase in macrophages throughout the body. Macrophage accumulation is tissue specific, and certain tissues simultaneously experience dearth of macrophages [17]. It is therefore evidently a problem of altered macrophage distribution than overall rise in macrophage activity. The redistribution is due to altered migration patterns of innate immune cells. The rate of macrophage and neutrophil migration is shown to be altered in diabetes [29–32]. Therefore my conjecture of altered phagocyte dynamics in diabetes has adequate support, although the complete picture of the dynamics is yet to emerge. From available data a number of mechanisms can be reconstructed by which the dynamics might shift.

1. Testosterone: One of the major players in the immune redistribution is likely to be testosterone. Traditionally testosterone is said to be immunosuppressive. Braude et al. [33] suggested that testosterone may be an immune redistributor rather than an immunosuppressor. According to their hypothesis, testosterone brings about movement of the immunological garrison to the periphery since testosterone is associated with aggression and therefore predicts injuries. Although there have been hardly any attempts to seriously test the Braude et al. [33] hypothesis, there is some circumstantial evidence in its favor. On the one hand serum testosterone levels are negatively associated with serum inflammatory markers and other components of the metabolic syndrome [34, 35]. Testosterone is shown to be protective against inflammatory complications of the metabolic syndrome [35]. On the other hand, although testosterone is generally considered immunosuppressive, it is shown to enhance inflammation (but not wound healing) of cutaneous wounds [36–38]. Thus the action of testosterone on inflammation appears to be in opposite directions in the context of skin and that of central organs. Testosterone is proinflammatory for cutaneous wounds and anti-inflammatory at the vascular and systemic level. Although testosterone does not help in wound healing, it stimulates secretion of EGF and other growth factors that are needed for wound healing. This is what we expect to be an adaptive response to aggression-provoked injury proneness.

2. Peroxisome proliferator activator receptors (PPARs): The PPARs appear to be another major class of players in a shift of balance. PPARs arrest inflammation on the one hand [39] and enhance wound healing and wound closure on the other [40, 41]. Epidermal injuries activate the expression of PPARs and especially that of PPAR-β [41]. Absence of injuries on adopting a diplomat way of life may shift the immune balance against healing and towards systemic inflammation by downregulating PPAR expression.
3. Sensory nerves and neuropeptides: Sensory nerves and neuropeptides also play a significant role in modulating inflammation and wound healing [42–45]. In diabetes there is a progressive degeneration of peripheral nerves [46], shifting the immune balance further away from the skin.
4. Adipokines: Perhaps the most important mechanisms of altering the dynamics of phagocytes in the body are through adipokines. Movement of immune cells is under a gradient of chemokines [47–50] secreted at the site of injury by resident tissue cells stimulated by injury. The chemokines are locally retained on extracellular matrix and cell surface heparan sulfate proteoglycans, establishing a stable concentration gradient around the stimulus. Chemokine signaling activates leukocyte integrins leading to extravasation. Cells, after leaving the vessels, migrate to the site of injury guided by the chemokine gradient. The magnitude of the gradient is determined by the difference between the local level and the baseline blood level.

This is where adipokines play a major role. If the basal levels of chemokines are high, the gradient will be weak, and migration will be downregulated. Adipocyte secretion of chemokines increases the basal level and thereby arrests the flow of inflammatory cells towards periphery. Fat accumulation is a sign of a lifestyle and behavioral change, and therefore, it is the right kind of tissue to bring about the redistribution of the immune system. Diabetes is known to reduce inflammatory response to cutaneous injuries [41], and this may be due to the raised basal levels of chemokines.

The action of chemoattractants and other signaling molecules largely depends upon their environment and their relative rather than absolute concentrations. Increased expression of a signaling molecule need not always mean increased action. The local topology of its relative concentrations can be of utmost importance. Experiments have shown that the effects of local application of TNF-α are substantially different from the effects of its systemic administration. In a mice study local application of 50–500 ng of TNF-α in collagen improved wound healing, but similar quantity in phosphate-buffered saline equivalent quantities intraperitoneally did not [51]. Presumably TNF-α in collagen was able to form a stable gradient, whereas that in saline dispersed systemically too rapidly. Systemic TNF-α was unable to produce any topological structure and therefore could not help the wound healing process at all .

We will now examine using diffusion kinetics the effects of raised basal chemokine levels on the gradient formation and the chemotactic migration of cells. If we assume simple diffusion of a chemoattractant, gradient formation can be computed using Fick's law of diffusion [43]:

$$X^2 = 4Dt \cdot ln\left(C_o / C_{xt}\right)$$

where X = distance from the origin of the chemoattractant, D = diffusion constant, t = time of diffusion, C_0 = concentration at the origin, and C_{xt} = concentration at a distance X at time t.

It can be seen that the concentration gradient thus formed is nonlinear and highly concave towards the lower end (Fig. 8.1). Let us assume that there is a threshold level of concentration difference that is necessary for the cells to recognize a gradient. A sphere (or hemisphere) of a radius equal to the distance between the source and the threshold forms the "catchment volume" of cell recruitment. The distance between the origin and threshold

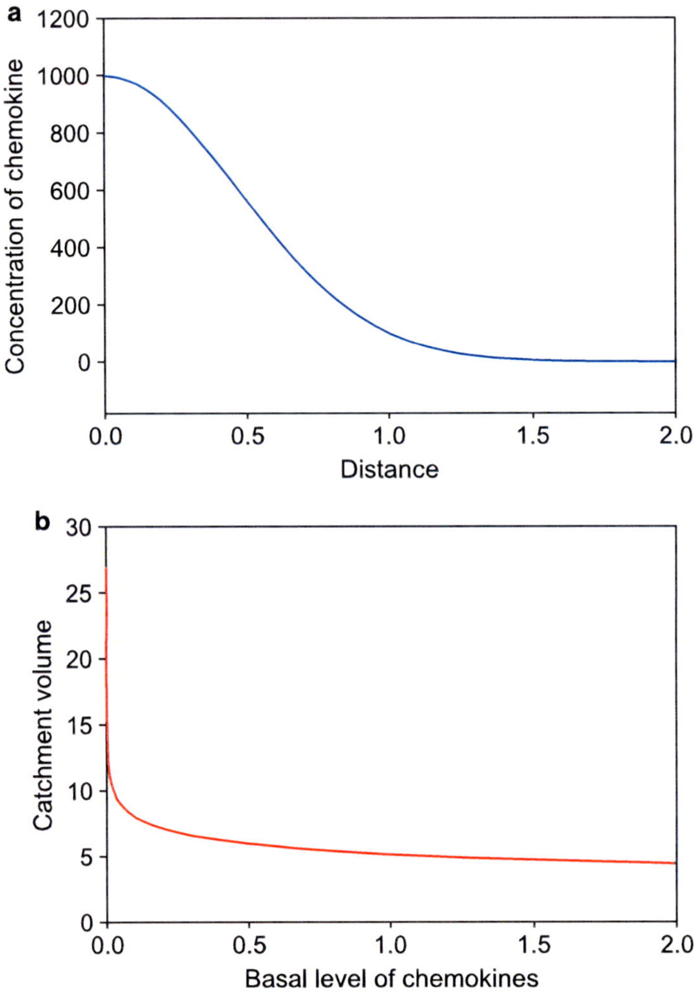

Fig. 8.1 Gradient of a chemoattractant from a site of injury and its effect on recruitment of inflammatory cells. (**a**) The expected diffusion gradient of a chemoattractant based on Fick's law of diffusion. The gradient will stop at the basal plasma level of the chemoattractant. Therefore the radius from which neutrophils and/or macrophages can be recruited is decided by the basal plasma level of chemoattractant. (**b**) The *curve* shows how the catchment volume for recruiting cells changes with basal chemokine levels. At small basal levels, a very small increase reduces the catchment substantially. At large basal levels it becomes relatively insensitive to changes. The *curves* illustrate the principle that a small increase in the basal plasma levels of a chemoattractant would result into substantial reduction in cell recruitment at the site of injury

concentration will be the radius of this hemisphere. Therefore the catchment volume will rise in cubic proportion of *X*. A larger catchment volume will result in greater infiltration of cells.

The threshold difference must depend upon the basal levels of the chemoattractant. If the basal levels rise, the threshold for chemotaxis should also increase. The concave nature of the concentration gradient results into a nonlinear reduction in the catchment volume such that when the threshold is small, even a little increment in it can cause a large decrement in the catchment. But when the threshold is high, a large change is required to cause the same decrement in the catchment (Fig. 8.1).

Data on normal and obesity-induced levels of TNF-α, IL-6, and MCP-1 show that there can be a two- to tenfold rise in their basal levels in obesity [52]. For example, Bastard et al. [52] found that the mean levels of TNF-α in obese patients were about double the mean for lean controls. Those for IL-6 were about tenfold. Data on their concentrations at the point of origin (C_0 of the above model) are difficult to obtain, but there are indications that the difference could be of many orders of magnitude as compared to the healthy basal levels [51]. Assuming a 1,000-fold difference between the source and threshold concentration, let us take the healthy basal level as 1 arbitrary unit and the source concentration as 1,000 units. Let us further assume that the threshold difference is directly proportional to the basal level with a proportionality constant of unity. With these assumptions, if the threshold levels are raised to 2 or 10 units, i.e., 0.2–1% of local levels, the model computes that there would be about 15 and 45% reduction in cell infiltration respectively. This calculation demonstrates that the reduction in infiltration is disproportionately larger. The values that we have assumed here need not match reality. As long as there is a large difference between local and systemic levels of chemokines, our conclusion will be valid. It means that a small rise in basal levels of chemokines can bring about substantial reduction in cell recruitment. Thus a small cost of raising the blood levels of chemokines can save a much greater investment in peripheral immunity. This is considerable immunological parsimony.

Interestingly, in diabetes, weakening of the gradient is brought about from the other end too. As the basal levels of chemokines are increased by adipocyte chemokine production from the other end, i.e., the site of injury, there is reduction and prolongation of chemokine production by local macrophages [17, 53–56]. Macrophage density in subcutaneous tissue is substantially reduced in diabetes [17]. This would be an autocatalytic response since a decreased density would decrease local chemokine production affecting further recruitment of macrophages. As the density of tissue macrophages goes down on the one hand, on the other, blood monocyte counts evidently go up in diabetes leading to macrovascular pathology [57–60].

A sedentary life devoid of physical activity and physical aggression creates a condition in which peripheral injuries are less likely. Obesity is prevalent under similar lifestyle. Therefore an immune reversal from the periphery under nonaggressive sedentary lifestyle mediated by adipose tissue could have evolved. The basal levels of chemokines cannot be expected to be stable. They will be subject to stochastic spatiotemporal fluctuations. The fluctuations are bound to increase in amplitude with rising basal levels. A stochastic peak exceeding the threshold can result into a nonspecific inflammatory trigger. This is a chance event that can occur anywhere in the body at any time. Therefore raised basal levels can result in stochastic inflammatory responses that can turn pathological. The modern urban lifestyle is much more sedentary and much less injury prone as compared to the conditions in which we evolved. It is possible therefore that this lifestyle acts as a supernormal stimulus for the naturally adaptive immune redistribution. An exaggerated immune redistribution can lead to vascular inflammation and its downstream pathological consequences.

There are three possible lines of work by which the hypothesis of adipokine-mediated redistribution of innate immune cells can be tested. First, it should be possible to test experimentally in animal models whether raising basal blood levels of chemokines by infusion decreases infiltration of immune cells to an experimental wound. This can be the most direct test of the hypothesis. As a long-term consequence of the altered basal levels, the distribution of immune cells in the body, mainly macrophages, is likely to change in obesity and related disorders. By appropriate labeling of macrophages followed by a whole-body scan, one could test whether the distribution of macrophages and other immune cells is different in healthy versus obese, insulin-resistant, or diabetic individuals. Also, on an epidemiological scale, it can be tested whether continued long-term exposure to minor cutaneous injuries is protective against obesity-induced inflammatory complications.

5. Signals from bone and muscle: It is well established by now that adipose tissue is an active endocrine organ and has effects on the immune system along with other systems. A less well-known but rapidly upcoming concept views bone and muscle too as endocrine organs. Osteocalcin, a molecule involved in strengthening bones, has multiple beneficial effects including insulin-sensitizing, β cell proliferative, and adiponectin-secreting actions [61–66]. Muscles also give out certain anti-inflammatory endocrine signals in the form of myokines [67–69]. This happens with every contraction of the muscle which means that the anti-inflammatory signals are directly proportional to muscle exercise. The effect of exercise on insulin resistance was earlier thought to be brought about through fat reduction. It is known now that exercise gives anti-inflammatory and insulin-sensitizing signals directly and independent of its effects on body fat. Similar to adipokines, the signal molecules released by muscle can be called "myokines." The concept of myokines is new, and not much research inputs have gone in. But we can expect them to play a significant role in the dynamics of innate immune cells.

   In addition to myokines there is a possibility of muscle-generated nervous signals as well. Muscles are rich in sensory nerve supply, and a sensory impulse is generated with every contraction of muscle. What happens to so much information being generated form muscle continuously? As argued earlier, it is possible that this information is used in the judgment of one's own muscle strength and activity and accordingly modifying behavioral and physiological strategies. Strong muscle signifies readiness for physical aggression and would program the immune system in anticipation of injuries. Thus exercise should stimulate the dynamics of macrophages and increase their density peripherally. This would have a systemic anti-inflammatory effect, and this is what is signaled by the myokines as well as neuronal signals given by every contraction of muscle. This is why exercise has immediate effects on the dynamics and distribution of innate immune cells of the body. It has been shown that with exercise, macrophage density and function in peripheral tissues are increased [70, 71] and that in adipose tissues [72] and lungs are decreased. This is in direct support of the macrophage redistribution hypothesis. Interestingly exercise-induced redistribution leads to increased susceptibility to respiratory infections [71, 73]. This gives plausible answers to why exercise has been reported to have contradictory effects on macrophage function [70, 74].

6. Cholesterol: Cholesterol is a marker of aggression deficiency. Since the need for cutaneous innate immunity is reduced in nonaggressive lifestyle, it would not be a surprise if cholesterol arrested movement of innate immune cells towards the periphery. Cholesterol, and LDL in particular, arrests the extravasation of macrophages at the endothelial level, and this may be viewed as a normal adaptive role of blood cholesterol, which may turn pathological under supernormal aggression loss.

We have thus seen that on adopting diplomat lifestyle, the distribution of innate immune cells of the body alters substantially. The changed distribution has a major role in the pathophysiology of diabetes and related disorders. There are three main mechanisms through which the immune redistribution contributes to the pathophysiology of diabetes and related disorders.

1. Atherosclerosis and CVD: The classical paradigm assumes LDL cholesterol to be solely responsible for the atherosclerotic changes in the arteries. What is not sufficiently appreciated is the role of macrophages in initiating these changes. Our hypothesis is that the reduced flow of monocyte–macrophages from blood to periphery leaves a larger number of macrophages within the blood vessels. Some of the excess macrophages accumulate somewhere in the vessel walls and trigger a mild localized inflammation-like response which is

the starting point of the fatty streaks and progressive pathophysiological processes leading to macrovascular pathology. The role of macrophages in initial stages of fatty streaks is clearly demonstrated [28, 75]. The novel suggestion coming out of our theory is only that the macrophages that normally should have migrated to peripheral tissues but fail to do so owing to lack of injury stimulation and raised basal chemokine levels are the ones that initiate the process. Therefore aggression and epidermal injury proneness can certainly be preventive and may even reverse the lesions in the blood vessels to some extent.

There is a paradox regarding inflammation and atherosclerosis and cardiovascular disease that the orthodox theory cannot explain but the new one may. Atherosclerosis and cardiovascular disease are said to be inflammatory disorders, and there is a wide-scale agreement on this [3, 75]. Glucocorticoids are known for their strong anti-inflammatory action and are successful therapeutic agents against a number of other inflammatory conditions [76–79]. However they do not appear to be useful in atherosclerosis and CVD. In fact high cortisol levels are positively associated with atherosclerosis and CVD and are considered risk factors [80–84]. Testosterone is also considered to be an immunosuppressive and anti-inflammatory agent, and it has protective effects against atherosclerosis and CVD [85–92]. If a generalized inflammatory tendency was responsible for initiating atherogenic changes, then any anti-inflammatory agent should have been sufficient to prevent it. But cortisol and testosterone appear to have opposite effects in this context, although both are said to be anti-inflammatory. This means that it is the immune redistribution and not generalized inflammation that is important. The altered distribution of macrophages can have proinflammatory effects in certain tissues and anti-inflammatory in others. This also implies that both testosterone and cortisol are not generalized anti-inflammatory agents as is usually believed. Both are differential immune redistributors, and as a result, they have differential pro- and anti-inflammatory effects in different tissues. The proaggression hormone testosterone drives the macrophages peripheral and reduces their vascular accumulation, whereas the anti-aggression cortisol is expected to have the opposite effect.

2. Oxidative stress: Oxidative stress is believed to be the single largest contributor to the pathological processes leading to diabetic complications. The most common source of reactive oxygen species (ROS) in healthy life is the phagocytes of the body. Unlike the popular perception, ROS production is not primarily a pathological process. It has important functions in normal metabolism and cell functions. In phagocytes of the body ROS are important weapons to kill bacteria that enter through an open injury [93]. When epidermal injuries become uncommon and the distribution of phagocytes changes, the distribution of ROS in the body is also bound to change. Since phagocytes remain within central circulation in greater numbers, there will be greater circulating concentrations of ROS. This is further enhanced by the chronic activation of these cells by inflammatory signals given by adipokines and partly by local inflammatory foci themselves. Therefore the increased oxidative stress in diabetes could be a direct consequence of the altered dynamics of the innate immune cells driven by a non-injury-prone diplomat lifestyle.
3. Dysfunctional wound healing and angiogenesis: Apart from their phagocytic role, macrophages have other important roles in the process of wound healing. Decreased macrophage density in the peripheral tissues is a likely cause of delayed wound healing [17]. A process of great importance during wound healing is the process of blood vessel formation called angiogenesis [94]. Macrophages have an important role in the initiation of angiogenesis [95]. Macrophages activated by inflammation or hypoxia can initiate neovascularization [96–98]. This suggests an interesting possibility that the adipocyte-derived

chemokines that attract macrophages have a selfish role. Growth of the adipose tissue needs angiogenesis [99, 100]. The macrophage population attracted and activated by adipokines may have a crucial role in the growth of adipose tissue by facilitating angiogenesis. Without angiogenesis the growing adipose tissue will not get sufficient supply of glucose and fatty acids to accumulate fat. A possible inference from this is that the primary role of adipokines may be in facilitating growth of adipose tissue itself through facilitated angiogenesis [101], and the systemic inflammatory changes is only a side effect seen under supernormal adipogenesis. However, the amount of adipokines produced and the number of macrophages attracted are substantially higher than what might be needed for adipose tissue angiogenesis. Adipokines could be simultaneously playing both the roles, that of facilitating a selfish local angiogenesis and that of achieving immunological parsimony for the whole body too.

Angiogenesis dysfunction is a major contributor to the pathophysiology of diabetes and, without the above logic, is full of apparent contradictions. One of the important factors causing impaired wound healing in diabetes is decreased angiogenesis, and any attempts to facilitate angiogenesis hasten the healing process [102–105]. It is interesting to note that most proaggression and sex signals have a proangiogenic action too, including testosterone, estradiol, and DHEA [106, 107]. This indicates that anticipation of injuries and preparation towards healing are inherently built in the process of aggression itself. Contrasting with this it is excessive and unwanted angiogenesis that leads to diabetic nephropathy, and attempts to arrest angiogenesis are helpful in arresting nephropathy [108–110]. The role of vascular endothelial growth factor (VEGF) in impaired wound healing and diabetic nephropathy is diametrically opposite. This paradox can be resolved by considering the macrophage redistribution. In tissues where the macrophage density is low, there is subnormal angiogenesis, and tissues where they accumulate, there is supernormal angiogenesis. Macrophage accumulation is shown to be one of the earliest events in diabetic nephropathy [111–113]. Thus, interestingly, both subnormal and supernormal angiogeneses contribute to diabetic complications simultaneously. The orthodox paradigm has not attempted to resolve this paradox, whereas the aggression deficiency hypothesis has logical and empirically supported explanation for both the extremes of angiogenesis simultaneously occurring in the body.

The central role of chronic low-grade systemic inflammation in T2D is quite well known. My interpretation of the causes and the process of inflammation is new. This is supported by a number of studies that show anti-inflammatory action of proaggression signals and vice versa. Although there are not many studies, brain serotonin, the mediator of aggression suppression, has proinflammatory effect in arthritis and depletion of serotonin ameliorates arthritis [114, 115] whereas dopamine appears to have anti-inflammatory actions [116, 117]. Melatonin also has proaggression action and is simultaneously anti-inflammatory and pain modulating [118, 119]. The anti-inflammatory properties of testosterone are already discussed, but even the female sex-related hormones estrogens are anti-inflammatory [120–123]. DHEA also has anti-inflammatory properties [124–126]. Effect of DHEA on angiogenesis is contradictory and dose dependent [127]. Although at higher concentrations DHEA inhibited in vitro endothelial cell proliferation, at concentration found in blood, it enhanced vascular endothelial proliferation [128]. Exogenous application of DHEA accelerated wound healing [129]. Apart from aggression signals, our hypothesis expects that frequent minor injuries or stimulation of the skin that generates minor but frequent local inflammations will have long-term systemic anti-inflammatory effects. Currently, we do not have data to test this, but there are some suggestive pieces of evidence. Many arthropod bites and stings have systemic anti-inflammatory effects; bee and scorpion venoms have been studied in some details, and a number of mechanisms for which are speculated [130–136]. Venom therapy has also been tried for inflammatory disorders. Acupuncture also has demonstrated anti-inflammatory effects, although the mechanisms may be debated [137, 138]. From our hypothesis' point of view, any injury-like skin stimulation would have systemic anti-inflammatory effects which are brought about by immune redistribution. When there is increased flow of phagocytes towards the skin, systemic inflammation presumably gets suppressed. This is certainly testable but has not been tested as yet.

We need to compare the two alternative interpretations of inflammatory tendency in T2D, that by the classical versus the new hypothesis. The classical view expects thrift to be accompanied by increased immunity. There appears to be no evidence for "thrift" being associated with increased immunity. The raised inflammatory markers in T2D do not signify higher levels of immunity. In fact the risk of certain types of infections appears to be significantly more. Interestingly not all types of infections seem to be more common in diabetics [13, 139]. The question why only certain systems become more infection prone and not others has not been addressed by the classical stream of thought.

The new interpretation does not talk about increased or decreased immunity. It rather postulates an altered distribution of immunity. In diabetes the innate immune mechanisms are partially withdrawn from the periphery. As a result there would be decreased immunity towards wound infections and other peripheral infections. The organs that undergo degenerative changes also may become more susceptible to infections in diabetes. However, there is no reason why upper respiratory or gastrointestinal infections should be more common in diabetics. In fact, although diplomats might be less prone to epidermal injuries, their more social and indoor life might make them more prone to respiratory infections. Therefore our hypothesis would expect that in anticipation, the respiratory immunity may be more enhanced in diplomats. In chimpanzees dominant individuals were found to have lower levels of circulating immunoglobulins and to be more susceptible to respiratory infections [140]. This is certainly testable in humans too, but testing this from published data is difficult. Positive associations between disorders are frequently reported, but there is a research bias as well as publication bias against negative associations. Also comparing frequencies of infections to reflect susceptibility to infections must be accompanied by careful control for factors affecting pathogen exposure. For example, individuals with sedentary lifestyle are more prone to diabetes, and we expect them to be less susceptible to respiratory infections, but their lifestyle may expose them more towards crowded indoors which are transmission-promoting environments for respiratory infections. Therefore it would be dangerous to compare from available literature. A study can be carefully designed to minimize confounding factors and then ask whether diabetics are more or less prone to upper respiratory infections. There are some data showing that they are more prone to lung infections [141–143], but lungs do undergo microangiopathy in diabetes [144–148] and that can be an added complication. No diabetes-related pathology in the upper respiratory tract is known, so it would be a good comparison. I can make a testable prediction here that although diabetics may get some specific infections more frequently, the frequency of upper respiratory infections would be at par or even lower than age-matched controls after correcting for other confounding factors. One large sample study has shown that in Asians, death due to respiratory infections decreases monotonically with obesity parameters. Unlike CVD and many other causes of death, it does not show a U-shaped curve with obesity [149]. Many other studies also show that upper respiratory infections in diabetics are not different or are less common than non-diabetic controls [141, 142]. However, inferring from published data being inadequate, one should wait for a specific study addressing this question. In addition to respiratory infections weight gain appears to have a negative association with gastrointestinal disease [150]. Whether obesity and insulin resistance offers better protection from gastrointestinal infections also needs to be investigated.

On the other hand wound and other skin infections are evidently more common in diabetics

[141, 142] which can be an effect of withdrawal of innate immunity from the periphery. Urinary infections are also more common in diabetics [141, 142], and this can possibly be explained better by the new interpretation. We have already talked about a positive association between sex and aggression. Sex increases the probability of transmission of urogenital infections in both sexes. In anticipation of this the urogenital mucosal immunity may be enhanced. There are no studies on humans or mammalian systems, but there is a study of gene expression in response to courtship signals in *Drosophila*. This study has shown that the most remarkable change is in the immunity-related genes presumably in anticipation of sexually transmitted infections [151–153]. This sounds logical and may work for mammals too [154], although this question has not been adequately addressed. If this is tested and turns out to be true, there is a possible reason for increased urinary infections in diabetics. Supernormal loss of aggression that accompanies reduced sexual activity may have resulted in supernormal disinvestment in urogenital mucosal immunity resulting in increased incidence of commensal bacteria turning pathogenic. The altered distribution of innate immune cells may be central to this change too. If that is true one should find lower densities of macrophages in uninfected urogenital mucosa of diabetics. This is a testable hypothesis and needs to be tested.

The hypothesis of behavior-mediated immune redistribution and a supernormal redistribution response leading to tissue-specific inflammatory reactions is certainly new and nontrivial. Admittedly a large body of research is needed to accept or reject it. My knowledge of immunology is very marginal. But I am convinced that the hypothesis has a promise and therefore should not be neglected by immunologists.

References

1. Tracy RP (1998) Inflammation in cardiovascular disease: cart, horse, or both? Circulation 97:2000–2002
2. Andersson SE, Edvinsson M-L, Edvinsson L (2003) Cutaneous vascular reactivity is reduced in aging and in heart failure: association with inflammation. Clin Sci 105:699
3. Libby P, Ridker PM, Maseri A (2002) Inflammation and atherosclerosis. Circulation 105:1135–1143
4. Festa A et al (2000) Chronic subclinical inflammation as part of the insulin resistance syndrome: the insulin resistance atherosclerosis study (IRAS). Circulation 102:42–47
5. Berg AH, Scherer PE (2005) Adipose tissue, inflammation, and cardiovascular disease. Circ Res 96:939–949
6. Lyon CJ, Law RE, Hsueh WA (2003) Minireview: adiposity, inflammation, and atherogenesis. Endocrinology 144:2195–2200
7. Weisberg SP (2003) Obesity is associated with macrophage accumulation in adipose tissue. J Clin Invest 112:1796–1808
8. Pickup JC (2004) Inflammation and activated innate immunity in the pathogenesis of type 2 diabetes. Diabetes Care 27:813–823
9. Fernández-Real JM, Ricart W (1999) Insulin resistance and inflammation in an evolutionary perspective: the contribution of cytokine genotype/phenotype to thriftiness. Diabetologia 42:1367–1374
10. Speakman JR (2006) Thrifty genes for obesity and the metabolic syndrome–time to call off the search? Diab Vasc Dis Res 3:7–11
11. Speakman JR (2008) Thrifty genes for obesity, an attractive but flawed idea, and an alternative perspective: the/'drifty gene/' hypothesis. Int J Obes 32:1611–1617
12. Dogra S, Kumar B, Bhansali A, Chakrabarty A (2002) Epidemiology of onychomycosis in patients with diabetes mellitus in India. Int J Dermatol 41:647–651
13. Joshi N, Caputo GM, Weitekamp MR, Karchmer AW (1999) Infections in patients with diabetes mellitus. N Engl J Med 341:1906–1912
14. Violante R et al (2005) Obesity risk factors in the ISAAC (International study of asthma and allergies in childhood) in Mexico city. Rev Alerg Mex 52:141–145
15. Plouffe JF, Silva J, Fekety R, Allen JL (1978) Cell-mediated immunity in diabetes mellitus. Infect Immun 21:425–429
16. Diepersloot RJ, Bouter KP, Beyer WE, Hoekstra JB, Masurel N (1987) Humoral immune response and delayed type hypersensitivity to influenza vaccine in patients with diabetes mellitus. Diabetologia 30:397–401
17. Maruyama K et al (2007) Decreased macrophage number and activation lead to reduced lymphatic vessel formation and contribute to impaired diabetic wound healing. Am J Pathol 170:1178–1191
18. Diaz BL et al (1996) Alloxan diabetes reduces pleural mast cell numbers and the subsequent eosinophil influx induced by allergen in sensitized rats. Int Arch Allergy Immunol 111:36–43
19. Trabucchi E et al (1988) The role of mast cells in wound healing. Int J Tissue React 10:367–372
20. Noli C, Miolo A (2001) The mast cell in wound healing. Vet Dermatol 12:303–313
21. Trautmann A, Toksoy A, Engelhardt E, Bröcker E, Gillitzer R (2000) Mast cell involvement in normal

human skin wound healing: expression of monocyte chemoattractant protein 1 is correlated with recruitment of mast cells which synthesize interleukin 4 in vivo. J Pathol 190:100–106
22. Hebda PA, Collins MA, Tharp MD (1993) Mast cell and myofibroblast in wound healing. Dermatol Clin 11:685–696
23. Weisberg SP et al (2003) Obesity is associated with macrophage accumulation in adipose tissue. J Clin Invest 112:1796–1808
24. Curat CA et al (2006) Macrophages in human visceral adipose tissue: increased accumulation in obesity and a source of resistin and visfatin. Diabetologia 49:744–747
25. Kanda H (2006) MCP-1 contributes to macrophage infiltration into adipose tissue, insulin resistance, and hepatic steatosis in obesity. J Clin Invest 116:1494–1505
26. Curat CA et al (2004) From blood monocytes to adipose tissue-resident macrophages. Diabetes 53:1285–1292
27. Hellman B, Larsson S, Westman S (1963) Mast cell content and fatty acid metabolism in the epididymal fat pad of obese mice. Acta Physiol Scand 58:255–262
28. Stary HC et al (1994) A definition of initial, fatty streak, and intermediate lesions of atherosclerosis. A report from the Committee on Vascular Lesions of the Council on Arteriosclerosis, American Heart Association. Arterioscler Thromb 14:840–856
29. Czepluch FS, Bergler A, Waltenberger J (2007) Hypercholesterolaemia impairs monocyte function in CAD patients. J Intern Med 261:201–204
30. Waltenberger J, Lange J, Kranz A (2000) Vascular endothelial growth factor-A-induced chemotaxis of monocytes is attenuated in patients with diabetes mellitus: a potential predictor for the individual capacity to develop collaterals. Circulation 102: 185–190
31. Sannomiya P, Pereira MAA, Garcia-Leme J (1990) Inhibition of leukocyte chemotaxis by serum factor in diabetes mellitus: selective depression of cell responses mediated by complement-derived chemoattractants. Agents Actions 30:369–376
32. Alba-Loureiro TC et al (2007) Neutrophil function and metabolism in individuals with diabetes mellitus. Braz J Med Biol Res 40:1037–1044
33. Braude S, Tang-Martinez Z, Taylor GT (1999) Stress, testosterone, and the immunoredistribution hypothesis. Behav Ecol 10:345–350
34. Laaksonen D et al (2003) Sex hormones, inflammation and the metabolic syndrome: a population-based study. Eur J Endocrinol 149:601–608
35. Muller M, Grobbee DE, den Tonkelaar I, Lamberts SWJ, van der Schouw YT (2005) Endogenous sex hormones and metabolic syndrome in aging men. J Clin Endocrinol Metab 90:2618–2623
36. Ashcroft GS, Mills SJ (2002) Androgen receptor-mediated inhibition of cutaneous wound healing. J Clin Invest 110:615–624
37. Gilliver SC, Wu F, Ashcroft GS (2003) Regulatory roles of androgens in cutaneous wound healing. Thromb Haemost 90:978–985
38. Gilliver SC, Ashworth JJ, Mills SJ, Hardman MJ, Ashcroft GS (2006) Androgens modulate the inflammatory response during acute wound healing. J Cell Sci 119:722–732
39. Moller DE, Berger JP (2003) Role of PPARs in the regulation of obesity-related insulin sensitivity and inflammation. Int J Obes Relat Metab Disord 27:S17–S21
40. Friedmann PS, Cooper HL, Healy E (2005) Peroxisome proliferator-activated receptors and their relevance to dermatology. Acta Derm Venereol 85:194–202
41. Tan N, Michalik L, Di-Poi N, Desvergne B, Wahli W (2004) Critical roles of the nuclear receptor PPARβ (peroxisome-proliferator-activated receptor β) in skin wound healing. Biochem Soc Trans 32:97–102
42. Quattrini C, Jeziorska M, Malik RA (2004) Small fiber neuropathy in diabetes: clinical consequence and assessment. Int J Low Extrem Wounds 3:16–21
43. Luger TA (2002) Neuromediators–a crucial component of the skin immune system. J Dermatol Sci 30:87–93
44. Misery L (1997) Skin, immunity and the nervous system. Br J Dermatol 137:843–850
45. Delgado AV, McManus AT, Chambers JP (2003) Production of tumor necrosis factor-α, interleukin 1-β, interleukin 2, and interleukin 6 by rat leukocyte subpopulations after exposure to substance P. Neuropeptides 37:355–361
46. Said G (2007) Diabetic neuropathy–a review. Nat Clin Pract Neurol 3:331–340
47. Luster AD (1998) Chemokines–chemotactic cytokines that mediate inflammation. N Engl J Med 338:436–445
48. Pelletier AJ et al (2000) Presentation of chemokine SDF-1 α by fibronectin mediates directed migration of T cells. Blood 96:2682–2690
49. Rothenberg ME (1999) Eotaxin. An essential mediator of eosinophil trafficking into mucosal tissues. Am J Respir Cell Mol Biol 21:291–295
50. Gillitzer R, Goebeler M (2001) Chemokines in cutaneous wound healing. J Leukoc Biol 69:513–521
51. Mooney DP, O'Reilly M, Gamelli RL (1990) Tumor necrosis factor and wound healing. Ann Surg 211:124–129
52. Bastard JP et al (2000) Elevated levels of interleukin 6 are reduced in serum and subcutaneous adipose tissue of obese women after weight loss. J Clin Endocrinol Metab 85:3338–3342
53. Chu X et al (1999) In vitro alteration of macrophage phenotype and function by serum lipids. Cell Tissue Res 296:331–337
54. Doxey DL et al (1998) Diabetes-induced impairment of macrophage cytokine release in a rat model: Potential role of serum lipids. Life Sci 63:1127–1136
55. Zykova SN et al (2000) Altered cytokine and nitric oxide secretion in vitro by macrophages from diabetic type II-like db/db mice. Diabetes 49:1451–1458
56. Naguib G, Al-Mashat H, Desta T, Graves DT (2003) Diabetes prolongs the inflammatory response to a

bacterial stimulus through cytokine dysregulation. J Invest Dermatol 123:87–92
57. Ford ES (2003) The metabolic syndrome and C-reactive protein, fibrinogen, and leukocyte count: findings from the Third National Health and Nutrition Examination Survey. Atherosclerosis 168: 351–358
58. Johnsen SH et al (2005) Monocyte count is a predictor of novel plaque formation: a 7-year follow-up study of 2610 persons without carotid plaque at baseline the Tromso study. Stroke 36:715–719
59. Chapman CML, Beilby JP, McQuillan BM, Thompson PL, Hung J (2004) Monocyte count, but not C-reactive protein or interleukin-6, is an independent risk marker for subclinical carotid atherosclerosis. Stroke 35:1619–1624
60. Zouaouiboudjeltia K et al (2006) Fibrinolysis and cardiovascular risk factors: association with fibrinogen, lipids, and monocyte count. Eur J Intern Med 17:102–108
61. Ferron M, Hinoi E, Karsenty G, Ducy P (2008) Osteocalcin differentially regulates β cell and adipocyte gene expression and affects the development of metabolic diseases in wild-type mice. Proc Natl Acad Sci USA 105:5266–5270
62. Ferron M et al (2010) Insulin signaling in osteoblasts integrates bone remodeling and energy metabolism. Cell 142:296–308
63. Kanazawa I et al (2009) Serum osteocalcin level is associated with glucose metabolism and atherosclerosis parameters in type 2 diabetes mellitus. J Clin Endocrinol Metab 94:45–49
64. Rached M-T et al (2010) FoxO1 expression in osteoblasts regulates glucose homeostasis through regulation of osteocalcin in mice. J Clin Invest 120: 357–368
65. Saleem U, Mosley TH, Kullo IJ (2010) Serum osteocalcin is associated with measures of insulin resistance, adipokine levels, and the presence of metabolic syndrome. Arterioscler Thromb Vasc Biol 30: 1474–1478
66. Fulzele K et al (2010) Insulin receptor signaling in osteoblasts regulates postnatal bone acquisition and body composition. Cell 142:309–319
67. Brandt C, Pedersen BK (2010) The role of exercise-induced myokines in muscle homeostasis and the defense against chronic diseases. J Biomed Biotechnol 2010:1–7
68. Pedersen BK, Åkerström TCA, Nielsen AR, Fischer CP (2007) Role of myokines in exercise and metabolism. J Appl Physiol 103:1093–1098
69. Pedersen BK (2006) The anti-inflammatory effect of exercise: its role in diabetes and cardiovascular disease control. Essays Biochem 42:105–117
70. Fehr H-G, Lotzerich H, Michna H (1989) Human macrophage function and physical exercise: phagocytic and histochemical studies. Eur J Appl Physiol Occup Physiol 58:613–617
71. Fehr H-G, Lötzerich H, Michna H (2008) The influence of physical exercise on peritoneal macrophage functions: histochemical and phagocytic studies. Int J Sports Med 09:77–81
72. Bruun JM, Helge JW, Richelsen B, Stallknecht B (2006) Diet and exercise reduce low-grade inflammation and macrophage infiltration in adipose tissue but not in skeletal muscle in severely obese subjects. Am J Physiol Endocrinol Metab 290:E961–E967
73. Davis JM et al (1997) Exercise, alveolar macrophage function, and susceptibility to respiratory infection. J Appl Physiol 83:1461–1466
74. Woods J, Lu Q, Lowder T (2000) Exercise-induced modulation of macrophage function. Immunol Cell Biol 78:545–553
75. Ross R (1999) Atherosclerosis–an inflammatory disease. N Engl J Med 340:115–126
76. Barnes PJ, Adcock I, Spedding M, Vanhoutte PM (1993) Anti-inflammatory actions of steroids: molecular mechanisms. Trends Pharmacol Sci 14: 436–441
77. Cline MJ, Melmon KL (1966) Plasma kinins and cortisol: a possible explanation of the anti-inflammatory action of cortisol. Science 153: 1135–1138
78. Saldanha C, Tougas G, Grace E (1986) Evidence for anti-inflammatory effect of normal circulating plasma cortisol. Clin Exp Rheumatol 4:365–366
79. Barnes PJ (1998) Anti-inflammatory actions of glucocorticoids: molecular mechanisms. Clin Sci 94:557–572
80. Walker BR, Soderberg S, Lindahl B, Olsson T (2000) Independent effects of obesity and cortisol in predicting cardiovascular risk factors in men and women. J Intern Med 247:198–204
81. Dekker MJHJ et al (2008) Salivary cortisol is related to atherosclerosis of carotid arteries. J Clin Endocrinol Metab 93:3741–3747
82. Smith GD et al (2005) Cortisol, testosterone, and coronary heart disease: prospective evidence from the Caerphilly study. Circulation 112:332–340
83. Nashel DJ (1986) Is atherosclerosis a complication of long-term corticosteroid treatment? Am J Med 80:925–929
84. Troxler RG, Sprague EA, Albanese RA, Fuchs R, Thompson AJ (1977) The association of elevated plasma cortisol and early atherosclerosis as demonstrated by coronary angiography. Atherosclerosis 26:151–162
85. Malkin C, Pugh P, Jones R, Jones T, Channer K (2003) Testosterone as a protective factor against atherosclerosis–immunomodulation and influence upon plaque development and stability. J Endocrinol 178:373–380
86. Jones RD, Nettleship JE, Kapoor D, Jones HT, Channer KS (2005) Testosterone and atherosclerosis in aging men: purported association and clinical implications. Am J Cardiovasc Drugs 5:141–154
87. Fukui M et al (2003) Association between serum testosterone concentration and carotid atherosclerosis in men with type 2 diabetes. Diabetes Care 26:1869–1873
88. Svartberg J et al (2006) Low testosterone levels are associated with carotid atherosclerosis in men. J Intern Med 259:576–582

89. Bhasin S (2003) Effects of testosterone administration on fat distribution, insulin sensitivity, and atherosclerosis progression. Clin Infect Dis 37: S142–S149
90. Simon D et al (1997) Association between plasma total testosterone and cardiovascular risk factors in healthy adult men: the Telecom study. J Clin Endocrinol Metab 82:682–685
91. Khaw K-T et al (2007) Endogenous testosterone and mortality due to all causes, cardiovascular disease, and cancer in men: European prospective investigation into cancer in Norfolk (EPIC-Norfolk) prospective population study. Circulation 116:2694–2701
92. Shabsigh R, Katz M, Yan G, Makhsida N (2005) Cardiovascular issues in hypogonadism and testosterone therapy. Am J Cardiol 96:67–72
93. Forman HJ, Torres M (2002) Reactive oxygen species and cell signaling: respiratory burst in macrophage signaling. Am J Respir Crit Care Med 166:S4–S8
94. Li J, Zhang Y, Kirsner RS (2003) Angiogenesis in wound repair: angiogenic growth factors and the extracellular matrix. Microsc Res Tech 60:107–114
95. Sunderkotter C, Steinbrink K, Goebeler M, Bhardwaj R, Sorg C (1994) Macrophages and angiogenesis. J Leukoc Biol 55:410–422
96. Koch A, Polverini P, Leibovich S (1986) Induction of neovascularization by activated human monocytes. J Leukoc Biol 39:233–238
97. Knighton D et al (1983) Oxygen tension regulates the expression of angiogenesis factor by macrophages. Science 221:1283–1285
98. Koch A et al (1992) Interleukin-8 as a macrophage-derived mediator of angiogenesis. Science 258:1798–1801
99. Rupnick MA et al (2002) Adipose tissue mass can be regulated through the vasculature. Proc Natl Acad Sci USA 99:10730–10735
100. Hausman GJ, Richardson RL (2004) Adipose tissue angiogenesis. J Anim Sci 82:925–934
101. Trayhurn P, Wood IS (2004) Adipokines: inflammation and the pleiotropic role of white adipose tissue. Br J Nutr 92:347–355
102. Galeano M et al (2004) Recombinant human erythropoietin stimulates angiogenesis and wound healing in the genetically diabetic mouse. Diabetes 53:2509–2517
103. Altavilla D et al (2001) Inhibition of lipid peroxidation restores impaired vascular endothelial growth factor expression and stimulates wound healing and angiogenesis in the genetically diabetic mouse. Diabetes 50:667–674
104. Galiano RD et al (2004) Topical vascular endothelial growth factor accelerates diabetic wound healing through increased angiogenesis and by mobilizing and recruiting bone marrow-derived cells. Am J Pathol 164:1935–1947
105. Rivard A et al (1999) Rescue of diabetes-related impairment of angiogenesis by intramuscular gene therapy with Adeno-VEGF. Am J Pathol 154:355–363
106. Häggström S, Lissbrant IF, Bergh A, Damber JE (1999) Testosterone induces vascular endothelial growth factor synthesis in the ventral prostate in castrated rats. J Urol 161:1620–1625
107. Belsare PV et al (2010) Metabolic syndrome: aggression control mechanisms gone out of control. Med Hypotheses 74:578–589
108. Yamamoto Y et al (2004) Tumstatin peptide, an inhibitor of angiogenesis, prevents glomerular hypertrophy in the early stage of diabetic nephropathy. Diabetes 53:1831–1840
109. Cha DR et al (2000) Role of vascular endothelial growth factor in diabetic nephropathy. Kidney Int 58:S104–S112
110. Nakagawa T (2007) Uncoupling of the VEGF-endothelial nitric oxide axis in diabetic nephropathy: an explanation for the paradoxical effects of VEGF in renal disease. Am J Physiol Renal Physiol 292:1665–1672
111. Chow F, Ozols E, Nikolic-Paterson DJ, Atkins RC, Tesch GH (2004) Macrophages in mouse type 2 diabetic nephropathy: correlation with diabetic state and progressive renal injury. Kidney Int 65:116–128
112. Nguyen D et al (2006) Macrophage accumulation in human progressive diabetic nephropathy. Nephrology 11:226–231
113. Chow FY, Nikolic-Paterson DJ, Atkins RC, Tesch GH (2004) Macrophages in streptozotocin-induced diabetic nephropathy: potential role in renal fibrosis. Nephrol DialTransplant 19:2987–2996
114. Harbuz MS, Marti O, Lightman SL, Jessop DS (1998) Alteration of central serotonin modifies onset and severity of adjuvant-induced arthritis in the rat. Br J Rheumatol 37:1077–1083
115. Harbuz MS et al (1996) The role of endogenous serotonin in adjuvant-induced arthritis in the rat. Br J Rheumatol 35:112–116
116. Bendele AM, Spaethe SM, Benslay DN, Bryant HU (1991) Anti-inflammatory activity of pergolide, a dopamine receptor agonist. J Pharmacol Exp Ther 259:169–175
117. Karanjia ND, Widdison AL, Lutrin FJ, Chang YB, Reber HA (1991) The antiinflammatory effect of dopamine in alcoholic hemorrhagic pancreatitis in cats. Studies on the receptors and mechanisms of action. Gastroenterology 101:1635–1641
118. Esposito E, Cuzzocrea S (2010) Antiinflammatory activity of melatonin in central nervous system. Curr Neuropharmacol 8:228–242
119. Ambriz-Tututi M, Rocha-González HI, Cruz SL, Granados-Soto V (2009) Melatonin: a hormone that modulates pain. Life Sci 84:489–498
120. Cuzzocrea S et al (2000) 17β-estradiol antiinflammatory activity in carrageenan-induced pleurisy. Endocrinology 141:1455–1463
121. Vegeto E et al (2003) Estrogen receptor-α mediates the brain antiinflammatory activity of estradiol. Proc Natl Acad Sci USA 100:9614–9619
122. Bodel P, Dillard GM Jr, Kaplan SS, Malawista SE (1972) Anti-inflammatory effects of estradiol on human blood leukocytes. J Lab Clin Med 80: 373–384

123. Josefsson E, Tarkowski A, Caristen H (1992) Anti-inflammatory properties of estrogen: I. In vivo suppression of leukocyte production in bone marrow and redistribution of peripheral blood neutrophils. Cell Immunol 142:67–78
124. Straub RH, Scholmerich JJ, Zietz B (2000) Replacement therapy with DHEA plus corticosteroids in patients with chronic inflammatory diseases - substitutes of adrenal and sex hormones. Z Rheumatol 59:II108–II118
125. Leowattana W (2001) DHEA(S): the fountain of youth. J Med Assoc Thai 84:S605–S612
126. Lopez-Marure R, Huesca-Gomez C, Ibarra-Sanchez Mde J, Zentella A, Perez-Mendez O (2007) Dehydroepiandrosterone delays LDL oxidation in vitro and attenuates several oxLDL-induced inflammatory responses in endothelial cells. Inflamm Allergy Drug Targets 6:174–182
127. Varet J et al (2004) Dose-dependent effect of dehydroepiandrosterone, but not of its sulphate ester, on angiogenesis. Eur J Pharmacol 502:21–30
128. Liu D et al (2008) Dehydroepiandrosterone stimulates endothelial proliferation and angiogenesis through extracellular signal-regulated kinase 1/2-mediated mechanisms. Endocrinology 149:889–898
129. Mills SJ, Ashworth JJ, Gilliver SC, Hardman MJ, Ashcroft GS (2005) The sex steroid precursor DHEA accelerates cutaneous wound healing via the estrogen receptors. J Invest Dermatol 125:1053–1062
130. Kwon YB et al (2002) The water-soluble fraction of bee venom produces antinociceptive and anti-inflammatory effects on rheumatoid arthritis in rats. Life Sci 71:191–204
131. Yoon S-Y et al (2007) Peripheral bee venom's anti-inflammatory effect involves activation of the coeruleospinal pathway and sympathetic preganglionic neurons. Neurosci Res 59:51–59
132. Chang Y-H, Bliven ML (1979) Anti-arthritic effect of bee venom. Agents Actions 9:205–211
133. Yoon S-Y et al (2008) Bee venom injection produces a peripheral anti-inflammatory effect by activation of a nitric oxide-dependent spinocoeruleus pathway. Neurosci Lett 430:163–168
134. Mirshafiey A (2007) Venom therapy in multiple sclerosis. Neuropharmacology 53:353–361
135. Petricevich V (2006) L. Balance between pro- and anti-inflammatory cytokines in mice treated with Centruroides noxius scorpion venom, Mediators Inflamm (2006)
136. Petricevich VL, Hernández Cruz A, Coronas FIV, Possani LD (2007) Toxin gamma from Tityus serrulatus scorpion venom plays an essential role in immunomodulation of macrophages. Toxicon 50:666–675
137. Zijlstra FJ, van den Berg-de Lange I, Huygen FJPM, Klein J (2003) Anti-inflammatory actions of acupuncture. Mediators Inflamm 12:59–69
138. Kavoussi B, Ross BE (2007) The neuroimmune basis of anti-inflammatory acupuncture. Integr Cancer Ther 6:251–257
139. Wheat LJ (1980) Infection and diabetes mellitus. Diabetes Care 3:187–197
140. Masataka N et al (1990) Dominance and immunity in chimpanzees (Pan troglodytes). Ethology 85: 147–155
141. Benfield T, Jensen JS, Nordestgaard BG (2006) Influence of diabetes and hyperglycaemia on infectious disease hospitalisation and outcome. Diabetologia 50:549–554
142. Muller LMAJ et al (2005) Increased risk of common infections in patients with type 1 and type 2 diabetes mellitus. Clin Infect Dis 41:281–288
143. Jeon CY, Murray MB (2008) Diabetes mellitus increases the risk of active tuberculosis: a systematic review of 13 observational studies. PLoS Med 5:e152
144. Ofulue AF, Kida K, Thurlbeck WM (1988) Experimental diabetes and the lung. I. Changes in growth, morphometry, and biochemistry. Am Rev Respir Dis 137:162–166
145. Kida K, Utsuyama M, Takizawa T, Thurlbeck WM (1983) Changes in lung morphologic features and elasticity caused by streptozotocin-induced diabetes mellitus in growing rats. Am Rev Respir Dis 128:125–131
146. Sandler M (1990) Is the lung a 'target organ' in diabetes mellitus? Arch Intern Med 150:1385–1388
147. Matsubara T, Hara F (1991) The pulmonary function and histopathological studies of the lung in diabetes mellitus. Nippon Ika Daigaku Zasshi 58:528–536
148. Marvisi M, Marani G, Brianti M, Della Porta R (1996) Pulmonary complications in diabetes mellitus. Recenti Prog Med 87:623–627
149. Zheng W et al (2011) Association between body-mass index and risk of death in more than 1 million Asians. N Engl J Med 364:719–729
150. Linares C, Su D (2005) Body mass index and health among Union Army veterans: 1891–1905. Econ Hum Biol 3:367–387
151. Lawniczak MK, Begun DJ (2004) A genome-wide analysis of courting and mating responses in *Drosophila melanogaster* females. Genome 47: 900–910
152. Peng J, Zipperlen P, Kubli E (2005) Drosophila sex-peptide stimulates female innate immune system after mating via the Toll and Imd pathways. Curr Biol 15:1690–1694
153. Fedorka KM, Linder JE, Winterhalter W, Promislow D (2007) Post-mating disparity between potential and realized immune response in Drosophila melanogaster. Proc R Soc B 274:1211–1217
154. McLean JM, Shaya EI, Gibbs ACC (1980) Immune response to first mating in the female rat. J Reprod Immunol 1:285–295

Why Population Density Matters 9

After having traveled from ecology to physiology, endocrinology, metabolism, and immunology, we will return once again to the basic drawing board of ecology. Ecology and evolution are the basic foundations of all these arguments, and one should not lose touch with the foundation at any time. The dynamics of populations is a very fundamental process of ecology and evolution, and it is certainly very important in physiology and medicine too, but its importance is not appreciated sufficiently in these areas of science. Population level processes and individual's behavior and physiology have subtle but far-reaching influences on each other, but this is an area that has not yet received sufficient attention. Ecologists have realized the importance of it [1–3] to some extent, but physiologists have not paid sufficient attention to it. I would attempt to make a case below for why and how population level processes are bound to affect individual behavior and physiology and why physiology and medicine need to understand population level processes. If I am able to convince the reader about this point, it would logically follow that today's unprecedented density of human population is a supernormal stimulus that could change human behavior as well as physiology substantially, ultimately having some consequences for health.

One of the basic models of population growth is called the logistic model of growth which describes the dynamics of growth when the resources are finite. Initially developed to describe population growth in a simplistic way, different versions and refinements of the model have been used in different contexts often going beyond population growth and predicting other trends such as rates of discoveries in a scientific field or the learning curves [4–8]. The baseline model can be described by a single equation:

$$\frac{\mathrm{d}X}{\mathrm{d}t} = rX\left(1 - \frac{X}{K}\right)$$

where X is the population size, r the intrinsic growth rate, and K the carrying capacity of the environment.

The equation is not difficult to interpret. The model has only two parameters, r, which is the intrinsic growth rate of the population, and K, which is the carrying capacity of the environment for the population of a given species. The term in the bracket denotes what fraction of the carrying capacity is yet unoccupied, and this gives a feedback to regulate the net growth rate. When the population is small, the feedback term is close to unity, and the equation simplifies to exponential growth. So when the population is small as compared to the carrying capacity of the environment, growth is nearly exponential. However, as the population increases, the feedback term starts becoming smaller and smaller which reduces the growth rate proportionately. When the population becomes equal to the carrying capacity, growth stops since the feedback term becomes zero. This dynamics results in a sigmoid growth curve (Fig. 9.1). The logistic curve forms the simple baseline model of population dynamics although in real life one sees situation-specific deviations

M. Watve, *Doves, Diplomats, and Diabetes: A Darwinian Interpretation of Type 2 Diabetes and Related Disorders*,
DOI 10.1007/978-1-4614-4409-1_9, © Springer Science+Business Media New York 2013

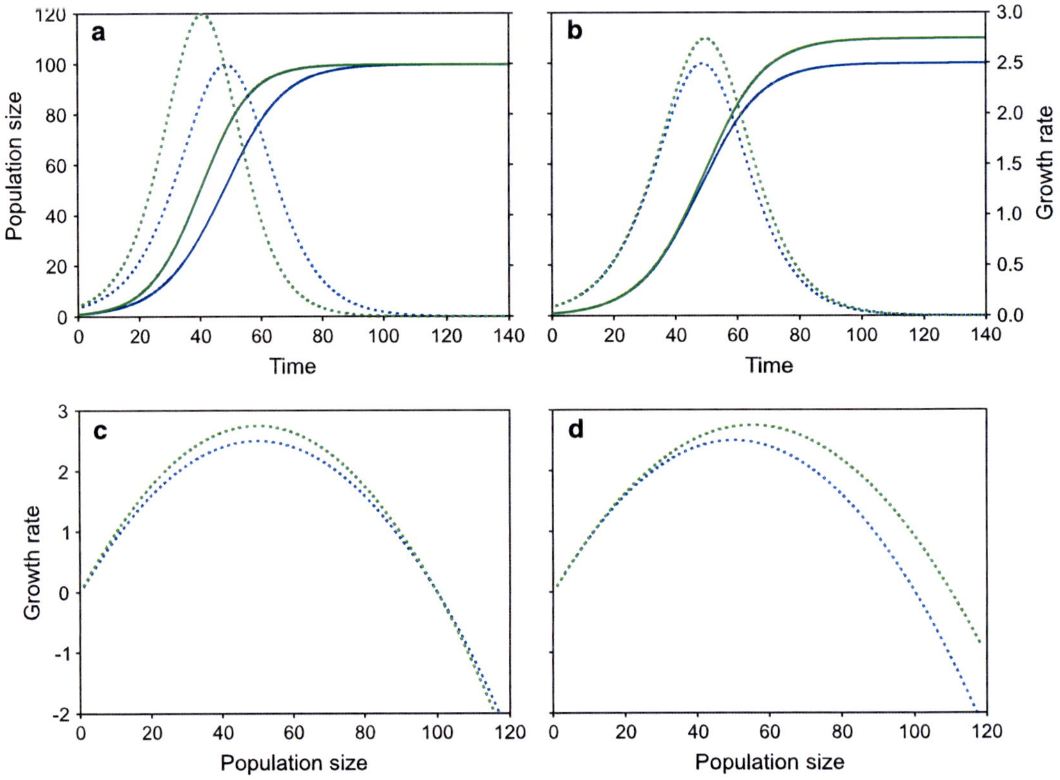

Fig. 9.1 The logistic growth curve: a fundamental model of population dynamics denoted by the equation $dX / dt = rX(1-(X / K))$, where r indicates intrinsic growth rate of the population and K the carrying capacity of the environment. *Solid lines* indicate population sizes (primary Y axis), and *dotted lines* represent net growth rates (secondary Y axis). (**a**) The effect of r on population growth, the *green line* has 10% higher r. (**b**) The effect of K on population growth, the *green line* having 10% higher K. (**c**) and (**d**) The relationship between population density and growth rates: a population with a higher r (**c**) has higher growth rate at moderate population densities but loses this advantage at high densities. A population with higher K (**d**) has a higher net growth rate at population densities close to saturation. This is an important point to realize. A common misunderstanding about logistic growth model is that a higher r represents all-time advantage in growth rate. Near the carrying capacity, even a small increase in K can give a significant growth rate advantage

from the assumptions of the model and the model can be refined to accommodate these complexities. For us, the baseline model is sufficient to build a conceptual framework.

Population ecologists and evolutionary biologists have used the model in different ways to cater for different classes of questions. While population ecologists have used it to predict growth trajectories, probabilities of extinctions, and effects of species interactions such as competition or predator prey and so on, evolutionary biologists have wondered how natural selection would shape the two parameters of the equation, namely, r and K. While it looks obvious that the population of a species with a higher reproductive rate r would grow faster and therefore outcompete one with smaller r, this is not unconditionally true. This can happen only under certain conditions. When much of the carrying capacity is still available so that the growth kinetics is close to exponential, a higher growth rate can give large returns. Therefore natural selection would favor larger r. However as the population approaches the carrying capacity, the importance of r is gradually reduced. Close to the carrying capacity, those that grow and reproduce faster are not the ones that have a larger r but those that make maximum efficient use of the limited resources available and ones that are highly competitive. In other words, those that make more efficient use of the limited resources have a higher effective K, and this is what selection will favor under crowded conditions. So at low population densities, there is selection for higher r, and at high population densities, there is selection for higher K. Then why can't a genotype evolve that has a high r as well as high K so that it is the fittest genotype under all conditions? This does not happen because there is an inevitable tradeoff between r and K. It is difficult to achieve both higher r and higher K simultaneously because the requirements for the two are conflicting and an individual has a finite resource. This has given rise to the concept of r and K selection [9–12].

In order to achieve a high reproductive rate, one has to increase the reproductive output. This can be done by increasing the number of offspring. Females can do this by increasing the number of eggs. Males can achieve this by mating with a greater number of females. However there is a cost in laying an extra egg or gaining an additional mating opportunity. Since every individual has a finite store of resources to be put in reproductive efforts, increasing the number of eggs might make reducing the size of eggs inevitable. A smaller egg may reduce the size and thereby the competitiveness of the emerging offspring. Competitiveness is an important component of K selection. Conversely, if egg size is kept large in spite of limited resources, either the number of eggs or longevity of the egg layer will have to be compromised. Perhaps both egg size and egg number may be increased by increasing the total fraction of resources going into reproduction. However, increasing reproductive effort might be accompanied by a compromise somewhere else such as longevity. This is an example of how there can be a conflict between r strategies of reproduction and K strategies of reproduction. It is difficult therefore to achieve both simultaneously. In reality, the precise form of expression of the r and K strategies and the nature of tradeoff is highly context dependent. For example, if saturating carrying capacity leads mainly to starvation, then a smaller body size and metabolic thrift could be the adaptive phenotype. However if it means increased aggressive competition, then a larger body size would be more adaptive. As a result, the characteristics that mark r and K strategies might be different in different species, but the robust principle is that the best reproductive strategies when a large fraction of carrying capacity is still available and those when it is saturated are different.

As a broad generalization, r strategy of reproduction is characterized by larger number of offspring with little investment in each and K by fewer offspring with greater investment in each. Some species are committed to the r strategy of reproduction and some others to the K strategy. Two extreme examples of r and K selection respectively are bacteria that may double the population every half an hour and elephants on the other hand that have a gestation period of 22 months ending in a single calf. Although the description so far sounds as if r and K are two discrete watertight compartments, in reality, they are the two ends of a continuum. Depending on the context, there can be an optimum position along the continuum that natural selection would favor. Originally the concept of r and K selection was applied across species, but now we know that even within a species, there can be a limited amount of variation between individuals with respect to their reproductive strategies [13–15]. Also for a given genotype, there could be some limited flexibility to shift towards r or towards K in one's lifetime [16]. In crowded conditions, animals are expected to shift more towards K strategy. This flexibility is evident since fecundity is shown to be suppressed by crowding [17–20]. Under what conditions a species may evolve to have this plasticity in reproductive strategies?

We have seen that there is a feedback term in the logistic model which regulates the growth rate based on the available fraction of the carrying capacity. It is assumed in the baseline logistic model that this feedback is instantaneous. In reality, this does not work quite often. For example, as the population grows, there may be relative scarcity of food. This should result in reduced reproductive output. But there can be a substantial

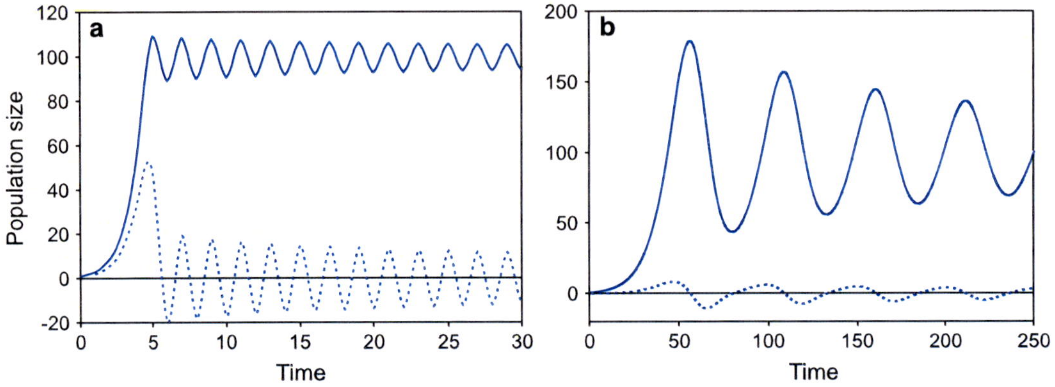

Fig. 9.2 Oscillations of populations around the carrying capacity: Populations can oscillate due to two main reasons. (**a**) If *r* is large, the oscillations are rapid typically completing one cycle in a few generations; (**b**) if there is a delay in feedback, there can be slower oscillations spanning over several generations. A combination of delayed sexual maturity and intrauterine programming of adult metabolic and reproductive functions can cause such a delay in feedback. Since both these factors are known to be predominant in humans, the prehistoric human populations are likely to have experienced (**b**)-type oscillations

delay in this process. Often, the nutrition in early developmental period decides the adult reproductive potential causing a time delay in the cause and effect. There can be many other reasons for such a delay. If there is a delay in the feedback, the growth rate does not reduce sufficiently in time to stabilize the population at the carrying capacity. In such a case, the population may actually overshoot the carrying capacity. This makes the effective growth rate negative in the following generation(s). The population then declines until it is substantially below the carrying capacity and then starts rising again (Fig. 9.2). Such oscillations in populations are not uncommon in real life for a wide variety of species [21]. If the population of a species oscillates with considerable amplitude, the selective pressure for reproductive strategies would also be fluctuating. If we assume that *r* and *K* strategies are genetically determined, then when the population is at its peak, *K* reproductive strategies should be selected for, and when at its trough, *r* should be selected. This might result in a coexistence of *r* and *K* genotypes. The endocrinological and physiological requirements of the two reproductive strategies are likely to be substantially different, and therefore, one expects to see a different set of genes or gene expression patterns under the two states. But what if some genome has both the reproductive capacities and can shift its strategies appropriately as the population density changes? It is not very difficult to visualize that such plasticity will have a selective advantage over a wide range of amplitudes and frequencies of oscillations. For a species to have this plasticity, it should have the ability to shift between two endocrinological and metabolic states supporting the two strategies, and highly coordinated mechanisms should have evolved to attain such plasticity. Such plasticity would also help adapting to a fluctuating environment. Even if the population does not oscillate, environments are known to fluctuate widely, and different reproductive strategies may be best under different environmental states.

We expect therefore that for a species which may have undergone population oscillations or wide environmental fluctuations for a sufficiently long evolutionary time, there could be plasticity in the reproductive strategy accompanied by a large number of coordinated changes in physiology. If such a phenomenon exists in the human species, it should be extremely relevant today since the human population is unprecedentedly high and therefore could serve as a supernormal stimulus for a *K* reproductive strategy.

We will now recall the insulin resistance syndrome and its associated sexual and reproductive states. The well-known consequences of insulin resistance for sex and reproduction are that on the one hand insulin resistance is associated with decreased ovulation [22–26] and decreased spermatogenesis [27–29] as well as subnormal sexual desire and function [30–35], all of which can result in reduced number of offspring. On the other hand, it is also well known that diabetic mothers have heavier babies [36–38]. This can be taken to mean that there is increased investment in the offspring whenever there is successful conception. The two effects together, namely, reduced chances of conception on the one hand and increased investment in offspring whenever conceived on the other, characterize *K* reproduction. We can suspect therefore that the reproductive component of insulin resistance syndrome is actually a *K* strategy of reproduction. The important question then is are the sexual and reproductive changes secondary to insulin resistance or are they the primary evolved plastic responses and is insulin resistance the accompanying physiological adjustment.

There appears to be evidence for both the directions. Type 1 diabetes in humans or streptozotocin-induced diabetes in rats results in sexual disorders indicating that diabetes comes first [30–35]. This is a more popular perception, and sexual dysfunction is perceived only as one of the many pathological consequences of diabetes. But there is also evidence on the contrary. Castration increased obesity and insulin resistance in rat as well as human studies [39, 40]. Contraceptives increase insulin resistance [41, 42]. Vasectomy induced atherogenic changes in some but not all primate studies [43–45] and fortunately not in humans [46, 47]. Erectile dysfunction is a common abnormality associated with diabetes [48–50]. Normalization of sugar does not appear to improve erectile dysfunction [51], but long-term treatment of erectile dysfunction with sildenafil is insulin sensitizing and restoring vascular function [52–55]. Diabetes does not affect menopause [56], but menopause leads to altered body fat distribution, impaired glucose tolerance, endothelial dysfunction, and systemic inflammation, all markers of insulin-resistant state [57]. These findings make us suspect that suppression of sex and reproduction could have a causal and not only a consequential role. For an evolutionary biologist, this certainly makes sense and is another example of an autocatalytic or positive feedback loop indicating a behavioral switch. If the possible number of offspring is reduced by any reason, the investment in each of the residual number of offspring should go up, and insulin resistance is the mechanism of doing so. On the other hand, if insulin resistance is primary and inevitably results in increasing the transplacental investment, it should be accompanied by a reduction in offspring number. The ultimate reasoning suggests that the association should be robust irrespective of the direction of the arrow of causation.

The other side of the story, that of diabetic mothers having heavier babies, is long known and undisputed. The mechanism by which this happens is also quite straightforward. Placental glucose uptake is insulin independent. Therefore, when maternal tissues develop insulin resistance, more glucose is diverted through the placenta. Macrosomia or large fetal size is directly related to maternal glucose levels. Although the phenomenon is long known, the interpretation of it being a mechanism of increasing investment in the fetus is new. Will it be possible for us to differentiate between the two views, namely, increased fetal weight as a coincidental consequence of maternal insulin resistance versus an evolved mechanism of increasing fetal investment? Although a clean choice between the two is difficult at this stage, there is some suggestive evidence. The fetal weight gain is not simply a consequence of the obesogenic physiological setup of the mother as evidenced by the study by Catalano et al. demonstrating that in normal mothers, the fetal weight gain was positively correlated with maternal weight gain, but in diabetic mothers, the increased weight of the fetus was independent of maternal weight gain [58]. Evolutionary biologists give substantial importance to parental investment. Parental investment appears to be central to the evolution of courtship displays, mating systems, and sexual dimorphism as well as biases in sex ratios at birth. Therefore, if the increased nutrient flow through the placenta in insulin-resistant mothers is an investment strategy, there should be other detectable correlated changes.

In species where monogamy is not obligate, a male's reproductive success depends upon access to females which in turn depend upon male's

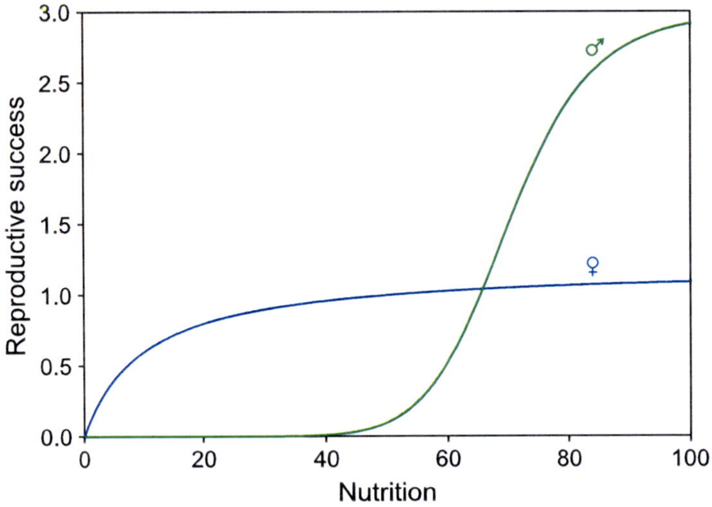

Fig. 9.3 The hypothetical differential fitness curves of males and females in promiscuous species. A female's reproductive success is decided by her own absolute fitness, whereas a male's reproductive success depends more on access to females which is decided by the social rank, that is, its fitness relative to other males. As a result, the difference between reproductive fitness of weaker and stronger males is very large as compared to females

relative social status. A high-ranking male may have a large reproductive success, and a weaker male may have little if any. A female's success depends on her absolute reproductive capacity and is little affected by the status of other females. As a result, a male's reproductive fitness is a sigmoid function of its physical fitness, whereas a female's curve is saturating (Fig. 9.3). It can be seen from this curve that at low level of physical fitness, females have better reproductive success than males, but at high levels of physical strength and fitness, males do better than females. Since early life development has lifetime consequences for the physique of an individual, a mother can potentially increase her net reproductive success by making careful investment choices at birth. Since a smaller or weaker male has very low chances of reproductive success, it makes no sense to make a weak male offspring. A weak female is not an outright loser like a weak male. She can reproduce as per her own absolute capacity. Therefore, a smart mother could maximize her net reproductive success by investing more in a male fetus. Alternatively, when the female is not in a capacity to invest sufficiently in the fetus, she should produce a female fetus preferentially. This is called the Trivers–Willard hypothesis of parental investment [59, 60]. If the Trivers–Willard hypothesis is true and if fetal macrosomia in diabetic mothers represents an evolved investment strategy, it should follow the predictions of the TW hypothesis. We can make two testable predictions based on this logic. The first is that the maternal insulin resistance or maternal blood sugar-induced fetal weight gain should be more pronounced for a male fetus than a female fetus. The other is that a female having higher nutritional reserves or higher plasma glucose at the time of conception should have a male-biased probability of fetal gender. The former has not been rigorously tested but is not difficult to test. There is already good amount of evidence for the latter [60–63]. Changes in glucose levels at conception have been associated with change in the sex ratios in mice. Excess glucose seems to choose male embryos [60]. One study shows that fetal sex is associated with diet at conception in humans. High preconceptional energy leads to higher number of boys suggesting a sexual dimorphism in the embryos based on the nutritional reserves [62–64]. Since humans have predominantly biparental care, the TW effect is expected to be very weak in humans. Therefore, it is unlikely that blood sugar makes any meaningful difference for a couple desirous of a baby girl or a boy, but it could be sufficient to show statistical significance to test the hypothesis.

The *r* and *K* selection and TW hypothesis work at the ultimate level. If the ultimate reasoning is correct, one should also find proximate mechanisms to bring about the expected adaptive changes. The next most relevant question then is at the level of endobolism: How are the *r* and *K* strategies executed? Which mechanisms lead to *r* reproduction and which ones lead to *K*? What triggers these mechanisms?

1. Adiponectin: Adiponectin appears to be the most important player in determining reproductive strategies. This is because adiponectin affects both fecundity and transplacental nutrient transfer. It is also associated with some other parameters of reproductive strategies such as litter size and weaning to estrus interval [65, 66]. Adiponectin is important in several steps of oogenesis as well as early embryonic development [67–70]. Not surprisingly, the common ovulation disorder of modern times, PCOS, is associated with low levels of adiponectin [71–74], and it may be suspected that low adiponectin may be causal to PCOS, although this is difficult to ascertain at this stage. Adiponectin receptors are abundant in testes indicating that in males, it is likely to be playing a crucial role in spermatogenesis [75–77]. Thus, adiponectin appears to enhance the probability of conception by multiple ways. As expected by the *r* and *K* selection theory, an agent that increases the number of offspring should bring down the investment in offspring simultaneously. Adiponectin does this by regulating the transplacental flow of nutrients. Evidence for this is not only correlational [78, 79] but also experimental, where rats infused with adiponectin showed reduction in fetal weights [80] and lower expression of glucose transporter in the placenta [81]. This is in addition to the insulin sensitizing effects of adiponectin. By increasing insulin sensitivity of maternal tissues, it shifts the budget allocation more towards maternal tissues and less towards offspring. Limiting the investment in a given fetus can be potentially helpful in providing for subsequent pregnancies. Under conditions such as transient maternal undernutrition where nutrients need to be preferentially transported to the fetus, adiponectin is downregulated [81]. The adiponectin response to caloric restriction is likely to have primarily evolved for this purpose.
2. Insulin and differential allocation of glucose: The placenta is an insulin-independent tissue. Therefore when maternal tissues become insulin resistant, there is increased flow of glucose through the placenta. This differential allocation of energy explains why insulin-resistant mothers have larger babies. High levels of leptin that are commonly associated with obesity also increase nutrient flow across the placenta by modulating system A transporters [82]. Not surprisingly, both obesity and insulin resistance are known to reduce fecundity. This, as opposed to adiponectin, marks an inclination towards *K* type of reproduction.
3. Active role of placenta: The placenta plays an active role in drawing more nutrients to it. A hormone called human chorionic somatomammotropin is secreted by the placenta in large quantities, and this hormone induces insulin resistance in maternal tissues [83]. As a result, less glucose is taken up by maternal tissues and thereby more is diverted through the placenta. To what extent this mechanism is important in gestational insulin resistance is not known. The placenta also actively secretes adiponectin [81] which has a diametrically opposite effect. The placenta appears to do a balancing or fine-tuning act. The dynamics of this in relation to sexual behavior and reproductive strategies would potentially be of great interest but currently remains an almost virgin area for research.
4. Investment in milk: Similar to transplacental investment, investment in lactation is also influenced by components of the insulin resistance syndrome. Adiponectin is negatively associated with the frequency and duration of lactation [84, 85]. IGF-1 which has insulin sensitizing effects is negatively associated

with milk yields in cows and positively with conception indicating that it is one of the mediators of a trade-off between number of offspring and investment in each [86]. Insulin and cortisol stimulate casein synthesis in the mammary gland and thereby increase the investment in milk [87–91]. Plasma glucose levels also enhance milk production [92]. More interestingly, milk glucocorticoids also affect hawk–dove behavior in future life of infants in a gender-specific manner [93]. All these findings are in support of the hypothesis of differential investment and *r* and *K* trade-off mediated by important players of the metabolic syndrome.

5. *K* strategy accompanied by longer life. Is there a contradiction? The *r* and *K* selection strategies in reproduction are also accompanied by differences in life span. The *r* strategists are typically short lived and *K* strategists long lived. In mammals, body size and longevity correlate positively with each other and negatively with the rate of reproduction [94, 95]. It makes sense therefore that insulin signaling is also connected to longevity. In the worm *Caenorhabditis elegans*, mutations of genes involved in insulin/IGF-1 signal transduction lead to extended life span [96, 97]. Mutations in genes homologous to the mammalian insulin/IGF-1 receptor signaling pathway, insulin receptor substrate homologue *chico*, or insulin-like receptor InR gene extend longevity in *Drosophila* [96]. In mammals and particularly humans, insulin is certainly connected to longevity but in a rather contradictory way. On the one hand, insulin resistance is associated with a series of fatal disorders, and centenarians are known to have high insulin sensitivity [96], but on the other, the *klotho* gene whose overexpression is known to increase life span has been shown to act by increasing insulin resistance [98]. FIRKO mice with fat cell-specific insulin resistance have a prolonged life span [99]. On the contrary, life extension in dwarf mice is associated with reduced secretion of insulin and increased hepatic sensitivity to insulin [100]. The key differences in the contradictory results might lie in differential tissue-specific resistance to insulin. Currently the association of insulin resistance with longevity in humans remains ambiguous. However, there is strong evidence that insulin signaling and insulin resistance are certainly involved in determining longevity in widely differing animals. We will come back to this paradox again in a later chapter with more clear assertion that actually insulin resistance has a pro-longevity effect, but certain problems arise in humans which make the change pathological.

There is one note of caution while using the *r* and *K* reproduction concepts in humans. Previously they have been applied to humans in a different context, with an argument that some human races are more *r* selected than others and that *K* selected races are more intelligent. This argument has a racist connotation. The argument above is driven by ecology and behavioral strategies, not by race or ethnicity. In this context, evidence that higher level of cognitive ability correlates with *K* strategy [101] has no racial implications. But why should *K* reproduction be accompanied by higher-level cognitive capacity? There is a logical reason which leads us back to the soldier–diplomat dichotomy.

Back to Soldier–Diplomat

Till the last chapter, we were talking about two alternative behavioral strategies, namely, soldier and diplomat, and arguing that diplomat behavior or deficiency of soldier behavior has a causative role in metabolic syndrome disorders. Here we are talking about two alternative reproductive strategies, namely, *r* and *K*, and contemplating that *K* strategy has a causal role in metabolic syndrome disorders and chronic crowding would trigger mechanisms of *K* reproduction. Are these two independent switches for the insulin-resistant state?

By all means, the two are not independent. Crowding or high population density affects reproductive strategies on the one hand which is more obvious. What is somewhat less obvious but well supported by both theory and empirical data is that crowding affects aggression to a considerable extent. A persistent and popular view holds that crowding leads to violence. Experimental data from free-ranging and captive animals is contradictory and so are data on humans. Some studies demonstrate increase in aggression upon crowding [102–109], and others report a net decrease with crowding [110–116]. Social isolation on the other hand is more consistently known to induce aggression and is a typical model of inducing aggression in rodent experiments [117–120]. This inconsistency comes at least partially from the different experimental designs and the history of animal groups under study. Crowding has a multitude of social effects too [121], and they may prevail over the actual effects of crowding on aggression. For example, if crowding is experimentally increased by increasing the number of animals, it inevitably means some unfamiliar individuals have to interact, and this creates a different social situation. The effects of a sudden change also need to be differentiated from the effects of crowding. For example, in one study, intense fighting and killing ensued when a group of macaques was released in a larger space. When after two and half years the same group was crowded back into a smaller pen, a similar bout of aggression was observed again [116]. This appears to be simply a response to change than a response to crowding.

Another important reason for the contradictory results is the lack of an appropriate null model for a fair comparison. The question relevant to us here is not whether the total incidents of aggression increase but whether individuals tend to behave more aggressively upon exposure to crowding. So the total frequency of crowding is not directly relevant. Many studies report mean per capita aggressive encounters. This also does not reflect on an individual becoming more aggressive. It should be noted that as crowding increases, the potential number of conflict situations would increase for an individual. If we assume that individuals interact with each other randomly, then the total number of encounters should increase in square proportion of crowding. The per capita encounters should increase in direct proportion to population size. In reality, this does not happen because there is some social and spatial viscosity. Any individual does not have equal probability of interacting with all other individuals. Therefore, in reality, the total number of potential encounters would increase in a saturation curve rather than in square proportion. This curve should be the null model in assessing the effects of crowding on aggression. Unfortunately, it is not easy to work out this saturation curve making it difficult to assess the effects of crowding on the frequency of aggression. In experiments where crowding appears to decrease the frequency of total aggressive encounters instead of increasing it, the interpretation is clear and robust. In studies where there is a marginal increase in the total number of aggressive encounters, after applying the null model curve even conceptually, the inference may reverse in most of the studies. This makes the evidence for pro-aggression effects of crowding substantially weaker [103, 122].

Unlike experimental crowding, which is a sudden change, in real life, population densities change gradually most of the times. Here reverse causation is also a likely important factor. Species that increase aggression upon crowding are most unlikely to reach high population densities as compared to ones that have evolved to control aggression. This effect of aggression modulation on population density has been shown in ecological and evolutionary context both empirically and with a mathematical model in cats [123, 124]. Therefore, the association of crowding with reduced aggression appears to be robust with a two-way causation.

But even more robust and consistent across studies, especially in primates, is the positive association between crowding and submissive or friendly displays. This is demonstrated with a number of species and has been interpreted as essential strategies for coping with crowding [115, 125]. Aggressive behavior may go up or down with crowding, but submissive displays and other means to defuse aggression certainly go up. If we extrapolate this behavior in the human context, we would expect diplomat behavior to increase with crowding, although whether soldier behavior would decrease may still be doubted.

In the human context, an anecdote is that of increased verbal and suppressed physical aggression with crowding. This can be easily witnessed in a highly crowded urban place; an ideal example could be local trains of Mumbai City. Here, owing to crowding, heat, humidity, and discomfort, people have higher irritability levels. As a result, minor conflicts can easily take the form of a fight. However, almost invariably, the fights remain elaborately verbal and do not translate into physical aggression, although threatening is commonplace. The large number of spectators appears to witness and perhaps even enjoy the fights as long as they are verbal. The moment they appear to turn physical, some of the spectators turn proactive and intervene to prevent physical aggression. Throughout these events, there is an amazing level of nonverbal signaling such that everyone knows what will remain at the level of verbal threats, when would the fighters "invite" intervention by spectators. On rare occasions, when real antisocial elements are involved in initiating overt physical aggression, their signals are very different. In such cases only a very small fraction of spectators would generally dare to intervene. Nevertheless in a large crowd, the probability of finding such daredevils is high. Therefore, antisocial individuals are also under pressure from the crowd and less likely to turn violent during crowded hours. Most of the antisocial violence takes place off-peak hours. In brief, if we clearly segregate physical versus verbal aggression, even in human societies, we would find physical aggression being suppressed with crowding, whereas verbal aggression may go up.

Let us examine the same problem with a theoretical framework. We apply the hawk and dove game model to test out the effect of population density on aggression (see Appendix III for mathematical framework). As the population increases, frequency of interactions between the individuals would increase. If two hawks are competing for a resource, then there is a physical fight, and one of the hawks wins the resource. During the physical fight, the competitors can get injured. There is a cost related to injury. As population increases, the number of fights/interactions increases. Then there is less time for the healing of the injury. The net cost of injury thus keeps on adding up like a saturating curve. We modified standard hawk and dove game to include a factor of interpersonal knowledge Q that represents the probability of knowing the other person's strength. This is important because we are considering the change in population density with time.

Both hawk and dove strategists compete for a resource/reward. Hawk strategists initiate aggressive behavior and fight until injured or until the opponent retreats, whereas dove strategists would display and retreat if the opponent initiates aggressive behavior. There is a cost related to injury. Crowding has three potential effects on the effective payoffs of hawks and doves:

(a) As the population increases, it becomes difficult to know and/or remember each potential opponent in terms of its physical strength. The knowledge about opponent's strength has some role in deciding an individual's strategy. One is unlikely to take an aggressive stand if the opponent is known to be stronger. In a small population, hierarchies are quickly established, and as a result, actual aggression is avoided. By this effect, aggression should increase with crowding.

(b) Since there is increased competition for the same set of resources, the reward per fight (v of the model) decreases. Either the reward itself gets shared or the number of fights one has to give to get the entire reward goes up so that there is substantial reduction in v per fight.

(c) The cost of injuries resulting from each fight is incorporated into the model. However, injuries heal with time. If on an average there is sufficient interval between fights, the cost of injuries can be substantially ameliorated. The classical hawk and dove model assumes that one fight has no influence on another. However, with increasing population density, the interval between conflicts can come down, and the assumption may not remain applicable. As the fight interval decreases, the cost of injuries will accumulate.

We incorporated the above-mentioned three factors in the hawk and dove model to check the effect of increasing population density on aggres-

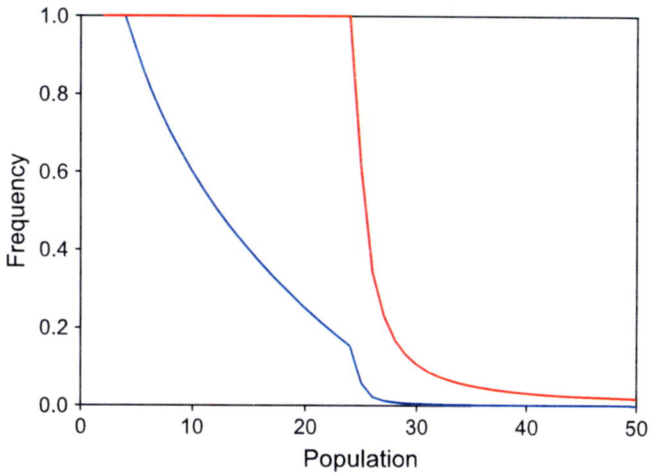

Fig. 9.4 The effect of population density on aggression as predicted by our model (see text and Appendix III). The *red line* represents the frequency of hawks in the population, whereas *blue line* represents frequency of aggressive encounters. Both decrease monotonically with population density but in a *complex shape*

sion, that is, the change in probability of hawk and dove strategists with change in population density (see Appendix III). Results of the model show that as the population density increases, the frequency of aggression in population reduces monotonically but in a complex curve. The frequency of hawks may remain high up to a threshold population density after which it declines very rapidly (Fig. 9.4). It is logical to assume therefore that at population densities close to saturation of the carrying capacity of the environment, when we expect a shift in strategies of reproduction from *r* to *K*, there should be an accompanying suppression of physical aggression.

The suppressing action of crowding on aggression links our two distinct lines of arguments. The physiological effects of soldier–diplomat behavioral dichotomy and *r*–*K* reproductive dichotomy are bridged by the common trigger of population density. High population density facilitates *K* reproductive strategy as well as diplomat behavior, both of which are pro-insulin resistance. In fact, one can argue that since the *r* to *K* and soldier to diplomat switches are likely to operate by common triggers, that is, the social and environmental stimuli for both the switches being frequently identical, common mechanisms for both the switches could have evolved. The common mechanisms constitute the wide variety of changes in the body that characterize the insulin resistance syndrome.

Does deficiency of physical aggression reduce ovulation and spermatogenesis? Does a correlation between female aggression and fecundity cut across taxa? Paper wasp queens are transiently highly aggressive when they become queens. Failure to become aggressive reduces lifetime fecundity [126]. Since female sex hormones are also involved in female aggression [127, 128], we can speculate that deficiency of physical aggression may be one of the triggering mechanisms behind PCOS. In some species of nonhuman primates, subordinate females undergo considerable suppression of reproduction. On the other hand, the role of testosterone in both male aggression and spermatogenesis is better known. T2D is associated with lower levels of testosterone which could be the proximate connecting link between aggression deficiency, reduced spermatogenesis, and erectile dysfunction.

There is one more effect of crowding that we need to worry about. Crowding anticipates competition for food and therefore possible starvation in near future. Therefore, our logical expectation would be that crowding should trigger thrift. There is some evidence that crowding does trigger thrift. In *C. elegans* crowding, caloric restriction or a combination of both results in a different morph called dauer that is a small, sluggish, and nonreproductive larval form that has very high lipid content and a prolonged life span [129, 130]. So crowding appears to increase lipid storage and decrease energy consumption in this species.

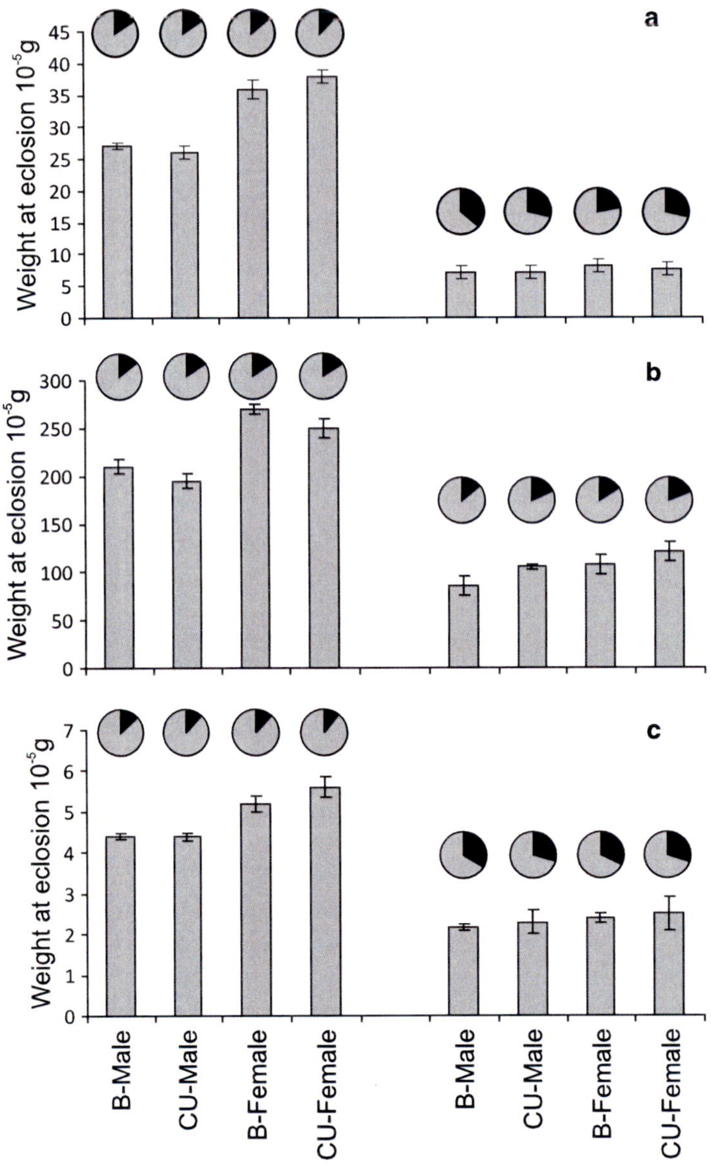

Fig. 9.5 The effect of larval crowding on body weight and lipid content in three species of *Drosophila* (data from [144]). Larvae of (**a**) *D. melanogaster*, (**b**) *D. nasuta*, and (**c**) *D. ananassae* were grown under crowded and uncrowded conditions, and the body weight and lipid content of the emerging adults were measured. The figures give mean body mass as bars and pie charts as relative lipid content, uncrowded larvae on the *left* and crowded on the *right*. Crowded larvae gave rise to lower mean body mass (bars) in all populations of the three species, but in spite of the lower body mass, the stored lipid (*dark sector* in the pie) was higher (significant in *D. melanogaster* and *D. ananassae*). In *D. nasuta*, where the difference in lipid content was not significant, the difference in population density was also smaller as compared to the two other species. Prior selection for crowded (CU) and uncrowded baseline (B) conditions did not make a significant difference in these parameters. It is possible that phenotypic plasticity towards metabolic response to crowding already existed, and therefore, selection did not make any difference. Larval density in uncrowded was 60–80 larvae per 8-dram vial with 6 ml food. Crowding in *D. melanogaster* consisted of 750–800 per vial, in *D. ananassae* 550–600 per vial, and *D. nasuta* 350–400 per vial. These researchers did not record aggression. My prediction is that crowded larval conditions will reduce aggression in adults

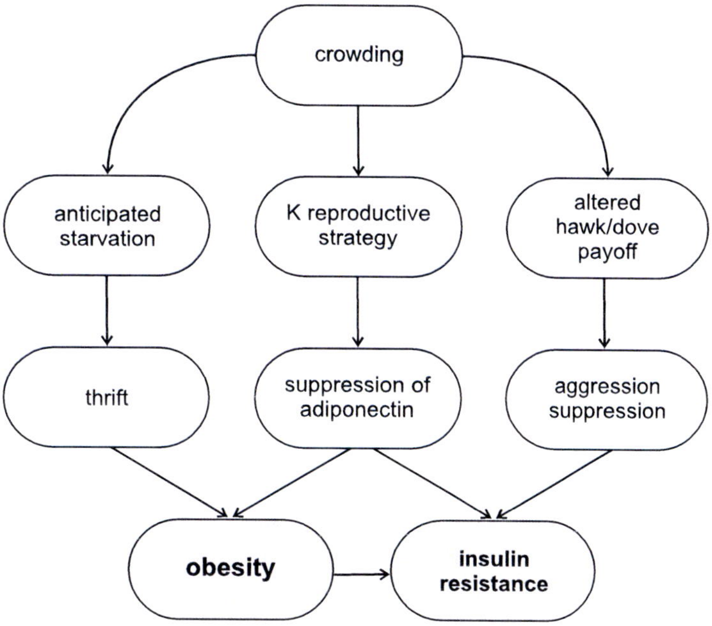

Fig. 9.6 Three different pathways by which high population density can affect obesity and insulin resistance

Larval crowding in drosophila results in adults that are smaller than the uncrowded controls but have a high proportion of lipids in the body (Fig. 9.5). One is easily tempted to compare the thin–fat phenotype in humans that is found in the most densely populated part of the world, the Indian subcontinent [131]. It is too premature to say whether there is something fundamentally similar between the small fat dauer worms or *Drosophila* and thin–fat Indian, but it is not impossible, in fact logically quite expected. Accumulating fat would be an appropriate adaptive biological response to crowding since crowding anticipates escalated competition for food leading to near-future starvation. A combination of crowding and availability of high-fat diet would presumably escalate the adaptive response and result in abnormally high adiposity. We can relate this now to the model of evolution of fetal programming for thrift discussed in Chap. 4. The Baig et al. [132] model showed that lifetime programming could evolve only if there was a correlation between birth time and lifetime nutritional conditions which make some anticipation possible. If such anticipation is possible, programming for thrift can evolve. Population density provides such a condition whether thrift can be adaptive, and this may be happening today because of the unprecedentedly high human population.

Summarily, we think that high population density is potentially an important risk factor driving our endobolic states towards metabolic syndrome. This can happen by three distinct mechanisms (Fig. 9.6): (1) by inducing thrift and lipid accumulation, (2) by shifting reproductive strategies from *r* to *K*, and (3) by suppressing physical aggression. If these links are included in the network model (Appendix II), it can be seen that crowding alone is sufficient to induce all the elements of metabolic syndrome. Suppression of adiponectin alone is also sufficient to induce all these changes. The prediction that crowding would suppress adiponectin needs to be tested. If this prediction is experimentally supported, the next natural question would be how it works. How does our body estimate population density and change the endobolic programs? We have little research in this direction so far. Both cognitive and noncognitive mechanisms may be involved. Visual perception of crowd noise can be perceived at conscious or subconscious levels. How much one worries about increasing population may matter on the one end, but on the other, it is possible that mechanisms completely inaccessible to conscious brain may operate involving pheromone sensing through olfactory or even vomeronasal route. Currently, we need to be more open minded and explore these possibilities with appropriate research tools.

An inevitable conclusion is that the increasing human population can be one of the reasons for the sudden burst of metabolic syndrome disorders throughout the world. Over the last century, population explosion and rise in metabolic syndrome disorders have gone hand in hand. But the trend alone does not tell us whether this is only a coincidence or there is a cause–effect relationship. There is some evidence that local crowding positively correlates with incidence of T2D [133–135]. For example, in the Indian city Chennai, along with increase in crowding, the prevalence of diabetes increased without a significant increase in anthropometric parameters [133]. The diabetic map of New York City demonstrates that the highest density of T2D lies in areas where there is a high proportion of people of tropical origin along with a high population density [136–140]. In almost all countries, the incidence of T2D in urban areas is substantially greater than rural areas [141]. The urban–rural difference cannot be completely explained by the differences in obesity alone [141], and crowding may be playing an important role. The regression between obesity and insulin resistance has a different slope and intercept in urban and rural areas, suggesting that some factor other than obesity is involved [142, 143]. This could very well be an effect of crowding. Since a causative effect of population density was never even suspected before, there have not been any epidemiological studies specifically testing crowding as a risk factor. But potentially, it should not be difficult to design specific epidemiological studies addressing this problem while controlling for other confounding factors. I hope epidemiological researchers realize the importance of population density and its effects on metabolic syndrome in the near future and attempt to test the hypothesis that population density can affect endocrine and metabolic states, a supernormal form of which can be pathological.

References

1. Sutherland W (1995) From individual behaviour to population ecology. Oxford University Press, Oxford
2. Matapurkar AK, Watve MG (1997) Altruist cheater dynamics in Dictyostelium: aggregated distribution gives stable oscillations. Am Nat 150:790–797
3. Levin PS, Tolimieri N, Nicklin M, Sale PF (2000) Integrating individual behavior and population ecology: the potential for habitat-dependent population regulation in a reef fish. Behav Ecol 11: 565–571
4. Tsoularis A, Wallace J (2002) Analysis of logistic growth models. Math Biosci 179:21–55
5. Bewley R (1988) A flexible logistic growth model with applications in telecommunications. Int J Forecast 4:177–192
6. Watve MG, Tickoo R, Jog MM, Bhole BD (2001) How many antibiotics are produced by the genus Streptomyces? Arch Microbiol 176:386–390
7. Weisstein EW. Logistic equation – from Wolfram MathWorld. http://mathworld.wolfram.com/LogisticEquation.html
8. Krebs C (2009) Ecology: the experimental analysis of distribution and abundance. NHBS Benjamin-Cummings
9. Parry GD (1981) The meanings of r- and K-selection. Oecologia 48:260–264
10. Pianka ER (1970) On r- and K-selecton. Am Nat 104:592–597
11. MacArthur R, Wilson E (2001) The theory of Island biogeography. Princeton University Press, Princeton
12. Watve MG et al (2000) The 'K' selected oligophilic bacteria: a key to uncultured diversity? Curr Sci 78:1535–1542
13. Richardson BJ (1975) r and K selection in kangaroos. Nature 255:323–324
14. Sinervo B, Svensson E, Comendant T (2000) Density cycles and an offspring quantity and quality game driven by natural selection. Nature 406: 985–988
15. Gillespie DOS, Russell AF, Lummaa V (2008) When fecundity does not equal fitness: evidence of an offspring quantity versus quality trade-off in pre-industrial humans. Proc Biol Sci 275:713–722
16. Chisholm JS, Burbank VK (2001) Evolution and inequality. Int J Epidemiol 30:206–211
17. Bruce HM (1963) Olfactory block to pregnancy among grouped mice. J Reprod Fertil 6:451–460
18. Morris EW (1977) Mobility, fertility, and residential crowding. Sociol Soc Res 61:363–379
19. Felson M, Solaun M (1975) The fertility-inhibiting effect of crowded apartment living in a tight housing market. Am J Sociol 80:1410–1427
20. Christian JJ, Lemunyan CD (1958) Adverse effects of crowding on lactation and reproduction of mice and two generations of their progeny. Endocrinology 63:517–529

21. Goulden CE, Hornig LL (1980) Population oscillations and energy reserves in planktonic cladocera and their consequences to competition. Proc Natl Acad Sci U S A 77:1716–1720
22. Norman RJ, Clark AM (1998) Obesity and reproductive disorders: a review. Reprod Fertil Dev 10:55–63
23. Pasquali R, Gambineri A, Pagotto U (2006) Review article: the impact of obesity on reproduction in women with polycystic ovary syndrome. BJOG Int J Obstet Gynaecol 113:1148–1159
24. Corbett SJ, McMichael AJ, Prentice AM (2009) Type 2 diabetes, cardiovascular disease, and the evolutionary paradox of the polycystic ovary syndrome: a fertility first hypothesis. Am J Hum Biol 21:587–598
25. Ibáñez L et al (2001) Sensitization to insulin induces ovulation in nonobese adolescents with anovulatory hyperandrogenism. J Clin Endocrinol Metab 86:3595–3598
26. Powers RW, Chambers C, Larsen WJ (1996) Diabetes-mediated decreases in ovarian superoxide dismutase activity are related to blood-follicle barrier and ovulation defects. Endocrinology 137:3101–3110
27. Altay B, Çetinkalp Ş, Do-anavşargil B, Hekimgil M, Semerci B (2003) Streptozotocin-induced diabetic effects on spermatogenesis with proliferative cell nuclear antigen immunostaining of adult rat testis. Fertil Steril 80:828–831
28. Cameron DF, Murray FT, Drylie DD (1985) Interstitial compartment pathology and spermatogenic disruption in testes from impotent diabetic men. Anat Rec 213:53–62
29. Arikawe AP, Daramola AO, Odofin AO, Obika LFO (2006) Alloxan-induced and insulin-resistant diabetes mellitus affect semen parameters and impair spermatogenesis in male rats/diabète sucré provoqué par l'alloxane aussi Bien que le diabète insulinorésistant influent sur les paramètres du sperme et entravent la spermatogenèse chez les rats mâles. Afr J Reprod Health 10:106–113
30. Doruk H et al (2005) Effect of diabetes mellitus on female sexual function and risk factors. Syst Biol Reprod Med 51:1–6
31. Rubin A, Babbott D (1958) Impotence and diabetes mellitus. J Am Med Assoc 168:498–500
32. Ellenberg M (1971) Impotence in diabetes: the neurologic factor. Ann Intern Med 75:213–219
33. Malavige LS et al (2008) Erectile dysfunction among men with diabetes is strongly associated with premature ejaculation and reduced libido. J Sex Med 5:2125–2134
34. Nakanishi S, Yamane K, Kamei N, Okubo M, Kohno N (2004) Erectile dysfunction is strongly linked with decreased libido in diabetic men. Aging Male 7:113–119
35. Lemone P (1996) The physical effects of diabetes on sexuality in women. Diabetes Educ 22:361–366
36. Neel JV (1999) Diabetes mellitus: a 'thrifty' genotype rendered detrimental by 'progress'? 1962. Bull World Health Organ 77:694–703, discussion 692–693
37. Schaefer-Graf UM et al (2003) Determinants of fetal growth at different periods of pregnancies complicated by gestational diabetes mellitus or impaired glucose tolerance. Diabetes Care 26:193–198
38. Moore TR (1997) Fetal growth in diabetic pregnancy. Clin Obstet Gynecol 40:771–786
39. Xu T et al (2002) Effect of surgical castration on risk factors for arteriosclerosis of patients with prostate cancer. Chin Med J 115:1336–1340
40. Holmäng A, Björntorp P (1992) The effects of testosterone on insulin sensitivity in male rats. Acta Physiol Scand 146:505–510
41. Godsland IF et al (1992) Insulin resistance, secretion, and metabolism in users of oral contraceptives. J Clin Endocrinol Metab 74:64–70
42. Diamanti-Kandarakis E, Baillargeon J-P, Iuorno MJ, Jakubowicz DJ, Nestler JE (2003) A modern medical quandary: polycystic ovary syndrome, insulin resistance, and oral contraceptive pills. J Clin Endocrinol Metab 88:1927–1932
43. Alexander N, Clarkson T (1978) Vasectomy increases the severity of diet-induced atherosclerosis in Macaca fascicularis. Science 201:538–541
44. Clarkson TB, Alexander NJ (1980) Long-term vasectomy: effects on the occurrence and extent of atherosclerosis in rhesus monkeys. J Clin Invest 65:15–25
45. Clarkson TB, Lombardi DM, Alexander NJ, Lewis JC (1986) Diet and vasectomy: effects on atherogenesis in cynomolgus macaques. Exp Mol Pathol 44:29–49
46. Manson JE et al (1999) Vasectomy and subsequent cardiovascular disease in US physicians. Contraception 59:181–186
47. Mullooly JP, Wiest WM, Alexander NJ, Greenlicki MR, Fulgham DL (1993) Vasectomy, serum assays, and coronary heart disease symptoms and risk factors. J Clin Epidemiol 46:101–109
48. Lu C-C et al (2009) Association of glycemic control with risk of erectile dysfunction in men with type 2 diabetes. J Sex Med 6:1719–1728
49. Awad H, Salem A, Gadalla A, El Wafa NA, Mohamed OA (2010) Erectile function in men with diabetes type 2: correlation with glycemic control. Int J Impot Res 22:36–39
50. Hidalgo-Tamola J, Chitaley K (2009) Review. J Sex Med 6:916–926
51. Yaman O, Akand M, Gursoy A, Erdogan MF, Anafarta K (2006) The effect of diabetes mellitus treatment and good glycemic control on the erectile function in men with diabetes mellitus-induced erectile dysfunction: a pilot study. J Sex Med 3:344–348
52. Ayala JE et al (2007) Chronic treatment with sildenafil improves energy balance and insulin action in high fat-fed conscious mice. Diabetes 56:1025–1033
53. Mammi C et al (2011) Sildenafil reduces insulin-resistance in human endothelial cells. PLoS One 6:e14542
54. De Young L, Chung E, Kovac JR, Romano W, Brock GB (2012) Daily use of sildenafil improves endothelial function in men with type 2 diabetes. J Androl 33:176–180

55. Sairam MR, Wang M, Danilovich N, Javeshghani D, Maysinger D (2006) Early obesity and age-related mimicry of metabolic syndrome in female mice with sex hormonal imbalances. Obesity (Silver Spring) 14:1142–1154
56. López-López R, Huerta R, Malacara JM (1999) Age at menopause in women with type 2 diabetes mellitus. Menopause 6:174–178
57. Rosano GMC, Vitale C, Marazzi G, Volterrani M (2007) Menopause and cardiovascular disease: the evidence. Climacteric 10:19–24
58. Catalano PM, Thomas AJ, Huston LP, Fung CM (1998) Effect of maternal metabolism on fetal growth and body composition. Diabetes Care 21(Suppl 2):B85–B90
59. Trivers RL, Willard DE (1973) Natural selection of parental ability to vary the sex ratio of offspring. Science 179:90–92
60. Cameron EZ (2004) Facultative adjustment of mammalian sex ratios in support of the Trivers-Willard hypothesis: evidence for a mechanism. Proc Biol Sci 271:1723–1728
61. Cameron EZ, Lemons PR, Bateman PW, Bennett NC (2008) Experimental alteration of litter sex ratios in a mammal. Proc Royal Soc B 275:323–327
62. Mathews F, Johnson PJ, Neil A (2008) You are what your mother eats: evidence for maternal preconception diet influencing foetal sex in humans. Proc Biol Sci 275:1661–1668
63. Larson MA, Kimura K, Kubisch HM, Roberts RM (2001) Sexual dimorphism among bovine embryos in their ability to make the transition to expanded blastocyst and in the expression of the signaling molecule IFN-τ. Proc Natl Acad Sci USA 98:9677–9682
64. Williams RJ, Gloster SP (1992) Human sex ratio as it relates to caloric availability. Soc Biol 39:285–291
65. Houde AA, Murphy BD, Mathieu O, Bordignon V, Palin MF (2008) Characterization of swine adiponectin and adiponectin receptor polymorphisms and their association with reproductive traits. Anim Genet 39:249–257
66. Chansomboon C, Elzo MA, Suwanasopee T, Koonawootrittriron S (2009) Genotypic polymorphisms for adiponectin and follicle stimulating hormone receptor genes related to weaning-to-first service interval and litter traits in a swine population in northern Thailand. Thai J Agric Sci 42:237–245
67. Ledoux S et al (2006) Adiponectin induces periovulatory changes in ovarian follicular cells. Endocrinology 147:5178–5186
68. Chappaz E et al (2008) Adiponectin enhances in vitro development of swine embryos. Domest Anim Endocrinol 35:198–207
69. Čikoš Š et al (2010) Expression of adiponectin receptors and effects of adiponectin isoforms in mouse preimplantation embryos. Hum Reprod 25: 2247–2255
70. Chappaz E (2006) Influence of adiponectin on porcine oogenesis. University of McGill, http://digitool.Library.McGill.CA:80/R/-?func=dbin-jump-full&object id=99329&silo library=GEN01
71. Groth SW (2010) Adiponectin and polycystic ovary syndrome. Biol Res Nurs 12:62–72
72. Sepilian V, Nagamani M (2005) Adiponectin levels in women with polycystic ovary syndrome and severe insulin resistance. J Soc Gynecol Investig 12:129–134
73. Carmina E et al (2005) Evidence for altered adipocyte function in polycystic ovary syndrome. Eur J Endocrinol 152:389–394
74. Panidis D et al (2003) Serum adiponectin levels in women with polycystic ovary syndrome. Hum Reprod 18:1790–1796
75. Ghazanfari S, Nobari K, Yamauchi T (2011) Adiponectin: a novel hormone in birds. Asian J Anim Vet Adv 6:429–439
76. Ocón-Grove OM, Krzysik-Walker SM, Maddineni SR, Hendricks GL, Ramachandran R (2008) Adiponectin and its receptors are expressed in the chicken testis: influence of sexual maturation on testicular ADIPOR1 and ADIPOR2 mRNA abundance. Reproduction 136:627–638
77. Caminos JE et al (2008) Novel expression and direct effects of adiponectin in the rat testis. Endocrinology 149:3390–3402
78. Shang L et al (2009) Relationship of adiponectin and visfatin with fetus intrauterine growth. Zhonghua Fu Chan Ke Za Zhi 44:246–248
79. Nanda S, Akolekar R, Sarquis R, Mosconi AP, Nicolaides KH (2011) Maternal serum adiponectin at 11 to 13 weeks of gestation in the prediction of macrosomia. Prenat Diagn 31:479–483
80. Schumacher MA (2009) Placental signaling mechanisms linking maternal obesity, high-fat diet, and adiponectin levels during pregnancy to fetal overgrowth. MS thesis, University of Cincinnati
81. Caminos JE et al (2005) Expression and regulation of adiponectin and receptor in human and rat placenta. J Clin Endocrinol Metab 90:4276–4286
82. Jansson N, Greenwood SL, Johansson BR, Powell TL, Jansson T (2003) Leptin stimulates the activity of the system A amino acid transporter in human placental villous fragments. J Clin Endocrinol Metab 88:1205–1211
83. Sladek C (2008) The effects of human chorionic somatomammotropin and estradiol on gluconeogenesis and hepatic glycogen formation in the rat. Horm Metab Res 7:50–54
84. Martin LJ et al (2006) Adiponectin is present in human milk and is associated with maternal factors. Am J Clin Nutr 83:1106–1111
85. Connor EE et al (2008) Effects of increased milking frequency on gene expression in the bovine mammary gland. BMC Genomics 9:362
86. Yong LC et al (1994) Relationship between dietary intake and plasma concentrations of carotenoids in premenopausal women: application of the USDA-NCI carotenoid food-composition database. Am J Clin Nutr 60:223–230
87. Feng Z et al (1995) Glucocorticoid and progesterone inhibit involution and programmed cell death in the mouse mammary gland. J Cell Biol 131:1095–1103

88. Berg MN, Dharmarajan AM, Waddell BJ (2002) Glucocorticoids and progesterone prevent apoptosis in the lactating rat mammary gland. Endocrinology 143:222–227
89. Terry PM, Banerjee MR, Lui RM (1977) Hormone-inducible casein messenger RNA in a serum-free organ culture of whole mammary gland. Proc Natl Acad Sci USA 74:2441–2445
90. Devinoy E, Houdebine L-M, Delouis C (1978) Role of prolactin and glucocorticoids in the expression of casein genes in rabbit mammary gland organ culture. Quantification of casein mRNA. Biochim Biophys Acta 517:360–366
91. Houdebine L-M, Devinoy E, Delouis C (1978) Stabilization of casein mRNA by prolactin and glucocorticoids. Biochimie 60:57–63
92. Linzell JL (1967) The effect of infusions of glucose, acetate and amino acids in hourly milk yield in fed, fasted and insulin-treated goats. J Physiol 190:347–357
93. Sullivan EC, Hinde K, Mendoza SP, Capitanio JP (2011) Cortisol concentrations in the milk of rhesus monkey mothers are associated with confident temperament in sons, but not daughters. Dev Psychobiol 53:96–104
94. Peters RH (1986) The ecological implications of body size. Cambridge University Press, Cambridge
95. Calder WA (1996) Size, function, and life history. Courier Dover Publications, Mineola, NY
96. Klöting N, Blüher M (2005) Extended longevity and insulin signaling in adipose tissue. Exp Gerontol 40:878–883
97. Kimura KD, Tissenbaum HA, Liu Y, Ruvkun G (1997) daf-2, an insulin receptor-like gene that regulates longevity and diapause in Caenorhabditis elegans. Science 277:942–946
98. Unger RH (2006) Klotho-induced insulin resistance: a blessing in disguise? Nat Med 12:56–57
99. Blüher M, Kahn BB, Kahn CR (2003) Extended longevity in mice lacking the insulin receptor in adipose tissue. Science 299:572–574
100. Bartke A, Brown-Borg H (2004) Life extension in the dwarf mouse. Curr Top Dev Biol 63: 189–225
101. Rushton JP (1996) Race, genetics, and human reproductive strategies. Genet Soc Gen Psychol Monogr 122:21–53
102. Southwick CH (1967) An experimental study of intragroup agonistic behavior in rhesus monkeys (macaca mulatta). Behaviour 28:182–209
103. Alexander BK, Roth EM (1971) The effects of acute crowding on aggressive behavior of Japanese monkeys. Behaviour 39:73–90
104. Nijman HLI, Rector G (1999) Crowding and aggression on inpatient psychiatric wards. Psychiatr Serv 50:830–831
105. Ng B, Kumar S, Ranclaud M, Robinson E (2001) Ward crowding and incidents of violence on an acute psychiatric inpatient unit. Psychiatr Serv 52:521–525
106. Owen C, Tarantello C, Jones M, Tennant C (1998) Violence and aggression in psychiatric units. Psychiatr Serv 49:1452–1457
107. Calhoun JB (1973) Death squared. Proc R Soc Med 66:80–88
108. Anderson BV, Elton RH (1977) The social behavior of a group of baboons (Papio anubis) under artificial crowding. Primates 18:225–234
109. Hazlett BA (1968) Effects of crowding on the agonistic behavior of the hermit crab pagurus bernhardus. Ecology 49:573–575
110. Wolkenten ML, Davis JM, Gong ML, Waal FBM (2006) Coping with acute crowding by cebus apella. Int J Primatol 27:1241–1256
111. Aureli F, De Waal FBM (1997) Inhibition of social behavior in chimpanzees under high-density conditions. Am J Primatol 41:213–228
112. Bercovitch FB, Lebrón MR (1991) Impact of artificial fissioning and social networks on levels of aggression and affiliation in primates. Aggress Behav 17:17–25
113. Harvey PW, Chevins PFD (1985) Crowding pregnant mice affects attack and threat behavior of male offspring. Horm Behav 19:86–97
114. Van Loo PLP, Mol JA, Koolhaas JM, Van Zutphen BFM, Baumans V (2001) Modulation of aggression in male mice: influence of group size and cage size. Physiol Behav 72:675–683
115. Sannen A, Elsacker LV, Eens M (2004) Effect of spatial crowding on aggressive behavior in a bonobo colony. Zoo Biol 23:383–395
116. de Waal FB, Aureli F, Judge PG (2000) Coping with crowding. Sci Am 282:76–81
117. Wongwitdecha N, Marsden CA (1996) Social isolation increases aggressive behaviour and alters the effects of diazepam in the rat social interaction test. Behav Brain Res 75:27–32
118. Matsumoto K, Pinna G, Puia G, Guidotti A, Costa E (2005) Social isolation stress-induced aggression in mice: a model to study the pharmacology of neurosteroidogenesis. Stress 8:85–93
119. Hodge GK, Butcher LL (1975) Catecholamine correlates of isolation-induced aggression in mice. Eur J Pharmacol 31:81–93
120. Sánchez C, Arnt J, Hyttel J, Moltzen EK (1993) The role of serotonergic mechanisms in inhibition of isolation-induced aggression in male mice. Psychopharmacology 110:53–59
121. Aiello JR, Nicosia G, Thompson DE (1979) Physiological, social, and behavioral consequences of crowding on children and adolescents. Child Dev 50:195–202
122. Regoeczi WC (2003) When context matters: a multilevel analysis of household and neighbourhood crowding on aggression and withdrawal. J Environ Psychol 23:457–470
123. Pontier D, Auger P, Bravo de la Parra R, Sánchez E (2000) The impact of behavioral plasticity at individual level on domestic cat population dynamics. Ecol Model 133:117–124

124. Auger P, Pontier D (1998) Fast game theory coupled to slow population dynamics: the case of domestic cat populations. Math Biosci 148:65–82
125. Judge PG, De Waal FM (1997) Rhesus monkey behaviour under diverse population densities: coping with long-term crowding. Anim Behav 54:643–662
126. Lamba S et al (2007) A possible novel function of dominance behaviour in queen-less colonies of the primitively eusocial wasp Ropalidia marginata. Behav Processes 74:351–356
127. Albert DJ, Jonik RH, Walsh ML (1992) Hormone-dependent aggression in male and female rats: experiential, hormonal, and neural foundations. Neurosci Biobehav Rev 16:177–192
128. Rubenstein DR, Wikelski M (2005) Steroid hormones and aggression in female Galápagos marine iguanas. Horm Behav 48:329–341
129. Hu PJ (2007) Dauer. In: Worm Book. doi:10.1895/wormbook.1.144.1
130. Fielenbach N, Antebi A (2008) C. elegans dauer formation and the molecular basis of plasticity. Genes Dev 22:2149–2165
131. Yajnik CS et al (2003) Neonatal anthropometry: the thin-fat Indian baby. The Pune Maternal Nutrition Study. Int J Obes Relat Metab Disord 27:173–180
132. Baig U, Belsare P, Watve M, Jog M (2011) Can thrifty gene(s) or predictive fetal programming for thriftiness lead to obesity? J Obes 2011:1–11
133. Ramachandran A, Snehalatha C, Latha E, Vijay V, Viswanathan M (1997) Rising prevalence of NIDDM in an urban population in India. Diabetologia 40:232–237
134. Riste L, Khan F, Cruickshank K (2001) High prevalence of type 2 diabetes in all ethnic groups, including Europeans, in a British inner city: relative poverty, history, inactivity, or 21st century Europe? Diabetes Care 24:1377–1383
135. Ellaway A, Macintyre S, Bonnefoy X (2005) Graffiti, greenery, and obesity in adults: secondary analysis of European cross sectional survey. Brit Med J 331:611–612
136. A Local National and Worldwide Scourge. http://berkeleycitizen.org/radiation/radiation2.htm
137. New Diabetes Report Documents Devastating Effects In New York City. New York city department of health and mental hygiene. http://www.nyc.gov/html/doh/html/pr2007/pr060-07.shtml
138. Then as now – New York's Shifting Ethnic Mosaic. http://www.nytimes.com/interactive/2011/01/23/nyregion/20110123-nyc-ethnic-neighborhoods-map.html
139. New York Population Map. http://en.wikipedia.org/wiki/File:New_York_Population_Map.png
140. Center for Urban Research – Long Island Index 2008. http://www.urbanresearch.org/projects/long-island-index-2008
141. Zimmet P, Dowse G, Finch C, Serjeantson S, King H (1990) The epidemiology and natural history of NIDDM–lessons from the South Pacific. Diabetes Metab Rev 6:91–124
142. Snehalatha C, Ramachandran A, Vijay V, Viswanathan M (1994) Differences in plasma insulin responses in urban and rural Indians: a study in southern-Indians. Diabet Med 11:445–448
143. Al-Nuaim AR (1997) Prevalence of glucose intolerance in urban and rural communities in Saudi Arabia. Diabet Med 14:595–602
144. Archana N (2009) The genetic architecture of fitness-related traits in populations of three species of drosophila subjected to selection for adaptation to larval crowding. PhD thesis, Jawaharlal Nehru Centre for Advanced Scientific Research, Bangalore

10 Time to Give Up Stress

A book on diabetes will remain grossly incomplete if we do not talk about stress. I suspect that many readers would be surprised that I did not even mention stress in the nine chapters so far. Some association between stress and health has been recognized for a long time by science, but till date, our understanding is rather naïve. I feel that the failure to understand stress and its effects clearly is owing to the lack of applying rigorous evolutionary logic to this field.

The history of stress-related health problems is interesting and has seen a number of ups and downs. From the beginning of the twentieth century, it has been speculated that diabetes and many other so-called "lifestyle" diseases are caused or aggravated by stress. These stress-related disorders included diabetes, hypertension, hypercholesterolemia, cardiovascular disease, gastric acidity, peptic ulcers, etc. After the World War I there was some setback to this argument due to evident reduction in some of the stress-related disorders, particularly diabetes and hypertension, during the Great War. It appeared that the stress of war not only failed to increase the incidence or severity but in fact reduced it dramatically [1–3]. Nobody doubted that there was tremendous stress on people during the Great War. But since there was no increase in stress-related disorders during the global stress period, there was a strong counterargument against stress having anything to do with these disorders. The antidiabetic effects of the stress of war were demonstrated once again during World War II. The story repeated itself in all subsequent long-drawn wars such as the Serbian or Sarajevan wars [3]. The war data are quite robust and consistent across wars. Clinical records, epidemiological surveys, and follow-up of known cases have consistently shown the ameliorating effects of war. Therefore it is unlikely to be a reporting or methodological bias. A common interpretation of the wartime suppression of diabetes and hypertension is that there is inevitable change in the diet during wartime. Unfortunately the diet hypothesis is more of hearsay and has not been pursued and tested with scientific rigor. Even if we get quantitative data on wartime diets, showing that there was a change in diet is not sufficient to conclude that the difference was caused by change in diet. A good test of the hypothesis would be to show that peacetime dietary interventions identical to the wartime diet produce the same effect. But this has never been done. Therefore all that we know is that war suppressed diabetes. Whether it was due to diet, stress of war, or some other effect is yet to be demonstrated.

While no one really knows why wartime stress decreased diabetes, in the last few decades, there has been an increasing belief in stress as the origin of an entire cluster of disorders, although some diseases such as peptic ulcers that were once thought to be stress induced are now excluded. The stress paradigm however is faced with a number of problems. The problems begin with the definition and measurement of stress itself.

Stress has been defined in a number of different ways, part of the difference being contextual.

M. Watve, *Doves, Diplomats, and Diabetes: A Darwinian Interpretation of Type 2 Diabetes and Related Disorders*, DOI 10.1007/978-1-4614-4409-1_10, © Springer Science+Business Media New York 2013

The word stress is used in animal, plant, and microbial physiology too. We would restrict ourselves to the use of the word in the field of medicine. Throughout the field of medicine, the definition and the methods (or the lack of them) of measuring stress have been elusive. Currently a disorganized set of concepts are being included under the name stress. Before we try to understand how and whether stress affects diabetes, we need to have a reasonable working definition of stress. A common and apparently reasonable definition is that of a challenge to one or more of the homeostatic mechanisms of the body or a departure from the normal homeostatic state in response to an environmental or behavioral stimulus. So, any environmental, social, or even an imaginary challenge that stretches the homeostatic mechanisms can be called a stress [4]. The definition sounds logical and should be acceptable as a working definition.

However there are a series of problems associated with this apparently simple definition. The homeostatic mechanisms of the body have substantial plasticity, and they appear to learn to accommodate a variety of challenges if faced repeatedly. Taking a simple example of a physical stress, if you run for a couple of kilometers, the mechanisms managing energy supply and energy balance are stretched rather too much, and you feel completely exhausted. However if you run 5 km every day, the mechanisms get adapted, and the systems adjust themselves such that you can easily run 2 km now without experiencing exhaustion. Something similar happens to psychological pressures too.

The corollary of it is that if you do not experience any challenge for sufficiently long time, the homeostatic mechanisms are slowly eroded. For example, a person who does not exercise for decades will be unable to run even 20 m. In this case, the homeostatic mechanisms appear to have deteriorated owing to chronic absence of challenge. Deterioration of a homeostatic system should also be considered a challenge to the homeostatic system. A challenge to the homeostatic system is stress by definition. This means that chronic lack of stress is a stress by itself. If lack of stress is a stress, then the definition of stress becomes meaningless. A definition of a concept needs to specify what it is and what it is not. If it is what it is not, the definition of definition is not satisfied, and the concept itself becomes meaningless.

This appears to reflect in the model stressors used by experimenters in different experiments too. For example, if rats are kept together, they indulge into fights and threat displays through which they establish and maintain a social hierarchy. This has been used as a model of "social stress" [5–8]. Other groups of experimenters addressing different questions have kept rats isolated in separate enclosures which they call "isolation stress" [9–12]. If being together is a stress and being isolated is also a stress, then one wonders whether there is a nonstress condition. Since various contradicting phenomena are dumped under a generic name of stress, there is contradiction in the effects of stress as well. Stress is immunosuppressive in certain context and enhancer of immunity in other [13–19]. Stress impairs learning and memory in some studies and enhances it in a few others [20–26]. Stress is proinflammatory in certain contexts [27, 28] and anti-inflammatory in others [29, 30]. The usefulness of stress as a generic term is therefore doubtful.

Another major problem lies in the circularity of stress-related arguments. If anything that challenges the homeostatic state is called stress, then stress changes the homeostatic state becomes a meaningless statement. Since inducing a physiological change is included in the definition of stress, the statement that stress affects physiology becomes a noninformative tautology. It effectively means that *anything that changes the homeostatic state changes the homeostatic state.* If you think that this statement is meaningful, then stress research is certainly meaningful. But if you feel this statement is meaningless, the logic behind entire stress research needs to be reexamined. A circularity problem of this type is not unique to stress. This philosophical problem is faced by many fundamental principles of science too. For example, the law of conservation of momentum states that the momentum of an object does not change unless an external force is applied. But force is measured by change in momentum, which means that *there is no change in momentum unless the momentum changes*.

Most researchers ignore this philosophical problem and indulge into the endocrinological nitty-gritty of the so-called stress-induced effects. For the time being we will follow this line of researchers and look at stress-related disorders without worrying about the definition of stress. All kinds of stresses are said to trigger a single pathway known as the HPA axis which stands for the hypothalamic–pituitary–adrenal axis [31–34]. This refers to the sequence in which a series of hormones are activated. The chain of events begins in the hypothalamus which receives nervous connections from many parts of the brain. After receiving certain kinds of signals which are believed to be "stress" signals, corticotropin-releasing factor (CRF) is produced by the paraventricular nucleus of the hypothalamus. CRF is carried to the anterior pituitary gland where it stimulates the production of adrenocorticotropic hormone (ACTH). ACTH release from the pituitary gland reaches the adrenocortical cells where it stimulates the production of glucocorticoids. This response takes a few minutes at the end of which raised levels of glucocorticoids can be detected in plasma. The rise can be as high as 20-fold of the basal levels. ACTH not only stimulates hormone secretion, it also induces hypertrophy and proliferation of the adrenocortical cells.

The end product of HPA activity is increased levels of glucocorticoids. In humans cortisol is the common form of the hormone, whereas in many animal models including rats, it is corticosterone. Either of them has a variety of actions on metabolism. What has been less appreciated is that they also have important effects on behavior. The metabolic effects consist of increased glucose production by liver utilizing amino acids that are mainly diverted from muscle. On the other hand the rate of muscle glucose utilization is reduced. Cortisol effectively depletes protein content of many tissues including muscle. It is also an effective immunosuppressor. The behavioral effects are less well known but nevertheless experimentally clearly demonstrated. Cortisol reduces risk-taking behaviors, increases submissive responses, and at the same time enhances learning, particularly learning in the context of coping strategies towards an environmental challenge [35]. The action of cortisol on hippocampus during sleep is important for memory consolidation [25, 36]. Cortisol enhances hippocampus-dependent declarative memory formation but suppresses amygdala-dependent emotional memory formation. The natural cortisol rise during late sleep may thus protect from overshooting emotional memory formation, a mechanism that could be potentially important for a diplomat who needs to take cool-headed decisions [25]. Although cortisol has many specific adaptive roles in fine manipulations of behavior, research on behavioral functions of cortisol appears to be largely overshadowed by the label of "stress hormone."

Stress researchers are so much obsessed by the association of stress and HPA axis that knowingly or unknowingly most of them define stress as anything that triggers the HPA axis. Some authors even overtly define stress by the HPA effect. Here lies the circularity of the argument in a more detailed form. The belief that all kinds of stresses stimulate the HPA axis is augmented by defining stress this way. Let us take an example. It can be argued that an alpha male in a primate group is under stress because with every potential competitor, its alpha position is at stake. A low-ranking male, on the other hand, has little to lose from losing another combat, and therefore, it can easily give a submissive display and avoid a conflict. Therefore alpha males must be under greater stress and uncertainty than low-ranking males. However, stress researchers will not agree with this argument because alpha males generally have lower corticosteroid levels [4]. This makes the researchers believe that low-ranking males are at a greater stress. Whenever alpha males were found to have higher corticosteroid levels, it was immediately concluded that alpha males are under greater stress [37]. This exemplifies how people use HPA response to define stress. If stress is defined this way, stress induces HPA activity is a meaningless statement. We will be able to test the hypothesis that stress activates HPA only if we have any independent quantitative measurement of stress. If stress itself is measured by corticosteroid levels, then stress increases corticosteroid levels cannot be the inference. And if stress is equated to corticosteroid levels, then there is no need to use the word stress. We can simply treat the phenomenon as raised corticosteroid levels. This would be a more objective working approach that does not involve any ill-defined terms such as stress.

The example of romantic love is even more entertaining. A number of hormonal changes are associated with the first phase of falling in love. One of the observations is that the corticosteroid levels are raised [38, 39]. Based on this observation some researchers have concluded that love is a stress. They do not appear to feel the need for any other justification as to why they want to call love as stress. A rise in cortisol is a sufficient indication of stress for them. This aptly demonstrates how HPA axis is used as a criterion to identify something as stress. This biased definition of stress has given rise to the highly flawed perception that all types of stressors trigger the HPA axis. This becomes a tautological argument once again saying *anything that triggers the HPA axis, triggers the HPA axis.*

The idea of all types of stresses triggering a single pathway appears very strange from the standpoint of a behavioral ecologist. Behavioral ecologists study the diversity of behavioral strategies displayed in response to a diversity of environmental and social challenges and therefore naturally expect an equally wide diversity in the physiological response of the body that accompanies a behavioral response. The notion that all types of challenges trigger the same stereotyped response therefore appears weird. How and why could have evolution favored such a stereotyped dumb response at the physiological level when there is a wide array of flexible and smart responses at the level of behavior? The answer to the question lies in the circularity of definition and the selective choice of stress models. Since knowingly or unknowingly HPA axis is used in defining stress and choosing model stressors, there is a false impression that all kinds of stresses trigger HPA axis. A number of stressors could be triggering different pathways, but stress researchers are generally reluctant to call them stressors because they do not see rising corticosteroids.

The evidence is on the contrary. As mentioned earlier wartime stress appears to have different effects than peacetime stress. Some stressors are reported to evoke hypocortisolemia rather than hypercortisolemia [40–42]. Particularly interesting is the fact that stress and depression caused by war exposure is associated with low rather than high cortisol levels, and in war veterans, cortisol levels are negatively associated with exposure to actual combat and violence [43, 44]. Evidence also points that the same stressor can elicit different hormonal responses in different individuals. For example, to the same stressor of a conspecific competitor, chimpanzees responded by increasing testosterone whereas bonobos responded by increasing cortisol [45]. This is in accordance with the known differences in the behavior of the two species. Chimpanzees are more aggressive and risk takers than bonobos [46], and they responded by testosterone, whereas bonobos, the more social cousins of chimpanzees, responded by cortisol. Even within a species, in response to a potential competitor, high-ranking individuals give a testosterone response, whereas low-ranking individuals give a corticosteroid response [47, 48]. Apart from social ranking differences in individuals' behavior and copying styles also affect stress responses [49]. It is possible therefore that in reality, different challenges evoke different responses and also the same challenge evokes different responses in different personalities, but it is the thinking trap of researchers that makes them believe that HPA is the only stress response pathway.

In fact some alternative responses are quite well known, and perhaps there are a number of others not so well known. Perhaps more important than HPA axis, there is another well-known stress response which is often referred to as a flight-or-fight response [50] with a quasi-evolutionary explanation. This is a response of the autonomous nervous system. In the field of medicine it is assumed that a typical natural cause of stress is say suddenly detecting a predator. The natural reaction to a predator is either to run away as fast as possible or to fight back with all wits and vigor.

This is called a flight-or-fight response. This response is mediated by activating the sympathetic nervous system which mobilizes a lot of energy in a short time by undertaking lipolysis, glycogenolysis, and liver gluconeogenesis resulting into release of more glucose and fatty acids that supply fuel to the fight-or-flight-related muscle activities. An old belief associated with this response is that although the stress and anxiety typical of modern urban life are of a very different nature, it triggers the same pathway as a predator response. So there is mobilization of fuel sources in the same way. Fatty acids and glucose levels in the plasma go up, but unlike classical predator situation, there is not any intense muscle activity to burn of the excess glucose and fatty acids. This eventually leads to the bad effects of stress including hyperglycemia, hyperlipidemia, hypercholesterolemia, atherosclerosis, and cardiovascular disease. The logic appears temptingly simple and convincing.

The only problem with the fight-or-flight hypothesis is that it is too naïve to be true. The first naïve assumption is that natural stressful situations are only predator or predator-like situations, and fight-or-flight response is the only adaptive response. This belief suits well for an urban biologist who does not know the complexities of life in the wilderness. The life of any social animal, primates in particular, is faced with a wide diversity of environmental and social challenges, and there is an equally wide diversity of behavioral responses to these challenges. Describing them would demand a separate book in itself, so I would avoid the temptation here. The only point relevant here is that fight or flight is not the only adaptive response to all types of challenges in the wilderness. This is recognized by many researchers, and alternatives are also suggested [51–53]. I would like to add two more "f"s to the response categories. They are *fox* and *freeze* in addition to *fight* or *flight*. *Fox* refers to the fox in Aesop's fables. It is equivalent to the diplomat strategy that we introduced earlier consisting of coping with a challenge by social manipulation rather than physical strength. *Fight* and *flight* together constitute our soldier strategy. Whereas *freeze* is a response of being helpless and seeking neither a soldier nor a diplomat solution. It would be somewhat counterintuitive, but in real life, a *freeze* response can give substantial returns in a number of situations. In fact, some species are specifically adapted to the *freeze* response. Quite well known to naturalists, physiologists appear to have largely ignored the importance of a nonsoldier–nondiplomat *freeze* response. A variety of species of quails inhabit grasslands, scrub, or forest floor. They are mainly ground-dwelling birds with relatively poor flight muscles. Having limited flight capacity, they need alternative strategies for escaping ground predators such as wild cats and jackals and areal predators such as falcons and hawks. They rely on a superb camouflage aided by a completely immobile crouched and frozen posture on suspecting predator presence. With the sharp vision of falcons and eagles, a combination of superb camouflage and complete freezing is essential, and no other active defense is likely to work.

It is not only for species that specialize in freeze response. To relate some brief anecdotes, I have experienced attacks by two dangerous species of animals in the wild, elephant and tiger. In both the cases standing firm on the ground facing the animal is the most successful strategy, while an attempt to run away increases the risk, and fighting back is simply out of question. One of my close friends and a well-known wildlife photographer tells me that when attacked by a lone tusker, he wanted to run away, but fear gripped him in such a way that he just could not move. He froze helplessly facing the tusker, and the tusker, after a threatening rush towards him, suddenly turned away from a distance of about 10 m and vanished into the thickets. My friend still could not move for the next few minutes. I am sure this is an evolved response which has a survival value. When fight, flight, or fox are unlikely to work, there is one more class of responses and there must be an elaborate neuronal and physiological set of mechanisms mediating it. But stress biologists are so obsessed by the fight-or-flight idea that they have a mental block towards other possible responses.

The picture with four "f"s instead of two is still very naïve and unable to fully capture the context-dependent diversity of responses. Nevertheless it is better than the classical presumption that the body gives a single stereotyped response to all kinds of stresses. The main point that needs to be driven is that different types of responses exist in nature, and each response has different physiological requirements. Therefore we expect that different neuroendocrine pathways must be used in different types of responses. The classical belief of a single pathway covering all stresses is not logical.

Let us then look into the details of the physiological requirements of the four "f"s (Table 10.1). The *fight*-or-*flight* response is a soldier response and needs fast action, muscle strength, and rapid mobilization of energy. It is also more risk prone and should prepare the body in anticipation of injuries. This is achieved by the sympathetic response in full bloom. The sympathetic response is the fastest of all, particularly in comparison with the HPA response which takes several minutes for execution. The sympathetic response begins in fractions of a second and is characterized by very quick mobilization of energy by increased glycogenolysis, lipolysis, and gluconeogenesis. This is accompanied by increased muscle efficiency and muscle utilization of glucose. A battery of other responses elicited by sympathetic stimulation is also supportive to muscle action or its anticipated consequences. There is increased arterial pressure, and blood flow is managed in such a way that there is increased blood flow to active muscles concurrent with decreased flow to organs such as the gastrointestinal tract that can be temporarily ignored. In addition, a number of changes occur in response to anticipation of injuries such as increase in the rate of blood coagulation, spurt of EGF and NGF release in the saliva [54–57], and suppression of pain responses. Testosterone and sympathetic response interact in a synergistic way, and testosterone is needed for many of the above pathways. For example, the expression of β adrenergic receptors on adipocytes which receive the sympathetic response is partially testosterone dependent [58]. If the receptors are depleted owing to testosterone deficiency, sympathetic stimulation of lipolysis is ineffective. Similarly testosterone facilitates synthesis of but not release of EGF. EGF synthesized under the influence of testosterone is released on sympathetic stimulation [59–61]. Thus the effects of the two are synergistic and interdependent and together are supportive of an explosive muscle action such as in fight or flight. However, sympathetic activity in the absence of testosterone might have a different set of effects. There would be glucose production but no lipolysis. As a result amino acids instead of fatty acids will be used for gluconeogenesis. Testosterone facilitates muscle protein synthesis. Testosterone deficiency and increased gluconeogenesis would mean more protein degradation and conversion to glucose. Most of this protein comes from muscle, so sympathetic response in the absence of aggression will have diametrically opposite consequences. It will lead to weakening of muscle and building of fat. Aggression is therefore a necessary part of the fight-or-flight response.

The *fox* or diplomat response need not be as rapid as the *fight–flight* response. It is substantially slower and thoughtful involving more of cognitive brain and less of muscle strength and rapid action. This is when the HPA axis is most appropriate. A comparison of the similarities and differences between the sympathetic versus HPA mechanisms demonstrates their differential use. As opposed to the sympathetic response, the HPA response is slower. Rapid explosive action is sustained for a short time, whereas diplomat activities are more long drawn. In accordance with this, adrenalin or epinephrine, a marker of sympathetic response, has a very short half-life, whereas half-life of cortisol is substantially greater. One of the practical consequences of this difference is that for an experimenter, it is easier to study the corticosteroid response. Studying the epinephrine response is more difficult since it gives little time to collect a sample and preserve/process the sample before the levels change. This is likely to be another practical reason why stress researchers have focused so much on HPA [4]. Convenience of work and availability of methods have often decided the direction of research throughout the history of science. Both sympathetic and HPA responses mobilize energy and increase plasma

Table 10.1 The four "f"s of alternative responses to challenges

	Fight	Flight	Fox	Freeze
Nature of response	Soldier response winner physiology	Soldier response loser physiology	Diplomat response	Learned helplessness neither soldier nor diplomat
Upregulated responses	Catecholamines, sex hormones, dopamine, EGF, other growth factors	Catecholamines, cortisol, EGF and other growth factors, CCK, BDNF, endorphin	Cortisol, cholesterol, insulin, serotonin, leptin, NGF, BDNF	Cortisol, cholesterol, serotonin
Suppressed responses	Cortisol, cholesterol, insulin	Cholesterol, insulin	Testosterone and other aggression hormones, EGF	Testosterone and other aggression hormones, insulin, EGF, NGF, BDNF, other growth factors, glut-1 in BBB
Insulin action	Sensitive	Sensitive	Resistant	Resistant

glucose by liver glucose production, but cortisol reduces muscle glucose uptake implying that this increased glucose is not meant for muscle. We can suspect then that it must have evolved to supply the brain instead.

EGF and NGF are secreted mainly by salivary glands. Both are involved in wound healing, but NGF is also important in brain function. Cortisol stimulates salivary NGF selectively and not EGF [62]. This also implies that cortisol response is in support of brain and not physical activity. In addition the behavioral and cognitive effects of corticosteroids including risk aversion and facilitated learning [35, 63] indicate that HPA response is specifically an adaptation for diplomat behavior. Success of diplomat behavior depends largely on social support. Therefore cortisol would be an appropriate response when seeking social support. If and when good social support is achieved, there should ideally be a feedback regulation of cortisol response since the target for the response is already achieved. This is apparently done by oxytocin. Oxytocin is stimulated by friendly social contact such as hugging, and it negatively regulates cortisol production [64]. All the facts together can be interpreted to mean that cortisol is not a marker of "stress" in general, but it is a marker of diplomat or *fox* response to an environmental or social challenge. In a wide variety of vertebrate species, where physical strength and aggression appear to determine social dominance, subordinate individuals need to suppress aggression, and they have higher levels of cortisol [4, 65–78]. Long-term cortisol treatment inhibited aggressive behavior in fish [79, 80]. ACTH, another player in the HPA, also has aggression suppression action [81]. Cortisol is negatively associated with aggression in humans in a number of studies, but there is some inconsistency in these data presumably due to different definitions and measures used for aggression [82–86]. This interpretation is further backed by the finding that in low-cortisol individuals, testosterone is positively correlated with dominance, but in high-cortisol individuals, testosterone is either not related or negatively related to dominance [87, 88]. It is clear that testosterone would help achieve dominance if physical strength and aggression are a determinant of dominance. But if dominance is decided by nonphysical means, testosterone would be irrelevant and perhaps counterproductive. This is the state of diplomat behavior marked by high cortisol. Therefore in high-cortisol individuals, testosterone is not related or negatively related to dominance. This might be the most logical explanation of the interaction between testosterone and cortisol. The role of corticosteroid as a marker of diplomat behavior is also illustrated by the observation that the ability to aggressively attack another individual (displacement aggression) reduces shock-induced corticosteroid response in rats [4]. It might be surprising in this connection that corticosteroids are immunosuppressive since subordinate individuals are under risk of attack by dominant individuals.

But interestingly there appears to be a built-in flexibility in the immunological effects of corticosteroids. In situations where the risk of injury is high, the immune system becomes resistant to the immunosuppressive action of corticosteroids [89–91]. It is possible therefore that in social situations where submissive displays can effectively inhibit aggression by dominant individuals, immunosuppressive effects are seen, whereas if submissive behavior does not prevent attacks, immunosuppressive action of corticosteroids is spared.

The *freeze* response shares some physiological requirements with *fox* response in terms of avoiding risk, suppressing aggression, and disinvesting from muscle. But the main difference is that here in addition to soldier functions, even the diplomat functions are switched off. Therefore even higher-order cognitive functions are not involved, and in addition to disinvestment from muscle, there would be a disinvestment from cognitive functions as well. Risk and aggression avoidance would need glucocorticoids and cholesterol, whereas disinvestment from muscle is mediated by insulin resistance. Since in the *freeze* response there is reduced demand on brain function, there should be disinvestment from brain energy supply too. This is particularly true for the *freeze* response since two main classes of brain functions are put off simultaneously. Complex nerve muscle coordination is not required, and many of the cognitive functions are also not required simultaneously. Therefore the activity and thereby energy requirement of the brain should come down substantially in comparison with both soldier and diplomats. Unlike muscle, disinvestment from brain is not brought about by insulin resistance since brain glucose uptake is insulin independent. The only way it can be brought down is by reducing the amount of glucose transporter glut-1 across the blood–brain barrier. This is a novel testable prediction of our model of multiple stress responses. There is little research input on glut-1 dynamics, but it is demonstrated in the rat model that in diabetes, glut-1 levels decrease substantially [92, 93]. It is also known that glut-1 levels are fine-tuned according to brain activity and energy demand [94]. But research so far is too scanty to be conclusive. Still we will explore the possible changes brought about by glut-1 dynamics in a later chapter.

Thus the different types of responses need different sets of physiological mechanisms. How frequently an animal gives these responses would slowly modulate the overall physiological and behavioral tendencies of it. What decides what type of response the body should give when faced with a challenge? It is partly decided extrinsically, i.e., by the nature of the challenge and partly by intrinsic factors that include the genetic constitution, developmental history, and all other factors that constitute the existing nature or personality of an individual. What happens when there is ambiguity in deciding which response is the most appropriate? For example, if a competitor is stronger, an HPA response is appropriate, and if it is weaker, then an aggressive response is more appropriate. But what if one is unable to judge the relative strength of the opponent? This is what happens when there is a conflict situation with a stranger or one of a comparable rank. In such a case as an interaction begins by giving threat displays, taking each other's judgment and perhaps actual fight, sooner or later it becomes clear who is stronger. Until then it would be ideal to be prepared for both possible outcomes. This is what actually happens in nature. Facing an aggressive encounter with an uncertain outcome, both sympathetic response and HPA response are simultaneously launched. When the winner–loser status becomes clear, the winner rapidly retracts the HPA response and loser retains it. Winner has an elevated testosterone level, and loser lowers its testosterone. This has been demonstrated across a number of species [4, 95]. It is well known that many types of stressors trigger both catecholamine and cortisol response, but some are dominated by catecholamine, and others by cortisol [4] demonstrating the flexibility and fine-tuning of the responses.

Ignoring the importance of the multitude of specific adaptive responses, the orthodox theory has tried to create a ghost phenomenon called stress and dumped all unexplained phenomena under it. I feel that our understanding of physiology would improve substantially if we completely give up the concept and the use of the word stress and instead differentially talk about how differential adaptive responses are given by different personalities to different types of environmental challenges. A number of phenomena that are grouped under the name stress response can be better understood if we interpret them without bringing in the word stress. I will try to illustrate this with a few specific examples where a logical alternative adaptive interpretation of the phenomenon is possible without bringing in the concept or the word stress anywhere:

1. Loneliness and mucosal immunity: A feeling of loneliness is shown to decrease mucosal immunity, and this has been interpreted as an effect of the "stress" of loneliness [96]. The alternative interpretation is more straightforward. Upper respiratory tract infections are more common in a social or crowded environment. A loner on the other hand is less likely to acquire them. As a result the perception of loneliness may result into disinvestment from mucosal immunity which is important for respiratory and gastrointestinal infections.
2. Acute versus chronic restraint and immunity: Not all types of stresses are immunosuppressive as admitted even by the orthodox school. In rats experimenters have claimed that short-term stress is immune enhancer and long-term stress is immunosuppressor [97, 98]. The stress they use in these experiments is restraint. The model of acute stress is one time restraint, and chronic stress is repeated restraint. There is no ultimate reason whatsoever why the effects of acute and chronic stress are diametrically opposite. An alternative interpretation is possible. Anyone who has handled animals knows that upon holding a naïve animal, it struggles to get free. This is an active and aggressive response. This is what would happen most probably on a short-term restraint. However after facing restraint repeatedly, realizing that struggling does not help, the animal starts giving a helpless response similar to the *freeze* above. In an aggressive response, injuries are anticipated for which immune system needs to be geared up. In a helpless response, there is no struggle, so no anticipation of injuries, and the immune system could be toned down. This is a clear-cut adaptive explanation of the response that does not need ambiguous concepts such as stress.
3. Early life maternal deprivation: Maternal deprivation in early infancy has been shown to induce anxiety-like behavior when adult [99, 100]. Here, again, the orthodox theory talks about the early life stress of maternal deprivation to be responsible for anxiety in adults. This example has dual ambiguity since along with stress there is this concept of anxiety which is only slightly better than stress in its clarity of definition. There is an alternative interpretation possible. It is natural that mother rats leave litter for some time and go out for foraging. However, the time taken for foraging reflects on the nutritional environment as well as the foraging capacity of the mother. Low-ranking females may take greater average time than dominant ones since they could be weaker, and their food could be snatched by stronger individuals. Therefore longer maternal deprivation means a low-ranking mother. Social ranks are often maternally transmitted, at least partially. Therefore offspring of low-ranking females are more likely to start with lower ranks. Also longer maternal deprivation means less breast-feeding opportunity which may leave the offspring weaker. Risk aversion is a typical behavioral strategy of a low-ranking individual, and for weaker individuals, it is adaptive too. The models of "anxiety" used in these experiments actually reflect risk avoidance, which they call "anxiety" since these models were developed to parallel human anxiety. But they actually measure the avoidance of the individual to exposures and potentially risky situations. In short, long-time maternal deprivation anticipates weaker and low-ranking status, and this anticipation induces risk-avoiding behavior which is adaptive for a low-ranking individual. When this interpretation is so clear, why do we need the words "stress" and "anxiety"?
4. Love as a stress? As mentioned earlier, romantic love increases cortisol levels [38, 39], and

therefore, love is interpreted as stress by some physiologists. There is a logical alternative interpretation. Love is often preceded by attempts to attract a potential mate, giving courtship displays which involve energetically intensive activity, aggression, and risk taking. The exact nature varies between as well as within species, but as a broad generalization, it is widely true that costly and risky behaviors are involved in mate-attracting displays and mate choice. However, once a pair bond is formed, there is a need for a rapid transition from aggression and risk taking towards tenderness, tolerance, and cooperation. During mate attraction phase, testosterone and other aggression-related hormones would be most active. The need for rapid change requires a strong antidote. Cortisol provides it since it is a strong suppressor of risk and aggression and enhancer of learning. This is a transient requirement though. Once the aggression levels are in check, tenderness and intimacy are achieved; high levels of cortisol are no more required. In reality, the love-induced cortisol response is transient and vanishes rapidly, although the pair would continue to love each other. Therefore rather than "stress" of love, it is the behavioral transition that requires high levels of cortisol.

5. Feeding as stress? Perhaps the most striking finding about cortisol response is that it is induced by feeding [101–105]. Carbohydrate

and fat as well as protein feeding increase cortisol response [103–105]. Do we conclude that feeding is stress? It will be obviously quite ridiculous to say so, although there are suggestions in this direction too [103]. There is a sensible alternative interpretation. In hunting or foraging exploration is required. Some risk taking is likely to be inevitable in exploration. However once a piece of rich food is obtained, it is maladaptive to be exposed, and therefore, the animal should withdraw and hide in a safe and secluded place. This is particularly relevant to species that can carry food to a safe refuge and eat. Cortisol brings about this change in mood from risk taking to risk avoiding. This logic is supported by an interesting finding. Rats feeding in light increased their corticosteroid levels during feeding but those feeding in dark did not [106]. This makes sense since already in a dark environment, the animal does not have to make special efforts to hide, but under bright light, one needs to be extra particular to hide and be safe while eating.

I do not intend to claim that the above explanations are *the* true ultimate causes of cortisol response. But it should be taken to mean that alternative interpretations are possible and need to be seriously considered and tested as hypotheses. Currently the use of the word stress is killing all possible alternative ways of thinking. If we decide to stop using the word stress, doors will be open to so many alternative interpretations, and soon we will be in a position to see which of the alternatives is true. The above interpretations are certainly more straightforward and logical than stress. But perhaps there could be even better ones. Once we remove our mental blocks and start looking at the phenomenon with more open minds, truth will start revealing itself. As our understanding of why and how different adaptive responses evolved to face different types of challenges deepens, the generic use of the word stress will become unnecessary. I feel that this time is not too far ahead. The word stress has served and for some more time will continue to serve a transient but useful function, just what "ether" did in physics. When the nature of light was first perceived as waves, people thought waves needed a medium to propagate, just as propagation of sound waves needs air. The thinking was perhaps dominated by the demonstration that sound waves do not propagate in vacuum. As a possible solution something called ether was assumed to be present everywhere through which light waves propagated. No one defined or demonstrated or characterized ether. Still it existed in all physics textbooks for several decades. Ultimately when quantum interpretation of light came in, ether, which was never there anyway, was formally thrown out from physics. Today it is only a historical entity. I feel confident to profess that as our understanding of behavioral flexibility and accompanying adaptive physiological correlates increases simultaneously at the proximate and ultimate level, "stress" will become as historical as ether of physics.

What is the relevance of the alternative interpretations of stress to our focus on diabetes and related disorders? One is to open up our minds for new possibilities. A number of possibilities have been killed under the name of stress. For example, there is good amount of evidence that exhaustive or muscle-damaging exercise increases insulin resistance [107–111] in spite of burning energy reserves. This needs a serious look and careful interpretation (which we have already discussed in Chap. 6). But instead, it is assumed that the stress of exercise increases insulin resistance. This paralyzes all other possible interpretations and gives a false satisfaction of having given an explanation.

There is an increased sympathetic tone in diabetes, the causes of which need to be carefully examined. However, if we are satisfied with the explanation that it is due to "stress," the process of scientific inquiry gets arrested. This leaves us with an incomplete understanding of diabetes. We need more specific reasoning for altered sympathetic activity in diabetes. This we will try in Chap. 12. In medical practice, whenever the cause of an ailment is not found, stress can be conveniently blamed since it is difficult to find a patient who can be certified as having zero stress. Removing the concept of stress will improve our understanding of diabetes and several other diseases by orders of magnitude.

Table 10.2 Soldier versus diplomat anxieties

Soldier	Diplomat
Predicting fights with rivals, predator attack, physical injuries, and discomfort	Predicting change in social status, financial position, strained relations
Solution involves intense physical action, often aggressive, involves risk taking	Solutions involve social manipulation, avoidance of physical risk
Stimulates testosterone, NGF, EGF	Stimulates corticosteroids, cholesterol
Enhances innate immunity	Suppresses innate immunity
Often but not always acute	Most likely chronic

Based on the "stress-free" logic that we developed above, a clearer interpretation of T2D and related conditions emerges. The *fox* and *freeze* responses are pro-insulin resistance, and *flight*-and-*fight* responses are insulin sensitizing. This might explain why the stress of war turned out to be antidiabetic. The wartime psychology is dramatically different from peacetime. One does not have to be an actual soldier fighting on the frontiers. Even for a civilian, the spirit of war catches inevitably. There are heightened levels of patriotism, mental preparation for quick action in an emergency situation, anticipation of attack, and anticipation of physical injuries which are typical soldier characteristics. In a way civilians become partial soldiers in wartime which may shift their physiological states substantially. Unfortunately owing to the biased definition of stress, there is little work on soldier stress or anxiety, but some exceptional data do exist. For example, in young soldiers, the stress of parachute jumping increased NGF levels. Even the thought of parachute jumping increased NGF by 84 %, and actual jump increased it further by 107 % [112]. If we have to use conventional terminology, we will have to say that the "anxiety" of anticipating a dangerous act and the "stress" of actually undergoing it induced NGF that has a protective role against diabetic complications as seen earlier.

Another possible example is tribes or forest villagers living in areas with dangerous animals. People collecting minor forest and mangrove produce from Sunderbans of Indian and Bangladesh are good examples. Here man-eating tigers have been common until recently, and there are sporadic incidences even now although considerably reduced in frequency [113, 114]. The threat of man eaters is a cause of severe chronic stress and anxiety for many [115], and since it is a question of life and death, its seriousness cannot be doubted. However, health surveys from this area have revealed practically nil records of diabetes among fin and shellfish collectors [116]. Diabetes is nevertheless prevalent in nearby areas among people doing indoor jobs and therefore not exposed to the man-eater anxiety. This is an example where people exposed to a source of serious anxiety are free of diabetes and hypertension, whereas people devoid of that anxiety have a high incidence. Similar is the picture in Masai and other African tribes who live among lion, leopard, hyena, elephants, and rhinos. The threat of potential attacks is associated with insulin sensitivity. This anxiety appears to be protective in spite of the chronic stress and high-fat diet in the case of Masai [117]. This point is of considerable interest. It has been argued that acute stress is "good," whereas chronic stress is "bad." I think that the difference lies not across acute–chronic axis but across the soldier–diplomat axis. A large number of soldier stresses are acute and diplomat stresses chronic. But here is an example of a soldier stress being chronic and still not having produced undesirable health effects of the metabolic syndrome type.

Let us now define the soldier–diplomat classification of stress and anxiety more clearly in Table 10.2. This is intended to trigger more research to either validate or reject this classification and its effects. Since this thinking was not there so far, there are not any experiments or data directly testing the effects of this dichotomy in stress and anxiety. I hope this situation would change in the near future, and we will have the hypothesis tested with unbiased experiments.

Since the physiological effects of soldier and diplomat stress are diametrically opposite, there is a simple message to clinicians as well as all middle-aged diabetes-prone and stress-prone individuals. If you have too much of diplomat stress (most of the stress of urban life is of this type anyway), there is simple antidote. It is impossible to get rid of stress, and it does not work. You should rather take on additional stress but that should be of a soldier type. Participate in martial arts, boxing, bungee jumping, paragliding, or any other sports involving aggression and adventure. Sportsmen scheduled to play a match against a reputed aggressive opponent are often restless and sleepless the previous night. This is anxiety, but it is soldier anxiety. Soldier stress and anxiety are most likely to restore the physiological balance upset by diplomat stress and anxiety and save you from its adverse effects!!

References

1. Hermanides J, Belqhazi L, Michels RPJ, Hoekstra JBL (2008) Lower incidence of type 2 diabetes mellitus with changes in lifestyle: clues from World War II. Ned Tijdschr Geneeskd 152:2415–2417
2. Westlund K (1966) Incidence of diabetes mellitus in Oslo, Norway 1925 to 1954. Br J Prev Soc Med 20:105–116
3. Kulenovic I, Robertson A, Grujic M, Suljevic E, Smajkic A (1996) The impact of war on Sarajevans with non-insulin-dependent diabetes mellitus. Eur J Public Health 6:252
4. Sapolsky RM (2004) Social status and health in humans and other animals. Annu Rev Anthropol 33:393–418
5. Albeck DS et al (1997) Chronic social stress alters levels of corticotropin-releasing factor and arginine vasopressin mRNA in rat brain. J Neurosci 17: 4895–4903
6. Blanchard RJ, McKittrick CR, Blanchard DC (2001) Animal models of social stress: effects on behavior and brain neurochemical systems. Physiol Behav 73:261–271
7. Krugers HJ, Koolhaas JM, Bohus B, Korf J (1993) A single social stress-experience alters glutamate receptor-binding in rat hippocampal CA3 area. Neurosci Lett 154:73–77
8. Rygula R et al (2005) Anhedonia and motivational deficits in rats: impact of chronic social stress. Behav Brain Res 162:127–134
9. Hatch A, Wiberg GS, Balazs T, Grice HC (1963) Long-term isolation stress in rats. Science 142:507
10. Parker LF, Radow BL (1974) Isolation stress and volitional ethanol consumption in the rat. Physiol Behav 12:1–3
11. Kehoe P, Shoemaker WJ, Arons C, Triano L, Suresh G (1998) Repeated isolation stress in the neonatal rat: relation to brain dopamine systems in the 10-day-old rat. Behav Neurosci 112:1466–1474
12. Weiss IC, Pryce CR, Jongen-Rêlo AL, Nanz-Bahr NI, Feldon J (2004) Effect of social isolation on stress-related behavioural and neuroendocrine state in the rat. Behav Brain Res 152:279–295
13. Dhabhar FS, McEwen BS (1999) Enhancing versus suppressive effects of stress hormones on skin immune function. Proc Natl Acad Sci USA 96:1059–1064
14. Hamilton DR (1974) Immunosuppressive effects of predator induced stress in mice with acquired immunity to hymenolepis nana. J Psychosom Res 18: 143–150
15. Dantzer R, Kelley KW (1989) Stress and immunity: an integrated view of relationships between the brain and the immune system. Life Sci 44:1995–2008
16. Dhabhar FS (2009) Enhancing versus suppressive effects of stress on immune function: implications for immunoprotection and immunopathology. Neuroimmunomodulation 16:300–317
17. Dorian B, Garfinkel PE (1987) Stress, immunity and illness ? A review. Psychol Med 17:393–407
18. Segerstrom SC, Miller GE (2004) Psychological stress and the human immune system: a meta-analytic study of 30 years of inquiry. Psychol Bull 130:601–630
19. Miller GE, Cohen S, Ritchey AK (2002) Chronic psychological stress and the regulation of pro-inflammatory cytokines: a glucocorticoid-resistance model. Health Psychol 21:531–541
20. Shors TJ (2004) Learning during stressful times. Learn Mem 11:137–144
21. Diamond DM, Fleshner M, Ingersoll N, Rose G (1996) Psychological stress impairs spatial working memory: relevance to electrophysiological studies of hippocampal function. Behav Neurosci 110: 661–672
22. Diamond DM, Park CR, Heman KL, Rose GM (1999) Exposing rats to a predator impairs spatial working memory in the radial arm water maze. Hippocampus 9:542–552
23. Taylor JA (1958) The effects of anxiety level and psychological stress on verbal learning. J Abnorm Psychol 57:55–60
24. Smeets T et al (2009) Stress selectively and lastingly promotes learning of context-related high arousing information. Psychoneuroendocrinology 34:1152–1161
25. Wagner U, Degirmenci M, Drosopoulos S, Perras B, Born J (2005) Effects of cortisol suppression on sleep-associated consolidation of neutral and emotional memory. Biol Psychiatry 58:885–893

26. Schwabe L, Böhringer A, Wolf OT (2009) Stress disrupts context-dependent memory. Learn Mem 16:110–113
27. Black PH, Garbutt LD (2002) Stress, inflammation and cardiovascular disease. J Psychosom Res 52:1–23
28. McDade TW, Hawkley LC, Cacioppo JT (2006) Psychosocial and behavioral predictors of inflammation in middle-aged and older adults: the Chicago health, aging, and social relations study. Psychosom Med 68:376–381
29. Iseri S (2006) Social isolation stress in the early life reduces the severity of colonic inflammation. Marmara Med J 19:65–72
30. Hermes GL, Rosenthal L, Montag A, McClintock MK (2006) Social isolation and the inflammatory response: sex differences in the enduring effects of a prior stressor. Am J Physiol Regul Integr Comp Physiol 290:273–282
31. Rosmond R (2003) Stress induced disturbances of the HPA axis: a pathway to Type 2 diabetes? Med Sci Monit 9:RA35–RA39
32. Kudielka BM, Kirschbaum C (2005) Sex differences in HPA axis responses to stress: a review. Biol Psychol 69:113–132
33. O'Brien JT, Ames D, Schweitzer I (1993) HPA axis function in depression and dementia: a review. Int J Geriatr Psychiatr 8:887–898
34. Chan O, Inouye K, Riddell MC, Vranic M, Matthews SG (2003) Diabetes and the hypothalamo-pituitary-adrenal (HPA) axis. Minerva Endocrinol 28:87–102
35. Thaker M, Lima SL, Hews DK (2009) Alternative antipredator tactics in tree lizard morphs: hormonal and behavioural responses to a predator encounter. Anim Behav 77:395–401
36. Rimmele U, Meier F, Lange T, Born J (2010) Suppressing the morning rise in cortisol impairs free recall. Learn Mem 17:186–190
37. Gesquiere LR et al (2011) Life at the top: rank and stress in wild male baboons. Science 333:357–360
38. Marazziti D, Canale D (2004) Hormonal changes when falling in love. Psychoneuroendocrinology 29:931–936
39. Esch T, Stefano GB (2005) The neurobiology of love. Neuro Endocrinol lett 3:175–192
40. Hellhammer J, Schlotz W, Stone AA, Pirke KM, Hellhammer D (2004) Allostatic load, perceived stress, and health: a prospective study in two age groups. Ann N Y Acad Sci 1032:8–13
41. Schuder SE (2005) Stress-induced hypocortisolemia diagnosed as psychiatric disorders responsive to hydrocortisone replacement. Ann N Y Acad Sci 1057:466–478
42. Oquendo MA et al (2003) Lower cortisol levels in depressed patients with comorbid post-traumatic stress disorder. Neuropsychopharmacology 28:591–598
43. Boscarino JA (1996) Posttraumatic stress disorder, exposure to combat, and lower plasma cortisol among Vietnam veterans: findings and clinical implications. J Consult Clin Psychol 64:191–201
44. Yehuda R et al (1995) Low urinary cortisol excretion in Holocaust survivors with posttraumatic stress disorder. Am J Psychiatry 152:982–986
45. Wobber V et al (2010) Differential changes in steroid hormones before competition in bonobos and chimpanzees. Proc Natl Acad Sci USA 107: 12457–12462
46. Heilbronner SR, Rosati AG, Stevens JR, Hare B, Hauser MD (2008) A fruit in the hand or two in the bush? Divergent risk preferences in chimpanzees and bonobos. Biol Lett 4:246–249
47. Sapolsky RM (1982) The endocrine stress-response and social status in the wild baboon. Horm Behav 16:279–292
48. Ray JC, Sapolsky RM (1992) Styles of male social behavior and their endocrine correlates among high-ranking wild baboons. Am J Primatol 28:231–250
49. Virgin CE Jr, Sapolsky RM (1997) Styles of male social behavior and their endocrine correlates among low-ranking baboons. Am J Primatol 42:25–39
50. Cannon WB (1932) The wisdom of the body. W.W. Norton & Company, Inc., New York
51. Bracha HS (2004) Freeze, flight, fight, fright, faint: adaptationist perspectives on the acute stress response spectrum. CNS Spectr 9:679–685
52. David DH, Lyons-Ruth K (2005) Differential attachment responses of male and female infants to frightening maternal behavior: tend or befriend versus fight or flight? Infant Mental Health J 26:1–18
53. Taylor SE et al (2000) Biobehavioral responses to stress in females: tend-and-befriend, not fight-or-flight. Psychol Rev 107:411–429
54. Byyny RL, Orth DN, Cohen S, Doyne ES (1974) Epidermal growth factor: effects of androgens and adrenergic agents. Endocrinology 95:776–782
55. Aloe L, Alleva E, Böhm A, Levi-Montalcini R (1986) Aggressive behavior induces release of nerve growth factor from mouse salivary gland into the bloodstream. Proc Natl Acad Sci USA 83: 6184–6187
56. Murphy RA, Saide JD, Blanchard MH, Young M (1977) Nerve growth factor in mouse serum and saliva: role of the submandibular gland. Proc Natl Acad Sci USA 74:2330–2333
57. Wallace LJ, Partlow LM (1976) α-Adrenergic regulation of secretion of mouse saliva rich in nerve growth factor. Proc Natl Acad Sci USA 73:4210–4214
58. Xu XF, De Pergola G, Björntorp P (1991) Testosterone increases lipolysis and the number of β-adrenoceptors in male rat adipocytes. Endocrinology 128: 379–382
59. Roberts ML (1974) Testosterone-induced accumulation of epidermal growth factor in the submandibular salivary glands of mice, assessed by radioimmunoassay. Biochem Pharmacol 23:3305–3308

60. Nexø E, Olsen PS, Poulsen K (1984) Exocrine and endocrine secretion of renin and epidermal growth factor from the mouse submandibular glands. Regul pept 8:327–334
61. Lamey PJ, Savage AP, Fisher BM, Bloom SR, Frier BM (1990) Secretion of epidermal growth factor in parotid saliva in diabetic patients: role of autonomic innervation. J Oral Pathol Med 19:351–354
62. Walker P, Weichsel ME, Hoath SB, Poland RE, Fisher DA (1981) Effect of thyroxine, testosterone, and corticosterone on nerve growth factor (NGF) and epidermal growth factor (EGF) concentrations in adult female mouse submaxillary gland: dissociation of NGF and EGF responses. Endocrinology 109:582–587
63. Thaker M, Lima SL, Hews DK (2009) Acute corticosterone elevation enhances antipredator behaviors in male tree lizard morphs. Horm Behav 56:51–57
64. Heinrichs M, Baumgartner T, Kirschbaum C, Ehlert U (2003) Social support and oxytocin interact to suppress cortisol and subjective responses to psychosocial stress. Biol Psychiatry 54:1389–1398
65. Louch CD, Higginbotham M (1967) The relation between social rank and plasma corticosterone levels in mice. Gen Comp Endocrinol 8:441–444
66. Davis DE, Christian JJ (1957) Relation of adrenal weight to social rank of mice. Proc Soc Exp Biol Med 94:728–731
67. Hardy MP et al (2002) Trends of reproductive hormones in male rats during psychosocial stress: role of glucocorticoid metabolism in behavioral dominance. Biol Reprod 67:1750–1755
68. Blanchard DC et al (1995) Visible burrow system as a model of chronic social stress: behavioral and neuroendocrine correlates. Psychoneuroendocrinology 20:117–134
69. Huhman KL, Moore TO, Mougey EH, Meyerhoff JL (1992) Hormonal responses to fighting in hamsters: separation of physical and psychological causes. Physiol Behav 51:1083–1086
70. Künzl C, Sachser N (1999) The behavioral endocrinology of domestication: a comparison between the domestic guinea pig (Cavia aperea f. porcellus) and its wild ancestor, the cavy (Cavia aperea). Horm Behav 35:28–37
71. Sachser N, Lick C (1989) Social stress in guinea pigs. Physiol Behav 46:137–144
72. Fox MW, Andrews RV (1973) Physiological and biochemical correlates of individual differences in behavior of wolf cubs. Behaviour 46:129–140
73. Carli G, Farabollini F, Di Prisco CL (1979) Plasma corticosterone and its relation to susceptibility to animal hypnosis in rabbits. Neurosci Lett 11: 271–274
74. Mallick J, Stoddart DM, Jones I, Bradley AJ (1994) Behavioral and endocrinological correlates of social status in the male sugar glider (Petaurus breviceps Marsupialia: Petauridae). Physiol Behav 55:1131–1134
75. Eberhart JA, Keverne EB, Meller RE (1983) Social influences on circulating levels of cortisol and prolactin in male talapoin monkeys. Physiol Behav 30:361–369
76. Martensz ND et al (1987) Relation between aggressive behaviour and circadian rhythms in cortisol and testosterone in social groups of talapoin monkeys. J Endocrinol 115:107–120
77. Manogue KR, Leshner AI, Candland DK (1975) Dominance status and adrenocortical reactivity to stress in squirrel monkeys (Saimiri sciureus). Primates 16:457–463
78. Magariños AM, McEwen BS, Flügge G, Fuchs E (1996) Chronic psychosocial stress causes apical dendritic atrophy of hippocampal CA3 pyramidal neurons in subordinate tree shrews. J Neurosci 16:3534–3540
79. Øverli Ø, Kotzian S, Winberg S (2002) Effects of cortisol on aggression and locomotor activity in rainbow trout. Horm Behav 42:53–61
80. Øverli Ø et al (2004) Stress coping style predicts aggression and social dominance in rainbow trout. Horm Behav 45:235–241
81. McGlone JJ (1984) Aggressive and submissive behavior in young swine given exogenous ACTH. Domest Anim Endocrinol 1:319–321
82. Yang S-J, Won Shin D, Sun Noh K, Stein MA (2007) Cortisol is inversely correlated with aggression for those boys with attention deficit hyperactivity disorder who retain their reactivity to stress. Psychiatr Res 153:55–60
83. Shoal GD, Giancola PR, Kirillova GP (2003) Salivary cortisol, personality, and aggressive behavior in adolescent boys: a 5-year longitudinal study. J Am Acad Child Adolesc Psychiatry 42:1101–1107
84. Gordis EB, Granger DA, Susman EJ, Trickett PK (2006) Asymmetry between salivary cortisol and α-amylase reactivity to stress: relation to aggressive behavior in adolescents. Psychoneuroendocrinology 31:976–987
85. Oosterlaan J, Geurts HM, Knol DL, Sergeant JA (2005) Low basal salivary cortisol is associated with teacher-reported symptoms of conduct disorder. Psychiatry Res 134:1–10
86. Cima M, Smeets T, Jelicic M (2008) Self-reported trauma, cortisol levels, and aggression in psychopathic and non-psychopathic prison inmates. Biol Psychol 78:75–86
87. Mehta PH, Josephs RA (2010) Testosterone and cortisol jointly regulate dominance: evidence for a dual-hormone hypothesis. Horm Behav 58:898–906
88. Popma A et al (2007) Cortisol moderates the relationship between testosterone and aggression in delinquent male adolescents. Biol Psychiatry 61: 405–411
89. Avitsur R, Stark JL, Dhabhar FS, Kramer KA, Sheridan JF (2003) Social experience alters the response to social stress in mice. Brain Behav Immun 17:426–437

90. Avitsur R, Stark JL, Dhabhar FS, Sheridan JF (2002) Social stress alters splenocyte phenotype and function. J Neuroimmunol 132:66–71
91. Avitsur R, Stark JL, Sheridan JF (2001) Social stress induces glucocorticoid resistance in subordinate animals. Horm Behav 39:247–257
92. Matthaei S, Horuk R, Olefsky JM (1986) Blood-brain glucose transfer in diabetes mellitus. Decreased number of glucose transporters at blood-brain barrier. Diabetes 35:1181–1184
93. Cornford EM, Hyman S, Cornford ME, Clare-Salzler M (1995) Down-regulation of blood-brain glucose transport in the hyperglycemic nonobese diabetic mouse. Neurochem Res 20:869–873
94. Zeller K, Rahner-Welsch S, Kuschinsky W (1997) Distribution of Glut1 glucose transporters in different brain structures compared to glucose utilization and capillary density of adult rat brains. J Cereb Blood Flow Metab 17:204–209
95. Overli O, Harris CA, Winberg S (1999) Short-term effects of fights for social dominance and the establishment of dominant-subordinate relationships on brain monoamines and cortisol in rainbow trout. Brain Behav Evol 54:263–275
96. Wawrzyniak AJ, Whitemn MCP (2011) Perceived stress, loneliness, and interaction with fellow students does not affect innate mucosal immunity in first year university students. Jpn Psychol Res 53:121–132
97. Dhabhar FS, Mcewen BS (1997) Acute stress enhances while chronic stress suppresses cell-mediated immunity in vivo: a potential role for leukocyte trafficking. Brain Behav Immun 11:286–306
98. Dhabhar FS (2000) Acute stress enhances while chronic stress suppresses skin immunity: the role of stress hormones and leukocyte trafficking. Ann N Y Acad Sci 917:876–893
99. Boccia ML, Pedersen CA (2001) Brief vs. long maternal separations in infancy: contrasting relationships with adult maternal behavior and lactation levels of aggression and anxiety. Psychoneuroendocrinology 26:657–672
100. Kalinichev M, Easterling KW, Plotsky PM, Holtzman SG (2002) Long-lasting changes in stress-induced corticosterone response and anxiety-like behaviors as a consequence of neonatal maternal separation in Long-Evans rats. Pharmacol Biochem Behav 73:131–140
101. Al damluji S et al (1987) Food induced cortisol secretion is mediated by central α 1 adrenoceptor modulation of pituitary ACTH secretion. Clin Endocrinol 26:629–636
102. Alexander LD, Evans K, Sander LD (1995) A possible involvement of VIP in feeding-induced secretion of ACTH and corticosterone in the rat. Physiol Behav 58:409–413
103. Tannenbaum BM et al (1997) High-fat feeding alters both basal and stress-induced hypothalamic-pituitary-adrenal activity in the rat. Am J Physiol Endocrinol Metab 273:E1168–E1177
104. Laugero KD (2001) A new perspective on glucocorticoid feedback: relation to stress, carbohydrate feeding and feeling better. J Neuroendocrinol 13:827–835
105. Gibson EL et al (1999) Increased salivary cortisol reliably induced by a protein-rich midday meal. Psychosom Med 61:214–224
106. Linda DS (1992) Circadian influences on feeding-induced changes in ACTH and corticosterone secretion in rats. Regul Pept 41:109–117
107. Asp S, Daugaard JR, Richter EA (1995) Eccentric exercise decreases glucose transporter GLUT4 protein in human skeletal muscle. J Physiol 482:705–712
108. Kirwan JP et al (1992) Eccentric exercise induces transient insulin resistance in healthy individuals. J Appl Physiol 72:2197–2202
109. Costill DL et al (1990) Impaired muscle glycogen resynthesis after eccentric exercise. J Appl Physiol 69:46–50
110. Del Aguila LF et al (2000) Muscle damage impairs insulin stimulation of IRS-1, PI 3-kinase, and Akt-kinase in human skeletal muscle. Am J Physiol Endocrinol Metab 279:206–212
111. Tuominen JA et al (1996) Postmarathon paradox: insulin resistance in the face of glycogen depletion. Am J Physiol Endocrinol Metab 270:336–343
112. Aloe L et al (1994) Emotional stress induced by parachute jumping enhances blood nerve growth factor levels and the distribution of nerve growth factor receptors in lymphocytes. Proc Natl Acad Sci USA 91:10440–10444
113. Neumann-Denzau G, Denzau H (2010) Examining certain aspects of human-tiger conflict in the Sundarbans forest, Bangladesh. Tigerpapers 37:1–11
114. Singh A, Bhattacharya P, Vyas P, Roy S (2010) Contribution of NTFPs in the livelihood of mangrove forest dwellers of Sundarban. J Hum Ecol 29:191–200
115. Chowdhury AN, Mondal R, Brahma A, Biswas MK (2008) Eco-psychiatry and Environmental Conservation: Study from Sundarban Delta, India. Environ Health Insights 2:61–76
116. Bhaumik U, Mitra A, Saha S, Paria T (2002) Professional health hazards among the fin and shellfish collectors of Sunderbans. Environ Ecol 20:514–519
117. Mann GV, Spoerry A (1974) Studies of a surfactant and cholesteremia in the Maasai. Am J Clin Nutr 27:464–469

11 Fat: Beyond Energy Storage

The rapidly increasing epidemic of obesity has stimulated substantial research focused on fat metabolism. For the alternative model of T2D and related disorders that this book is developing, obesity is not as important and essential as the classical model presumes. By now the new argument is sufficiently clear on its stand on the relationship between obesity and T2D although a part of it is yet to be discussed. In the new emerging picture obesity is not causal to insulin resistance and T2D, but the behavioral factors that cause obesity and those that cause insulin resistance are largely overlapping. Therefore the association between obesity and insulin resistance leading to T2D is not surprising. But it is also not obligate. Physiologically it is possible (and also common in some populations) to be insulin resistant and eventually diabetic without ever being obese. Similarly it is also possible to be obese and still insulin sensitive. The two are independent but largely overlapping etho-physiological states. We will now examine this statement more elaborately starting from the basic functions of fat and why fat metabolism is linked with a diversity of other functions in the body.

It is quite well known that fats or lipids are storage molecules, and fat or adipose tissue serves as an energy storage tissue. Fortunately or unfortunately the word fat is used to denote both a kind of macromolecule and a kind of tissue that stores fat. I will make use of this ambiguity below while describing the seven functions of fat in our body. Some of the functions described below are of lipid molecules and some of adipose tissue. This distinction will be obvious as I elaborate on these functions. The important point to be emphasized here is that fat for the body is much more than storage of excess calories, and while talking about the causes and consequences of obesity, it is important to keep all of them in mind with due "weighting." This is not intended to be an extensive and comprehensive review of the existing literature. Much comprehensive and scholarly work on obesity by several authors is available [1, 2]. I will avoid repeating what has already been said. Instead I will focus on aspects that these authors appear to have missed and those that are directly relevant to the synthesis in this book.

There is a very fundamental function of lipids, phospholipids in particular, in any living organism, and that is they form the outer and inner membranes of cells. Without this structural role of lipids, life would have been just impossible. But this is very basic for all forms of life, and it is taken for granted. I will not include this fundamental structural role of lipids while talking about the seven functions of fat. The seven functions of fat are:

1. As an energy reserve
2. As the most abundant macromolecular constituent of the brain
3. As an insulator against cold temperatures
4. As an impact buffer
5. As a modulator of aggression, sex, and reproduction
6. As an immune modulator
7. As a social signal

M. Watve, *Doves, Diplomats, and Diabetes: A Darwinian Interpretation of Type 2 Diabetes and Related Disorders*, DOI 10.1007/978-1-4614-4409-1_11,

Some of them are quite well known, others are less well known, and some are recent surprise findings. I will elaborate on each of them soon and show how each one is related in a different way to the four fundamental processes in fat dynamics, i.e., (1) lipogenesis, (2) lipolysis, (3) distribution of fat tissue in the body, and (4) effects of fat on other functions of the body.

1. Fat as an energy reserve: Excess energy left after meeting the current needs of the body is said to be stored, the common energy stores being fat. In this statement there is one fundamental assumption that appears to be uncritically accepted by everyone. That is, there is something like a fixed current requirement of the body above which it will be treated as excess and stored. However, the concept of current requirement of the body can be challenged. The current requirement is not a physiological constant but is a behavioral decision. An individual has a choice of what to do with energy. It is possible to spend an extra quantum of energy to go and chase more females, take another round of the territorial boundary to ensure there are no intruders, go and challenge one's competitors, or just tease other group-mates. Storing as fat is only one of the many available options of what to do with energy. Which option is the best bet to increase one's lifetime success is decided by ecological and social circumstances as well as by the personality of the decision maker. In terms of evolutionary optimization the available energy can be best utilized for the present if the situation is favorable to grow and reproduce and/or the future is relatively uncertain. On the other hand if the present circumstances are not so encouraging, there is a chance of the future being better and the probability for surviving till the next possible opportunity is fair; saving rather than spending is a better strategy. For example, if one's reproductive opportunities are blocked by other more dominant individuals, it is no use spending time and energy in attempting to reproduce; it would rather pay to save energy and wait for a future opportunity. That is why lipogenesis is more adaptive under depressing or subordinate conditions. However, the most important condition necessary for storing for the future is that there is a good chance of surviving sufficiently long to make use of the stored energy. If predator or other environmental risks are large, it is better to spend as much energy as possible at the present and maximize immediate reproductive success. If the prospects for the future are to be counted, one needs to ensure survival before storing fat, and in order to do so, one should avoid all risk-taking behaviors till then. Therefore adiposity and risk-taking behavior have to develop an inverse association. Since physical aggression is necessarily coupled with risk of injury and perhaps death, obesity and aggression also need to be negatively related. Thus the very definition of excess energy on which the concept of energy storage is based is itself decided by ecology, social status, and behavior. Therefore believing in the popular notion that excess energy is stored as fat and ignoring the ecology and ethology of fat storage will not lead us to any insightful understanding of fat metabolism.

Leaving aside this consideration for the time being, let us return to the orthodox view and examine the concept of obesity as the result of net positive energy balance, i.e., calorie intake minus calorie burnt. Calorie burnt has two distinguishable components, the basic or resting metabolic rate and that due to physical activity or exercise. The former and even partly the latter are dependent on the total body mass. The greater the amount of respiring tissue, the greater the total fuel burnt. As a result, the body mass stabilizes at an equilibrium value of M. This is given by a simple baseline equation $dM = I - aM - b$, where I is energy intake, M is the body mass, a is the basal metabolic rate, and b the expenditure due to physical activity [3]. This equation can explain why at a given food intake and lifestyle, different people can have different body weights.

Unfortunately the reality is not as simple as the equation. What the equation does not incorporate is the influence of M on I. I is taken as an independent variable by this equation which is not true in reality. A large number of different mechanisms of the body regulate food intake and fat storage. There are two parallel lines of thinking and research which have failed to come together effectively. One is the energy balance equation given above which appears to neglect that a large number of mechanisms of intake regulation exist in the body and simply assumes that either a decrease in energy expenditure or increase in intake or both together are responsible for the obesity epidemic. If one or many of the intake regulation mechanisms were effective, neither of them could ever lead to obesity. And not only one but a series of mechanisms of food intake regulation exist which work at different levels (see Table 3.1). If there has to be a positive energy imbalance, not one but all the mechanisms would have to fail or be overridden, and as yet there is no suggestion as to how and why this is happening with increasing frequency only in the last couple of generations in a large and increasing fraction of the population.

We have seen earlier that a variety of at least partially independent mechanisms regulate food intake in the short [4–16] and long run [17, 18]. The body thus has adequate mechanisms to keep the short-term and long-term energy balance. Unless there is something wrong simultaneously with all these mechanisms, we will not eat excessively and will not accumulate fat. The short-term and long-term mechanisms of energy intake control appear to form a multilayered and robust system of regulation. It is difficult to perceive how such a system with a series of backups can fail to regulate food intake. If one system fails, others are there to compensate, and therefore, a single defect is most unlikely to cause an eating disorder. It is more likely that overeating represents an adaptive response. Being an evolved adaptive response, there would be mechanisms to modify the entire series of regulation mechanisms simultaneously. It is possible that the evolved mechanisms of overeating are facing supernormal stimuli today and therefore overexpressing now. In order to understand this we need to know under what conditions overeating would be adaptive in natural settings.

We have seen earlier that food is one of the major causes of aggression. This is certainly true for species having a patchy distribution of available food. For grass eaters, there need not be food-related aggression since the food is abundantly and widely distributed. For large carnivores a successful kill becomes a focus of intense competition and aggression. Frugivores also have a patchily distributed food although competition and aggression may be somewhat less intense than carnivores. On a food patch there is a complex interplay of aggression and food intake, and it is not a coincidence that common signal molecules such as serotonin are involved in regulating both the processes [11, 19]. We can visualize four possible scenarios on a food patch.

1. There is competition and the opponents are weaker than you: Here there are clear advantages of being aggressive. You can drive away others and eat. Serotonin is expected to be low in this case. However, as you eat sufficiently, serotonin levels rise and aggression levels come down. At this stage, the competitors will become more active once again, and you retreat since your desperation for food decreases and you want to avoid risk of getting injured unnecessarily. Therefore you suppress aggression by increasing serotonin, and this suppresses hunger as well. This process ends up in moderate eating.
2. There is competition and opponents are stronger: You need to suppress aggression by increasing serotonin and thereby partially suppressing hunger too. After waiting for variable time, you might get some share of the food, but you are unlikely to overeat.
3. No competition: As a rare chance, you have no competitors on the food patch. This is a rare opportunity, and you should take maximum advantage of it. Since there is no competition, there is no need to suppress aggression. As a result serotonin remains low and satiety response is delayed. This is the only situation where overeating is possible and there is nothing wrong in overeating since such a situation is unlikely to come very frequently in a highly competitive environment. Therefore overeating response could be a natural adaptive response when food is abundant and there is no perceived chance of aggression related to food. As a result absence of aggressive competitors could have evolved a link with overeating response in species that normally have food-related aggression. It is unlikely to work

the same way in species in which naturally there is no food-related aggression.

Another related factor is the risk of predators. In the wild, one has to forage for food. For species having predators, foraging often increases the risk of predators. Therefore there would be a trade-off between the nutritional benefits of foraging and the increasing risk of predation. The trade-off point is largely decided by the perceived risk of predation. If it is too high, then one should stop foraging after having the minimum intake needed for survival. Predators generally have a large home range and move over a large area. As a result, there is wide day-to-day variation in the risk of predation in any locality. When an animal perceives high risk, it is adaptive to arrest foraging efforts. When a predator is not perceived, it is equally adaptive to compensate by overeating. Thus eating behavior is intimately linked to perceived risk from predators and aggressive competitors. At a proximate level, this is mediated by aggression, anxiety, and fear signals in the brain. The list of regulators inhibiting food intake is considerably longer than that of appetite stimulators. Most of the peptides inhibiting food intake are also the ones mediating fear reactions, whereas the majority of the agents reducing anxiety responses stimulate appetite [20]. Of particular interest are the brain signal molecules, α melanocyte-stimulating hormone (α-MSH) [21–23], histamine [24–28], and cocaine- and amphetamine-regulated transcript (CART) [20, 29, 30], all the three being related to fear, anxiety, and risk avoidance on the one hand and decreased food intake on the other. Inhibition of histaminergic pathway is suspected to be important in hyperphagia and obesity [31]. It is very likely therefore that signaling by CART and histamine in the brain has evolved to fine-tune the trade-off between energy gain by foraging and predator or competitor aggression risk increasing with foraging.

When a higher risk is perceived, higher levels of these signals are generated which will reduce foraging attempts and thereby food intake. Appetite control is a complex interplay between short-term metabolic signals such as insulin, long-term signals such as leptin, and behavioral signals such as CART and α MSH. The action of leptin is at least partially CART dependent, and CART levels are fine-tuned by perceived fear and aggression. It has been shown that the difference between diet-induced obese rats and diet-resistant rats (those that fail to become obese even when kept on a high-fat diet) lies in the CART and α MSH levels in the hypothalamus [32]. If CART and α MSH levels are low, food intake regulation fails and animals become obese. This may be the key

to behavior-induced suppression of appetite. On the other hand if an animal is starving, or if leptin signaling is low indicating little stored energy, CART levels are depleted [33]. This impairs the fear response, and the animal is prepared to take greater risk for foraging. It is unlikely to be a coincidence that molecules involved in anxiety are also involved in appetite control. The leptin, CART, and α MSH interactions appear to have evolved in such a way to achieve an optimum

A seventeenth-century Indian folktale has an interesting reference to body weight regulation. The emperor Akbar floated a riddle. He gave a goat to each village headman and ordered that the goat should be fed well, but after 6 months it should weigh exactly the same. Birbal, a famous wise man of his times, appears to have helped a village headman solve the riddle by an advice to tie the goat in front of a caged tiger at night and then feed the goat well during day time. He said the fear of the tiger will prevent the goat from being fat. We know today that fear induces peptides including CART, histamine, and alpha MSH which regulate food intake. Birbal obviously did not know the fear peptides but perhaps knew by observations and traditional wisdom that fear prevents one from overeating.

trade-off between the metabolic demands of the body and the risk associated with foraging for food. In order to understand the interaction of these molecules in appetite regulation, it is necessary to know what is an optimum trade-off of food intake driven by perceived fear. It is necessary to make it clear at this point that this fear is necessarily physical. We have differentiated between soldier and diplomat anxieties in the last chapter whose physiological effects are expected to be different. Histamine, CART, and α MSH are shown to be driven by physical anxiety. So far, there is little data showing that diplomat anxiety triggers the same response, but this needs further investigation. We will continue with physical fear and anxiety in the discussions below.

The effects of predator fear and aggression anxiety can be incorporated in a formal model that gives good insights into what optimum food intake is from an integrated ecological and metabolic point of view. The model begins with the relationship of net energy intake per unit time such as a day and the evolutionary fitness contributed by this. In calculating net energy intake it is assumed that the energy expenditure in foraging, i.e., obtaining the food, is already subtracted. This relationship is most likely to go in a saturation curve with a maximum fitness that we will assume to be in unity. It is much like the famous Michaelis–Menten curve that every biochemistry student learns in the first year, with two differences. One being that it does not start from the origin. There is some basic energy needed for maintenance above which it starts contributing towards reproductive fitness. The other is that at very high intake the fitness may actually come down. Ignoring the latter for the time being, the curve will look as in Fig. 11.1. An equation for this curve can be written by adding a constant for maintenance energy I_0 to each of the terms in a Michaelis–Menten type of equation:

$$R = \frac{I - I_0}{(K - I_0) + (I - I_0)}$$

Since animals have to forage for food and the risk of predation or injuries from aggressive competitors increases with time spent in foraging, the risk will be positively correlated with energy intake. This is shown in the figure by the straight line with a slope c. The optimum food intake would be one at which the difference between the benefit and the risk is maximum. This happens at a point along the fitness curve where the slope becomes equal to c. Since the derivative of a curve is its slope, the optimum food intake I_{opt} is obtained when

$$\frac{dR}{dI} = c$$

This is the optimum food intake per day that maximizes the net evolutionary fitness. It can be seen from Fig. 11.1 that c, the risk of foraging, is an important determinant of optimum foraging and thereby food intake. A small decrease in c can cause a disproportionately large increase in I_{opt}. An assumption behind the model is that evolution will fine-tune the homeostatic mechanisms of the body to achieve this optimum. This is a calculation for a short-term optimum assuming that the animal has no reserve food. If there is a reserve biomass M, a part of M can be made available per day if the animal is undernourished on a given day. As a result the curve will shift to the left by a difference I_m where I_m is the reserve fat that can be mobilized in a day. This itself would be a saturating function of total body fat with an upper limit j. If the curve shifts to the left, the optimum food intake would also shift to the left. Thus there is a feedback from energy stores to food intake. Food intake I will in turn affect M by the well-known energy balance equation,

$$dM = I - aM - b$$

where a is the energy consumption per unit body mass for maintenance (basic resting metabolic rate) and b is energy expenditure related to activities other than foraging. The three equations above make an integrated dynamics, and it can be easily seen in this dynamics that a stable M is reached very soon and this steady state M is decided by the parameters of the equations namely, a, b, c, and j. Figure 11.2 shows the relative effects of changes in the four parameters. It is interesting to note that b, which represents physical activity or exercise, has

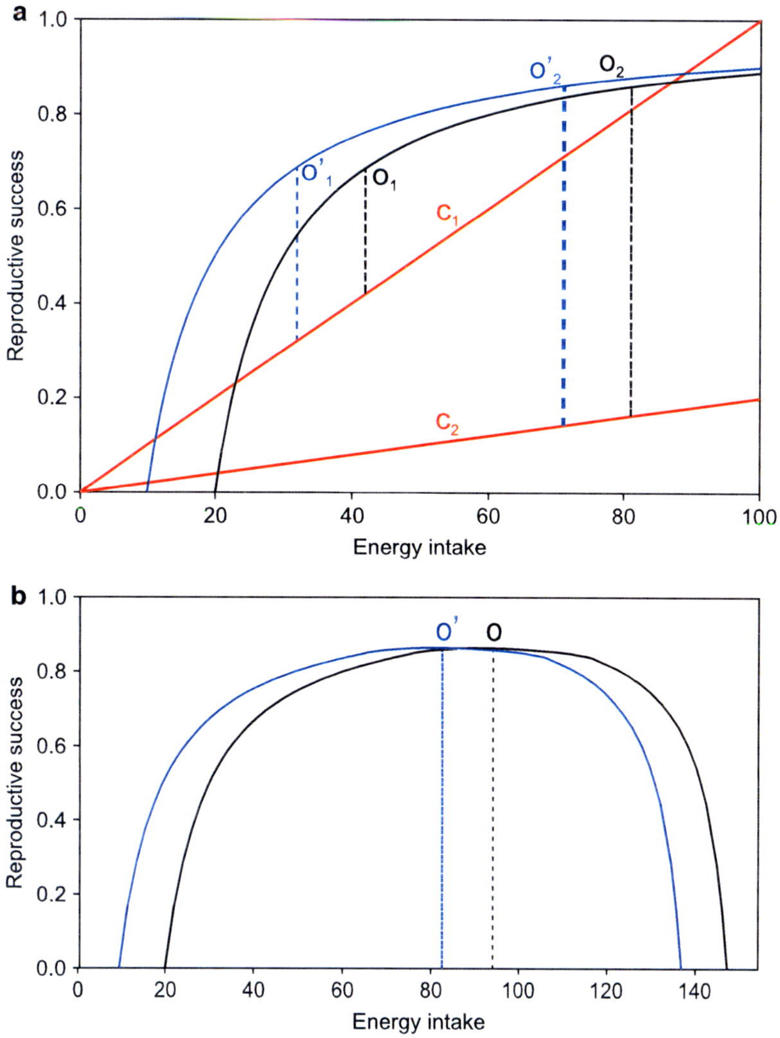

Fig. 11.1 (**a**) A foraging theory-based model of food intake optimization. The saturation curves represent the contribution of net energy intake to reproductive fitness. The *straight lines* with slope *c* denote risks associated with foraging. The optimum food intake *O* is one where the net benefit, i.e., the distance between the curve and the line, is maximum. This happens where the slope of the curve equals the slope of the line. It can be easily seen that when risk decreases ($c1$ to $c2$), the optimum food intake increases ($O1$ to $O2$). If the body has a preexisting energy reserve, the curve shifts left (*blue line*) and so does the optimum ($O'1$ to $O'2$). (**b**) Food intake optimum (O) without considering foraging risk: This is likely to develop for animal species like elephants that have neither predators nor food-related aggression. Here metabolic regulation is expected to prevail over behavioral regulation, and such species of animals are expected to be relatively resistant to developing obesity

the least effect on body mass. Fat oxidation rate is the most effective fine-tuner, whereas foraging risk and basic metabolic rate per unit body mass have almost the same effect.

The difference between the classical energy balance equation and our model is that we have incorporated the feedback effect of M on I. It is also important to note that in our model c, the risk or the nonenergy cost of obtaining food is one of the most important determinants of food intake. In nature c is not constant and varies according to seasonal cyclic changes such as vegetation, visibility, migration, as well as purely stochastic factors. As a result individuals need to adjust the trade-off point

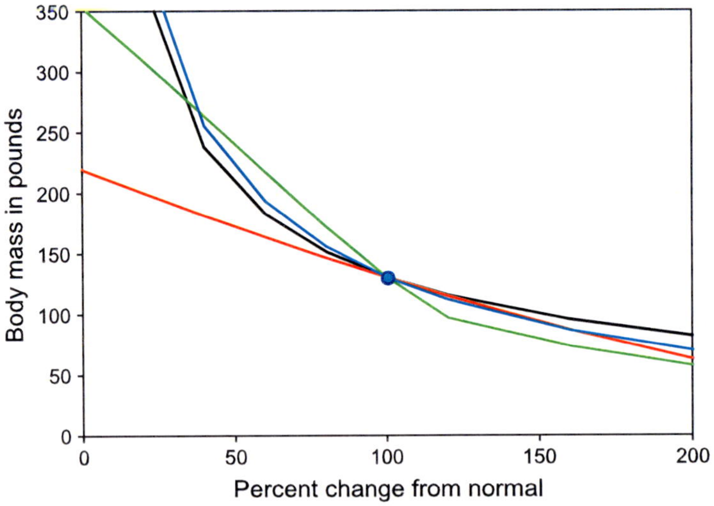

Fig. 11.2 The relative effects of four possible factors that can influence equilibrium body weight. The figure is based on simulation results from the model. An equilibrium point at moderate values of all parameters was set as standard. Percent change in each parameter and its effect on the equilibrium body weight are represented. Close to the normal body weight, exercise (*red line*) appears to have the minimum effect on body weight and the rate of fat oxidation (*green line*) has maximum. Risk of foraging (*blue line*) and basal metabolic rate (*black line*) also have large effect sizes

between benefit and risk accordingly. They would eat more when predator risk is less and vice versa. Since long-term feedbacks in terms of effects of M are there in the system, the average c decides the mean M. However, what could we expect if the mean c changes? With decreasing nonenergy costs of food, there would be a dramatic rise in M. This is not a failure of homeostasis; instead, the desired set point of homeostatic regulation is shifted to the right, and the body has programs hardwired to attain the shifted desired level. This is one possible reason for the apparent "failure" of the homeostatic mechanisms in regulating obesity. This model suggests that this can happen when food-related risks disappear. This is the first level of ecological optimization of food intake that we will call risk optimization control. In the case of animals that naturally have high risk associated with foraging, removal of the risk will lead to obesity if food availability is not limiting.

What about species that naturally have a very low mean c? This may happen for animals such as elephant or rhino that are almost free of predators and food-related aggression. Many large ungulate species of the savanna that have predators but do not have a safe predator-free refuge will also have a small c since the difference in predator risk during foraging and non-foraging is not very large. In such animals c is unlikely to be important in the evolution of homeostatic mechanisms. Here we would expect that the deleterious effects of large M would shape the homoeostatic mechanisms to regulate food intake to keep an optimum M directly. This can happen by the short-term and long-term metabolic signals mentioned above. The short-term regulation works through gut and pancreatic signals including insulin, whereas long-term regulation works through leptin and other adipose-generated molecules. Let us call it metabolic regulation. In species where the behavioral regulation is frequently triggered by predation and aggression, metabolic control would not be needed most frequently. As a result metabolic control may not evolve to be strong. Moreover there is a reason why metabolic control will directionally evolve to be weaker. If predator risk is high, the behavioral optimum is always lower than the metabolic optimum. This will eventually lead to loss of weight. However predator risk is variable. Therefore whenever no risk of predator or aggressive competitor is perceived, an animal can make up by eating more. For such a compensatory hyperphagia to work, the metabolic regulation needs to be weaker and overridable. It appears necessary therefore that in species in which behavioral control of food intake prevails, metabolic control would evolve to be weaker and surmountable. The environmental cue for overriding the control mechanisms would be absence of fear and aggression. The mediator of this overriding is likely to be downregulation of CART, which can weaken the action of normal food intake regulators such as leptin.

In species where behavioral control never works in its ecological setting, metabolic control would be strong and almost infallible. A testable prediction of this would be that animal species that have neither predators nor food-related aggression will not show a tendency to become obese even when food availability is high. A possible example is elephants. So far, I have not been able to put together elephant morphometry data from various sources along with dietary and ecological data for each individual elephant. But such data do exist scattered over various sources, and therefore, the prediction is testable. Having spent a few years in the company of wild and captive Asiatic elephants, I have an impression that morphometric standards in elephants are highly reproducible and the variance around them is relatively small both in the wild as well as in captivity. Although a number of elephants in captivity are given food with high sucrose, other soluble carbohydrates, and oils, they do not drift substantially from the morphometric standards. There is no published data on obesity in elephants, but I have not seen rapid and uncontrolled weight gains in elephants, comparable to what happens to carnivores or primates in captivity. Obesity in carnivores and primates in captivity and occasionally even in the wild or semiwild state is not uncommon. Zookeepers know that the amount of food provided for carnivores needs to be carefully controlled, but excess of food does not matter so much for grass eaters. It does matter for primates that are predominantly fruit eaters. The distinction may not lie in diet since elephants in captivity are given fruits, sugarcane, and coconut which do not appear to lead to obesity.

The ancestral humans had a patchy distribution of food, whether scavenged, hunted, or gathered; therefore, a carnivore-like response is more likely to have evolved. Human ancestors had predators until recently, and the cultural development of death rituals is likely to have discouraged predators according to one theory [34]. Speakman also considers predators to be responsible to arresting obesity for a different reason [35]. Obese individuals are likely to be more susceptible as well as more attractive to predators, and therefore, evolution would strengthen energy regulation mechanisms for species with predators. This reasoning is compatible with our model with some differences. Speakman thinks that after being free of predators, a process of genetic drift increased obesity in humans. The above model says that being free of predation would reduce the risk of foraging c and thereby increase the optimum body weight. In human ancestors food-related aggression was also present to varying degrees almost throughout history. Food could have been one of the causes of wars and fights throughout human history. In hunter-gatherers foraging often involved driving away competing animals. Protection of crops, stored grain, and livestock from wild animals is a major aggressive activity even today in areas of residual wildlife. However, in modern society, an urban consumer is almost never exposed to food-related aggression. Therefore the c of the model in today's society is almost zero. This makes the risk optimization-related food intake regulation driven by CART and histamine completely ineffective. We have seen above that in species where behavioral regulation is important, metabolic regulation would be partially degenerate. It is possible therefore that the human species is primarily a behavioral regulation species, not a metabolic regulation species. In modern lifestyle behavioral regulation factors have vanished, and metabolic regulation was always weak. Therefore energy intake regulation is failing in modern life.

Another result of the model is of considerable interest to us since it matches very well with reality. The parameter I_m which reflects the maximum energy that can be made available for the body from the stored energy per unit time is an important determinant of obesity. The prediction of the model is that the basal metabolic rate or fat oxidation rate has similar effects on M but exercise has a much smaller impact on obesity (Fig. 11.2). For a long time a popular belief has been that obese and non-obese individuals differ in the basal metabolic rate, i.e., *a* of this model. Although the model supports this, evidence does not appear to give unanimous support to this idea [36, 37]. Exercise is always advocated as an antidote on obesity, but according to the model, exercises have limited efficiency in reducing body weight. Individuals differ dramatically in their tendencies, some rapidly gaining weight in spite of exercises and others remaining lean without any exercises. More robust data relate fat oxidation rate with obesity. Low rate of fat oxidation appears to be the main difference between people of obese versus lean tendencies [38–42]. The traditional models of energy balance equation or homeostasis have not been able to incorporate the rate of fat oxidation in their models which our model does effectively.

The rate of fat oxidation is also aggression dependent by a mechanism that is quite well worked out. One of the main mechanisms that cause rapid lipolysis is sympathetic stimulation [43–45]. Sympathetic stimulation is triggered in a flight-or-fight response. We have seen while discussing stress that flight or fight is typically a soldier response. Moreover the sympathetic activation is received by the adipose tissue by β adrenergic receptors [43], and the formation and distribution of these receptors is testosterone dependent [46]. Therefore testosterone arrests visceral fat accumulation [47–49]. Typical diplomat personalities have lower levels of testosterone [50], and therefore even if they happen to activate a sympathetic response, it does not lead to lipolysis as much as in a high-testosterone individual. Sympathetic nerves also appear to regulate the proliferation of preadipocytes since denervation of fat pad has been shown to increase the number of preadipocytes and weight of fat pad in rats [51–54]. This is another possible mechanism by which absence of aggression could facilitate obesity.

This also partially explains the distribution of fat in the body. The question as to why the distribution of fat is different in different persons remains largely unanswered although the differential effects of fat accumulation in different parts of the body are well known [55–61]. Since the same blood carrying the same concentration of fatty acids and hormones is circulated throughout the body, why does fat deposit in one adipose depot more than another? The answer certainly does not lie in hormones and metabolites since the same blood carrying these molecules circulates to all fat depots. A possible logical answer is nerve supply. Different adipose depots are innervated differently [62]. For example, abdominal visceral fat has an intricate sympathetic supply through the vagus nerve. Vagus does not innervate subcutaneous fat [63]. As a result, a combination of testosterone and sympathetic activation, both triggered by aggression, keeps on mobilizing visceral fat. This is how aggression changes the ratio of visceral to subcutaneous fat. There is no wonder then that sumo wrestlers have little visceral fat in spite of having large amount of total body fat. This is true as long as they practice wrestling. On retirement, their fat may go visceral rapidly, and they might become insulin resistant [64].

This is not the first time that a model is being used to explain obesity. However, the uniqueness of this model is that for the first time it brings together three different concepts, namely, energy balance equation, homeostatic control on energy intake, and foraging theory. Somehow, researchers had so far either not recognized the importance of bringing the three concepts together or had failed to do so, the former being the most likely case. But now we appear to be closer to the reality of how and why the homeostatic mechanisms evolved under ancestral conditions are failing in the modern lifestyle. There are a few more implications of the model. Both predation and aggression are population regulation mechanisms. Removal of predation increases prey population [65, 66], and reduced aggression also enables higher population densities [67]. Therefore if the predatory pressure as well as intraspecific aggression is suddenly removed in a species, there can be anticipation of crowding and thereby decreased food availability. Therefore it would be adaptive to increase thrift on removal of a fear signal. It is possible that because of this reason too, evolution might have built in a program to increase food intake on loss of predator fear and signs of population growth. For wild populations crowding is almost invariably associated with increased food competition, exerting a negative feedback control that would effectively arrest obesity. The problem with the current human condition is that the natural fear signals associated with foraging are absent, eating has lost its connection with foraging associated risks, there is increasing crowding which could be stimulating a thrift response, and there are no signs of reduction in food availability that naturally accompanies these conditions. If vanished risk of predation and food-related aggression anticipating crowding is at the root of these changes, it is logical then that reproductive strategies also change in this situation, and the reduction in fecundity with thrift is likely to be a response evolved for this reason.

Obese persons are often known to be lethargic. In the light of the model described above, it can be easily seen why lethargy is adaptive. Lethargy is not an inevitable result of storing fat. Extra energy makes one lethargic is paradoxical. Lethargy is more likely to have evolved as a feedback mechanism in energy homeostasis and risk minimization. If an individual has sufficient stored energy, it is unprofitable to keep on foraging and being exposed to risks. So lethargy increases the chances of survival in a wilderness environment by reducing foraging-related risks. Simultaneously lethargy decreases foraging and thereby food intake, contributing to effective feedback in energy homeostasis. If fat causes lethargy then the foraging feedback is almost infallible. It is only in the modern human society that feeding is detached from foraging. As a result, in spite of being lethargic, food intake may not be affected. In our hunter-gatherer ancestors the lethargy feedback must have been very effective.

There are other possible means by which evolved responses might be leading to overeating in modern life. There are reward centers in the brain that are important in regulating food intake [17, 68]. In the wilderness, the most common natural rewards activating reward centers are food, sex, and presumably occasional social rewards such as achieving dominance. The reward centers when activated by getting a sumptuous food reward initiate a satiety pathway and stop food intake [17]. The alternative rewards also trigger the same reward centers, and in most of these rewarding situations, temporary suppression of hunger is essential. For example, while having sex, hunger would be an interfering drive, and therefore, reward center activation by sex should also give satiety signals. Such an adaptive loop must have evolved. However, it is also important to put a limit on such nonfood reward suppression of hunger. So if the reward center is suppressed by nonfood rewards too often, there would be a temporary desensitization of the reward center. This would bring back the hunger response and, as a result of the partial desensitization, result into compensatory overeating.

It is very likely that this is happening in today's monetary economics. In modern life money is perceived as a reward and is perhaps the most common nonfood reward of modern life. Money keeps on stimulating the reward center repeatedly. Neurobiological studies have shown that the brain areas activated by getting prize money and those activated by a sweet taste are the same [69–73]. Experimental psychologists have also demonstrated that there is a neuronal cross talk between food and money such that hunger affects food-related decisions and thinking about money affects food intake [74]. It is very likely that repeated activation of the reward centers by money over a long time results into chronic desensitization of the reward center resulting into habitual overeating. A downregulation of dopamine signaling is indeed demonstrated in obesity [75, 76]. There is a link with aggression here too. Aggression is linked to the dopamine reward pathway, and aggression increases dopamine activity [77]. This is perhaps one more mechanism by which aggression arrests obesity.

In an attempt to test the hypothesis that nonfood rewards affect obesity, we carried out a survey of people working as full-time cashiers in different types of organizations. Some of the cashiers were owners or partners in the endeavor, and others were only salaried cashiers. Since they were all doing physically very similar jobs, the levels of on-job physical activities could be assumed to be similar. There was an obvious difference between the reward values of the money being handled. For the owner cashiers the money had a strong reward value, whereas for the salaried cashiers, the reward value of the money they handled would have been at the most weak. The survey showed that owners had a greater BMI, waist circumference, and waist-to-hip ratio than salaried cashiers. Also, two other variables, namely, the amount of money handled per day and the duration of doing a cashier's job, were also positively related with obesity, but the duration of doing a sedentary job and self-reported exercise or absence of it was not. The most important part of the study, in spite of its small sample size, was that this single behavioral factor explained about 20% of the population variance [16]. This is interesting on the background that genome-wide fishing for obesity-related genes has so far explained not more than 2% of population variance in obesity [78–83]. Yet researchers as well as laymen have a strong belief that obesity is genetic and are reluctant to believe that behavior could be a more important determinant of it.

The purpose of the above discussion was not to claim to have identified all factors and mechanisms responsible for overeating but to highlight that neurobehavioral processes are perhaps the most important contributors to it. Unless the neurobehavioral mechanisms that have evolved to face social and ecological conditions are clearly identified and addressed, all efforts to mitigate obesity are unlikely to work.

One more important point to be recognized before ending a discussion on fat as energy storage is that energy is not only stored for future starvation. Energy storage is needed for a variety of other functions including migration, breeding, maternal care, and immunity. This aspect is insightfully discussed by Pond [1], and I will avoid repeating it.

In a nutshell, all components of the energy balance, namely, energy intake, lipogenesis, lipolysis, and fat distribution, are affected by personality and behavior, particularly physically aggressive behavior and exposure to physical risks. We will see below that most other functions of fat also have significant evolutionary and behavioral components.

2. Fat as insulation: People in extreme cold weather need a layer of subcutaneous fat that acts as an effective insulator and prevents heat loss from the body. As a result ethnic groups evolved in colder climates for sufficient number of generations should have a more peripheral distribution of fat. Tropical people, on the other hand, want to dissipate heat as rapidly as possible and therefore have little subcutaneous fat. When these people accumulate fat, it tends to be abdominal. Not surprisingly colder climate people such as Tibetans do not show much tendency towards central obesity [84]. Although this looks more of a common sense, analytical inputs to examine this question are rare [1].

3. Fat as an impact buffer: The important function of fat as a subcutaneous impact buffer and its relevance to aggression and to metabolic syndrome has almost never attracted the attention of obesity researchers. A subcutaneous layer of fat takes a major part of the impact in a blunt trauma [85, 86] and prevents or reduces injury to inner parts. Therefore subcutaneous fat is adaptive for an aggressive individual. This cushion effect is purely physical and may not involve any hormonal or biochemical processes. Therefore if subcutaneous fat is mainly evolved for this purpose, it need not be metabolically or endocrinologically active. Nevertheless, what drives fat subcutaneously is an important question, and neuroendocrine mechanisms are likely to have evolved to drive fat subcutaneously in anticipation of injurious impacts. One possibility is that the dissolution of visceral fat by a combination of testosterone and sympathetic activation is sufficient to deposit the same fat subcutaneously. But perhaps more complex signaling involving impact sensing by the subcutaneous nerves may be at work.

Both wrestlers and body builders need strong muscle. However, there is a major difference in the requirement and function of fat in the two. For wrestlers, subcutaneous fat is helpful to tolerate blows, punches, kicks, and other impacts that they receive frequently. Body builders, on the other hand, should make all-out efforts to avoid subcutaneous fat, because they want to make every muscle curve conspicuous and a layer of fat over it will mask them. As a result body builders will not be able to make good wrestlers, and wrestlers will not show every minute muscle although they need to be muscular. They signal strength by overall body size and muscle mass but not by showing every curve. It would be adaptive for a fighter to have his fat distribution more subcutaneous and less visceral, but the mechanisms that do so are only partially known.

4. Fat in neuronal development: As opposed to muscle tissue which is chemically predominantly proteins, the brain and nervous tissue is predominantly fat. The brain is about 70% fat by dry weight. Lipids are therefore important in the development as well as maintenance of brain and peripheral nervous system. Lipid metabolism and certain classes of lipids, such as omega three fatty acids, are particularly known to help brain development, maintenance, and function [87–92]. It would be logical to expect therefore that a muscle-dependent lifestyle will be associated with a protein anabolic metabolism and a brain-dependent lifestyle with a lipid anabolic bias in the metabolism. T2D is known to be associated with a slow decay of muscle strength and build up of fat, but a quantitative model of this shift in bias or fine-tuning of metabolism to suit the requirement has not been attempted so far.

In an interesting study, rats malnourished for several generations showed a highly insulin-resistant phenotype. In these rats the relative brain weight was substantially higher, and the absolute brain weight was marginally higher than controls. Accompanying the brain-biased development, they had a higher fat content, although total body weight was reduced [93]. Greater percentage of fat in malnourished individuals is a contradiction for the energy-centric view but not for the wider view that understands the greater relative importance of brain during muscle-wasting malnourishment and role of fat in brain.

Brain development specifically needs certain classes of fatty acids which are not synthesized in the body and have to be supplied by dietary intake. In a diplomat lifestyle this requirement might be higher. Now, if diet is deficient in these fatty acids, brain will encourage fat intake until the demand is fulfilled. This will be less intense in a soldier life if the requirement is relatively less. Deficiency of omega 3 fatty acids may thus interact with diplomat strategy to result in obesity.

5. Fat as a modulator of immunity: The adipose tissue secretes a number of proinflammatory and anti-inflammatory signals, which modulate the dynamics of the innate immune cells. The role of adipose tissue in initiating a state of low-grade

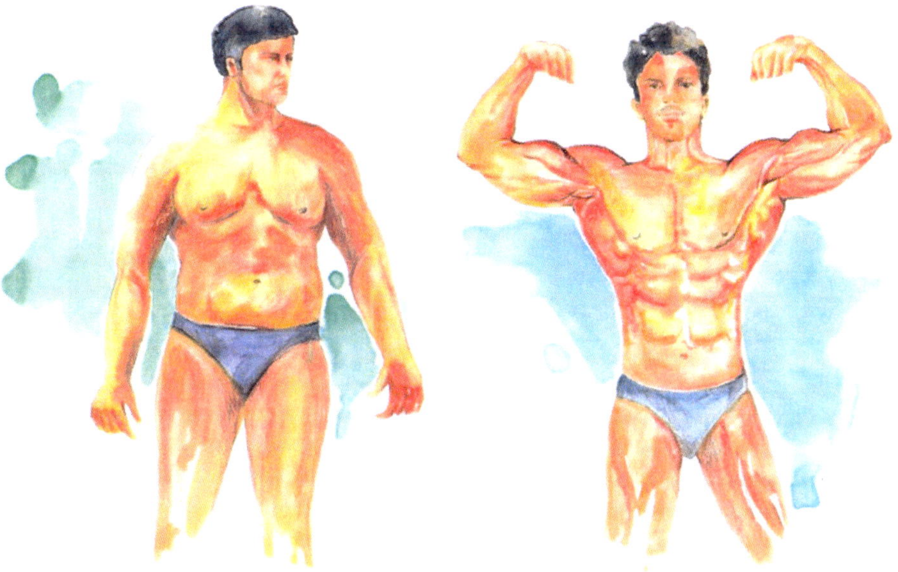

chronic systemic inflammation is well known. We have seen this in sufficient details in Chap. 8. The prevalent paradigm only blames the adipose tissue for the state of this systemic inflammation which underlies many of the pathological consequences of obesity and T2D. What is less appreciated is that the same adipose tissue also gives anti-inflammatory signals too. The adipose-derived molecules adiponectin and sFRP5 have significant anti-inflammatory effects [94, 95], and the role of visfatin is debated [96–99]. This means that the normal role of adipose tissue in the fine-tuning of innate immunity must be more subtle and balancing. In obesity-related disorders there is a suppression of the anti-inflammatory signals and overexpression of the proinflammatory ones [100]. Why this happens is not clearly known. A possible player in this shift of balance is likely to be testosterone, which is known to suppress the systemic inflammatory response. It is likely therefore that in the absence of physical aggression, the balance between pro- and anti-inflammatory signals is shifted. This is a likely parsimonious measure in anticipation of a less injury-prone lifestyle as discussed in Chap. 8. It is also likely that adiponectin, which is a marker of *r* reproductive strategy, is suppressed by crowding, and this leads to the shift in immune balance as well as greater accumulation of fat. Both are speculative at present but provide a logically sound explanation to the shift of balance. In contrast there does not appear to be any alternative explanation coming from the classical school that works at both ultimate and proximate level.

Adipose tissue has another interesting connection with lymphoid tissue. Most lymph nodes are physically intimately associated with small masses of adipose tissue. These are likely to be important energy sources for activated lymph nodes [101–105]. Lymphocytes appear to prefer fatty acids as fuel sources [103]. It is possible therefore that the dispersed adipose masses associated with lymph nodes have a critical role in acquired immunity. However, in an obese person, the fraction of the total adipose tissue in association with lymph nodes is quite small as compared to total body fat. Therefore it is unlikely that proneness to obesity may have evolved primarily to serve an immune function, although this is interesting hypothesis and needs to be investigated [106].

Summarily it appears that fat has a negative role in innate immunity, in the context of injuries, but a positive role in acquired immunity which is more important in infectious diseases. This is compatible with diplomat life which is less prone to physical injuries but presumably more to certain types of infections such as respiratory infections. But our understanding of the interaction between adipose and immune cells is still in infancy. More efforts are needed to understand the normal role of adipose tissue in immune modulation before we understand the pathological role of excess adiposity. This is another major area of unanswered questions.

6. Fat as a modulator of sexual, reproductive, and social behavior: The interaction between obesity, particularly abdominal obesity, and sexual and reproductive performance is well documented, but the mechanisms linking the two are not yet very clearly known. Abdominal obesity is associated with reduced sexual desire [107], loss of libido [108], substantially compromised sexual attractiveness [57], as well as reduced fertility [109]. On the other hand fat is not always bad for sex, fertility, and reproduction. In fact an optimum level of fat is essential for fertility [110].

As fat affects sexual and reproductive function, we can expect that a change in sexual and reproductive strategy, capacity, or function would in turn affect fat deposition. There is some evidence in this direction. Castration has been shown to induce obesity in rats [111]. Also, adiponectin, whose primary function is to facilitate *r* reproductive strategy, is differentially associated with body fat distribution [112, 113] and is likely to be a mediator rather than an effect of central fat distribution. We have seen time and again that testosterone, a major player in both sex and aggression, has important effects on lipolysis. Sex and reproductive life therefore is evidently a major player in the dynamics of adiposity.

7. Fat as a social signal: Until recently, the scientific community had almost entirely ignored this important social function of the fat tissues and particularly fat distribution in the body. It is a

relatively recent suggestion that fat serves as a social signal [1] signaling recent nutritional history [114] but more important than that, signaling biological as well as personality characteristics [115]. The female body form, particularly waist-to-hip ratio, is believed to reflect fertility [1]. In males the role of central obesity is of particular interest. It is well known that central obesity is a better predictor of obesity-related disorders and that visceral fat is metabolically more active than subcutaneous fat. If the metabolic and behavioral role of adipose tissue and fat distribution as a social signal coevolved, it makes sense that metabolically active fat should be deposited abdominally. Subcutaneous fat is difficult to differentiate from muscle mass and therefore can be of little signal value. Abdominal fat, on the other hand, changes the body proportions substantially and therefore stands out quickly. For a person approaching from a distance, body proportions can be perceived much before facial expressions. Further the theory of honest signaling or the handicap principle states that only costly signals can be evolutionary-stable honest signals [116–118]. Fat has a high energy cost, and therefore, signaling by fat can evolve to be honest.

People are known to make personality judgments very quickly based on facial characters. The judgments are made instantaneously, and thinking for a longer time appears to make little difference [119, 120]. Furthermore subjects may not be able to state how they made these judgments suggesting that the judgments are not always made at a conscious level. Most of the earlier studies are restricted to facial features, whereas we showed in an earlier study that when faces are not shown, people make use of body proportions to read personality characteristics. Faceless drawings of three male body forms, namely, lean, muscular, and feminine, each with and without abdominal obesity (Fig. 11.3), were shown in a randomized order to a group of 222

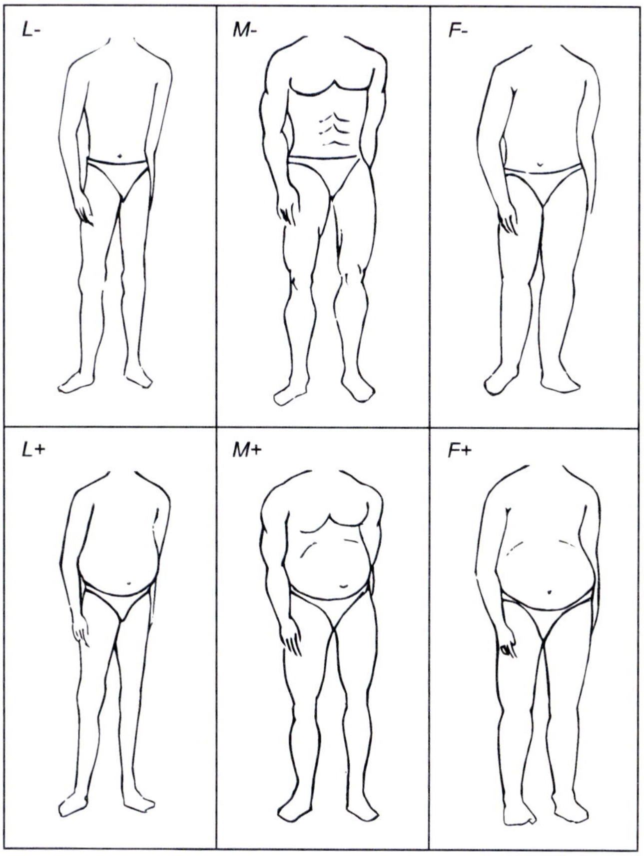

Fig. 11.3 The six body forms used for examining whether people from different genetic and cultural backgrounds associate personality with body forms consistently. If body proportions have some evolved social signal value, they should be perceived similarly by different cultures

respondents. A list of 30 different adjectives or short descriptions of personality traits was given to each respondent, and they were asked to allocate the most appropriate figure to each of them independently. The traits included those directly related to physique, those related to nature, attitude, and moral character, and also those related to social status. For 29 out of the 30 adjectives, people consistently attributed specific body forms. Based on common choices, the 30 traits could be clustered into distinct "personalities" which were strongly associated with particular body forms (Fig. 11.4). A centrally obese figure was positively associated with the adjectives lethargic, greedy, political, money-minded, selfish, and rich. It was also negatively associated with physical aggression and swiftness.

The respondents were asked whether they could reason out their choices of figures for each trait. For physical traits, the proportion of people choosing with reason was significantly higher as expected. For traits related to nature, attitude, moral character, and social status, there was a high proportion of "just felt like" responses. However, the high level of concordance showed that these responses were highly nonrandom.

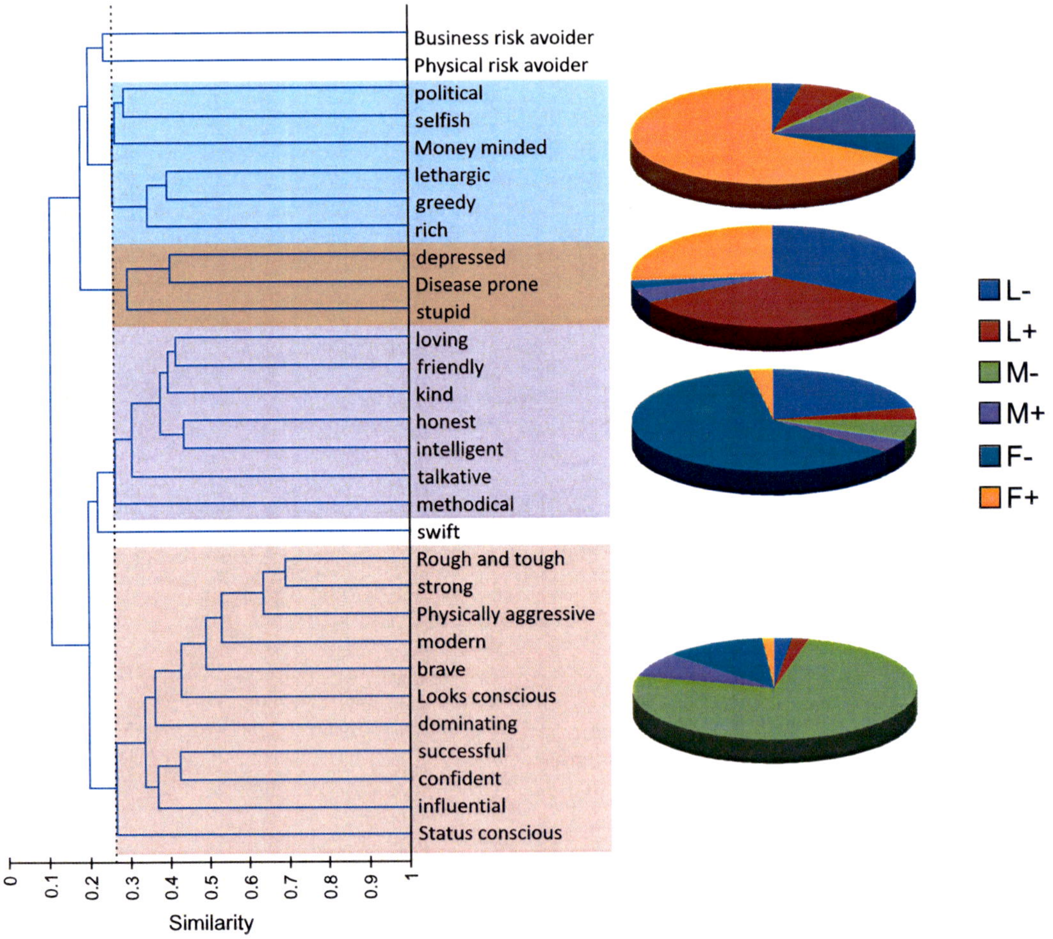

Fig. 11.4 From perceived associations of body forms with personality characteristics, distinct personality clusters emerged with predominant association with certain body forms. The similarity values between personality characteristics used for the clusters represent what fraction of individuals choose the same body form for the two personality characteristics. The body forms chosen are seen in the accompanying pie charts. Data comes from a study of university students in India by Mankar et al. [115]

This indicates that most of these choices could have been made at a subconscious level, and although respondents largely converged on their choices, they were not able to give explicit reasons. In a further and yet unpublished study, the same questionnaire was used on a group of university students in Germany and in the Caribbean islands. Although the precise meaning of the set of adjectives associated with the figures could be somewhat different in different cultures, there was some convergence in the responses. The negative association between central obesity and physical aggression was consistent across cultures (Fig. 11.5). There appears to be a cross-cultural consistency in many if not all associations of body form with personality characteristics (Fig. 11.6). An interesting inference of this study is that people already seem to know what I am trying to argue in several chapters of this book. Most researchers may not have yet suspected that there is a consistent negative association between obesity and aggression. But cross-culturally people seem to know this relation subconsciously.

If fat indeed evolved to work as a social signal, in addition to its metabolic and other functions, it explains why there is a difference in male and female abdominal fat distribution. We have seen earlier that food is one of the main natural causes of aggression. A hungry individual having greater desperation for food is more aggressive, and one with a full stomach avoids aggression. The link is mainly through serotonin. It is possible therefore that instinctively a full stomach is taken as a signal of nonaggression. When a large-sized and dominant animal is at a kill, others are most likely to keep distance. But when the dominant animal has a full stomach, the waiting animals would start advancing since there is less risk of aggression from the dominant individual. The dominant animal would generally retract from the kill at this stage. A distended stomach is the most likely signal that brings about this change. This is frequently observed during the fights between carnivores and scavengers over a kill. Abdominal fat mimics the same signal and gives the same message. This is true for both genders, but for females there is an additional context for controlling aggression. That is pregnancy. A pregnant female needs to be risk averse and therefore nonaggressive. Perception of this signal also appears to have evolved. Female abdominal obesity therefore mimics pregnancy. This is likely to be the reason why male (android) and female (gynoid) abdominal obesity differs in form (Fig. 11.7)

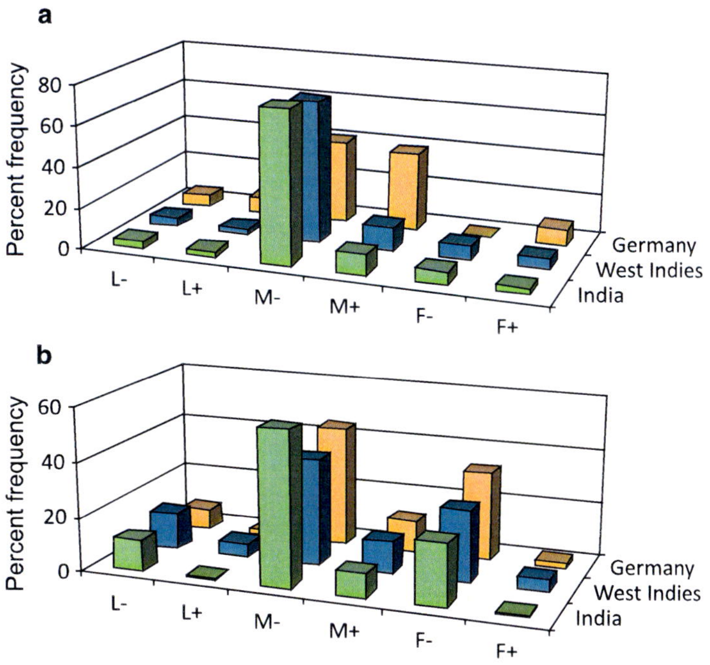

Fig. 11.5 When the study represented in Figs. 11.3 and 11.4 was extended to two other countries (similar age group and education), namely, Germany and one of the Caribbean islands, most of the associations were found to be culture invariant. Shown here are responses to the adjectives physically aggressive (**a**) and brave (**b**). Unpublished data by Meenakshi Prabhune (Germany) and Krystal Philip (Caribbean)

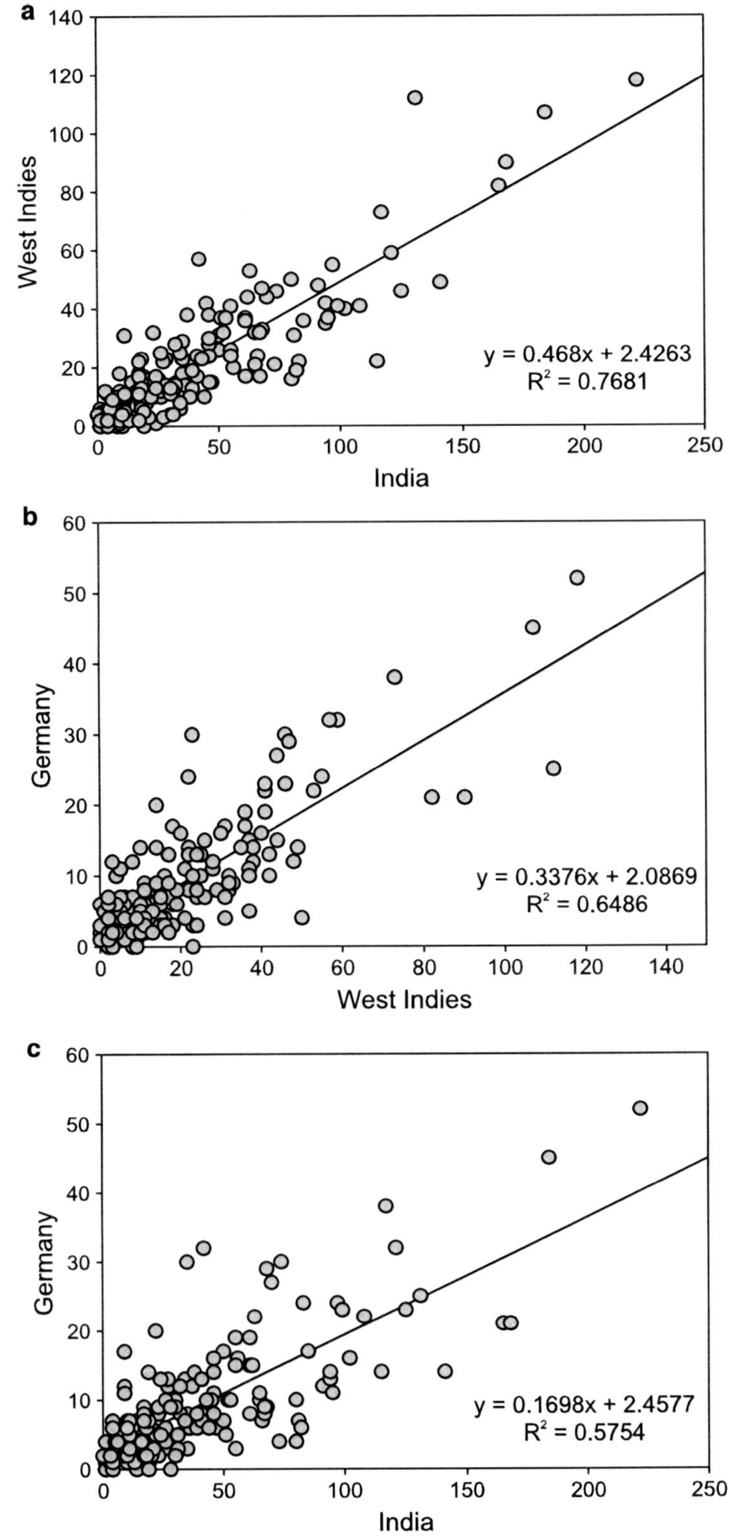

Fig. 11.6 Cross-cultural correlations in associations of body forms with personality characters. The *points* represent the number of persons in a cultural group that associated a given body form with a given personality character. Since the number of body forms is limited for each character, there are degree of freedom problems with these correlations. But all correlations are highly significant ($p<0.01$ for all the three) after working out probabilities for appropriate degrees of freedom. The sample sizes were for India, for Germany, and for the Caribbean

Fig. 11.7 Male and female forms of abdominal obesity have noticeable differences and are likely to carry different social signals

although biochemically it is accumulation of fat. Male obesity is more in the upper abdomen which resembles a full stomach, whereas female abdominal obesity covers the lower abdomen and resembles the appearance of pregnancy. Apart from aggression, fat distribution is also likely to signal other social and behavioral strategies and status. They may include fertility [121–124], parity status, or cognitive abilities [61, 125].

Appreciating the multiple functions of fat in the body itself resolves many unresolved paradoxes and questions. So far, a large volume of research is centered on mainly a single function of fat, that of energy storage. All others have been looked at as pathophysiological by-products of excess fat. We need to come out of this thinking trap and appreciate that normally lipids and the adipose tissue have a wide variety of functions in the body [1, 2] and no single function can be studied in isolation. One of the common threads linking the different functions appears to be aggression or the absence of it. However, this is not intended to deny other links. The multifunctionality makes it likely that a change in any one or more of them can have an effect on fat accumulation and distribution and that will inevitably alter other functions too. For example, castration in rats and rabbits is shown to increase obesity [111] and insulin resistance [126]. A change in one's social role may induce changes in fat distribution with associated metabolic effects. Owing to the heavy energy balance bias in obesity research, so far we have left so many loose ends and information voids. Unless obesity research adequately covers the virgin areas and information voids, our understanding of obesity and its effects on health can never be sound and useful.

What is the exact nature of association of obesity and type 2 diabetes? Historically there have been arguments in both directions. Obesity causes insulin resistance, and insulin resistance or a diabetic tendency in some form causes obesity. We are now leading to a novel viewpoint that neither obesity nor insulin resistance is the causal agent of the other. It is behavioral syndrome that precedes both, and the neuroendocrine processes driven by specific behavioral strategies initiate pathways leading to obesity on the one hand and T2D and associated pathophysiology on the other. We have already seen a number of different pathways by which behavior affects hyperinsulinemia, insulin resistance, systemic inflammation, and angiogenesis dysfunction. Behavior could be related to the origins of obesity by a number of possible links.

An interesting observation that provides additional support to the neurobehavioral hypothesis is that it is relative rather than absolute obesity that appears to be related to insulin resistance. There are four reasons why I say relative rather than absolute obesity is more closely related to insulin resistance parameters:

1. The obesity–insulin resistance relationship between high-obesity countries versus low-obesity countries: The prevalence and intensity of the obesity epidemic in the USA is substantially greater than that in India. A comparison of data from the two countries shows that in the USA where obesity is more prevalent, the slope of the regression line between obesity parameters and insulin resistance parameters is substantially lower than that in India. In other words where obesity is less, people appear to become diabetic at lower BMI. Where obesity is more they need greater BMI to become diabetic.
2. A similar pattern exists between rural and urban populations within a country. Obesity is generally more prevalent in urban areas where the slope of the line is less than that in rural areas where fewer obese individuals are found (Fig. 11.8). This pattern has been noted across many countries [127, 128].
3. The mortality rate is a U-shaped function of body weight. There is an optimum BMI that ensures maximum health. In the USA over 100 years, the mean BMI increased substantially, and along with it the optimum also shifted substantially towards the right (Fig. 11.9).
4. In an unpublished rat experiment in Pune, a colony of chronically malnourished rats showed high levels of insulin resistance. These rats, which we will call Hardikar rats after the experimenter, were smaller in size but had larger brains. In comparison with normal rats, the Hardikar rats had lower body weights and higher insulin resistance, leading to a negative association between body weights and insulin resistance between groups. But within both the groups, there was a positive correlation between body weights and insulin resistance [93]. The malnourished

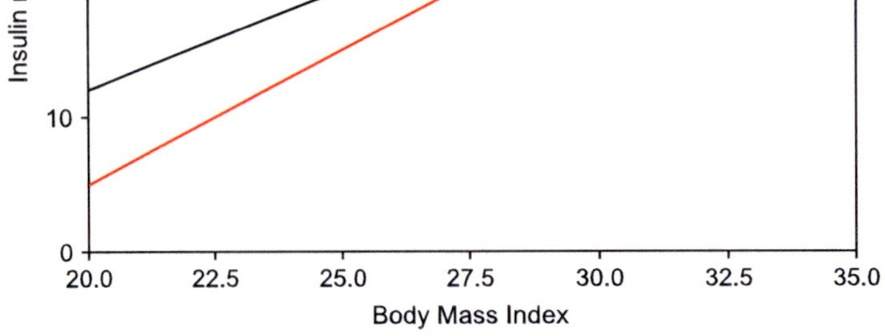

Fig. 11.8 A conceptual pattern in the slope of the regression between obesity parameters and insulin resistance parameters. In populations with lower mean obesity, the slope of the line is steeper although its position may be lower. A number of studies [127–130] indicate that the urban–rural difference cannot be accounted by obesity alone. Not only obesity levels are different in urban and rural environments, the relationship of obesity with insulin resistance is also different

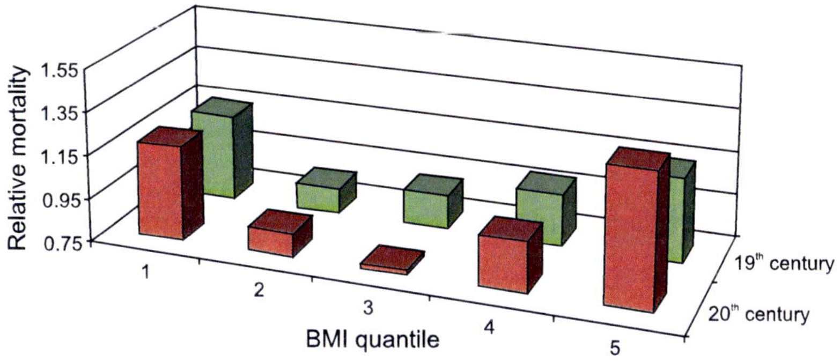

Fig. 11.9 Mortality is known to have a U-shaped relationship with BMI with an optimum BMI ensuring minimum risk. A comparison of data on American males in the late nineteenth century and late twentieth century shows that as the mean BMI increased over a century, the optimum BMI also shifted to the right. A BMI of 20–22 was optimum earlier, and that of 24–26 appears to be optimum after nearly 100 years (based on data from [131, 132])

group that had lower mean body weight appears to become insulin resistant at much lower body weight.

All the four observations support my speculation that it is relative obesity that correlated with insulin resistance than absolute obesity. So even in a group that has lower mean obesity, the relatively more obese are more susceptible to insulin resistance. If obesity had a direct connection with insulin resistance metabolically, then only one's own obesity should have mattered. What appears to matter is one's obesity in comparison with others in the group. This can happen only if perception of relative obesity is involved somewhere. An individual's perception of where he stands among the obesity ranking of the society could be affecting the metabolic actions of obesity. This is consistent with the observation above that people perceive obesity as a social signal. This is an indication that obesity has a perceptional and neuronal connection rather than a direct metabolic connection with insulin resistance.

There is one more piece of evidence that relates perception of calories to metabolism, this time from *Drosophila*. Caloric restriction is shown to increase life span in widely different species, and *Drosophila* is not an exception. In an interesting experiment the smell of food was shown to partially reverse the effects of caloric restriction [133]. There is a similar finding in *Caenorhabditis elegans* too [134]. In both the experiments actual intake of calories was not necessary to bring about the metabolic change that marked the reversal of the effects of caloric restriction. Perception of calories was sufficient. These experiments also indicate the involvement of neuronal mechanisms in mediating the metabolic effects of calories.

If relative obesity matters, relative muscle mass and perhaps more importantly muscle strength should matter even more. There are no studies in this direction. But this could explain certain perplexing known patterns. South Asians who migrated to Western (Caucasoid-dominated) countries show a very high prevalence of T2D which is not matched by the urban populations of their native places. Even if we assume that there is a genetic difference across ethnic groups, why should the ones who migrated to Western societies be more susceptible? A plausible answer worth investigating is that South Asians have a lower proportionate muscle mass as compared to Caucasoids. In their own country they are not inferior in muscle mass, but when they migrate to another society that is dominated by an ethnic group having larger muscle mass on an average, their perceived ranking in muscle strength is low, and that may matter in shaping behavioral and metabolic strategies.

Beginning from Chap. 5 till now, we have seen a large number of factors and processes linking obesity to insulin resistance. In this light obesity has a strong association with insulin resistance without being an inevitable biochemical cause of insulin resistance. We can strengthen this statement using an alternative method. It can be seen that increasing food intake in the network model (Appendix II) does not result into all components of metabolic syndrome if behavioral changes are blocked. On the other hand, if behavioral changes are primary, then through altered adiponectin, leptin, and insulin-related mechanisms, the system becomes sensitive to diet-induced obesity.

I will briefly recapitulate now the origins of obesity and its links with insulin resistance syndrome according to the new line of thinking:

(a) Overeating and obesity are not possible as long as the food intake regulation mechanisms of the body are functional. Therefore the origin of obesity does not lie in greater availability of food or macronutrient composition of food. It lies in disruption of the regulatory mechanisms.
(b) In species that have predators or food-related aggression, behavioral regulation of food intake works more frequently than metabolic regulation; also in these species, metabolic regulation evolves to be overridable by behavioral cues. Human species belongs to this category; therefore, the key to obesity lies in behavior rather than metabolism. Absence of food-related aggression and freedom from predation may deplete the levels of anxiety-related peptides which are crucial in food intake regulation.
(c) A transition from soldier to diplomat behavior is accompanied by hyperinsulinemia, and sustained hyperinsulinemia needs to be accompanied by weakening of food intake regulation since insulin itself is anorectic. A hyperinsulinemic individual will die of undernutrition unless the food intake regulation mechanisms are partially disrupted. Sustained hyperinsulinemia must also be accompanied by insulin resistance, without which hypoglycemia will threaten life. Therefore insulin resistance and propensity to be obese have a common origin.
(d) Other concurrent changes accompanying soldier to diplomat transition are testosterone deficiency, lipid anabolic bias, and early activation of central fatigue mechanisms. These change body composition and pattern of fat deposition.
(e) Obesity reinforces diplomat behavior by making some components of soldier behavior such

as agility more difficult and also by enhancing cognitive function through the agency of leptin and cholesterol.

(f) Energy storage and risk-taking behavior have conflicting life history consequences, because of which obesity has evolved to be associated with decreases in risk-taking behavior, weakening the soldier component of behavior further.

(g) Adiponectin downregulation as an intrinsic part of *K* like reproduction in response to high population density or social subordination can also trigger a chain of events similar to the above.

It is no surprise therefore that obesity and insulin resistance are so tightly linked, but I would still maintain that obesity is neither necessary nor sufficient for the development of insulin resistance. This relationship of obesity with insulin resistance is different from what is traditionally perceived. Brain and behavior appear to be involved in almost every step in some way or the other. This is only a small part of the central point that I want to emphasize on a much broader scale throughout the book. It is impossible to understand metabolism without understanding behavioral ecology.

References

1. Pond CM (1998) The fats of life. Cambridge University Press, Cambridge
2. Wells JCK (2009) The evolutionary biology of human body fatness. Cambridge University Press, Cambridge
3. Summary of SIAM talk « Scientific Clearing House. http://sciencehouse.wordpress.com/2010/07/23/summary-of-siam-talk/
4. Gibbs J, Young RC, Smith GP (1973) Cholecystokinin decreases food intake in rats. J Comp Physiol Psychol 84:488–495
5. Pappas TN, Melendez RL, Debas HT (1989) Gastric distension is a physiologic satiety signal in the dog. Dig Dis Sci 34:1489–1493
6. Myers R, McCaleb M (1980) Feeding: satiety signal from intestine triggers brain's noradrenergic mechanism. Science 209:1035–1037
7. Gibbs J, Smith GP (1986) Satiety: the roles of peptides from the stomach and the intestine. Fed Proc 45:1391–1395
8. Levin BE, Dunn-Meynell AA, Routh VH (1999) Brain glucose sensing and body energy homeostasis: role in obesity and diabetes. Am J Physiol 276:R1223
9. Debons A, Krimsky I, From A (1970) A direct action of insulin on the hypothalamic satiety center. Am J Physiol 219:938–943
10. Rodin J, Wack J, Ferrannini E, DeFronzo RA (1985) Effect of insulin and glucose on feeding behavior. Metabolism 34:826–831
11. Leibowitz SF, Alexander JT (1998) Hypothalamic serotonin in control of eating behavior, meal size, and body weight. Biol Psychiatry 44:851–864
12. Benoit SC et al (2002) The catabolic action of insulin in the brain is mediated by melanocortins. J Neurosci 22:9048–9052
13. Schwartz MW et al (1992) Inhibition of hypothalamic neuropeptide Y gene expression by insulin. Endocrinology 130:3608–3616
14. Schwartz MW, Figlewicz DP, Woods SC, Porte D Jr, Baskin DG (1993) Insulin, neuropeptide Y, and food intake. Ann N Y Acad Sci 692:60–71
15. Loftus TM, Maggs DG, Lane MD (1997) The adipose tissue/central nervous system axis. Diabetologia 40(Suppl 3):B16–B20
16. Karve S et al (2011) Money handling and obesity: a test of the exaptation hypothesis. Curr Sci 100:1695–1700
17. Schwartz MW, Woods SC, Porte D Jr, Seeley RJ, Baskin DG (2000) Central nervous system control of food intake. Nature 404:661–671
18. Munzberg H, Myers MG (2005) Molecular and anatomical determinants of central leptin resistance. Nat Neurosci 8:566–570
19. Duman EA, Canli T (2010) Social behavior and serotonin. In: Handbook of the behavioral neurobiology of serotonin, vol. 21, pp. 449–456
20. Kask A, Schiöth HB, Mutulis F, Wikberg JES, Rägo L (2000) Anorexigenic cocaine- and amphetamine-regulated transcript peptide intensifies fear reactions in rats. Brain Res 857:283–285
21. File SE (1981) Contrasting effects of org 2766 and α MSH on social and exploratory behavior in the rat. Peptides 2:255–260
22. Kokare DM, Dandekar MP, Chopde CT, Subhedar N (2005) Interaction between neuropeptide Y and α melanocyte stimulating hormone in amygdala regulates anxiety in rats. Brain Res 1043:107–114
23. Rao TL et al (2003) GABAergic agents prevent α-melanocyte stimulating hormone induced anxiety and anorexia in rats. Pharmacol Biochem Behav 76:417–423
24. Hasenöhrl RU, Weth K, Huston JP (1999) Intraventricular infusion of the histamine H 1 receptor antagonist chlorpheniramine improves maze performance and has anxiolytic-like effects in aged hybrid Fischer 344 × Brown Norway rats. Exp Brain Res 128:435–440
25. Privou C, Knoche A, Hasenöhrl RU, Huston JP (1998) The H1- and H2-histamine blockers chlorpheniramine and ranitidine applied to the nucleus basalis magnocellularis region modulate anxiety and reinforcement related processes. Neuropharmacology 37:1019–1032
26. Ookuma K et al (1993) Neuronal histamine in the hypothalamus suppresses food intake in rats. Brain Res 628:235–242

27. Ookuma K, Yoshimatsu H, Sakata T, Fujimoto K, Fukagawa K (1989) Hypothalamic sites of neuronal histamine action on food intake by rats. Brain Res 490:268–275
28. Mercer LP, Kelley DS, Humphries LL, Dunn JD (1994) Manipulation of central nervous system histamine or histaminergic receptors (H1) affects food intake in rats. J Nutr 124:1029–1036
29. Stanek LM (2006) Cocaine- and amphetamine related transcript (CART) and anxiety. Peptides 27:2005–2011
30. Asakawa A et al (2001) Cocaine-amphetamine-regulated transcript influences energy metabolism, anxiety and gastric emptying in mice. Horm Metab Res 33:554–558
31. Teff KL, Kim SF (2011) Atypical antipsychotics and the neural regulation of food intake and peripheral metabolism. Physiol Behav 104:590–598
32. Tian D-R et al (2004) Changes of hypothalamic α MSH and CART peptide expression in diet induced obese rats. Peptides 25:2147–2153
33. Kristensen P et al (1998) Hypothalamic CART is a new anorectic peptide regulated by leptin. Nature 393:72–76
34. Watve M (1993) Why man has no predator. Curr Sci 65:120–122
35. Speakman JR (2008) Thrifty genes for obesity, an attractive but flawed idea, and an alternative perspective: the 'drifty gene' hypothesis. Int J Obes 32:1611–1617
36. Bandini LG, Schoeller DA, Dietz WH (1990) Energy expenditure in obese and nonobese adolescents. Pediatr Res 27:198–203
37. Dietz WH, Bandini LG, Schoeller DA (1991) Estimates of metabolic rate in obese and nonobese adolescents. J Pediatr 118:146–149
38. Frisancho AR (2003) Reduced rate of fat oxidation: a metabolic pathway to obesity in the developing nations. Am J Hum Biol 15:522–532
39. Zunquin G, Theunynck D, Sesboüé B, Arhan P, Bouglé D (2009) Comparison of fat oxidation during exercise in lean and obese pubertal boys: clinical implications. Br J Sports Med 43:869–870
40. Zurlo F et al (1990) Low ratio of fat to carbohydrate oxidation as predictor of weight gain: study of 24-h RQ. Am J Physiol Endocrinol Metab 259:650–657
41. Rogge MM (2009) The role of impaired mitochondrial lipid oxidation in obesity. Biol Res Nurs 10: 356–373
42. Sawaya AL, Verreschi I, Tucker KL, Roberts SB, Hoffman DJ (2000) Why are nutritionally stunted children at increased risk of obesity? Studies of metabolic rate and fat oxidation in shantytown children from São Paulo, Brazil. Am J Clin Nutr 72:702–707
43. Weiss B, Maickel RP (1968) Sympathetic nervous control of adipose tissue lipolysis. Int J Neuropharmacol 7:395–403
44. Yoshimatsu H et al (2002) Histidine induces lipolysis through sympathetic nerve in white adipose tissue. Eur J Clin Invest 32:236–241
45. Bartness TJ et al (2010) Sensory and sympathetic nervous system control of white adipose tissue lipolysis. Mol Cell Endocrinol 318:34–43
46. De Pergola G (2000) The adipose tissue metabolism: role of testosterone and dehydroepiandrosterone. Int J Obes Relat Metab Disord 24:S59–S63
47. Mrin P, Arver S (1998) Androgens and abdominal obesity. Baillières Clin Endocrinol Metab 12:441–451
48. Mårin P, Odén B, Björntorp P (1995) Assimilation and mobilization of triglycerides in subcutaneous abdominal and femoral adipose tissue in vivo in men: effects of androgens. J Clin Endocrinol Metab 80:239–243
49. Mårin P et al (1993) Androgen treatment of abdominally obese men. Obes Res 1:245–251
50. Sellers JG, Mehl MR, Josephs RA (2007) Hormones and personality: testosterone as a marker of individual differences. J Res Pers 41:126–138
51. Cousin B et al (1993) Local sympathetic denervation of white adipose tissue in rats induces preadipocyte proliferation without noticeable changes in metabolism. Endocrinology 133:2255–2262
52. Penicaud L, Cousin B, Leloup C, Lorsignol A, Casteilla L (2000) The autonomic nervous system, adipose tissue plasticity, and energy balance. Nutrition 16:903–908
53. Shi H, Song CK, Giordano A, Cinti S, Bartness TJ (2005) Sensory or sympathetic white adipose tissue denervation differentially affects depot growth and cellularity. Am J Physiol Regul Integr Comp Physiol 288:R1028–R1037
54. Youngstrom TG, Bartness TJ (1998) White adipose tissue sympathetic nervous system denervation increases fat pad mass and fat cell number. Am J Physiol Regul Integr Comp Physiol 275: R1488–R1493
55. Larsson B et al (1984) Abdominal adipose tissue distribution, obesity, and risk of cardiovascular disease and death: 13 year follow up of participants in the study of men born in 1913. Br Med J (Clin Res Ed) 288:1401–1404
56. Risica PM, Ebbesson SO, Schraer CD, Nobmann ED, Caballero BH (2000) Body fat distribution in Alaskan Eskimos of the Bering Straits region: the Alaskan Siberia Project. Int J Obes Relat Metab Disord 24:171–179
57. Singh D (1993) Body shape and women's attractiveness. Hum Nat 4:297–321
58. Landin K, Krotkiewski M, Smith U (1989) Importance of obesity for the metabolic abnormalities associated with an abdominal fat distribution. Metab Clin Exp 38:572–576
59. Berman DM et al (2001) Racial disparities in metabolism, central obesity, and sex hormone-binding globulin in postmenopausal women. J Clin Endocrinol Metab 86:97–103
60. Kerwin DR et al (2010) The cross-sectional relationship between body mass index, waist-hip ratio, and cognitive performance in postmenopausal women enrolled in the Women's Health Initiative. J Am Geriatr Soc 58:1427–1432

61. Lassek WD, Gaulin SJC (2008) Waist-hip ratio and cognitive ability: is gluteofemoral fat a privileged store of neurodevelopmental resources? Evol Hum Behav 29:26–34
62. Bartness TJ, Song CK (2007) Thematic review series: adipocyte biology. Sympathetic and sensory innervation of white adipose tissue. J Lipid Res 48:1655–1672
63. Kreier F et al (2002) Selective parasympathetic innervation of subcutaneous and intra-abdominal fat–functional implications. J Clin Invest 110: 1243–1250
64. Nesto RW (2005) Obesity. Tex Heart Inst J 32: 387–389
65. Pech RP, Sinclair ARE, Newsome AE, Catling PC (1992) Limits to predator regulation of rabbits in Australia: evidence from predator-removal experiments. Oecologia 89:102–112
66. Tapper SC, Potts GR, Brockless MH (1996) The effect of an experimental reduction in predation pressure on the breeding success and population density of grey partridges Perdix perdix. J Appl Ecol 33:965–978
67. Auger P, Pontier D (1998) Fast game theory coupled to slow population dynamics: the case of domestic cat populations. Math Biosci 148:65–82
68. Morton GJ, Cummings DE, Baskin DG, Barsh GS, Schwartz MW (2006) Central nervous system control of food intake and body weight. Nature 443:289–295
69. Elliott R, Newman JL, Longe OA, Deakin JFW (2003) Differential response patterns in the striatum and orbitofrontal cortex to financial reward in humans: a parametric functional magnetic resonance imaging study. J Neurosci 23:303–307
70. Johnson PM, Kenny PJ (2010) Dopamine D2 receptors in addiction-like reward dysfunction and compulsive eating in obese rats. Nat Neurosci 13: 635–641
71. Wang G-J et al (2004) Exposure to appetitive food stimuli markedly activates the human brain. NeuroImage 21:1790–1797
72. Elliott R, Newman JL, Longe OA, William Deakin JF (2004) Instrumental responding for rewards is associated with enhanced neuronal response in subcortical reward systems. NeuroImage 21:984–990
73. Stice E, Yokum S, Burger KS, Epstein LH, Small DM (2011) Youth at risk for obesity show greater activation of striatal and somatosensory regions to food. J Neurosci 31:4360–4366
74. Briers B, Pandelaere M, Dewitte S, Warlop L (2006) Hungry for money: the desire for caloric resources increases the desire for financial resources and vice versa. Psychol Sci 17:939–943
75. Wang G-J, Volkow ND, Fowler JS (2002) The role of dopamine in motivation for food in humans: implications for obesity. Expert Opin Ther Targets 6:601–609
76. Wang G-J et al (2001) Brain dopamine and obesity. Lancet 357:354–357
77. Couppis MH, Kennedy CH (2008) The rewarding effect of aggression is reduced by nucleus accumbens dopamine receptor antagonism in mice. Psychopharmacology 197:449–456
78. Sladek R et al (2007) A genome-wide association study identifies novel risk loci for type 2 diabetes. Nature 445:881–885
79. Scott LJ et al (2007) A genome-wide association study of type 2 diabetes in Finns detects multiple susceptibility variants. Science 316:1341–1345
80. Li S et al (2010) Cumulative effects and predictive value of common obesity-susceptibility variants identified by genome-wide association studies. Am J Clin Nutr 91:184–190
81. McCarthy MI, Zeggini E (2009) Genome-wide association studies in type 2 diabetes. Curr Diab Rep 9:164–171
82. Thorleifsson G et al (2009) Genome-wide association yields new sequence variants at seven loci that associate with measures of obesity. Nat Genet 41:18–24
83. Rankinen T et al (2006) The human obesity gene map: the 2005 update. Obesity (Silver Spring) 14:529–644
84. Vaz M et al (1999) Body fat topography in Indian and Tibetan males of low and normal body mass index. Indian J Physiol Pharmacol 43:179–185
85. Arbabi S et al (2003) The cushion effect. J Trauma 54:1090–1093
86. Wang SC et al (2003) Increased depth of subcutaneous fat is protective against abdominal injuries in motor vehicle collisions. Annu Proc Assoc Adv Automot Med 47:545–559
87. Haag M (2003) Essential fatty acids and the brain. Can J Psychiatry 48:195–203
88. Willatts P, Forsyth J, DiModugno M, Varma S, Colvin M (1998) Effect of long-chain polyunsaturated fatty acids in infant formula on problem solving at 10 months of age. Lancet 352:688–691
89. Stevens LJ, Zentall SS, Abate ML, Kuczek T, Burgess JR (1996) Omega-3 fatty acids in boys with behavior, learning, and health problems. Physiol Behav 59:915–920
90. Hibbeln JR et al (1998) A replication study of violent and nonviolent subjects: cerebrospinal fluid metabolites of serotonin and dopamine are predicted by plasma essential fatty acids. Biol Psychiatry 44:243–249
91. Reisbick S, Neuringer M, Hasnain R, Connor WE (1994) Home cage behavior of rhesus monkeys with long-term deficiency of omega-3 fatty acids. Physiol Behav 55:231–239
92. Kitaysky AS, Kitaiskaia EV, Piatt JF, Wingfield JC (2006) A mechanistic link between chick diet and decline in seabirds? Proc Biol Sci 273:445–450
93. Hardikar A (1999) Role of environmental factors in induction, prevention and reversal of diabetes mellitus. PhD thesis, University of Pune
94. Ouchi N, Walsh K (2007) Adiponectin as an anti-inflammatory factor. Clin Chim Acta 380:24–30
95. Ouchi N et al (2010) Sfrp5 is an anti-inflammatory adipokine that modulates metabolic dysfunction in obesity. Science 329:454–457
96. Sonoli SS et al (2011) Visfatin–a review. Eur Rev Med Pharmacol Sci 15:9–14

97. Tilg H, Moschen AR (2008) Role of adiponectin and PBEF/visfatin as regulators of inflammation: involvement in obesity-associated diseases. Clin Sci 114:275
98. Sethi JK, Vidal-Puig A (2005) Visfatin: the missing link between intra-abdominal obesity and diabetes? Trends Mol Med 11:344–347
99. Stephens JM, Vidal-Puig AJ (2006) An update on visfatin/pre-B cell colony-enhancing factor, an ubiquitously expressed, illusive cytokine that is regulated in obesity. Curr Opin Lipidol 17:128–131
100. Bastard J-P et al (2006) Recent advances in the relationship between obesity, inflammation, and insulin resistance. Eur Cytokine Netw 17:4–12
101. Pond CM, Mattacks CA (1995) Interactions between adipose tissue around lymph nodes and lymphoid cells in vitro. J Lipid Res 36:2219–2231
102. Pond CM (2003) Paracrine interactions of mammalian adipose tissue. J Exp Zool A Comp Exp Biol 295A:99–110
103. Wells AS, Read NW, Laugharne JD, Ahluwalia NS (1998) Alterations in mood after changing to a low-fat diet. Br J Nutr 79:23–30
104. Pond CM, Mattacks CA (1998) In vivo evidence for the involvement of the adipose tissue surrounding lymph nodes in immune responses. Immunol Lett 63:159–167
105. Caroline MP (2005) Adipose tissue and the immune system. Prostaglandins Leukot Essent Fatty Acids 73:17–30
106. Wells JCK (2009) Ethnic variability in adiposity and cardiovascular risk: the variable disease selection hypothesis. Int J Epidemiol 38:63–71
107. Kolotkin RL et al (2006) Obesity and sexual quality of life. Obesity (Silver Spring) 14:472–479
108. Hammoud A et al (2009) Effect of Roux-en-Y gastric bypass surgery on the sex steroids and quality of life in obese men. J Clin Endocrinol Metab 94:1329–1332
109. Gesink Law DC, Maclehose RF, Longnecker MP (2007) Obesity and time to pregnancy. Hum Reprod 22:414–420
110. Mitchell M, Armstrong DT, Robker RL, Norman RJ (2005) Adipokines: implications for female fertility and obesity. Reproduction 130:583–597
111. Hausberger FX, Hausberger BC (1966) Castration-induced obesity in mice. Acta Endocrinol 53:571–583
112. Staiger H et al (2003) Relationship of serum adiponectin and leptin concentrations with body fat distribution in humans. Obesity 11:368–376
113. Park K-G et al (2004) Relationship between serum adiponectin and leptin concentrations and body fat distribution. Diabetes Res Clin Pract 63:135–142
114. Pond CM (2001) Long-term changes in adipose tissue in human disease. Proc Nutr Soc 60:365–374
115. Mankar M, Joshi RS, Belsare PV, Jog MM, Watve MG (2008) Obesity as a perceived social signal. PLoS One 3:e3187
116. Zahavi A, Zahavi A, Balaban A, Ely MP (1999) The handicap principle: a missing piece of Darwin's puzzle. Oxford University Press, New York, NY
117. Lachmann M, Szamado S, Bergstrom CT (2001) Cost and conflict in animal signals and human language. Proc Natl Acad Sci USA 98:13189–13194
118. Gintis H, Smith EA, Bowles S (2001) Costly signaling and cooperation. J Theor Biol 213:103–119
119. Ambady N, Rule NO (2008) Brief exposures: male sexual orientation is accurately perceived at 50 ms. J Exp Soc Psychol 44:1100–1105
120. Elfenbein HA, Ambady N (2002) On the universality and cultural specificity of emotion recognition: a meta-analysis. Psychol Bull 128:203–235
121. Rilling JK, Kaufman TL, Smith EO, Patel R, Worthman CM (2009) Abdominal depth and waist circumference as influential determinants of human female attractiveness. Evol Hum Behav 30:21–31
122. Singh D, Renn P, Singh A (2007) Did the perils of abdominal obesity affect depiction of feminine beauty in the sixteenth to eighteenth century British literature? Exploring the health and beauty link. Proc Biol Sci 274:891–894
123. Renato P (2006) Obesity, fat distribution and infertility. Maturitas 54:363–371
124. Diamanti-Kandarakis E, Bergiele A (2001) The influence of obesity on hyperandrogenism and infertility in the female. Obes Rev 2:231–238
125. Lassek WD, Gaulin SJC (2006) Changes in body fat distribution in relation to parity in American women: a covert form of maternal depletion. Am J Phys Anthropol 131:295–302
126. Georgiev IP et al (2009) Evaluation of insulin resistance in obese castrated New Zealand white rabbits. Rev Méd Vét 160:335–340
127. Al-Nuaim AR (1997) Prevalence of glucose intolerance in urban and rural communities in Saudi Arabia. Diabet Med 14:595–602
128. Snehalatha C, Ramachandran A, Vijay V, Viswanathan M (1994) Differences in plasma insulin responses in urban and rural Indians: a study in Southern-Indians. Diabet Med 11:445–448
129. Yajnik CS et al (2008) Adiposity, inflammation and hyperglycaemia in rural and urban Indian men: Coronary Risk of Insulin Sensitivity in Indian Subjects (CRISIS) Study. Diabetologia 51:39–46
130. Zimmet P, Dowse G, Finch C, Serjeantson S, King H (1990) The epidemiology and natural history of NIDDM–lessons from the South Pacific. Diabetes Metab Rev 6:91–124
131. Linares C, Su D (2005) Body mass index and health among Union Army veterans: 1891–1905. Econ Hum Biol 3:367–387
132. Su D (2005) Body mass index and old-age survival: a comparative study between the Union Army Records and the NHANES-I Epidemiological Follow-Up Sample. Am J Hum Biol 17:341–354
133. Libert S et al (2007) Regulation of drosophila life span by olfaction and food-derived odors. Science 315:1133–1137
134. Smith ED et al (2008) Age- and calorie-independent life span extension from dietary restriction by bacterial deprivation in Caenorhabditis elegans. BMC Dev Biol 8:49

12 Why Blood Sugar Goes Up

Readers are bound to wonder why in a book on diabetes we have not discussed blood sugar levels at sufficient length so far and why it is delayed till the 12th chapter. The common perception of diabetes begins with blood sugar and often ends with blood sugar. We saw in Chap. 3 that there are a series of paradoxes associated with the existing theory of glucose homeostasis and the conventionally perceived etiology of hyperglycemia. The alternative picture needs to resolve all the paradoxes. We can attempt to do so at this stage since by now I think sufficient background about the new interpretation has been built on which we can construct a new model of normal glucose regulation and altered glucose dynamics in T2D. Discussing sugar so late in this book also serves as a gesture intended to emphasize that sugar levels in blood are actually not as important in diabetes as generally believed. Raised blood sugar to diabetes is just what fever is to typhoid. It is only one of the many symptoms. However, it is one that is easiest to monitor and therefore the first indication in diagnosis and one that can also be conveniently used to follow the progression of the illness as well as any recovery from it. But that is its limitation. Just as fever is not everything in typhoid, blood sugar is not everything in T2D. Typhoid is an infection, and fever is only a symptom. If typhoid is cured, fever will certainly vanish, and the body temperature will return to normal. The reverse is not true. Fever can be reduced by antipyretic drugs or other physical means. Keeping the body on ice can also reduce the body temperature temporarily. But fever gone is not typhoid gone. The relation of blood sugar to T2D is very similar. Bringing blood sugar back to normal is not an indication of diabetes gone.

There is an important difference in the two analogous situations though. We know the cause of typhoid quite well, and treatment that directly attacks the infective agent is available. Symptomatic treatment with antipyretics may be given as a supplement to the main anti-infective treatment but that is not the main course of the treatment strategy. This has not been so with T2D so far. Since we have not understood the underlying disease process as yet, conventional medicine has caricatured the symptom as the cause. Treatments to bring blood sugar to normal have been the focus of almost the entire research and clinical practice. This is partly a burden of history. Sugar was among the first markers of diabetes to be discovered. Therefore, all the thinking was stuck on to the only thing known. Now, since we know a lot more about the variety of changes in different systems of the body, the thinking needs to be reoriented. But we are still haunted by the ghost of history which makes us think that sugar is everything to diabetes. There is considerable amount of success in bringing the sugar down too, at least in the short run. But this approach does nothing to the roots of diabetes itself. It is not surprising that sugar control by drugs is also short lived and glycemic control worsens slowly.

Any theory of diabetes cannot be complete without explaining the changes in glucose homeostasis that take place in diabetes. I will argue in

M. Watve, *Doves, Diplomats, and Diabetes: A Darwinian Interpretation of Type 2 Diabetes and Related Disorders*, DOI 10.1007/978-1-4614-4409-1_12, this chapter that the central nervous system is an important component of the mechanisms of glucose homeostasis and that without understanding the role of brain and behavior, we cannot understand glucose homeostasis. This argument is currently based more on theoretical and mathematical analysis. Currently experimental data are fragmentary but still sufficient to try and join dots to make a picture. The strength of the new model that I will describe in this chapter lies in its bold predictions that can account for most observed patterns that the orthodox paradigm is unable to account for.

More than 100 years ago, the celebrated physiologist Claude Bernard showed that puncture of the floor of the forth cerebral ventricle induced diabetes. This was a clear demonstration of the brain's direct role in glucose homeostasis. But after the discovery of insulin, the role of brain was largely neglected. Although researchers are aware of it and there is revival of interest in possible central regulation of peripheral glucose [1], the picture is not yet sufficiently clear to bring about a substantial change in the prevalent thinking. Although the molecular mechanisms involved in signaling by glucose, insulin, and leptin in the brain are partially revealed, it is not yet clear in what way the signaling is impaired in diabetes and why.

At the level of clinical practice, brain's role is practically denied, and in drug discovery, there are only feeble attempts to identify and use central drug targets for possible next-generation antidiabetic drugs. Any such attempts are faced with many problems so that a central action antidiabetic drug is unlikely to be in the market in the near future unless there is a radical change in the line of thinking [2]. There are some signs of change on the frontiers of research, but it has not gathered sufficient gravity, which is presumably due to the lack of appropriate paradigm rather than any methodological or implementational problems. The traditionally known circuit of glucose regulation on which the current clinical practice is based consists of the interplay between glucose, insulin, and glucagon. Raised plasma glucose induces insulin release by the pancreatic β cells, which facilitates glucose uptake by muscle and many other tissues. Insulin also arrests hepatic glucose production. This brings back the raised level of glucose. If glucose level goes below normal, glucagon release is stimulated which increases liver glucose production and thereby restores blood glucose levels. This sounds like a simple, nice, and efficient way of regulating glucose levels in blood.

But the reality is much more complex. The glucose–insulin–glucagon-dominated paradigm excludes or neglects a number of quite well-demonstrated phenomena. These include the role of parasympathetic inputs in insulin secretion, the cephalic phase of insulin response, direct vagus control of liver glucose production, the hepatic portal glucose sensors, the role of glucose-stimulated and glucose-sensitive neurons in the brain, and so on. By the orthodox view, the neuronal mechanisms are important only in the counterregulatory response to hypoglycemia and not otherwise [3, 4]. This is demonstrably not true. It has been shown, for example, that even for the "normal" regulatory function of insulin on liver glucose production, neuronal signaling is essential [5–7]. The orthodox view also fails to explain a number of "inconvenient" findings that we discussed in Chap. 3 which I will briefly reiterate now because any new theory needs to explain most or all the odd facts. The questions that the new theory needs to answer are as follows: (1) Why in an HIIR state suppressing insulin production does not lead to rise in plasma glucose, instead the insulin sensitivity increases? (2) Why in MIRKO mice high muscle insulin resistance and lack of compensatory hyperinsulinemia fail to increase blood sugar? (3) Why in LIRKO mice in spite of liver insulin resistance the fasting blood glucose spontaneously returns to normal but in diabetes it does not? (4) Why perfusion of glucose directly into the brain results in peripheral hypoglycemia? or (5) Why in some patients plasma glucose fails to respond to insulin, and further why infusion of insulin paradoxically increases blood sugar levels in some patients (Somogyi paradox)? (see Chap. 3 for details of all these paradoxes). We need a model of glucose homeostasis that accommodates all the known sensors of glucose levels in the body and the brain and all known effectors of glucose dynamics in such a way that the so far unexplained patterns in glucose homeostasis get adequate explanation. Of particular importance are the changes in the glucose and insulin curves during GTT. There are four different features of altered GTT curve in early diabetes: (1) delayed insulin response, (2) increased height of the glucose peak from the fasting glucose level, (3) increased fasting glucose level, and (4) a long right-hand tail, that is, a delay in returning the glucose level back to fasting level. Any theory of diabetes should explain all the four together. It should also explain differential occurrence of some of them, that is, IGT in the absence of IFG and vice versa. Any new theory needs to show how it explains these features of the altered curve better than the classical theory. The new model also needs to either find support in the existing literature or make experimentally testable predictions. Let us try our luck to find such a theory of diabetic hyperglycemia now.

The mechanisms involved in glucose sensing can be grossly classified in two categories, namely, (1) peripheral, consisting of direct effects of glucose on the α and β cells of the islet, and (2) central or neuronal, consisting of glucose transport to the brain, glucose utilization by the brain, glucose sensing neurons, information processing, neuronal signals to pancreas to regulate insulin and glucagon production, and direct neuronal signals to liver to regulate glucose production. If two parallel circuits of homeostasis exist in the body, we can theoretically construct four possible models of glucose homeostasis: (1) one in which only the peripheral mechanisms work, which is the traditional model, (2) one in which only the central mechanisms work, (3) one in which both work together all the time, and (4) one in which the two operate conditionally, that is, under one set of conditions only the peripheral control works and in other set of conditions the central mechanisms take over the control. A detailed formal mathematical model considering the possible outcomes of all the four combinations is described in Appendix IV. Here it is represented by a schematic diagram (Fig. 12.1), and we will discuss in this chapter the intuitive outcomes of the model alone without worrying much about the mathematics.

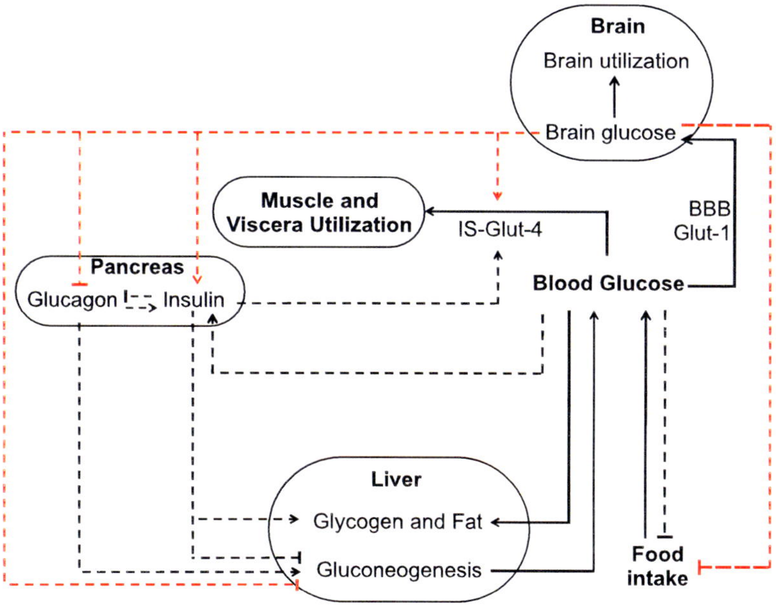

Fig. 12.1 A schematic diagram depicting the peripheral and central regulation of glucose homeostasis (*IS* insulin sensitivity, *BBB* blood–brain barrier). *Solid black* lines depict glucose flow, *black dotted* lines peripheral regulation signals, *red dotted* lines central regulation signals

It is not difficult to understand that a peripheral mechanism alone would be more efficient in maintaining plasma glucose level and bringing it back after any perturbation. This is because peripheral control has less number of steps in the feedback loop and therefore minimum delay between sensing a change and mobilizing effectors in response to the change. A mechanism that involves central sensing of brain glucose levels has more number of steps that are likely to cause a delay in response. Glucose in plasma is transported to the brain through specific transporters in the blood–brain barrier called glut-1 [8]. The brain level of glucose or the rate of energy production in the brain is sensed by the glucose sensing neurons; they send signals to the arcuate nucleus of the hypothalamus, where the information is processed and neuronal signals are given to the pancreas to secrete insulin or glucagon and also directly to the liver to regulate glucose production [9]. The role of neuronal signals in the function of endocrine pancreas is long known, but its importance is being realized only recently [10–13]. The neuronal steps in this process are rapid and therefore will not cause any significant delay in the circuit. However the single process that can cause substantial delay is the transport or diffusion of glucose from plasma to the brain. So if brain glucose is being sensed instead of plasma glucose, it will result in some delays in regulation. It is well known in cybernetics that a delayed feedback causes oscillations. It is not surprising therefore that peripheral control is more efficient than central control in minimizing deviation in plasma glucose under any positive or negative peripheral perturbation such as food intake or muscle exercise. We can perceive three types of perturbations in steady-state plasma glucose which are commonly experienced in everyday life. One is energy intake which transiently increases plasma glucose, the other is muscle exercise that causes rapid uptake of plasma glucose by muscle, and the third is increased brain demand that can be triggered by intensive mental activity or mental "stress." The stress can be acute or chronic. In all these types of perturbations, peripheral control is more efficient in maintaining stable plasma glucose and returning to it quickly after the perturbation.

Why does the central control exist then? The prevalent thinking is that the central control is an emergency measure that would prevent brain damage if glucose level in the brain suddenly reduces. This is called the counterregulatory response [14, 15]. The current perception is that the counterresponse operates in case of severe hypoglycemia, that is, if the brain glucose levels dangerously reduce as a result of reduction in plasma glucose levels. This, however, does not make much sense from the system design point of view. If a drop in brain glucose levels is caused by a drop in plasma glucose levels since that is the only source of glucose for the brain, sensing a drop in plasma glucose level would work more efficiently than waiting for the brain level to drop and then act. For a sensitive and important organ like the brain, if it was to be protected against hypoglycemia, sensing plasma levels would have been the best strategy. Hence peripheral glucose sensing would have been more efficient even for a counterregulatory response to hypoglycemia. But surprisingly no neuronal mechanism of sensing plasma glucose appears to have evolved. Neuronal sensing of glucose happens only in the brain. Therefore either the system is badly designed or it is evolved for a purpose other than protecting the brain from hypoglycemia.

There is an alternative scenario when sensing brain glucose would be more efficient than sensing plasma glucose and that is if the demand for glucose in brain tissue goes up acutely or chronically. The steady-state concentration of glucose in the brain, in the absence of perturbations, is decided by the rate of glucose diffusion from blood and the rate of consumption by the brain tissue. The brain consumes disproportionately high quantities of glucose. Although the total weight of the brain is only about 2–3% of the body, it consumes about 20% of total body glucose at rest [16]. If glucose consumption by brain tissue increases, the steady-state glucose level in the brain will fall. The only means to correct it is by increasing transport from plasma since the only source of glucose to the brain is plasma. If brain glucose levels are depleted by explosive brain activity, the primary change is in the brain and not in the plasma, and therefore, sensing brain sugar would be specifically needed and sensing plasma sugar would be useless. Therefore, from a system design point of view, we can argue that the central mechanisms of glucose regulation must have evolved not to protect brain from a drop in plasma glucose but to ensure adequate glucose supply under conditions of increased brain demand. In other words, if the cause of brain glycopenia is hypoglycemia, then central mechanisms would be highly inefficient in dealing with the situation. However, if the cause of neuroglycopenia lies within the brain, central mechanisms would be highly efficient in bringing glucose supply of the brain back to normal. Therefore it would be logical to assume that the central mechanism evolved primarily to handle neuroglycopenia caused by explosive brain activity. But the same mechanism may also work to protect the brain from hypoglycemia, although it will not be too efficient in this task.

When do we expect a higher glucose demand in the brain? This can happen when facing acute or chronic environmental or social challenges that need computation of a complex solution. Such a rise in brain glucose demand has been demonstrated [17, 18]. Here we return to our soldier–diplomat dichotomy. Based on the soldier–diplomat hypothesis, I am going to make a rather bold speculation here. On adopting a diplomat lifestyle over soldier, not only the brain glucose demand increases, the brain glucose levels become more important than plasma glucose levels, and the primary focus of glucose homeostasis shifts from maintaining plasma glucose levels to maintaining brain glucose levels. The idea that the true target for glucose homeostasis is not plasma glucose but brain glucose is not new [19]. Over the last decade, a group of researchers have been promoting the concept of "selfish brain" that exerts an active energy pull. According to this theory, glucose homeostasis, and particularly glucose supply to the brain, is not governed by a push from glucose availability but rather an active pull from the brain [20–26]. The proponents of the selfish brain theory postulate further that a healthy state is marked by a "competent brain pull," whereas an "incompetent brain pull" leads to obesity and diabetes.

While I am happy with the idea of a brain pull and in some sense the central regulation used in the analysis below can be taken to be equivalent to a brain pull, I differ from these ideas in proposing that there is flexibility in the homeostasis priority. In soldier life, muscle glucose supply for explosive muscle activity is a priority, whereas in diplomat life, long-term sustained supply to the brain is the priority. This shift in priority is expected to be adaptive because the nature of challenge for glucose homeostasis in the two lifestyles is likely to be different. In soldier life, explosive muscle respiration is the most frequently expected perturbation. Brain activity is certainly involved in complex nerve muscle coordination during a physical encounter, but this is typically of a short duration. On the other hand, muscle building and maintenance are needed over sustained time scale. In contrast, in diplomat life, intensive brain activity may be more frequent and prolonged. Simulations show that when intensive muscle glucose utilization perturbs glucose homeostasis, peripheral control is highly efficient in restoring normality of plasma glucose. Central control is less efficient. On the contrary, if the primary challenge consists of increased brain glucose utilization, then central control is more efficient in restoring brain glucose homeostasis at the cost of raised plasma glucose (Fig. 12.2).

The suggestion therefore is that in a diplomat lifestyle the central mechanisms of glucose homeostasis should be more active than the peripheral ones. This central shift is of quantitative nature on a continuous scale which can be mediated by a change in the sensitivity of pancreas and liver to peripheral versus central signals. In soldier life, islets should be more responsive to peripheral glucose stimulation than to autonomic stimulation, and in diplomat life, autonomic activation should be more important for insulin secretion than stimulation by peripheral glucose. Similarly liver glucose production should be more under feedback control directly by glucose and insulin in soldier life and more under neuronal control in diplomat life. This idea is certainly testable, and there is already some evidence in this direction. Pancreatic islets are

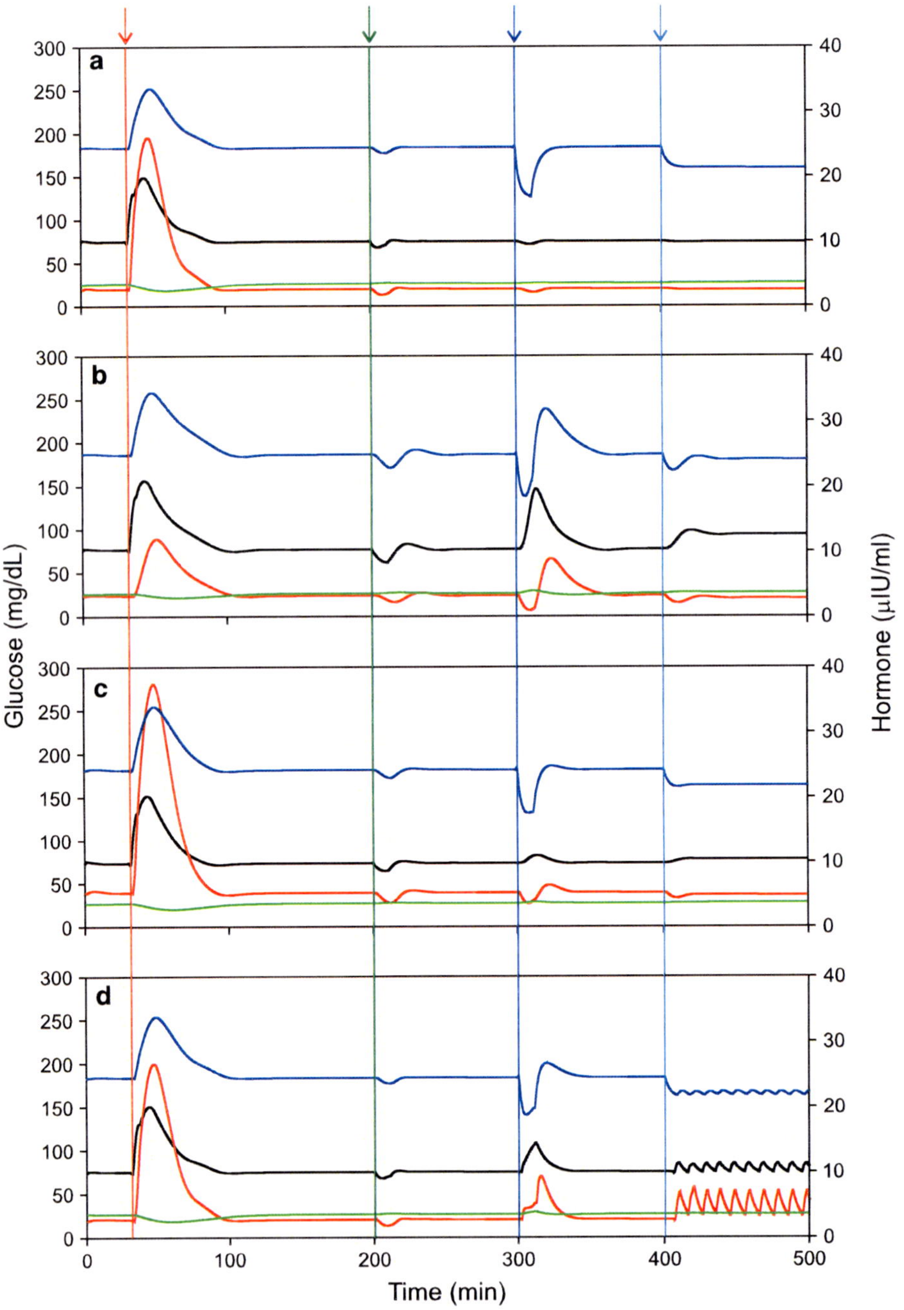

Fig. 12.2 Four possible ways in which peripheral and central regulation of glucose may interact and the relative performance of the four systems against four different perturbations: (**a**) peripheral control alone, (**b**) central control alone, (**c**) joint central and peripheral control with $x=0.5$ (see Appendix IV), and (**d**) switchover control where central mechanisms take over if brain glucose levels reduce below a threshold (22 mg/dL in these simulations). Four types of perturbations were introduced at four time points shown by *vertical lines* from *left* to *right* respectively: food intake, acute muscle exercise, acute brain exercise, and sustained brain exercise. It can be seen that peripheral control is most efficient in maintaining stable plasma glucose with all possible types of perturbations, but it allows the brain glucose to reduce to much lower levels during sustained brain hyperactivity. The central control is not very efficient in maintaining plasma glucose but maintains stable brain glucose particularly during sustained brain hyperactivity. The switchover control is good at handling acute brain exercise but may lead to rapid oscillations under sustained increase in brain activity. The oscillations are caused by alternately switching to central control when brain glucose goes below the threshold and switching back to peripheral when it returns to normal. *Black line*: plasma glucose, *red line*: plasma insulin, *green line* plasma glucagon and *blue line* (secondary axis) brain glucose. The same color code is used in figures 12.8, 12.9, 12.11 and 12.13

richly innervated by parasympathetic and sympathetic neurons [27]. These play some role in the regulation of insulin secretion under normal healthy state and under all conditions including meal-stimulated insulin release [28] or exercise-induced insulin suppression [29]. The autonomic system is involved in determining the glucose sensitivity of β cells [30]. Vagotomy can partially abolish the altered responsiveness of β cells exposed to chronic hyperglycemia [30]. Our hypothesis expects that in the obese, insulin-resistant, prediabetic, or early diabetic state insulin secretion should become less responsive to peripheral glucose and more responsive to central stimuli. Such a shift in response is demonstrated by the comparative response to glucose versus that to the cholinergic agonist carbachol. In fat-fed C57BL/6J strain of mice, after 12 weeks of fat feeding the glucose-induced insulin secretion was markedly impaired but carbachol-stimulated insulin secretion was potentiated [31]. This indicates a possible shift in the responsiveness of β cells from peripheral to central stimuli and more generally a change in emphasis of the system from peripheral to central mechanisms. In a human study, cholinergic enhancement of insulin secretion was stronger in obese than in non-obese subjects [32], indicating that obesity is associated with increased sensitivity of β cells to autonomic signals. In diabetics without autonomic neuropathy, the basal and meal-stimulated levels of pancreatic polypeptide are higher than normal [33, 34]. The pancreatic polypeptide is a marker of autonomic activity in the pancreas, and therefore, higher levels of this peptide in diabetes suggest greater role of autonomic inputs in pancreatic functions. It has been suggested that stress-related suppression of insulin is actively mediated by the brain [22], and this is likely to happen through autonomic control. In diabetic rats, β cells showed enhanced sensitivity to sympathetic signals [35]. The liver cell membranes of ob/ob mice have threefold higher β adrenergic receptor binding sites and show threefold increase in response to catecholamines than controls [36]. This again indicates that liver and pancreatic β cells might be becoming more sensitive to central regulation, along with losing sensitivity to peripheral insulin and glucose signals, respectively. The demonstration that the increased response of liver cells to central stimuli is mediated by corticosteroids [36] raises the possibility that HPA might be an important mediator of the central shift. Since HPA is a marker of diplomat life (Chap. 11), our notion of a central shift in diplomat life is supported.

PPAR gamma coactivator-1(PGC-1) is an important modulator of gluconeogenesis in the liver [37, 38], and PGC-1 expression is shown to be under sympathetic influence [39, 40]. Interestingly PGC-1 expression is downregulated in muscle and adipose tissue where its downregulated state is correlated to insulin resistance [41–44] and it is upregulated in liver where its upregulation correlates with liver insulin resistance [37, 38]. This paradox (one of the several dozen associated with insulin resistance) can be resolved by the differential action of the sympathetic nervous system. In a nonaggressive diplomat lifestyle, there is lowered sympathetic input to muscle and adipose tissue. According to the above argument, it should be increased in the liver. In diplomat life, visceral fat lipolysis and muscle activity are downregulated. At the same time, sympathetically stimulated liver glucose production is upregulated. Both seem to be achieved through mediation by PGC-1 but in opposite directions. This is a nice demonstration of how the sympathetic system through the agency of PGC-1 mediates fine-level changes in muscle, fat, and liver. All the changes are in different directions but bring about a consorted and organized action. All this is more than preliminary evidence in support of my speculation that diplomat life is characterized by increased importance of autonomically mediated central regulation of glucose homeostasis and decreased importance of peripheral regulation.

It is necessary to clarify here the relationship of this central shift with counterregulatory response. If we visualize a zero to one scale, zero represents complete peripheral control as visualized by the current working clinical model. The argument above states that in a soldier life, the system may be say at a score of 0.2 or 0.3, and on adopting a diplomat life, it may gradually shift to say 0.6 or 0.7. The all-out counterregulatory response stands at 1. This zero to one conceptual scale is incorporated explicitly in the model in the parameter x (Appendix IV). Thus, even after a soldier to diplomat transition, a counterregulatory response is possible, but the more the shift, the less will be the difference between the basal standing state and the all-out counterregulatory response. This will reflect into an apparently blunted counterregulatory response in diabetes.

There are two alternative ways in which the central mechanisms influence plasma glucose. One way is by modulating pancreatic hormones insulin and glucagon and the other by direct neuronal modulation of liver glucose production. If glucose utilization by the brain is increased owing to increased cognitive demands, higher level of plasma glucose is needed. This can be achieved by suppressing insulin and enhancing glucagon production. But suppressing insulin may not be desired since insulin itself is needed for cognitive functions. Therefore, the other pathway, that of direct neuronal regulation of liver glucose production, could have evolved. With two alternative pathways available within the central regulation mechanisms, it becomes possible to fine-tune glucose and insulin levels independently. If the involvement of insulin was mandatory in the regulation of glucose, suppression of insulin would have been crucial to increase brain supply of glucose. However, since the cognitively active brain needs both glucose and insulin simultaneously, stimulation of insulin production along with suppression of insulin action on the liver and direct regulation of glucose production by autonomic control would be a better strategy. This will result in a high fasting HOMA-IR index, although there is nothing wrong with the liver. On the other hand, whenever cognitive demand for insulin is not very high but liver glucose production is needed (such as during starvation [45]), suppression of insulin and stimulation of glucagon production might be a more desirable pathway. The presence of these two alternative central pathways to regulate glucose production brings this flexibility, and the brain must be smart enough to make contextually differential use of them.

For a cognitively intensive diplomat life, direct neuronal regulation of liver glucose production would be more important than regulating pancreatic hormones. There is increasing evidence that hypothalamic neuronal signals through the vagus nerve play an important role in regulating liver glucose production [46–49]. Moreover, it is also possible that the "direct" action of insulin on liver glucose production is not really "direct" but depends upon its interaction with neuronal inputs [50]. This hypothesis expects that a direct neuronal control over liver glucose production or contribution of neuronal mechanisms in its regulation will be more important in a diplomat than in a soldier personality. Since the neuronal regulation of liver glucose production is under sympathetic control, there would be increased sympathetic activity in a diplomat's glucose homeostasis. Glucose levels in the brain appear to be finely monitored and modulated depending upon the activity and respiration rate of brain tissue [51–54]. In effect, in a diplomat lifestyle, there could be activation of mechanisms that modulate the plasma glucose in such a way to maintain steady and adequate brain glucose level.

It is possible that the central effects on pancreas and liver take two different time courses. One consists of active and ongoing neuronal modulation of insulin production and insulin action on the liver. The other could be long-term programming of liver and β cells. The latter has a possible adaptive value. For a chronically altered behavior, a chronic change in pancreatic and hepatic response is likely to be more economical than ongoing active modulation by neurons. Further, if the peripheral system is kept active, there can be conflict of interests. For example, if the brain needs chronically elevated plasma sugar, it is necessary to reduce β cell responsiveness to glucose. Otherwise, the central and peripheral systems will work against each other and waste much energy in the conflict. Similarly, along with neuronal regulation of liver glucose production, it is necessary to reduce insulin sensitivity of the liver; otherwise, the two systems will be in conflict. The suggestion logically emerging from this is that the loss of β cell responsiveness and liver insulin resistance could be strategies to reduce the conflict between the two homeostatic systems. It is also likely that β cells are programmed neuronally [30] on a prolonged exposure to hyperglycemia to reduce insulin production since chronic hyperglycemia could be a marker of altered homeostatic strategies. In effect, the new hypothesis states that the diplomat behavioral strategy is supported by a shift in the glucose homeostatic mechanisms from peripheral to the central ones which is actively mediated by the autonomic nervous system.

One can clearly appreciate at this level that a central shift in homeostatic mechanisms combined with increased brain glucose demand is sufficient to give rise to hyperglycemia along with reduced sensitivity of β cells to glucose and reduced sensitivity of liver to insulin. This can happen without β cell degeneration. This is a strong candidate competing with IR-RII for explaining hyperglycemia in diabetes. But this is not the complete story. There are more players in the act of central regulation.

Vagus control of hepatic function not only regulates glucose production by liver but also fine-tunes muscle insulin sensitivity. Interruption of the hepatic parasympathetic nerves by surgical denervation, atropine, or blockade of hepatic NO synthase induces peripheral insulin resistance [55]. In a response to the synergistic action of insulin, parasympathetic signal, and hepatic NO production, a hormone called hepatic insulin sensitizing substance (HISS) is produced which increases postmeal insulin sensitivity of specifically muscle tissue [55, 56]. This means that the central system is also capable of manipulating muscle insulin resistance. Muscle insulin action has HISS-dependent and HISS-independent components, and the HISS-dependent component is responsible for about 56% of muscle glucose uptake. The proportion of insulin action under central control is substantially large to affect peripheral glucose dynamics.

Another important player influencing central regulation is the glucose transporter in brain capillaries called glut-1. An alternative possible way of increasing the glucose supply of the brain would be to increase the amounts of glut-1 across the blood–brain barrier (BBB). Glut-1 action is insulin independent unlike the muscle glucose transporter glut-4. We have seen earlier that muscle insulin resistance is a mechanism by which investment in muscle is decreased by reducing glut-4 activity. By the same logic, if the investment in brain is to be increased, increasing glut-1 activity would be appropriate. Since glut-1 is insulin independent, the only way this can be done is by increasing the levels of glut-1 itself.

However, there is a serious problem here. The plasma glucose levels are subject to wide fluctuations normally caused by food assimilation after meals, muscle uptake following exercise, and other physiological perturbations. The brain glucose levels are relatively constant, and glut-1 has an important role in maintaining steady glucose levels in the brain. It can be shown mathematically that the lower the levels of glut-1, the steadier will be the brain glucose levels. Simulations (Fig. 12.3) showing the effect of glut-1 levels on the ratio of the amplitude of fluctuations in plasma and brain illustrate this. The principle can be intuitively appreciated through a cartoon (Fig. 12.4) which helps us imagine a high-amplitude wave being forcibly pushed through a narrow slit so that the wave coming out on the other end has compressed amplitude. If a stable glucose supply to the brain is desired in diplomat life, then to protect the brain from fluctuations in plasma glucose, a decrease rather than an increase in glut-1 will help. However, a decreased glut-1 will also result in a decreased mean brain glucose level. The solution to this problem is to bring about a compensatory increase in the plasma glucose. This is possible only if the central regulation is in operation. The model shows that rather than increasing glut-1 levels, increasing mean plasma glucose and slightly reducing glut-1 ensure a more steady supply of glucose to the brain (Fig. 12.3). A moderate reduction in glut-1 can therefore be adaptive for a diplomat lifestyle in coordination with central regulation mechanisms. However, a supernormal reduction in glut-1 can be detrimental as shown by these simulations.

Now, along with central shift of the homeostatic mechanisms and increased brain glucose demand, glut-1 downregulation is an additional factor that can contribute to raising plasma glucose. This logic raises the possibility that it is the brain

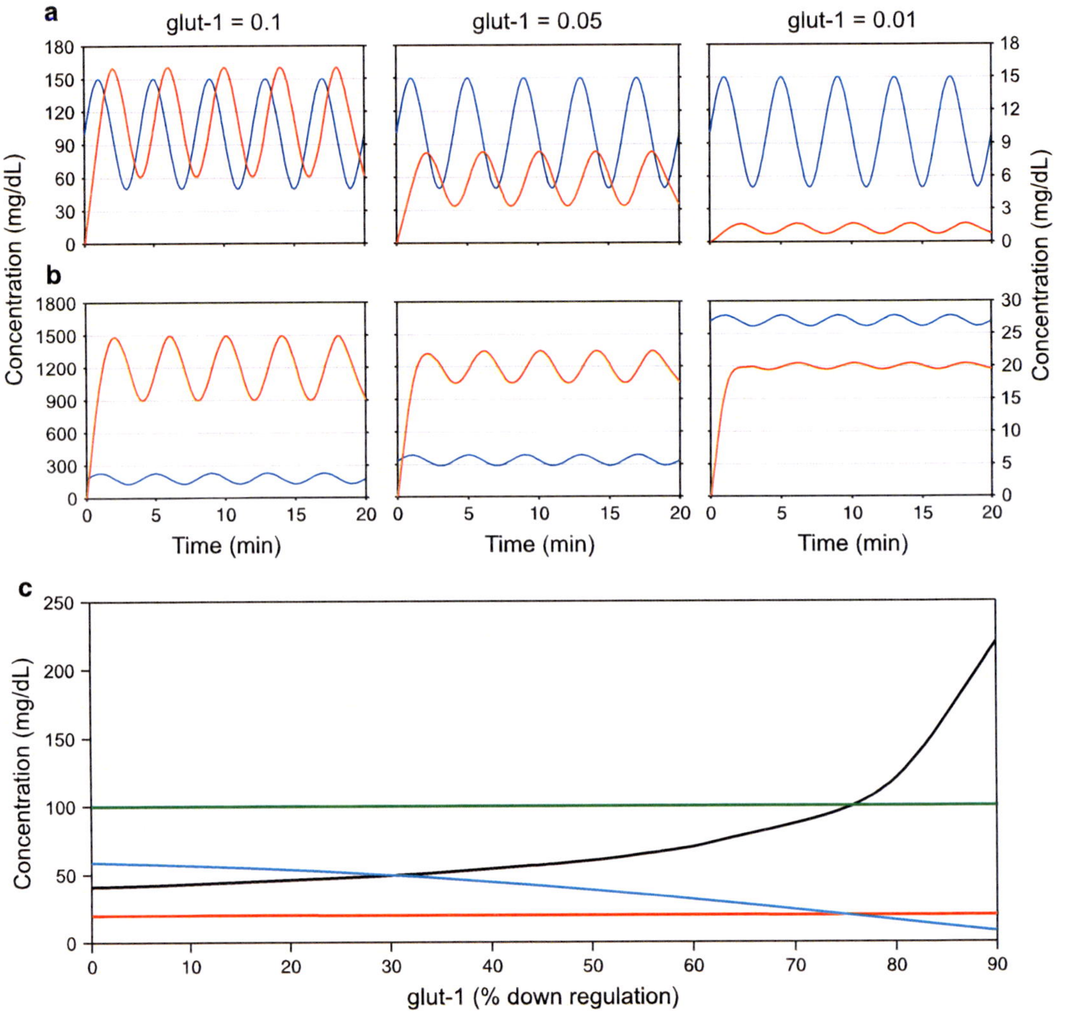

Fig. 12.3 Effect of the width of glut-1 window on stability of blood glucose levels: Here, in a model system, plasma glucose (*blue line*, primary *Y* axis) is given an oscillation of constant amplitude in a sine curve. The resulting oscillations in brain glucose are monitored (*red line* secondary *Y* axis). (**a**) As glut 1 decreases from 0.1 to 0.01 (arbitrary units), the amplitude of oscillation of brain sugar decreases but so does the mean. (**b**) If the mean brain glucose is to be kept constant and the amplitude is to be decreased, decreasing glut-1 and increasing mean plasma glucose are the only possible solution. (**c**) The behavior of mean and amplitude of plasma and brain glucose with different levels of glut-1. In these simulations, the amplitude of plasma sugar oscillations is kept constant (*green line*). The mean plasma glucose (*black line*) is adjusted to keep a constant mean brain glucose (*red line*). With greater suppression of glut-1, the amplitude of fluctuation of brain glucose decreases (*blue line*), but the mean plasma glucose needed to achieve this increases disproportionately. This is what is expected if brain glucose stability is the main homeostatic target, and there is reduction in glut-1 levels

glucose dynamics that primarily underlies diabetic hyperglycemia and IR-RII is incidental. This argument is elaborated more with some mathematics in Appendix IV, and readers that are not averse to mathematics and expect more rigorous argument may visit Appendix IV. For mathematics-averse readers, the set of arguments above should be sufficient to convince that at least a logical possibility of alternative reason for diabetic hyperglycemia exists and needs serious investigation.

The levels of glut-1 in brain capillaries indeed go down in diabetes, but the data are fragmentary.

Fig. 12.4 A small glut-1 window can effectively dampen fluctuations in glucose concentration

There are no studies on glut-1 levels across the brain capillaries in human diabetes to the best of my knowledge. In rat models of diabetes, glut-1 levels have been repeatedly shown to reduce, but the experiments lack consistency [57] presumably due to methodological problems and therefore are somewhat difficult to interpret. There can be several reasons for the inconsistency across experiments. The nonlinearity of glucose transport, difference between different models of diabetes, methods of estimating glut-1 levels or glucose transport rates, failure to make observations at steady-state concentrations, need for sufficiently long period for detection of a chronic response, regional differences in glucose dynamics [58, 59], and a number of other factors need to be taken care of in such experiments. In spite of these problems, a general pattern emerging from many experiments is that glut-1 levels are downregulated in response to hyperglycemia and upregulated in response to hypoglycemia [8, 60–68]. As perceived currently, glut-1 reduction is a consequence rather than the cause of hyperglycemia. Another general and robust trend in all the studies appears to be that the alterations in glut-1 levels are slow. Chronic hyperglycemia, at least over several weeks, is needed for detectable downregulation of glut-1. Upregulation by hypoglycemia could be a little faster but still requires several days to show detectable effects.

Several questions are raised by the demonstration of altered glut-1 levels in diabetic rat models, and I consider two of them to be of prime importance. The first is what causes reduction in glut-1 and why. The second is what the implications of reduced glut-1 are for glucose homeostasis. We will look at the latter first since that emphasizes the importance of altered glut-1 levels. Once we know how important it is, the pursuit of the first question will get more weighting. When we start examining the possible consequences of reduced glut-1, we find explanations for a number of observed patterns in diabetes. This leads to a strong alternative reasoning for diabetic hyperglycemia along with a number of characteristic changes associated with it.

1. Raised fasting plasma glucose: The lower the glut-1 expression, the slower will be the rate of glucose transport across the blood–brain barrier (BBB). This would lead to a lower steady-state brain glucose level if the rate of uptake by brain tissue is assumed constant. However, if a desired steady-state level has to be maintained in spite of reduced glut-1, the only way is to increase plasma sugar. This can happen only when the central regulation is sufficiently active, that is, x of the model is sufficiently large. The lower the levels of glut-1, the greater would be the plasma glucose needed to maintain a given level in the brain. Therefore, reduced glut-1 can be a cause of hyperglycemia. If this level falls, brain glucose level will fall, and the autonomic control will increase blood sugar by activating liver glucose production. For this to happen, any defect in β cell is unnecessary. The lower level of brain glucose as sensed by glucose sensing neurons will suppress insulin production in spite of β cells being normal and/or increase liver glucose production in spite of liver cells being normal.

2. Failure of adjustment of insulin sensitivity to decreased insulin production: We have seen in Chap. 3 that in an HIIR state, if insulin production is artificially suppressed, instead of glucose level rising, insulin sensitivity increases such that plasma glucose remains normal. This well-demonstrated phenomenon is not taken into account by the IR-RII paradigm. Experiments demonstrate the existence of a mechanism which brings about compensatory increase in insulin sensitivity when insulin production is suppressed. If a compensatory rise in insulin sensitivity is possible in experimental suppression of insulin production, what prevents it from happening in diabetic hyperglycemia? Lowered level of glut-1 is a promising answer to this question. According to this hypothesis, the transition from insulin-resistant to hyperglycemic state is not caused by β cell dysfunction but by altered brain glucose dynamics. Under central shift and reduced glut-1 state, it is in the interest of the brain to raise plasma glucose, and therefore, central signals reduce insulin secretion and/or enhance liver glucose production so that a higher plasma glucose level is maintained. Here, the primary cause is not insulin resistance, and therefore, even if insulin sensitivity is restored following insulin suppression, it cannot prevent hyperglycemia.
3. Beta cell dysfunction? In this model, the central system decides what the plasma glucose level should be to attain the desired steady-state brain glucose. Insulin and glucagon secretions and sympathetic stimulation of liver for glucose production are neuronally regulated to achieve these levels. When neuronal control takes on a predominant role, insulin secretion is no more directly responsive to glucose, independent of neuronal inputs. If brain needs increased supply of glucose, it will suppress insulin production and/or enhance glucagon secretion and liver glucose production. As a result, β cells will produce less amount of insulin than what is expected from them at the standing glucose level. This could give an impression of β cell dysfunction when nothing is actually wrong with them. Researchers unaware of neuronal control or choosing to ignore it are bound to think that as β cells are not responding to glucose as expected, there must be some kind of functional β cell defect. Further, if insulin production is chronically suppressed by the neuronal system, disuse atrophy may slowly reduce the β cell mass. Chronic suppression of insulin secretion should be accompanied by chronic suppression of amylin secretion too since both are cosecreted. But amylin has much longer half-life than insulin, and amylin disposal is presumably more difficult than insulin. As a result, excess amylin that is not secreted in time could slowly accumulate and form amyloid deposits in the islet tissue. The amyloid deposits which are long known to be associated with islet degeneration [69–72] could be a consequence rather than a cause of reduction in β cell secretary activity. Alternatively, once amyloid deposition begins, islet degeneration may begin.
4. Diminished AIR and delayed insulin response in GTT: After a meal or oral glucose intake, there is a rapid and short-lived burst of insulin production which is called acute-phase insulin response (AIR). In diabetes AIR is suppressed and even undetectable. This is often suspected to be the prime reason of impaired GTT. We have seen before (see Chap. 3) that neuronal inputs have a major role in AIR [73–77]. Even in ivGTT, there is a neuronal component in AIR, although it may be less important than in oral GTT [78, 79]. For ivGTT, neuronal inputs presumably come when the glucose-responsive neurons in the brain respond to the rising concentration of glucose in the brain. With reduced glut-1 and thereby reduced rate of diffusion to the brain, the glucose-responsive neurons may respond with a delay, or if they require some threshold rate of rise in glucose concentration for AIR which is not achieved with very low levels of glut-1, AIR may be completely absent. Currently, the blame for the absence of AIR is placed on β cell "defect," but the nature of the defect or

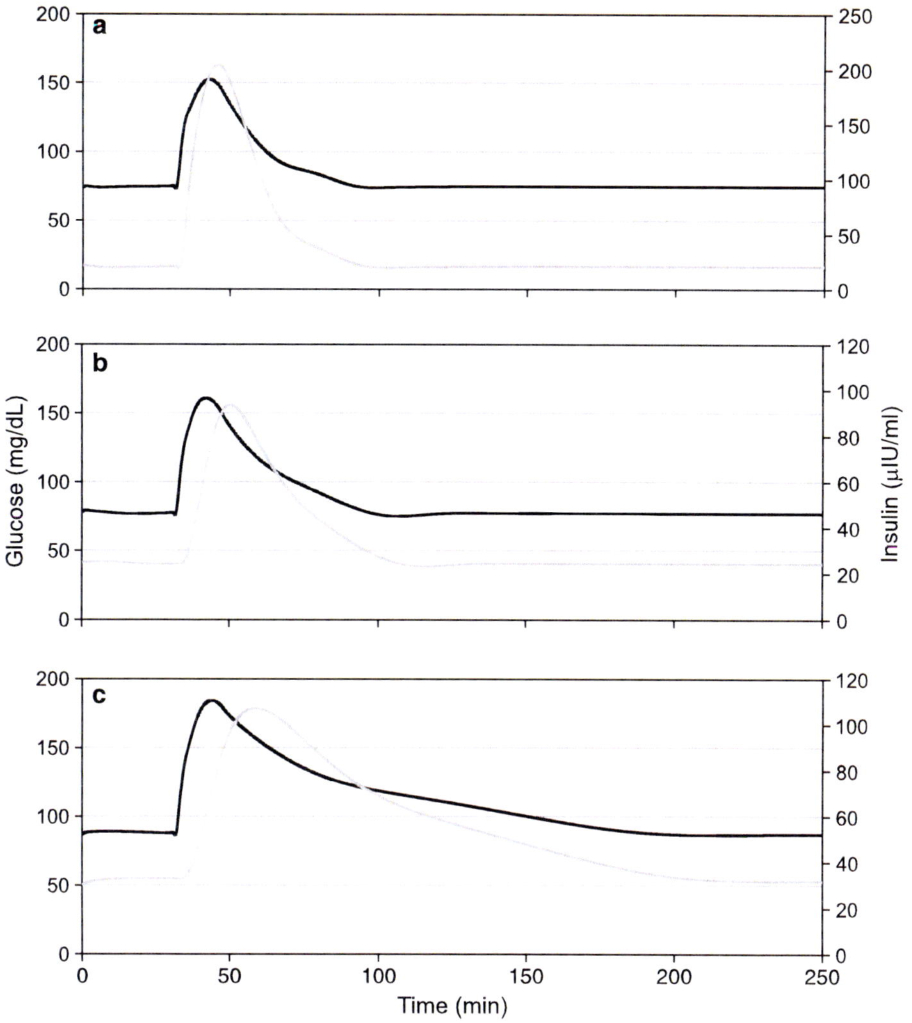

Fig. 12.5 Shift to central regulation and glut-1 reduction causes delay in insulin response to glucose. (**a**) OGTT with peripheral control, (**b**) OGTT with central control, and (**c**) OGTT with central control and reduced glut-1. Note the progressively rightward shift in the insulin peak. Only models involving central control lead to delayed insulin response. The peripheral control model does not give this feature of altered OGTT under any parameter values

mechanism by which β cells respond with a delay is not known. Glut-1 reduction offers an alternative hypothesis for this phenomenon which appears more logical, and the possibility needs to be tested experimentally.

In IGT and early diabetics, the total amount of insulin produced by the pancreas is not reduced; in fact, it may have increased substantially. But there is a delay in the beginning of the insulin response. This is likely to be an effect of central shift and glut-1 reduction which delays the rise in the levels of brain glucose (Fig. 12.5). Assuming that neuronal inputs become more important for insulin release in a diplomat state, glut-1 downregulation rather than β cell defect is a better model for explaining the delayed insulin curve. Since the total amount of insulin produced is not reduced, β cell capacity to produce insulin cannot be doubted. Again, what kind of "defect" makes the β cells work with a delay but without loss of capacity is not known. In in vitro experiments, altered responsivity of β cells to glucose has been shown, but I failed to find an experiment demonstrating the phenomenon of delayed response. This suggests that delayed response is not a property of β cells. Therefore we need to examine glut-1 reduction and central shift as an alternative hypothesis for the altered AIR and GTT.

5. Triggering counterregulatory response at higher plasma glucose: If there is glut-1 reduction, there would be lower levels of brain glucose per unit plasma glucose at steady state. Since the glucose sensing mechanisms in the brain sense the brain glucose levels and do not have a direct access to plasma levels, the counterregulatory response may be triggered at a higher plasma glucose level which is known to happen in diabetics [80–82].
6. Why the counterregulatory response in diabetics is delayed and less glucagon dependent: It is known that in long-term diabetics, the glucagon response to hypoglycemia is blunted [83]. Since the rate of diffusion from blood to brain is reduced because of reduced glut-1, a sudden drop in blood sugar will also be sensed more slowly by the brain and the counterregulatory response will be delayed. This is compatible with the observations [84]. It is possible that as a compensatory measure against the delay, the hormonal glucagon response which is independent of sympathetic activation [85] is bypassed and replaced by direct autonomic stimulation of liver glucose production [86] which is a faster response. The delay in sensing could thus be partially compensated by reduction in response time.
7. Shift in baseline autonomic tones: Both sympathetic and parasympathetic systems have different roles in glucose homeostasis. Sympathetic activation suppresses insulin and enhances liver glucose production. Parasympathetic on the other hand enhances insulin production and also, through the agency of HISS, facilitates muscle glucose uptake. Logically, if the brain is short of glucose, the sympathetic system should be activated, and if there is excess glucose, parasympathetically activated insulin production and diverting glucose to muscle would help in regulating brain sugar. If along with diabetic hyperglycemia brain sugar also had gone up, we would expect parasympathetic activation. On the contrary, sympathetic activity predominates, and parasympathetic is suppressed in diabetes. This pattern gets an explanation if glucose transport to the brain is assumed to be reduced by reduced glut-1, reduced capillary blood flow, or increased glucose consumption in the brain.
8. Rise in stress-related sugar: When the glucose demand of the brain goes up which may happen in emotional arousal, worries, or anxiety, the increased demand can be met by slightly increasing the plasma glucose levels. Figure 12.6 demonstrates the interesting dynamics of this under high and low levels of glut-1. The hypothetical diagram is not based on data since data on brain glut-1 levels in humans under different physiological states is conspicuously absent, but it illustrates conceptually what may be happening with glut-1 reduction. Glucose transporters operate by facilitated diffusion and therefore follow Michaelis–Menten kinetics. At high level of the transporter, an increased demand can be met with by a small rise in plasma glucose since in this region, the slope of the line is steep. A reduction in glut-1 changes the nature of the curve such that the slope of the line in this region is much smaller, and as a result, a large change in plasma glucose is needed to fulfill a small increase in brain demand. This is an alternative explanation for stress- or anxiety-induced rise in plasma sugar. Again, there is a need to test this model in comparison with the conventional model of cortisol-mediated rise in sugar. The two are not mutually exclusive and may work in harmony.
9. Why HOMA-based and other indices may not reflect insulin resistance in advanced T2D: There are a large number of indices for measuring peripheral insulin resistance. In normoglycemic conditions, most of them correlate quite well with the more elaborate "gold standard" measurement of insulin-mediated glucose uptake, namely, the euglycemic hyperinsulinemic clamp. The correlations become substantially weaker as

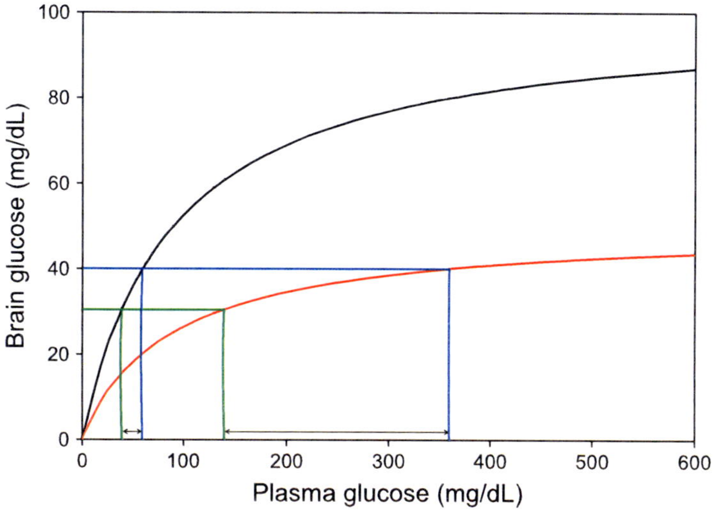

Fig. 12.6 Sugar response to mental stress at two different levels of glut-1: The dynamics of glut-1-mediated glucose transport follows saturation kinetics. If we assume that the normal glucose requirement of the brain (*green line*) is fulfilled by some steady-state plasma glucose levels, a slight increase in brain glucose utilization (*blue line*) necessitates a marginal increase in plasma glucose if glut-1 level is high. The same requirement puts greater demand on plasma glucose if glut-1 is reduced (*red curve*)

diabetes advances [87–91]. My interpretation of this phenomenon is that as diabetes advances brain glucose dynamics and central regulation become increasingly important in determining blood glucose. Direct neuronal action rather than glucose feedback insulin action decides fasting glucose levels in advanced stages. Since the causal role of insulin and insulin sensitivity in deciding plasma glucose is substantially undermined, the correlations become weak. Central shift and glut-1 reduction can increase HOMA-IR without increasing insulin resistance because of which HOMA-IR does not faithfully reflect insulin resistance.

10. Making exogenous insulin ineffective: The model shows that even if hyperglycemia is caused by glut-1 reduction, it may still respond to exogenous insulin as long as the shift in emphasis from peripheral to central regulation is moderate. It is likely that in the supernormal diplomat lifestyle of modern society, there is a supernormal shift towards central regulation. If this happens, clinically feasible doses of exogenous insulin will be unable to bring glucose to normal levels. This indeed happens in many advanced diabetics for whom normal glucose levels are not attained even after exceedingly high doses of insulin. Going by traditional thinking, these patients must have extremely high insulin resistance. But this is not supported by empirical findings. MIRKO mice have demonstrated that muscle insulin resistance does not raise fasting plasma glucose levels even if there is no compensatory rise in insulin [92]. In human studies, it is estimated that peripheral insulin resistance reaches a plateau at FPG levels of 160–180 mg/dL and does not contribute to further rise in glucose. Thus there is a limit to insulin resistance and its effects on the rise in blood glucose level. Any rise beyond this is mainly contributed by excessive liver glucose production rather than reduced glucose uptake [93]. But liver-specific insulin receptor knockouts have demonstrated that even after complete liver insulin resistance, fasting sugar increases moderately and returns to normal in a few months. Restoration of normoglycemia is because of central regulation compensating for liver insulin resistance independent of insulin action [94, 95]. This means that neither muscle nor liver insulin resistance can

account for extreme hyperglycemia not responding to insulin. If the picture is upside down, that is, increased glucose production is caused by central mechanisms, independent of insulin, exogenous insulin is unlikely to have much effect on it.

Further it is known for a long time that in a class of patients, large doses of insulin paradoxically increase hyperglycemia which is ameliorated only after reducing insulin dose (Somogyi phenomenon) [96]. This appears highly counterintuitive. It was initially suspected that large doses of insulin first lead to hypoglycemia, and then the counterregulatory response increases plasma glucose. However, continuous monitoring of glucose in such patients has failed to detect a hypoglycemic phase [97]. This paradox is best explained by the glut-1 model and by the glut-1 model alone. In patients with very low glut-1, the difference in plasma and brain glucose levels is large, and as a result, a high dose of insulin reduces brain glucose levels below the threshold even when the plasma levels are above normal. In effect, insulin stimulates the counterregulatory response even without hypoglycemia. Withdrawing insulin in such patients improves glycemic control.

11. Glucose infusion in the brain reduces plasma glucose: Perhaps the most direct evidence of how brain glucose dynamics affects plasma glucose levels comes from the experiments that infused glucose in the intracerebroventricular space [98]. This infusion of small quantity of glucose induced a rapid reduction in plasma glucose, indicating that plasma glucose is controlled by the brain glucose sensing and regulating mechanisms even under normoglycemic conditions. Another important inference of this experiment is that this mechanism works both ways. It is well accepted that lowering brain glucose triggers the counterregulatory sympathetic mechanisms. But the perception has mostly been that this is only an emergency-triggered mechanism which otherwise remains silent. If increasing brain sugar exerts a regulatory effect too, it means that the central mechanism is not only turned on to combat hypoglycemic emergency, it is all-time active although not in the all-out mode. Unfortunately, there is little research in this direction to confirm the reproducibility of these results.
12. Hyperphagia, type 1 thrift: glucose sensing neurons in the brain have an important role in regulation of food intake [99, 100]. In the hypothalamic arcuate nucleus, glucose has excitatory actions on anorexigenic POMC neurons, while the appetite-promoting NPY neurons may be directly inhibited by glucose. If the rate of diffusion from blood to brain is slowed down, the "stop feeding" response would be substantially delayed leading to overeating. This could be a mechanism of thrift, type 1 thrift in particular. Since type 1 thrift could be adaptive for socially subordinate individuals, we may expect that social subordination would reduce BBB glut-1. This has never been tested but is also not very difficult to test in appropriately designed experiments combining social behavior and brain chemistry.
13. The adaptive importance of ketone bodies: The glut-1 hypothesis also suggests a possible functional role for diabetic ketosis. Ketone bodies are made by the liver in advanced diabetes or during starvation. They are derived from fatty acids. A traditional explanation for ketosis is that when tissues are unable to utilize glucose as a fuel due to starvation, insulin deficiency, or insulin resistance, fatty acids are utilized and ketosis is either an inevitable product of fatty acid metabolism or is produced as alternative fuel for tissues. This cannot be true for many reasons. Most tissues can utilize fatty acids as fuels but do not generate ketone bodies. The enzymes required for ketogenesis are specifically located in liver cell mitochondria [101–103]. This appears to be a specific adaptation to provide an alternative fuel for the brain since the brain is unable to utilize fatty acids as fuel but can use keto acids efficiently. Although other tissues can also utilize keto acids, for the brain, in the absence of glucose, it is the only option available. However, there is a paradox here. If keto acids are crucial for the brain in the absence of glucose, why are they produced

when there is excess glucose in the blood? Brain is not dependent on insulin for glucose uptake and therefore should be able to take up glucose as a fuel even in an insulin-resistant state. In diabetes, ketone bodies will be needed by the brain only if it does not get sufficient glucose supply. This can happen either in extreme hypoglycemia or when glucose transport to the brain is impaired. If ketosis takes place when there is high or normal blood sugar, we can infer that glucose transport to the brain must be impaired. It may be a specific adaptive response when glut-1 levels are lowered since it results in reduced glucose transport to brain and fatty acids cannot be utilized by the brain tissue [104]. To the best of my knowledge, there is no other ultimate cause for a ketogenic response to evolve.

A possible additional factor that might drive ketoacidosis is as follows. At extremely low glut-1 levels, any other stress or disturbance that can cause sudden increase in brain glucose consumption or drop in availability triggers the sympathetic response. This increases liver glucose production to compensate for low glucose availability to the brain. However, the rate of gluconeogenesis has a physical limit. Also, since glut-1-mediated transport follows a saturation curve, much higher rise in blood glucose would be needed to bring about a small rise in transport to the brain (Fig. 12.6). Simultaneously ketogenesis is stimulated as an alternative rescue mechanism. Both the responses are mediated by the counterregulation machinery. Ketogenesis is able to give some rescue. If ketogenesis fails to start for any reason, another condition called hyperglycemic, hyperosmolar non-ketotic state (HHNKS) ensues. If this model is correct, we can expect that HHNKS should be more dangerous than ketosis. This indeed appears to be true as shown by a comparative study that had a 5% mortality rate in ketoacidosis and 15% for HHNKS [105]. Also, as required by the glut-1 hypothesis, both the conditions are characterized by intense expression of the components of counterregulatory response [106–109]. The traditional line of thinking does not adequately explain why the sympathetic counterregulatory response is triggered when blood sugar is high.

There is one more reason to believe that the ketosis and HHNKS states are driven by central and not peripheral mechanisms. The extremely high levels of plasma sugar reached in these states cannot be explained by insulin resistance and/or insulin deficiency. This is demonstrated by LIRKO mice which have extreme liver insulin resistance but only moderate hyperglycemia that ultimately returns to normal and also by liver parenchyma cells in culture. The difference between gluconeogenesis of liver cells in vitro in the presence (physiological concentrations) versus absence of insulin is of the order of 20% only [110]. This means that even extreme liver insulin resistance or insulin deficiency cannot give the exceedingly high levels of plasma sugar seen in these conditions. The mechanisms have to be other than IR-RII.

If ketogenesis is adaptive, why does insulin, which suppresses ketogenesis, help in such cases? The answer may lie not in ketogenesis but in the hemodynamic action of insulin. Insulin has both short-term and long-term effects enhancing blood flow to tissues which is reduced in insulin resistance [111–113]. Infusion of extra insulin is likely to be helpful in increasing cerebral blood flow and thereby glucose as well as keto acid transport to the brain. This would restore energy flow to the brain and rescue the ketotic condition.

A very important interpretation of the model above is that the pathological effects correlated with hyperglycemia may not be the effects of hyperglycemia but the effects of the relative neuroglycopenia that is causative to hyperglycemia. A condition like HHNKS does not precipitate due to extreme insulin resistance. It happens because of a counterregulatory response to acute neuroglycopenia, and the neuroglycopenia itself could be responsible for the fatal consequences and not hyperglycemia.

Although the glut-1 hypothesis potentially explains over a dozen observed phenomena unexplained by the traditional theory, so far, glut-1 dynamics has not been demonstrated in humans, and therefore, it can be said to be purely speculative. Yet there is some evidence pointing in this direction. Many elements of this story have been demonstrated, although the complete story is yet to be demonstrated. High plasma insulin, which has been shown to enhance cognitive function, increases the rate of brain glucose utilization [114], and this rise is mostly marked in the cortical areas, whereas in the cerebellum, known to be more important in muscle coordination, it is marginal [114]. This supports the notion that brain glucose demand may be increased specifically for cognitive function in diplomat life marked by hyperinsulinemia. Availability of glucose to the brain via the plasma affected cognitive functions in a number of studies [115–119] supporting our notion of the importance of brain glucose in a cognitive brain-dependent diplomat life.

If maintaining a desired level of brain sugar is the objective of regulating plasma sugar, we would expect the brain sugar to remain constant in diabetic patients, although their blood sugar may vary substantially. This has not been tested extensively, but there are some suggestive data. In a rat study, brain glycogen levels did not change in diabetic rats in spite of high plasma glucose, whereas infusing plasma glucose in normal rats increased brain glycogen. Since brain glucose uptake is insulin independent, this difference is not explained by insulin resistance. The difference in the effect of high plasma sugar in normal and diabetic rats suggests that diabetes makes some difference in glucose transport to the brain so that even with rise in plasma glucose, there is no excess glucose in the brain to increase glycogen synthesis [120]. In another comparative study of Zucker obese and Zucker diabetic fatty rats versus lean rats, there was lower brain glucose metabolism, TCA cycle, and glycogen content in obese and diabetic rats than in lean rats [121]. In human studies, there are no data on glut-1 levels, but the ratio of glucose concentration in various regions of the brain to fasting plasma levels can be expected to be lower in diabetics than healthy controls if the BBB glucose transport is slower in diabetics. This is seen in a study where the mean plasma sugar levels of diabetics and nondiabetics were markedly different (9 and 5 mmol/l, respectively), but the brain concentrations were not significantly different [122].

It is very likely that a number of researchers might have suspected some role of brain glucose and looked at brain glucose levels in diabetes. But they did not find anything wrong there and therefore thought that brain glucose was not important in the pathology of T2D. I suspect many researchers might have given up investigating brain sugar in T2D thinking that it was futile, often keeping their results unpublished. This reminds me of the famous Sherlock Holmes story the "Silver Blaze" and the famous dialogue:

> Gregory (Scotland Yard detective): "Is there any other point to which you would wish to draw my attention?"
> Holmes: "To the curious incident of the dog in the night-time."
> Gregory: "The dog did nothing in the night-time."
> "That" said Holmes, "was the curious incident."

If the brain glucose does not change substantially in diabetic hyperglycemia, there are important implications of the dog not barking. Minimal changes in brain glucose in spite of diabetes are a significant clue in itself and can be taken as an evidence that brain glucose is the real target of homeostasis. Perhaps there are many mechanisms that buffer brain glucose dynamics from fluctuations in plasma glucose dynamics [59]. Manipulating plasma glucose and restricting BBB glucose transport are among the possible mechanisms of this buffering effort. It can be calculated from the Heikkilä et al. [122] data that under fasting conditions, the ratio of brain sugar to plasma sugar in different parts of the brain was markedly lower in diabetics than nondiabetics (cerebellum 0.51 and 0.86, cortex 0.29 and 0.42, white matter 0.25 and 0.29, and thalamus 0.25 and 0.36, respectively). In another human study, the authors claim to contradict this finding, but the actual data do not. In this study ($n = 14$ each), after keeping the blood glucose level constant, glucose levels in the occipital cortex of diabetics were lower than that of controls, but the difference was not significant [123]. Since four out of five uppermost brain glucose concentrations belonged to controls and four out of five lowermost to diabetics in these data, it may be suspected that the failure to see significant difference may be due to small sample size and large individual variation. The authors' conclusion that diabetes does not affect brain glucose dynamics may therefore be too premature. In another study (7 diabetic 11 control), when plasma glucose was increased by 12 mmol/l, brain glucose rise was less in diabetics as compared to controls in all three regions monitored, namely, cortex (2.7±0.9 mmol/l in control versus 2.0±0.7 mmol/l in diabetics), thalamus (2.3±0.7 mmol/l in control versus 1.1±0.4 mmol/l in diabetics), and white matter (1.7±0.7 mmol/l in control versus 1.3±0.7 mmol/l in diabetics), but only thalamus was significantly different [124]. Although individual comparisons were not statistically significant in these studies, the consistent trend of brain sugar rise in diabetics being less than control warrants more attention. In the other direction, there is evidence that similar to rats, even in humans, hypoglycemia is associated with increased rate of glucose transport to the brain [125–127]. So, although there are no data directly on glut-1 levels in humans, the rate of glucose transport across BBB appears to hold an inverse relationship with plasma glucose levels similar to that in rats.

Even if we assume that glut-1 is not affected, indirectly, glucose transport to the brain can be affected by reduced blood flow to the brain, which is demonstrated in diabetes and is more marked in T2D [128, 129]. There is a general decrease in capillary density in diabetes [130–133] presumably owing to disinvestment in angiogenic mechanisms which may be responsible for reduced blood flow to the brain too. More clear evidence exists for decreased brain glucose transport in Alzheimer's disease [134–136]. Thus, although studies on glut-1 expression are scanty, available evidence points towards decrease in glucose transport to the brain in diabetes. If vascular pathologies reduce glucose supply to brain and brain reacts to this by activating central stimulation of liver glucose production, diabetic hyperglycemia can be said to be of vascular origin.

Although altered glut-1 and brain glucose dynamics explain a number of patterns unexplained or only partially explained by traditional theory, there is one piece of evidence going strongly against the glut-1 hypothesis. In a certain disorder characterized by frequent seizure, genetic deficiency of glut-1 has been detected [137, 138]. There is a mouse model of glut-1 haploinsufficiency too [139]. Glut-1 deficiency results in impaired supply of glucose to the brain leading to frequent seizures, delayed development, and many other symptoms. This is an example where glut-1 deficiency does not lead to hyperglycemia. The brain glucose levels are substantially below normal in the glut-1 deficiency syndrome, but surprisingly, counterregulatory response does not seem to operate in the glut-1 deficiency syndrome. There are two possible solutions to this riddle. One possibility is that this is a diagnosis bias. Reduction in brain glucose levels should normally trigger counterregulatory mechanisms as widely known. When the counterregulatory mechanisms are normal, glut-1 deficiency will result in hyperglycemia rather than low brain glucose leading to seizures. This would then be diagnosed as diabetes, and the diagnostic process is most likely to stop there. Only in cases where for some reason the counterregulatory mechanisms do not work, diagnosis would be pursued to detect the glut-1 deficiency syndrome. Otherwise, we diagnose it as diabetes, and since diabetes is "well known," nobody pursues the case sufficiently to see whether there is glut-1 deficiency. The other possible reason is that in glut-1 deficiency of genetic origin, glut-1 levels are low throughout the development period, and in type 2 diabetes, glut-1 deficiency develops in late adulthood. The consequences of the two could be substantially different. We already know that in response to chronic or very frequent hypoglycemia, the counterregulatory response gets progressively blunted leading to hypoglycemia unawareness [140–142]. If neuroglycopenia is faced from early development, there could be complete blunting of the response. In such a physiological state, brain metabolism is likely to shift completely from glucose to keto acids as fuels. Not surprisingly, ketogenic diet is helpful in glut-1 deficiency syndrome [143, 144]. The most important thing to note is that peripheral glucose homeostasis returns to normal in spite of having neuroglycopenia. This is comparable to and a counterpart of LIRKO mice in which in spite of complete liver insulin resistance, the glucose levels eventually return to normal. In the LIRKO example, a problem in peripheral homeostasis is compensated by central control. In the glut-1 deficiency syndrome, the reverse seems to happen. When there is a serious problem in the central mechanisms of glucose homeostasis, there are compensatory adjustments which supply brain with an alternative fuel, switch off the counterregulatory response part of the central control and normalize peripheral glucose homeostasis. Since LIRKO and glut-1-deficient models have demonstrated efficient mutually compensatory effects of peripheral and central control, diabetic hyperglycemia is possible only if both peripheral and central mechanisms fail or are altered simultaneously, and one needs to explain why both might fail or alter together.

The important unanswered question now is why and under what conditions glut-1 levels decrease. We are stuck here due to conspicuous absence of conclusive data. Nevertheless, there are some possible reasons and some empirical support to many of them.

1. Response to hyperglycemia: The prevalent thinking is that glut-1 levels reduce in response to hyperglycemia, and this is an adaptive response to avoid high brain glucose levels. On the other hand, hypoglycemia increases glucose transporter [8, 60–68]. Alteration in glut-1 level is a chronic response and may need several weeks of hyperglycemia for a detectable reduction in glut-1. Although well supported by evidence, there are two potential problems with this hypothesis. The first is that it implies that glut-1 reduction is only consequential and not causative to type 1 or type 2 diabetes. If glut-1 involvement is only consequential and has no causative role, one should not see glut-1 polymorphisms to be related to any of the diabetes types. But this is not true. Glut-1 polymorphism is significantly associated with both type 1 and type 2 diabetes and some of the complications [145–149] indicating that it may have some causal role in diabetes.

The other problem is related to the structure of the control loops. Hyperglycemia reducing glut-1 levels becomes a negative feedforward control in the brain glucose dynamics. This negative feedforward lies within a negative feedback loop comprising brain glucose sensing and neuronal control of liver glucose production. A negative feedforward within a negative feedback is generally not seen anywhere in biological systems since it can quickly lead to escalating cascades crusading to collapse of the system. In the following hypothetical loop (Fig. 12.7), if accumulation of B arrests conversion of B to C and deficiency of D stimulates the production of B, B will keep on accumulating and C and D will go on reducing. This is a badly designed control loop, and there is no wonder that it is almost never found in living systems. Now substitute this by glucose dynamics assuming that hyperglycemia reduces glut-1 levels. Here,

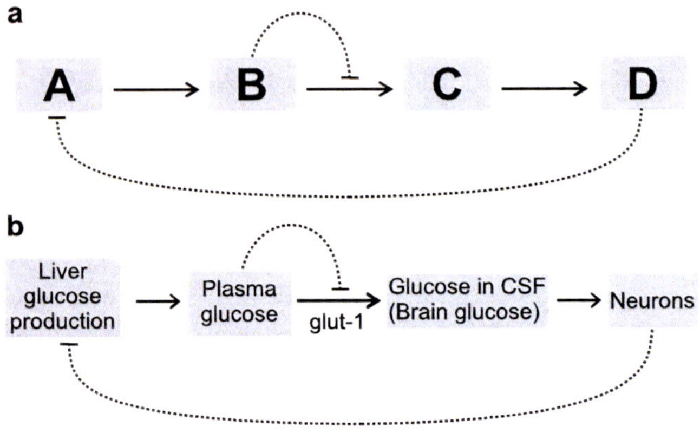

Fig. 12.7 There is evidence pointing to reduction in glut-1 in response to sustained hyperglycemia. If true, this gives rise to a strange regulatory circuit consisting of a negative feedforward within a negative feedback, a loop that can enter a vicious cycle. (**a**) a hypothetical generalized case (**b**) the case of glucose homeostasis in the brain

B stands for plasma glucose and C for brain glucose and D for glucose sensing mechanisms in the brain which trigger hepatic glucose production A that leads to B. A slight perturbation of such a system can lead to a progressive escalation of plasma glucose and simultaneous depletion of brain glucose ultimately leading to death. Therefore, the postulate that glut-1 reduction is a compensatory response to hyperglycemia is a problematic one. However, there is a possible way out of this paradox. This vicious cycle operates only if the central control loop is sufficiently active. If glucose homeostasis is under peripheral control, the vicious cycle does not begin, and hyperglycemia-induced reduction in glut-1 can be a reasonable measure to protect brain from hyperglycemia. The model helps us here again and shows that along the continuum between completely peripheral and completely central regulation, that is, x going from zero to one, there is a threshold level of x beyond which the system enters a vicious cycle. It is possible that during hunter-gatherer life, the threshold was seldom exceeded. As a result, there was no problem in the evolution of a mechanism for hyperglycemia-induced glut-1 downregulation. It is only in the supernormal diplomat lifestyle that the systems get unusually biased towards central regulation and exceed the threshold x. The model shows that the threshold is affected by muscle exercise as well as brain work. Chronically increased muscle glucose consumption protects the system from going into the vicious cycle of glut-1 lowering (Fig. 12.8) which is compatible with the empirical finding that exercise helps in keeping glut-1 levels high [150]. A certain amount of muscle exercise increasing muscle glucose consumption can save the system from entering a vicious cycle. However, once the system enters the vicious cycle, it becomes more difficult to reverse it. The amount of exercise that could prevent entry into the vicious cycle is unable to stop the cycle once started. Much higher increment in muscle exercise is needed to bring the system back to a stable state (Fig. 12.9). If this model is correct, it explains the observed difficulty in reversing T2D.

An inquiry into the parameter space over which there is a risk of entering the vicious cycle gave a major surprise. As expected, the vicious cycle begins when muscle glucose utilization is low. I expected the reverse trend with brain glucose utilization, that is, very high brain glucose demand, will trigger the vicious cycle. However, the relationship with brain glucose demand turned out to be highly non-monotonic. Vicious cycle was more likely at high demand and least likely at normal average demand as expected. But abnormally reduced demand also triggered the cycle (Fig. 12.10). This unexpected result was due to a different sequence of events. The extremely low brain consumption increased brain glucose levels, and the high brain glucose stimulated excessive insulin production (for $x>0$). With insulin, amylin production also increased, and because amylin has a longer half-life and has a nonlinear relationship with insulin sensitivity, it induced high and sustained insulin resistance. This increased blood sugar eventually downregulated glut-1 and the cycle continued. The role of amylin was crucial, and removal of amylin from the model stopped the vicious cycle. These unexpected results raise certain novel possibilities. It explains the origin of obesity and insulin resistance in certain forms of mental retardation. It also means that departure from normal brain glucose dynamics in either direction will lead to similar outcomes through different pathways. It also supports the possibility of double deficiency (simultaneous soldier and diplomat deficiency) leading to T2D.

It is necessary to point out another possibility here. The reason for increased average glucose consumption in the brain is not increased cognitive demand alone. There is one more important reason. Sleep is an important determinant of total daily glucose consumption by the brain. Brain glucose consumption decreases by 30–40% in non-REM sleep [151–153]. Therefore loss of sleep can be a potential reason for entering the vicious cycle area. Insomnia has an association with T2D and related disorders, and the pathway could be through altered brain glucose dynamics. An important inference is that high-level cognitive activity such as that by the brain of a mathematician or a scientist is unlikely to trigger the vicious cycle if sleep is adequate and sound. On the other hand, worries that affect sleep are likely to be more dangerous.

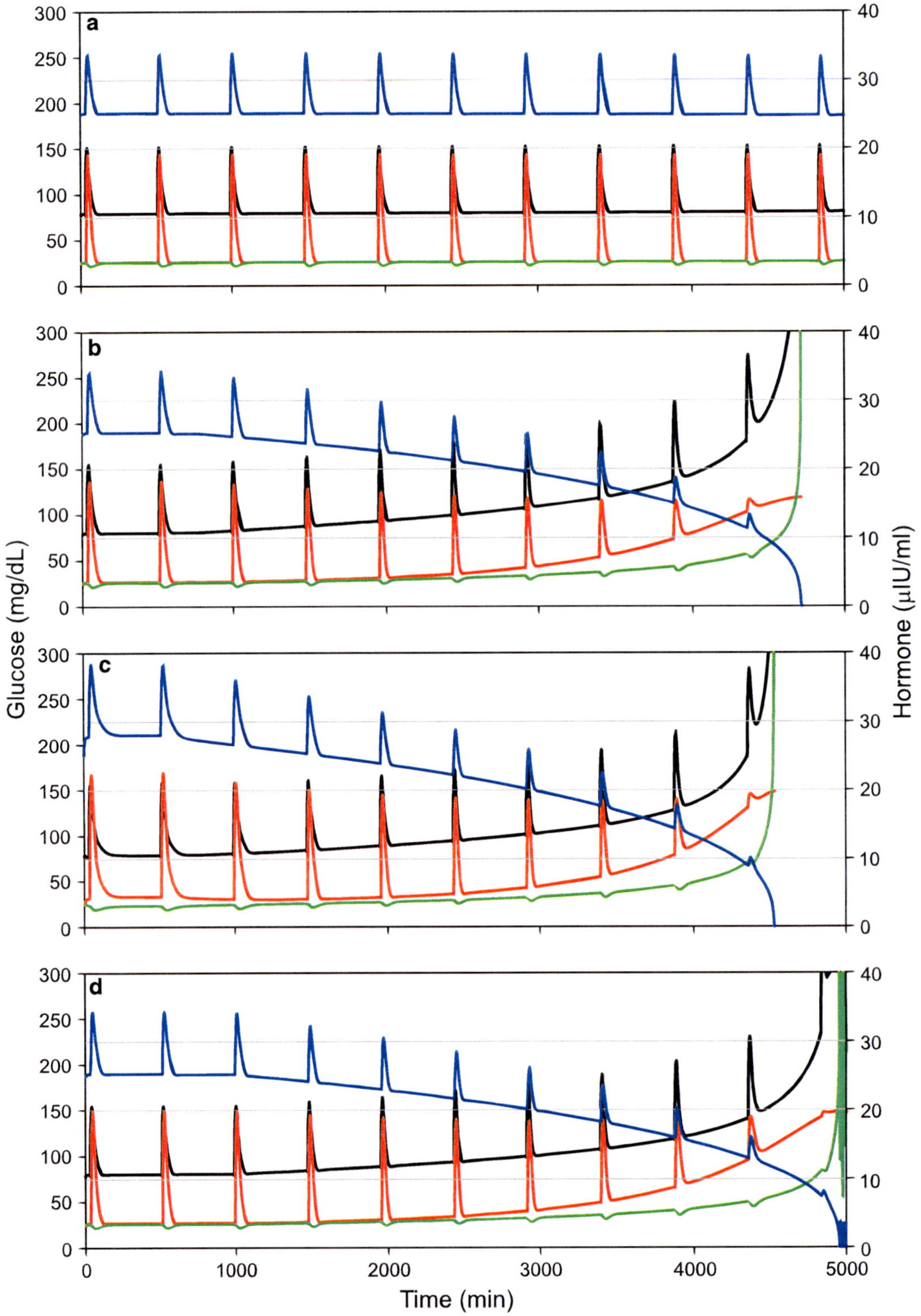

Fig. 12.8 A model incorporating chronic changes (see Appendix IV) illustrates the potential vicious cycle and its risk factors. In a long-term simulation that includes periodic meals, a chronic condition is introduced in which if the mean blood sugar averaged over time T is greater than a threshold, there is reduction in glut-1 by a very small quantum Δglut-1. For demonstration purpose, in these simulations, $T = 500$ min. In reality, it will be

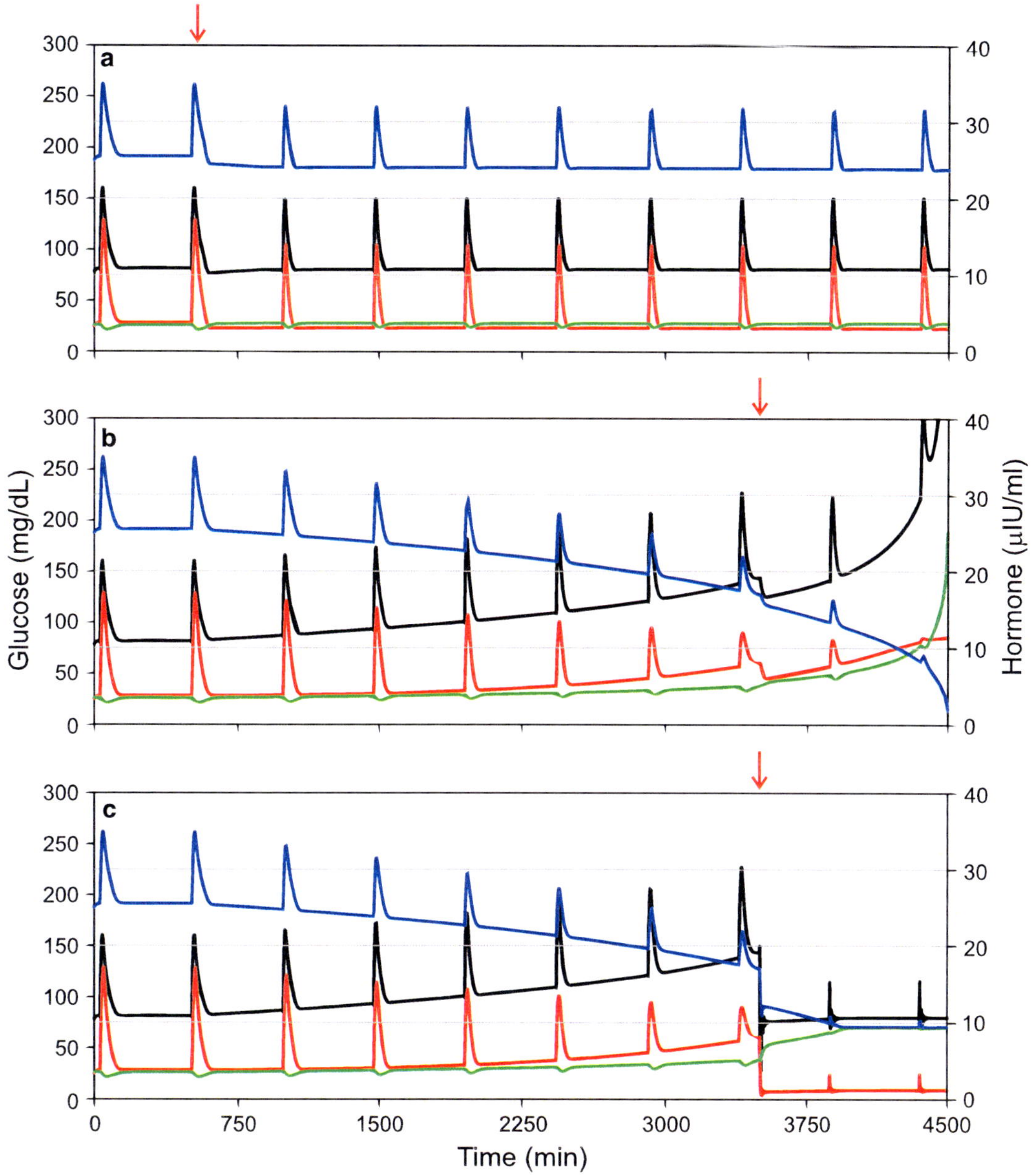

Fig. 12.9 Effect of exercise before and after entering a vicious cycle: Here, simulations are run at a set of parameters that lead to vicious cycle, but (**a**) exercise (sustained increased in *MU*) is introduced in the beginning which can prevent the cycle. (**b**) The same amount of exercise started at a much later stage after the cycle progresses is ineffective in arresting the detrimental effects; (**c**) about 40 times greater increment in *MU* was needed to come out of the cycle in these simulations. This suggests that the amount of exercise needed to prevent T2D will not be sufficient to reverse T2D once established, a finding that could be realistic

Fig. 12.8 (continued) orders of magnitude greater than this value. Therefore, the changes that take place over a short time in the demonstration simulations may take years in real life. Simulations show that the system enters a vicious cycle if x is above a threshold. *BU* and *MU* (Appendix IV) affect this threshold. Vicious cycle can begin with a change in any one or more of the three parameters. (**a**) Stable state without entering vicious cycle. The system entering a vicious cycle at a higher value of x (**b**), higher *BU* (**c**), and lower *MU* (**d**)

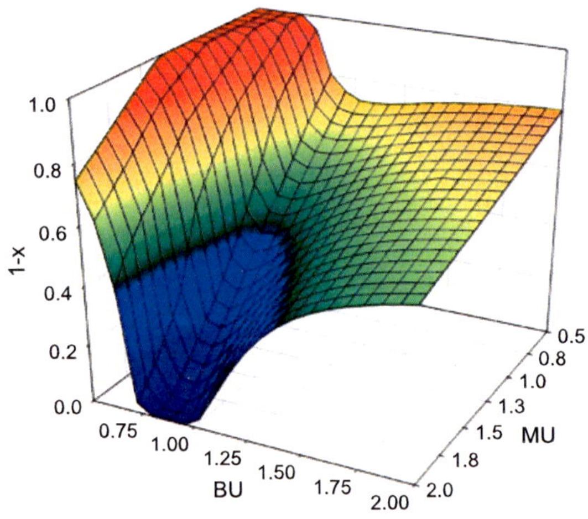

Fig. 12.10 The parameter space in which vicious cycle begins (*below the surface*). This is characterized by large x and small MU. Surprisingly, the effect of BU is non-monotonic. At very high BU, the system enters a vicious cycle as expected by the soldier–diplomat hypothesis, but the same is also observed at very low BU. The chain of causation is likely to be different at low and high BU, but the results are similar

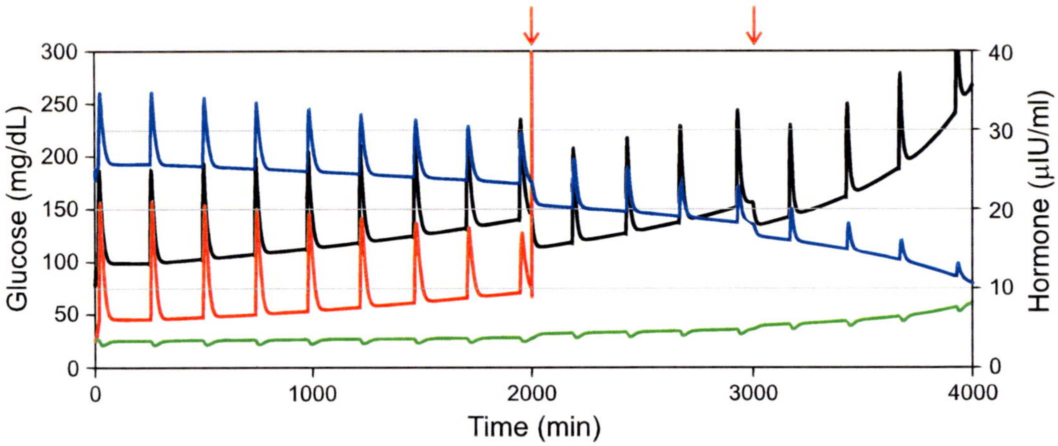

Fig. 12.11 The honeymoon exposed: The model suggests a plausible reason why the honeymoon period of insulin treatment ends eventually. Exogenous insulin is started at *first arrow* which can bring down plasma sugar effectively and rapidly but is unable to change glut-1 rapidly. As a result, the central stimulation of liver glucose production resumes and the vicious cycle continues. Further down the line, as glut-1 is depleted further, the same dose of insulin becomes ineffective, and an increase in the dose (*second arrow*) is needed to show the same effect. This dose eventually turns ineffective too. The glut-1 model thus explains the gradually losing glycemic control without involving insulin resistance

Although central to the vicious cycle is glut-1 level and brain glucose dynamics which is insulin independent, the model shows that exogenous insulin does exert a short-term effect. However it does not help in the long run since reduced glut-1 levels are not restored by insulin treatment [154]. If the main cause of hyperglycemia is left untouched, any attempt to reduce plasma sugar will only have short-term effects. It is quite well known that all the current pharmaceutical treatments for T2D have excellent short-term effects, but the honeymoon period ends in a while, and the sugar levels increase in spite of continued treatment. A model combining insulin resistance and glut-1 reduction explains this behavior very well (Fig. 12.11). If insulin treatment is given to a patient who is already in a slow chronic vicious cycle, regular administration of insulin can bring down the glucose levels but is unable to increase glut-1 levels in the BBB and therefore unable to stop the cycle. As a result, glut-1 keeps on reducing slowly further, and after a while, the current dose of insulin becomes inadequate. An increased dose shows an effect in the short run, but again the cycle continues to operate, and this dose proves inadequate subsequently.

Thus hyperglycemia-induced glut-1 downregulation explains a number of observed patterns in the pathophysiology of diabetic hyperglycemia. But there are other possible causes of glut-1 downregulation too, and we must keep these possibilities open until there are experiments elucidating the true causes of glut-1 downregulation.

2. To protect from fluctuations in plasma glucose: I have introduced this concept already, and this is my favored hypothesis although I am open to all others. A small transport window has a dampening effect on oscillations. This is simple to show mathematically as well as visualize intuitively (Figs. 12.3 and 12.4). It is possible therefore that the soldier to diplomat transition itself results in moderate decrease in glut-1 levels as an anticipatory strategy. But it is also possible that glut-1 reduction is induced by large amplitude of fluctuation of blood glucose. When the first signs of IGT appear due to IR-RII, the amplitude of daily glucose fluctuations goes up, and glut-1 reduction is a protective response to chronically increased amplitude of fluctuating plasma glucose. This response could then increase fasting glucose levels. With the progression of diabetes, the daily fluctuations might increase in amplitude, and therefore, glut-1 may decrease further. Again there have been no studies so far to test this possibility, but it is not difficult to test it in animal models. Glut-1 reduction in response to plasma glucose fluctuations can give rise to a vicious cycle similar to the one described above.

It also explains in an alternative way why in insulin-treated diabetics insulin becomes progressively ineffective. This is because insulin treatment leads to higher amplitude of fluctuations in plasma glucose. If glut-1 responds to the amplitude of fluctuations, it will reduce further in response to insulin treatment. This raises the mean plasma glucose in spite of insulin to which the diabetologist responds by increasing insulin dose, and the vicious cycle continues. It is possible that the observed decrease in glut-1 in streptozotocin-induced diabetic rats is induced not by hyperglycemia but by the wider fluctuations in plasma glucose which are inevitable after loss of insulin control.

3. As a mechanism of thrift: As argued above, low glut-1 can lead to overeating. This is type 1 thrift. Since we had predicted earlier that subordinate individuals should be thrifty because of lower food security, we can expect that subordinate or nonaggressive individuals will have lower glut-1 levels. This has not been tested anytime but is not difficult to test in a socially reared and behaviorally monitored rodent colony.

4. Deficiency of $n-3$ polyunsaturated fatty acids: The microvessels and cortex of rats fed $(n-3)$ PUFA-deficient diet had 25–30% reduction in glut-1 levels [155–157], although glut-1 mRNA was not altered. Although research in this direction is scanty, it raises an interesting possibility that the main pathophysiological mechanism of deficiency of $(n-3)$ PUFAs may be through downregulation of glucose transport in the brain [158]. This along with the previous possibility makes coherent sense. Since $(n-3)$ PUFAs are needed by the brain and they cannot be synthesized by the body, they have to come through the food intake. If the concentration in food is low, increasing the total food intake is the only option left. This can be achieved by reduced glut-1 levels. Therefore if glut-1 expression is $(n-3)$ PUFA dependent, it serves as a control mechanism ensuring adequate supply of $(n-3)$ PUFAs.

5. Obesity: Obesity by itself may downregulate glut-1. This has not been clearly demonstrated as yet, but there are two pieces of suggestive evidence. Caloric restriction has been shown to increase glut-1 levels in one study [159]. In an unpublished experiment in my lab, a student observed lower levels of glut-1 in female rats fed high-fat diet and subject to chronic restraint. The most physically inactive animals had lowest glut-1 levels. However, in these experiments, glut-1 levels negatively correlated with plasma glucose as well, and therefore, the cause–effect relationship between plasma glucose and glut-1 remained obscure.

6. Response to higher levels of brain glucose: An alternative possibility to hyperglycemia-induced glut-1 lowering is that glut-1 reduces in response to high brain glucose. The reason why we see glut-1 reduction in chronic hyperglycemia

is that even the brain glucose levels are raised. Although this might look very similar to the earlier argument, it makes a substantial difference for the operation of the control system. Glut-1 reduction in response to chronically raised brain glucose becomes a negative feedback, and such a system can achieve considerable stability and robustness towards perturbations. The reduction in glut-1 observed in streptozotocin-induced diabetic rats can be a response to higher brain glucose than higher plasma glucose. Although this is more logical, currently, there is no direct evidence to differentiate between glut-1 responses to blood glucose versus brain glucose. But the idea can be tested with a simple and feasible experiment. One has to infuse calculated quantities of glucose directly into the brain or CSF of rats chronically such that the brain levels are maintained above normal but plasma levels are not increased. In fact, they may decrease in response to high brain sugar levels if our hypothesis is correct. If glut-1 levels or glut-1 transcription levels reduce in response to chronically elevated brain sugar accompanied by normal or lowered blood sugar, the hypothesis can be validated. This has not been attempted as yet, but if results comply with the expectation, it can increase our understanding of brain glucose dynamics substantially. Currently this remains an open possibility. Unlike 1 and 2 above, this mechanism does not lead to a vicious cycle.

7. Response to reduced brain activity: As a broad generalization, the body invests more in organs and functions used more frequently or being more crucial for survival and disinvests from unused and unimportant organs and functions. We have argued earlier that insulin resistance is a mechanism of disinvesting from muscle when not in use. Since brain glucose pickup is insulin independent, investment in brain cannot be reduced by an insulin-dependent mechanism; it can be only reduced by lowering the transport protein glut-1. Although studies testing this are few, there is some evidence that brain activities increase glucose utilization rate which stimulates glucose transport too [51, 58, 160]. Glut-1 levels appear to be so sensitive to glucose demand that differences in activity of different brain regions appear to affect local glut-1 levels [161, 162]. There is a positive correlation between activity of different regions and the local glut-1 levels [163]. Also, frequent seizures, which acutely increase brain glucose utilization, are shown to increase glut-1 content of brain capillaries [164]. BDNF is shown to increase glut-3 activity in the brain, and although a direct effect of BDNF on glut-1 has not been shown, it is shown to increase glucose turnover in the brain [165]. So although of a preliminary nature, there is positive evidence that glut-1 levels are fine-tuned according to brain activity. This is in contradiction with the possibility 2 above. However we need to keep all possibilities open until more evidence eliminates one of them.

We had made a prediction in an earlier chapter that a chronic helpless response such as depression should downregulate glut-1 (Chap. 10). This prediction also remains to be tested. Another testable epidemiological prediction is that professions and personalities that involve high cognitive brain activities should show lower incidence of fasting hyperglycemia since their glut-1 levels are unlikely to be low. Such highly brain-dependent but physically sedentary professions might become insulin resistant but will not become hyperglycemic as long as they are involved in frequent and intensive cognitive brain functions.

Physical inactivity has been well recognized as a characteristic of modern lifestyle. What is less appreciated is that even cognitive brain work is considerably reduced today because of calculators, computers, ready online information, GPS aid in finding directions during driving, and so on. These aids have demonstrable effects on human cognitive demands [166]. If both physical and mental activities are simultaneously reduced, glut-1 may be going down substantially leading to hyperglycemia in the long run.

Hypotheses 2, 6, and 7 above together suggest a novel possibility. If we incorporate a combination of central shift in glucose homeostasis mechanisms and low brain demand in the model (Appendix IV), we get an interesting chain of events. Low glucose utilization in the brain will increase steady-state brain glucose which if hypothesis 6 is true will decrease glut-1 levels. A combination of central shift, lower brain activity, and lowered glut-1 has interesting effects on fasting and OGTT response. Simulations show that under these conditions, the fasting glucose is normal, but fasting insulin is increased, glucose-stimulated insulin response is delayed, the glucose peak is elevated, the right-hand tail is extended, and area under the glucose curve is increased and area under insulin curve might be increased too. If fasting glucose is normal but the peak height is increased, it means the amplitude of fluctuation in plasma glucose is increased. As a slow and chronic response to increased amplitude, glut-1 may decrease further and then the fasting plasma glucose starts rising. All these are typical characteristics of impaired glucose tolerance in a prediabetic state progressing to overt diabetes. This single model explains all the four characteristics of altered GTT. Interestingly no other model explains all of them together as we will see soon. It is possible therefore that central shift and altered brain glucose dynamics are responsible for the characteristics of altered glucose tolerance curve. The only possible weak link in this argument is that if brain glucose demand is lower, why should there be a central shift in homeostatic mechanisms. A possible answer is that loss of physical activity may be a sufficient trigger for the central shift. Evolved responses of the body may assume that when physical activity and aggression go down, demand for cognitive functions is bound to increase, and central shift is an anticipatory response. Anticipatory responses are not new to physiology. An alternative possibility is that central shift is an adaptive response not only to reduced brain glucose levels but also to abnormally high glucose levels. If brain glucose consumption decreases, the steady-state levels would go up, and central shift can effectively bring it back. This suggests that central shift could be an evolved response to any departure from normal brain glucose dynamics. If this is true, it explains almost all observable elements of normal and altered glucose dynamics.

A fairly robust conclusion is that altered brain glucose dynamics, central shift in glucose homeostatic mechanisms, and lowering of glut-1 by any of the above reasons are sufficient to give typical picture of diabetic hyperglycemia. This works irrespective of whether the brain glucose demand goes up or down.

If glut-1 and altered brain glucose dynamics have an important role in diabetic hyperglycemia as our model suggests, it is extremely important to identify the factors that affect standing levels of glut-1 in brain capillaries since that can have serious clinical implications. In the absence of data, we need to be open for all the above possibilities and perhaps even more. This is the most important missing link in the new model of diabetic hyperglycemia and needs to be addressed at the earnest. If glut-1 levels or glucose transport to the brain is demonstrated to be always lowered in T2D and the reasons for this alteration are revealed, almost every mystery and paradox in diabetes would be resolved.

We have now two competing schools of thought for explaining fasting hyperglycemia. One is the traditional school of IR-RII which is aware of the existence of central control but largely ignores its role. The other stated above consisting of central shift of homeostatic mechanisms, altered brain glucose utilization, and gut-1 reduction. In order to be fair to both the hypotheses, we need to use the same treatment for both. Therefore we will now try to take a new approach of examining the expected shape of GTT given by each of the schools and the sub-hypotheses. Figure 12.12 gives the expected glucose and insulin curves during OGTT by a large number of possible combinations of processes.

Using this approach it can be realized that:

1. None of the peripheral mechanisms are able to give delayed insulin response typical of early T2D. All models involving central mechanisms give a delayed insulin response, with or without altered glut-1. With reduced glut-1, there is a greater delay.
2. In a peripheral model, if we reduce insulin production capacity, there is no effect on glucose curve until about 85% loss of the capacity. This is very close to the partial pancreatectomy experiments in which at least 85% removal was needed to cause hyperglycemia.
3. If we assume that hyperinsulinemia is primary and insulin resistance is a compensatory

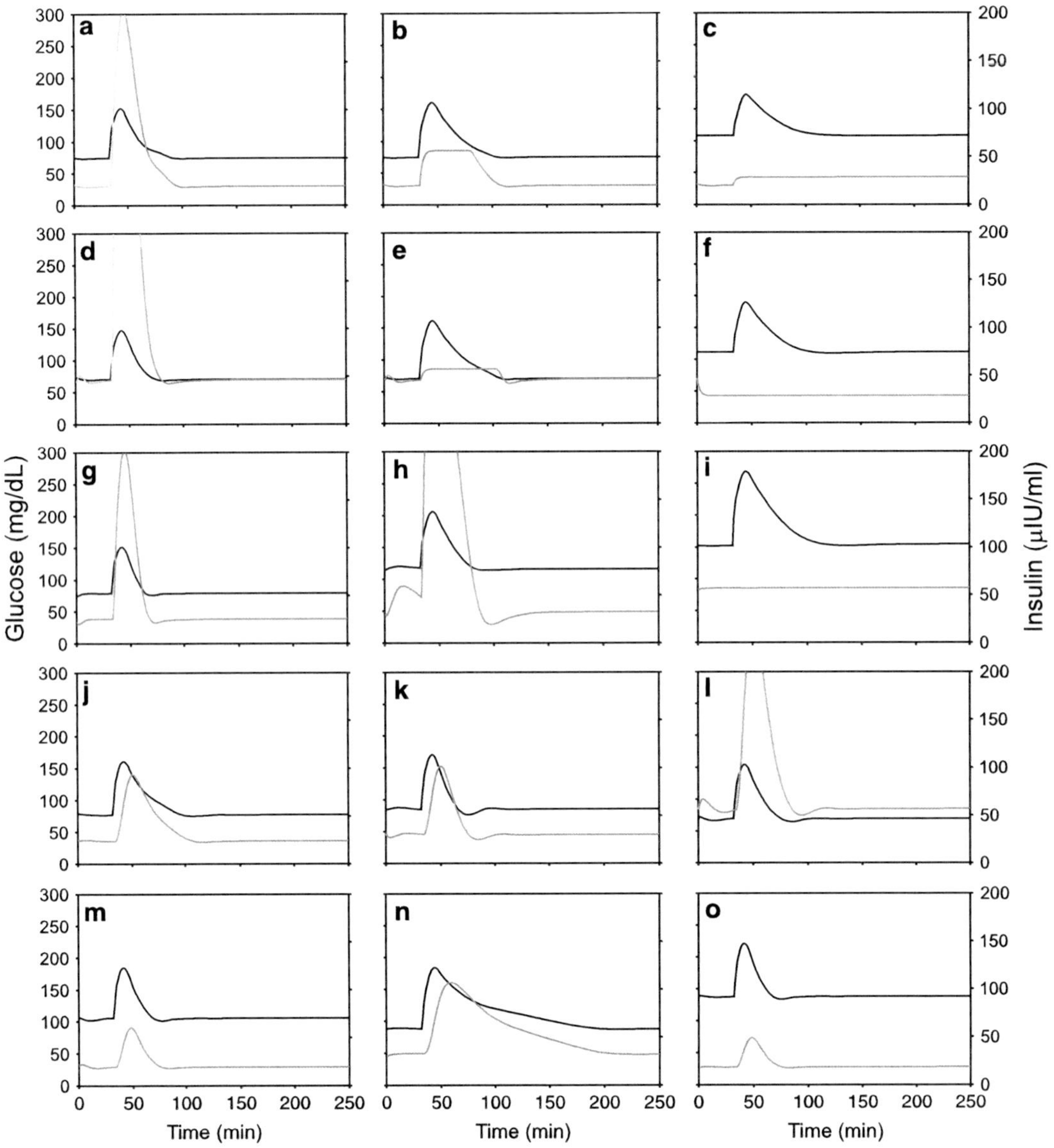

Fig. 12.12 Simulated OGTT curves with different sets of assumptions. Glucose is administered at 30 min. The simulations are intended to identify the conditions under which the four characteristics of impaired OGTT, namely, delayed insulin response, raised fasting glucose, increased height of glucose peak, and extended right-hand tail, are qualitatively observed. In the results shown, only extreme values of x ($x=0$ for (**a–i**) and $x=1$ for (**j–m**) are used. The general inference is based on several other simulation runs (results not shown) with intermediate values of x and exploring the parameter space widely. (**a**) OGTT with peripheral control normal insulin sensitivity. (**b**) Same as (**a**) with 80% reduction in the beta cell capacity. The glucose curve hardly changes despite much reduction in insulin (**c**) same as (**a**) with a 90% reduction in beta cell capacity. Here, the glucose curve changes substantially, but there is no glucose-stimulated insulin peak. (**d**) OGTT of an HIIR state: For achieving HIIR state, insulin resistance and insulin responsivity are proportionately increased. Here, the glucose curve is normal, whereas both fasting and glucose-stimulated insulin are high (**e**) same as (**d**) with 80% suppression of beta cell capacity, glucose curve only marginally altered (**f**) same as (**d**) with 90% suppression of beta cell capacity. Here, the glucose curve changes substantially, but there is no glucose-stimulated insulin peak. (**g–i**) Amylin action is disabled, otherwise (**g**) is similar to (**a**). (**h**) Insulin resistance is increased without altering beta cell responsivity. (**i**) Further to (**h**), beta cell capacity is reduced by 90%. In these and several other simulations over wider parameter space, delayed insulin response was not seen at $x=0$. (**j**) Central control active and normal insulin sensitivity. Note the delayed and somewhat suppressed insulin peak although glucose curve is normal. (**k**) is the same as (**j**) with the effect of amylin disabled. Reactive hypoglycemia is more likely under these conditions. (**l**) Central control with increased insulin resistance and insulin responsivity. Both fasting and glucose-stimulated insulin

response, for which there is more evidence (see Chaps. 3, 5 and 6), the model is unable to give hyperglycemia under any conditions except when insulin production is substantially below the normal healthy insulin-sensitive state. This is compatible with a number of experiments of suppressing insulin production in an HIIR state which have demonstrated that insulin sensitivity increases on suppressing insulin production so that the sugar level remains normal.

4. If amylin and its action on insulin sensitivity are incorporated into the model, hyperglycemia is extremely difficult to achieve in spite of suppressing insulin production in an insulin-resistant state. On the other hand, it is equally difficult to achieve hypoglycemia on stimulation of insulin production. Amylin thus has a buffering action on plasma glucose levels against extreme insulin levels. The IR-RII model does not give hyperglycemia when amylin is included in the model except at extremely low insulin levels. This highlights the role of amylin which is largely ignored by the classical school.
5. If we ignore the action of amylin and also assume that insulin effect on muscle glucose uptake is linear and go back to the classical assumptions that insulin resistance is primary and hyperinsulinemia is a compensatory response, then hyperglycemia can result under the following circumstances.

   There is an upper limit to insulin production capacity of the pancreas, and insulin resistance increases beyond this capacity so that compensation is no more possible. This condition can produce postmeal hyperglycemia but cannot achieve fasting hyperglycemia unless the difference between the two is too large. If the maximum insulin-producing capacity is lowered substantially to simulate islet degeneration, fasting hyperglycemia can be achieved in this model only when the meal-stimulated insulin production is completely blunted. In no case, fasting hyperglycemia and meal-stimulated insulin peak can coexist. This means that if in reality diabetic patients at any stage show IFG and are still able to give postmeal insulin peak (even if delayed), this model needs to be rejected. Since early diabetics show this condition, this set of assumptions cannot be true. However, this set of assumptions gives three out of the four characteristics of impaired GTT curve: raised fasting levels, increased peak, and extended right-hand tail. But delayed insulin peak is never seen in this model.

   If we keep the nonlinearity of insulin action and effects of amylin in the model but make an assumption that a chronic factor such as sedentary life and/or high-calorie diet gradually induces insulin resistance on the one hand and slowly reduces insulin-producing capacity of β cells on the other, fasting hyperglycemia can be achieved along with a GTT pushed up. But in this case, it is pushed up in the same shape. The shape of the curve does not change.

In short, any set of assumptions under the peripheral control and IR-RII model does not give us all the four characteristics of the glucose and insulin curve typical of early T2D. On the other hand, central control models with increased or decreased brain demand and reduced glut-1 give all the characteristics. Although using these models we can make only qualitative predictions, the predictions go clearly against the RI-RII model and favor the central shift, glut-1 drift, and brain thrift model.

However, I will still like to keep the IR-RII model alive because this is the model believed by generations of diabetologists. It is good to be con-

Fig. 12.12 (continued) levels are high, but glucose-stimulated peak is delayed, and there is no significant change in glucose curve. (**m**) Central control with normal insulin sensitivity but increased brain glucose utilization. Note suppressed insulin peak and increased fasting sugar. (**n**) Reduced glut-1 levels accompanied by reduced brain glucose demand. Note the substantial delay in insulin peak and altered glucose curve in spite of normal fasting glucose. These features of the curve are unaffected by the HIIR state. (**o**) Central control with reduced glut-1 and highly increased brain glucose demand. Note the increased fasting glucose as well as height of the glucose peak. The general inference is that IR-RII is neither necessary nor sufficient to get all the characteristics of the altered course of OGTT, but it can contribute to some of them. On the other hand, a combination of central shift (large x), glut-1 reduction, and altered brain utilization is both necessary and sufficient to give all the characteristics. The delay in insulin response is unique to central control (i.e., large x) and can be taken as a diagnostic marker of large x

servative while dealing with evidence against existing and widely accepted theories. Let us therefore keep alive the possibility that both the alternative mechanisms of diabetic hyperglycemia exist in reality and may even work in synergy.

Interaction of the Two Alternative Models of Hyperglycemia

Since the two are not mutually exclusive it is likely that (1) one mechanism is primary and gives way to the other in due course of pathological progress, (2) the two mechanisms work simultaneously and synergistically to increase plasma sugar, (3) one is predominant in one class of patients and the other predominant in others, and (4) only one of them is true. It is possible to experimentally resolve between the four possibilities if the two mechanisms have differential markers or differential testable predictions. The two mechanisms certainly make differential testable predictions apart from the evidence discussed above.

The most important part of the hypothesis to test is whether there really is a central shift, whether the role of central mechanisms is increased, and whether that of peripheral ones decreased in diabetes. It should be possible for experts to visualize ways to do so. I am fascinated by the carbachol experiments by Ahrén et al. [31] and think that this line of work needs to develop further.

The BBB glucose transporter is difficult to study quantitatively unlike blood parameters for which sampling is easy. So far all studies directly looking at glut-1 levels have used brain tissue after death or sacrifice of the experimental animal. Owing to this limitation, it is not possible to follow the time course of changes in glut-1 levels. It may be possible to use some modern techniques such as magnetic resonance spectroscopy [123, 167, 168] to monitor brain glucose levels in live animals. However, this approach could still be useless because it does not make differential testable predictions. By the glut-1 model, the brain glucose to blood glucose ratio should be lower in diabetes, but the same can happen by the peripheral model too with an added component of consequential hyperglycemia-induced glut-1 lowering. Therefore, demonstration of altered ratio remains inconclusive.

Therefore it is necessary to rely on indirect evidence. The model helps us here. In real life, it is difficult to segregate the effects of insulin resistance and glut-1 reduction since they would most likely co-occur. In the model, we have the freedom to alter one at a time and see the effect. Therefore we can use the model to make one or more differential testable predictions. Since the models are not based on empirically estimated realistic values of parameters, we cannot make quantitative predictions and will have to rely on qualitatively different testable predictions.

An easy-to-test prediction arising from the two models is about the result of insulin sensitizing treatment. If hyperglycemia is caused by glut-1 lowering and brain glucose dynamics and if we reduce insulin resistance in the model, insulin levels go down quickly, but sugar levels remain almost the same for quite a long time. On the other hand, if IR-RII is the cause of hyperglycemia, reducing insulin resistance in the model reduces plasma glucose and subsequent decrease in insulin production (Fig. 12.13). This particular prediction of the model is extremely important since it can qualitatively differentiate between the two alternative causes of hyperglycemia. If hyperglycemia is caused primarily by insulin resistance and relative insulin insufficiency, then upon increasing insulin sensitivity, plasma glucose levels should reduce first and insulin levels may reduce in response to it. If the delay is not detectable, then both will appear to come down simultaneously. If this happens, it can be safely inferred that hyperglycemia was caused by IR-RII. However, if insulin level reduces first without affecting glucose level for a substantial period of time, which can be predicted to be several days to a few weeks, we should infer that hyperglycemia is caused by altered brain glucose dynamics. We should also take into account the possibility that both work simultaneously. Figure 12.13c shows the qualitatively expected patterns if only one of the two factors works or if both work simultaneously. Even if both work simultaneously, the pattern of reversal on giving insulin sensitizing treatment is substantially different.

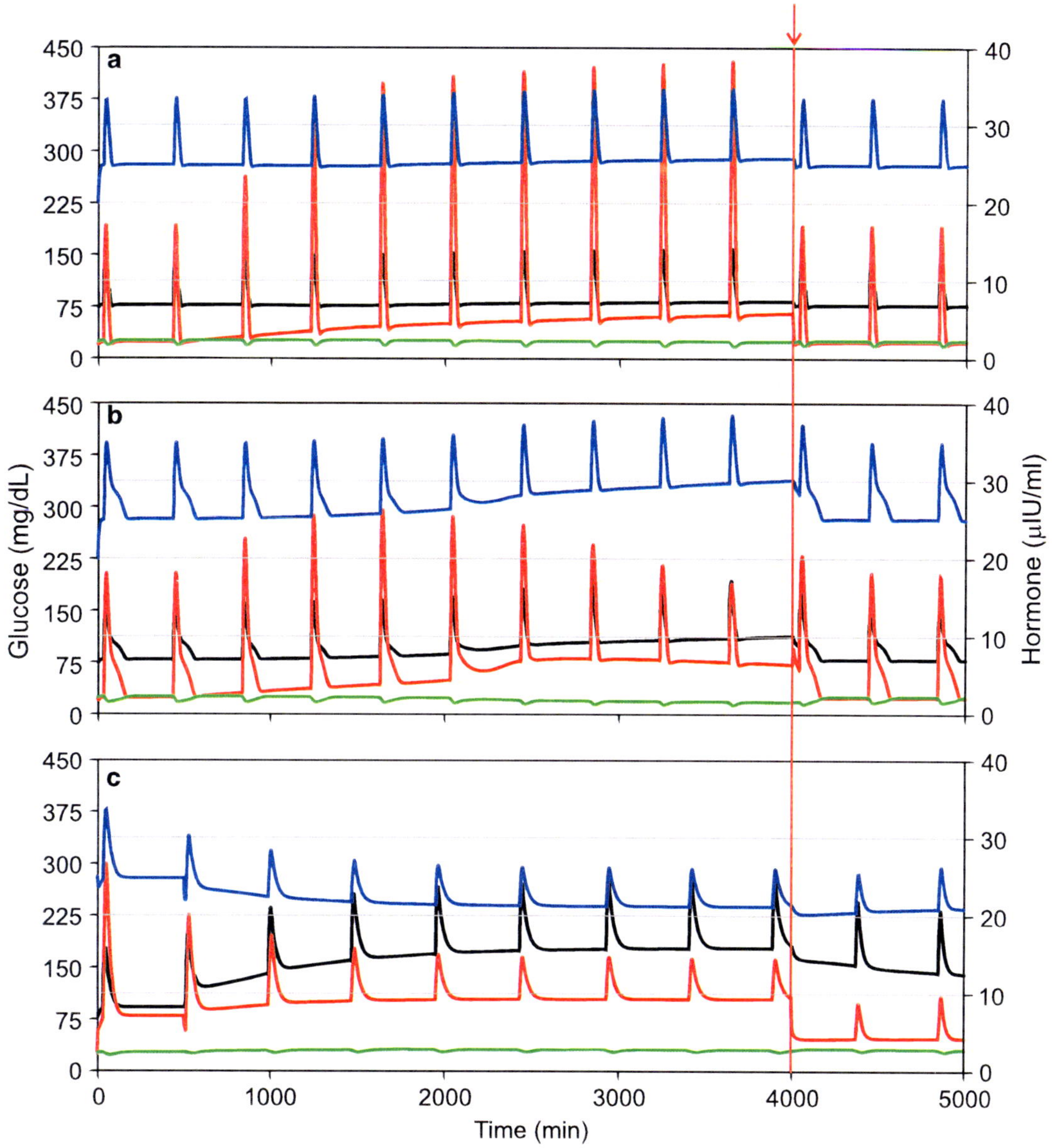

Fig. 12.13 How to differentiate between hyperglycemia of peripheral (IR-RII) origin versus central (central shift and glut-1 depletion) origin, if any: Simulations help in differentiating the two by a test that can be performed at the clinical level. (**a**) HIIR normoglycemic state: If insulin sensitivity is increased by exercise, insulin levels reduce but sugar levels remain unaltered. (**b**) In a hyperglycemic state caused by IR-RII, if insulin sensitizing treatment is given, glucose and insulin levels reduce simultaneously. (**c**) If HIIR state is accompanied by hyperglycemia caused by central shift and glut-1 depletion, insulin sensitizing treatment brings down insulin rapidly, but glucose level comes down only gradually after a considerable time lag or it may not come down at all. The detectable difference in the insulin and glucose response to insulin sensitizing treatment can be used to differentiate between hyperglycemia of peripheral and central origin

It is likely that in some T2D patients, one cause prevails and in others the second cause predominates. In that case, the reversal pattern can potentially be used as a diagnostic test to know whether hyperglycemia is caused by IR-RII or by altered brain glucose dynamics.

Data to test this may already exist in clinical trials of insulin sensitizing drugs or exercise trials.

However, we are unlikely to find it in published literature perhaps owing to the fact that reduction in insulin levels without reducing glucose is most inconvenient to publish. I had access to some unpublished data of insulin sensitizing treatment where I could see that in a substantial proportion of patients, fasting insulin levels come down first before any detectable change in fasting glucose levels. However, since in publications only the mean levels are reported, the fact that such a class of patients exists does not get revealed. Also typically results are reported only after 12 weeks or more by which time reduction in sugar is noticeable. One should look at the pattern every week or twice a week to know whether insulin or glucose reduces first. There is some published data that indirectly indicates this pattern (see Chap. 3), but let us treat this as a testable prediction rather than a demonstrated pattern and test it again. This will be an acid test for the two alternative models.

As it stands today, although there exist a logically coherent theory now, evidence is fragmentary and there are gaps in the data, and therefore, I will only conclude that we are far from understanding how the peripheral and central regulations interact and why blood sugar goes up in diabetes. The models discussed above cannot be treated as conclusive but certainly generate many new and interesting possibilities that are testable.

At the end, in spite of gaps in data and some uncertainties, I will try to join the available pieces together to make a logically coherent picture of the origin and progress of diabetic hyperglycemia. Part of this is speculative, but these speculations join the missing pieces of the jigsaw puzzle and therefore should be accepted as a hypothesis and put to test:

Phase 1: The soldier to diplomat transition needs higher levels of insulin for cognitive brain function. This is mediated by parasympathetic stimulation of pancreatic islets. At this stage, β cells are responding to both peripheral glucose and parasympathetic inputs, and therefore, their glucose sensitivity appears to be increased. In response to hyperglycemia, peripheral insulin resistance develops so that plasma sugar remains normal.

Phase 2: In response to the altered balance between glucose demands by muscle versus by the brain, there is a progressive shift of the glucose homeostasis mechanisms from peripheral to central regulation. As a result, the islets become less sensitive to peripheral glucose and more to autonomic inputs. Also, the liver glucose production becomes less sensitive to inhibition by peripheral insulin and glucose and more sensitive to autonomic inputs. This shift in the homeostatic emphasis from peripheral to central is what is currently being called "liver insulin resistance" and "β cell dysfunction." At this stage, any sudden increase in brain glucose consumption can cause transient rise in plasma glucose, a condition that we may equate to "stress diabetes." If there is sustained increase in brain demand for glucose, fasting plasma glucose will show a sustained rise.

Phase 3: Overlapping with phase 2 slight downregulation of glut-1 levels might begin as an adaptive response to stabilize brain glucose levels. Reduction in glut-1 is accompanied by raised plasma glucose, particularly FPG. Reduced rate of glucose transport to the brain may delay insulin response to meals leading to altered OGTT. Up to this point, the changes are adaptive, and there is a shift in the homeostatic target but without any pathological consequences.

Phase 4: This is the phase of supernormal reduction in glut-1, the reasons for which are not clearly known. The most likely reason is crossing of the threshold where the vicious cycle of our model begins. As seen earlier, this can happen by a combination of decreased muscle activity, increased role of central regulation, and altered brain demand for glucose. Increased amplitude of plasma glucose fluctuation may also matter. If the system bias towards central regulation increases and if glut-1 levels and/or brain capillary density is reduced substantially, the system becomes progressively less responsive to exogenous insulin or insulin sensitizing agents.

Phase 5: More serious pathological changes begin with deficiency of EGF and other growth factors preventing β cell regeneration. This leads to progressive loss of β cell mass. Almost simultaneously, deficiency of NGF and BDNF results in progressive autonomic degeneration. Both the deficiencies arise because of deficiency of soldier lifestyle components. As a result, both peripheral and central homeostatic mechanisms become progressively dysfunctional. Normally, central and peripheral mechanisms can compensate each other's dysfunction as demonstrated by normalization of glucose in LIRKO and glut-1 deficiency models. However, since both fail more or less simultaneously in T2D, the compensation is not seen.

It can be appreciated that phases 1–3 are adaptive responses to a transition in behavioral strategy from soldier to diplomat. Phases 4 and 5 are truly pathological which are triggered by supernormal behavioral shifts. Each of the proposed chain of events has some support in literature as discussed earlier. But evidence for every step is not necessarily evidence for the ladder. Therefore, much research is needed to put the proposal and its components to experimental tests. It is important and urgent to test these ideas because the clinical implications could be radical. All the changes above are reversible. The only possible irreversibility would be if neuropathy in phase 5 involves neuronal deaths. All the phases originate in behavioral states, and therefore, appropriate behavioral intervention should be the best strategy to reverse them. This will be discussed more elaborately in a later chapter.

Implications for Clinical Practice, Research, and Drug Development

The theoretical prediction of the role of altered brain glucose dynamics in regulating plasma glucose levels is still open to experimental confirmation. In physics, it is common to make theoretical predictions first and then look for experimental confirmation. In medicine, this has been a red herring. But things are changing rapidly, and therefore, we need to take theoretical predictions seriously. Some evidence supporting the theory is already at hand, but substantial research inputs are needed to reveal the details. If we assume at this stage that experiments confirm the theoretical prediction, the implications are clear and important. When glut1 levels in the blood–brain barrier are reduced, a higher level of plasma glucose is needed to maintain normal brain glucose dynamics. I am reluctant to call this a pathological rise in plasma glucose. I am more inclined to say that the normal range of plasma glucose is shifted upwards when glut-1 reduces. This is different from the current perception of pathological rise in plasma glucose. A limited and stable rise in plasma glucose may be caused by a shift in the normal range rather than a pathological increase. Dysregulation of glucose begins only in phases 4 and 5, but plasma sugar could be raising much before that. Conceptually, the two conditions need to be differentiated. A practical way to differentiate the two at diagnostic level would evolve with more research inputs, and we will have some discussion about it in the following chapter too. This distinction is important because when glucose transport across the BBB is slowed down, insistence on normalizing blood sugar can potentially result in reduced glucose supply to brain. This has been demonstrated as a principle [61, 169], but to what extent it is clinically important is not yet known. Although it has been demonstrated that glut-1 levels rise in response to chronic hypoglycemia [8, 66, 170, 171], brain damages can be acute and the speed of adjustment of glut-1 relatively slow. This means that rapid "normalization" of blood sugar can be dangerous even when peripheral blood sugar does not reach "dangerously low" levels. Good clinical practice might have to realize that the normal range plasma glucose of a diabetic has shifted upwards, and one should try to keep the altered range as a target rather than the normal range of a nondiabetic person. The problem is that, currently, we have no tools to identify and differentiate between shifted normal range versus pathological shift in the level. Conceptually it is not difficult to do so, but intensive research inputs would certainly be needed to evolve strategies and tools to differentiate between the two.

Currently, for patients not responding to insulin sensitizing drugs, infusing insulin is the only possible solution, which gradually goes on losing its effect, and in response, the advocated insulin doses slowly keep on increasing. If glut-1 levels are reduced and there is extreme central shift, increasing insulin doses would be ineffective even theoretically. Attempts to increase glut-1 could be the only possible solution. But as yet, we do not know how to increase glut-1. Perhaps, it could be achieved by chronically keeping low glucose levels since glut-1 is shown to increase on facing hypoglycemia. But this is a potentially dangerous practice. Currently, we have no proven way as to how to achieve this without risking brain damage. Long-term exercise may be of some help in increasing glut-1 [150], but more studies are needed to reveal this. Although we are much in the dark currently, conceptually realizing the ineffectivity of insulin in low glut-1 patients and not recommending higher ineffective doses of insulin would also serve a great deal to the patient.

If the hypothesis that acquired glut-1 deficiency can shift the normal range of blood sugar upwards stands experimental tests, what we need clinically is diagnostic tools to differentiate between hyperglycemia caused by peripheral mechanisms versus one induced by glut-1-mediated central mechanisms. The patient can be asked to take an exercise course and be monitored for fasting insulin and sugar levels. As the HOMA index falls in response to exercise, we can look at the relative contributions of insulin and sugar to change in HOMA. If sugar drop is substantial in comparison with drop in insulin, it indicates hyperglycemia of peripheral origin. If on the other hand, fasting insulin drops significantly without appreciable decrease in fasting sugar, it indicates hyperglycemia of central origin. Another possible diagnostic approach for the future could be to monitor brain glucose levels using some advanced noninvasive tools such as magnetic resonance spectroscopy [123, 167, 168]; the ratio and response patterns between fasting blood and brain sugars can give a differential diagnosis of hyperglycemia of peripheral versus central origin. This sounds feasible for diagnostic laboratories in the near future since these techniques have already started finding applications in neurosurgery [172]. If we can differentially diagnose hyperglycemia of peripheral versus central origin, it would be of great value in a pinpointed diagnosis and a better guideline for treatment.

In pharma R&D, if the hypothesis about brain glucose dynamics being central to hyperglycemia is correct, there would be a shift in focus from pancreas, β cell, and insulin action to focusing on the brain. The next-generation drugs for treating hyperglycemia could be glut-1 transcription enhancers or any other compounds targeting important steps in brain glucose dynamics. These drugs could be more effective in normalizing blood sugar. Patents have already started getting filed for CNS targets of antiobesity and antidiabetic drugs [173, 174], but these seem to be hitting in the dark in the absence of a basic understanding of glucose dynamics in the brain. It needs theoretical development and experimental work simultaneously. At the moment, it can be said with a high level of confidence that the brain is the most important player in glucose regulation, and unless we understand its role, we will not understand hyperglycemia in diabetes.

But understanding the brain's role and finding new targets and new effective drugs to regulate sugar are not enough. Even if a new drug discovered this way turns out to be extremely efficient in normalizing glucose, I am not prepared to call it the next-generation antidiabetic drug. It would only treat hyperglycemia, not diabetes. Diabetes is not mainly about blood sugar, it is about so many different and simultaneous changes in the body; blood sugar is only one of the changes, certainly not central to diabetes as I will elaborate once again now in the following chapter using an even wider canvas.

References

1. Schwartz MW, Porte D (2005) Diabetes, obesity, and the brain. Science 307:375–379
2. Prodi E, Obici S (2006) Minireview: the brain as a molecular target for diabetic therapy. Endocrinology 147:2664–2669
3. Cryer PE (1994) Banting lecture. Hypoglycemia: the limiting factor in the management of IDDM. Diabetes 43:1378–1389
4. Mitrakou A et al (1991) Hierarchy of glycemic thresholds for counterregulatory hormone secretion, symptoms, and cerebral dysfunction. Am J Physiol 260:67–74
5. Obici S, Zhang BB, Karkanias G, Rossetti L (2002) Hypothalamic insulin signaling is required for inhibition of glucose production. Nat Med 8:1376–1382
6. Matsuhisa M et al (2000) Important role of the hepatic vagus nerve in glucose uptake and production by the liver. Metabolism 49:11–16
7. Xue C et al (2000) Isolated hepatic cholinergic denervation impairs glucose and glycogen metabolism. J Surg Res 90:19–25
8. Simpson IA et al (1999) Blood–brain barrier glucose transporter. J Neurochem 72:238–247
9. Yamaguchi N (1992) Sympathoadrenal system in neuroendocrine control of glucose: mechanisms involved in the liver, pancreas, and adrenal gland under hemorrhagic and hypoglycemic stress. Can J Physiol Pharmacol 70:167–206
10. Gilon P, Henquin J-C (2001) Mechanisms and physiological significance of the cholinergic control of pancreatic β-cell function. Endocr Rev 22:565–604
11. Mitrani P, Srinivasan M, Dodds C, Patel MS (2007) Autonomic involvement in the permanent metabolic programming of hyperinsulinemia in the high-carbohydrate rat model. Am J Physiol Endocrinol Metab 292:1364–1377
12. Mitrani P, Srinivasan M, Dodds C, Patel MS (2007) Role of the autonomic nervous system in the development of hyperinsulinemia by high-carbohydrate formula feeding to neonatal rats. Am J Physiol Endocrinol Metab 292:1069–1078
13. Rossi J, Santamäki P, Airaksinen MS, Herzig K-H (2005) Parasympathetic innervation and function of endocrine pancreas requires the glial cell line-derived factor family receptor α2 (GFRα2). Diabetes 54:1324–1330
14. Cryer PE, Gerich JE (1985) Glucose counterregulation, hypoglycemia, and intensive insulin therapy in diabetes mellitus. N Engl J Med 313:232–241
15. Borg WP et al (1994) Ventromedial hypothalamic lesions in rats suppress counterregulatory responses to hypoglycemia. J Clin Invest 93:1677–1682
16. Benton D, Parker PY, Donohoe RT (1996) The supply of glucose to the brain and cognitive functioning. J Biosoc Sci 28:463–479

17. Madsen PL, Cruz NF, Sokoloff L, Dienel GA (1999) Cerebral oxygen/glucose ratio is low during sensory stimulation and rises above normal during recovery[colon] excess glucose consumption during stimulation is not accounted for by lactate efflux from or accumulation in brain tissue. J Cereb Blood Flow Metab 19:393–400
18. Lund Madsen P et al (1995) Persistent resetting of the cerebral oxygen/glucose uptake ratio by brain activation: evidence obtained with the kety-schmidt technique. J Cereb Blood Flow Metab 15:485–491
19. Lu G, Hidenori K (2009) A mathematical model of brain glucose homeostasis. Theor Biol Med Model 6:26
20. Peters A et al (2004) The selfish brain: competition for energy resources. Neurosci Biobehav Rev 28: 143–180
21. Langemann D (2007) Selfish-brain theory: challenges in the top-down analysis of metabolic supply chains. In: Grundy J, Hartmann S, Laender AHF, Maciaszek L, Roddick JF (eds) Proceedings of the tutorials, posters, panels and industrial contributions at the 26th international conference on conceptual modeling – ER 2007. Auckland, New Zealand, CRPIT, 83, pp. 39–49
22. Hitze B et al (2010) How the selfish brain organizes its supply and demand. Front Neuroenergetics 2:7
23. Peters A (2011) The selfish brain: competition for energy resources. Am J Hum Biol 23:29–34
24. Peters A, Langemann D (2009) Build-ups in the supply chain of the brain: on the neuroenergetic cause of obesity and type 2 diabetes mellitus. Front Neuroenergetics 1:1–15
25. Chung M, Göbel B (2012) Mathematical modeling of the human energy metabolism based on the selfish brain theory. Adv Exp Med Biol 736:425–440
26. van Dyken R et al (2010) Low plasma lactate concentration as a biomarker of an incompetent brain-pull: a risk factor for weight gain in type 2 diabetes patients. Psychoneuroendocrinology 35:1287–1293
27. Prentki M (2006) Islet cell failure in type 2 diabetes. J Clin Invest 116:1802–1812
28. D'Alessio DA, Kieffer TJ, Taborsky GJ, Havel PJ (2001) Activation of the parasympathetic nervous system is necessary for normal meal-induced insulin secretion in rhesus macaques. J Clin Endocrinol Metab 86:1253–1259
29. Jarhult J, Holst J (1979) The role of the adrenergic innervation to the pancreatic islets in the control of insulin release during exercise in man. Pflugers Arch 383:41–45
30. N'Guyen JM et al (1994) Involvement of the autonomic nervous system in the in vivo memory to glucose of pancreatic β cell in rats. J Clin Invest 94:1456–1462
31. Ahrén B et al (1997) Dissociated insulinotropic sensitivity to glucose and carbachol in high-fat diet-induced insulin resistance in C57BL/6 J mice. Metabolism 46:97–106
32. Del Rio G et al (1997) Cholinergic enhancement by pyridostigmine increases the insulin response to glucose load in obese patients but not in normal subjects. Int J Obes Relat Metab Disord 21:1111–1114
33. Glasbrenner B, Dominguez-Munoz E, Riepl RL, Vetsi A, Malfertheiner P (1995) Cholecystokinin and pancreatic polypeptide release in diabetic patients with and without autonomic neuropathy. Dig Dis Sci 40:406–411
34. Buysschaert M, Donckier J, Dive A, Ketelslegers JM, Lambert AE (1985) Gastric acid and pancreatic polypeptide responses to sham feeding are impaired in diabetic subjects with autonomic neuropathy. Diabetes 34:1181–1185
35. Kurose T et al (1992) Glucagon, insulin and somatostatin secretion in response to sympathetic neural activation in streptozotocin-induced diabetic rats. A study with the isolated perfused rat pancreas in vitro. Diabetologia 35:1035–1041
36. Begin-Heick N (1994) Liver β-adrenergic receptors, G proteins, and adenylyl cyclase activity in obesity-diabetes syndromes. Am J Physiol 266:1664–1672
37. Koo S-H et al (2004) PGC-1 promotes insulin resistance in liver through PPAR-α-dependent induction of TRB-3. Nat Med 10:530–534
38. Yoon JC et al (2001) Control of hepatic gluconeogenesis through the transcriptional coactivator PGC-1. Nature 413:131–138
39. Terada S et al (2002) Effects of low-intensity prolonged exercise on PGC-1 mRNA expression in rat epitrochlearis muscle. Biochem Biophys Res Commun 296:350–354
40. Puigserver P, Wu Z, Park CW, Graves R, Wright M, Spiegelman BM (1998) A cold-inducible coactivator of nuclear receptors linked to adaptive thermogenesis. Cell 92:829–839
41. Mootha VK et al (2003) PGC-1α-responsive genes involved in oxidative phosphorylation are coordinately downregulated in human diabetes. Nat Genet 34:267–273
42. Hammarstedt A, Jansson P-A, Wesslau C, Yang X, Smith U (2003) Reduced expression of PGC-1 and insulin-signaling molecules in adipose tissue is associated with insulin resistance. Biochem Biophys Res Commun 301:578–582
43. Crunkhorn S et al (2007) Peroxisome proliferator activator receptor gamma coactivator-1 expression is reduced in obesity: potential pathogenic role of saturated fatty acids and p38 mitogen-activated protein kinase activation. J Biol Chem 282:15439–15450
44. Liang H, Ward WF (2006) PGC-1α: a key regulator of energy metabolism. Adv Physiol Educ 30:145–151
45. Evans ML, Sherwin RS (2002) Brain glucose metabolism and hypoglycaemia. Diabetes Nutr Metab 15:294–296, discussion 296
46. Buettner C, Camacho RC (2008) Hypothalamic control of hepatic glucose production and its potential role in insulin resistance. Endocrinol Metab Clin North Am 37:825–840

47. German J et al (2009) Hypothalamic leptin signaling regulates hepatic insulin sensitivity via a neurocircuit involving the vagus nerve. Endocrinology 150: 4502–4511
48. Rother E, Könner AC, Brüning JC (2008) Neurocircuits integrating hormone and nutrient signaling in control of glucose metabolism. Am J Physiol Endocrinol Metab 294:E810–E816
49. Obici S et al (2001) Central melanocortin receptors regulate insulin action. J Clin Invest 108:1079–1085
50. Könner AC et al (2007) Insulin action in AgRP-expressing neurons is required for suppression of hepatic glucose production. Cell Metab 5:438–449
51. Silver I, Erecinska M (1994) Extracellular glucose concentration in mammalian brain: continuous monitoring of changes during increased neuronal activity and upon limitation in oxygen supply in normo-, hypo-, and hyperglycemic animals. J Neurosci 14:5068–5076
52. Gjedde A, Rasmussen M (1980) Pentobarbital anesthesia reduces blood-brain glucose transfer in the rat. J Neurochem 35:1382–1387
53. Hargreaves RJ, Planas AM, Cremer JE, Cunningham VJ (1986) Studies on the relationship between cerebral glucose transport and phosphorylation using 2-deoxyglucose. J Cereb Blood Flow Metab 6: 708–716
54. Cremer JE, Cunningham VJ, Seville MP (1983) Relationships between extraction and metabolism of glucose, blood flow, and tissue blood volume in regions of rat brain. J Cereb Blood Flow Metab 3:291–302
55. Lautt WW et al (2001) Hepatic parasympathetic (HISS) control of insulin sensitivity determined by feeding and fasting. Am J Physiol Gastrointest Liver Physiol 281:G29–G36
56. Correia NC, Guarino MP, Raposo J, Macedo MP (2002) Hepatic guanylyl cyclase inhibition induces HISS-dependent insulin resistance. Proc West Pharmacol Soc 45:57–58
57. Pelligrino DA, LaManna JC, Duckrow RB, Bryan RM Jr, Harik SI (1992) Hyperglycemia and blood-brain barrier glucose transport. J Cereb Blood Flow Metab 12:887–899
58. McNay EC, McCarty RC, Gold PE (2001) Fluctuations in brain glucose concentration during behavioral testing: dissociations between brain areas and between brain and blood. Neurobiol Learn Mem 75:325–337
59. Herzog RI, Sherwin RS, Rothman DL (2011) Insulin-induced hypoglycemia and its effect on the brain. Diabetes 60:1856–1858
60. McCall AL, Millington WR, Wurtman RJ (1982) Metabolic fuel and amino acid transport into the brain in experimental diabetes mellitus. Proc Natl Acad Sci U S A 79:5406–5410
61. Gjedde A, Crone C (1981) Blood-brain glucose transfer: repression in chronic hyperglycemia. Science 214:456–457
62. Matthaei S, Horuk R, Olefsky JM (1986) Blood-brain glucose transfer in diabetes mellitus. Decreased number of glucose transporters at blood-brain barrier. Diabetes 35:1181–1184
63. Pardridge WM, Triguero D, Farrell CR (1990) Downregulation of blood-brain barrier glucose transporter in experimental diabetes. Diabetes 39: 1040–1044
64. Cornford EM, Hyman S, Cornford ME, Clare-Salzler M (1995) Down-regulation of blood-brain glucose transport in the hyperglycemic nonobese diabetic mouse. Neurochem Res 20:869–873
65. Kumagai AK (1999) Glucose transport in brain and retina: implications in the management and complications of diabetes. Diabetes Metab Res Rev 15:261–273
66. Koranyi L et al (1991) Glucose transporter gene expression in rat brain: pretranslational changes associated with chronic insulin-induced hypoglycemia, fasting, and diabetes. Mol Cell Neurosci 2:244–252
67. Vorbrodt A, Dobrogowka D, Meeker H, Carp R, Tarnawski M (2007) Quantitative immunogold study of glucose transporter (GLUT-1) in five brain regions of scrapie-infected mice showing obesity and reduced glucose tolerance. Acta Neuropathol 102:278–284
68. Wertheimer E, Sasson S, Cerasi E, Ben-Neriah Y (1991) The ubiquitous glucose transporter GLUT-1 belongs to the glucose-regulated protein family of stress-inducible proteins. Proc Natl Acad Sci USA 88:2525–2529
69. Ehrlich JC, Ratner IM (1961) Amyloidosis of the islets of langerhans. Am J Pathol 38:49–59
70. Maloy AL, Longnecker DS, Robert Greenberg E (1981) The relation of islet amyloid to the clinical type of diabetes. Hum Pathol 12:917–922
71. Kahn SE, Andrikopoulos S, Verchere CB (1999) Islet amyloid: a long-recognized but underappreciated pathological feature of type 2 diabetes. Diabetes 48:241–253
72. Hull RL, Westermark GT, Westermark P, Kahn SE (2004) Islet amyloid: a critical entity in the pathogenesis of type 2 diabetes. J Clin Endocrinol Metab 89:3629–3643
73. Teff KL (2010) Cephalic phase pancreatic polypeptide responses to liquid and solid stimuli in humans. Physiol Behav 99:317–323
74. Berthoud HR, Bereiter DA, Trimble ER, Siegel EG, Jeanrenaud B (1981) Cephalic phase, reflex insulin secretion. Neuroanatomical and physiological characterization. Diabetologia 20:393–401
75. Mattes RD, Engelman K, Mattern J, Teff KL (1993) Cephalic-phase insulin in obese and normal-weight men: relation to postprandial insulin. Metabolism 42:1600–1608
76. Berthoud HR, Jeanrenaud B (1982) Sham feeding-induced cephalic phase insulin release in the rat. Am J Physiol Endocrinol Metab 242:280–285

77. Rohner-Jeanrenaud F, Jeanrenaud B (1983) The central nervous system-endocrine pancreas axis. Ann Endocrinol (Paris) 44:217–227
78. Calhoun P et al (1986) Evaluation of insulin secretion after pancreas autotransplantation by oral or intravenous glucose challenge. Ann Surg 204:585–593
79. Succurro E et al (2008) Insulin secretion in metabolically obese, but normal weight, and in metabolically healthy but obese individuals. Obesity 16:1881–1886
80. Spyer G, Hattersley AT, MacDonald I, Amiel S, MacLeod KM (2000) Hypoglycaemic counter-regulation at normal blood glucose concentrations in patients with well controlled type-2 diabetes. Lancet 356:1970–1974
81. DeFronzo RA, Hendler R, Christensen N (1980) Stimulation of counterregulatory hormonal responses in diabetic man by a fall in glucose concentration. Diabetes 29:125–131
82. Boyle PJ et al (1988) Plasma glucose concentrations at the onset of hypoglycemic symptoms in patients with poorly controlled diabetes and in nondiabetics. N Engl J Med 318:1487–1492
83. Bolli G et al (1983) Abnormal glucose counterregulation in insulin-dependent diabetes mellitus. Interaction of anti-insulin antibodies and impaired glucagon and epinephrine secretion. Diabetes 32:134–141
84. Simonson D, Tamborlane WWV, De Fronzo R, Sherwin RS (1985) Intensive insulin therapy reduces counterregulatory hormone responses to hypoglycemia in patients with type I diabetes. Ann Intern Med 103:184–190
85. Palmer JP, Henry DP, Benson JW, Johnson DG, Ensinck JW (1976) Glucagon response to hypoglycemia in sympathectomized man. J Clin Invest 57:522–525
86. Popp DA, Shah SD, Cryer PE (1982) Role of epinephrine-mediated β-adrenergic mechanisms in hypoglycemic glucose counterregulation and post-hypoglycemic hyperglycemia in insulin-dependent diabetes mellitus. J Clin Invest 69:315–326
87. Monzillo LU, Hamdy O (2003) Evaluation of insulin sensitivity in clinical practice and in research settings. Nutr Rev 61:397–412
88. Brun J-F, Raynaud E, Mercier J (2000) Homeostasis model assessment and related simplified evaluations of insulin sensitivity from fasting insulin and glucose. Diabetes Care 23:1037–1038
89. Mari A, Pacini G, Murphy E, Ludvik B, Nolan JJ (2001) A model-based method for assessing insulin sensitivity from the oral glucose tolerance test. Diabetes Care 24:539–548
90. Avignon A, Boegner C, Mariano-Goulart D, Colette C, Monnier L (1999) Assessment of insulin sensitivity from plasma insulin and glucose in the fasting or post oral glucose-load state. Int J Obes 23:512–517
91. Matsuda M, DeFronzo RA (1999) Insulin sensitivity indices obtained from oral glucose tolerance testing: comparison with the euglycemic insulin clamp. Diabetes Care 22:1462–1470
92. Brüning JC et al (1998) A muscle-specific insulin receptor knockout exhibits features of the metabolic syndrome of NIDDM without altering glucose tolerance. Mol Cell 2:559–569
93. DeFronzo RA, Ferrannini E, Simonson DC (1989) Fasting hyperglycemia in non-insulin-dependent diabetes mellitus: contributions of excessive hepatic glucose production and impaired tissue glucose uptake. Metabolism 38:387–395
94. Kahn CR (2003) Knockout mice challenge our concepts of glucose homeostasis and the pathogenesis of diabetes. Exp Diabesity Res 4:169–182
95. Michael MD et al (2000) Loss of insulin signaling in hepatocytes leads to severe insulin resistance and progressive hepatic dysfunction. Mol Cell 6:87–97
96. Campbell IW (1976) The somogyi phenomenon. A short review. Acta Diabetol Lat 13:68–73
97. Shanik MH et al (2008) Insulin resistance and hyperinsulinemia: Is hyperinsulinemia the cart or the horse? Diabetes Care 31:262–268
98. Ono T, Steffens AB, Sasaki K (1983) Influence of peripheral and intracerebroventricular glucose and insulin infusions on peripheral and cerebrospinal fluid glucose and insulin levels. Physiol Behav 30:301–306
99. Burdakov D, Luckman SM, Verkhratsky A (2005) Glucose-sensing neurons of the hypothalamus. Philos Trans R Soc Lond B Biol Sci 360:2227–2235
100. Levin BE, Dunn-Meynell AA, Routh VH (1999) Brain glucose sensing and body energy homeostasis: role in obesity and diabetes. Am J Physiol 276:R1223–R1231
101. Clinkenbeard KD, Reed WD, Mooney RA, Lane MD (1975) Intracellular localization of the 3-hydroxy-3-methylglutaryl coenzyme A cycle enzymes in liver. Separate cytoplasmic and mitochondrial 3-hydroxy-3-methylglutaryl coenzyme A generating systems for cholesterogenesis and ketogenesis. J Biol Chem 250:3108–3116
102. Williamson DH, Bates MW, Krebs HA (1968) Activity and intracellular distribution of enzymes of ketone-body metabolism in rat liver. Biochem J 108:353–361
103. Chapman MJ, Miller LR, Ontko JA (1973) Localization of the enzymes of ketogenesis in rat liver mitochondria. J Cell Biol 58:284–306
104. Laffel L (1999) Ketone bodies: a review of physiology, pathophysiology and application of monitoring to diabetes. Diabetes Metab Res Rev 15:412–426
105. Kitabchi AE et al (2001) Management of hyperglycemic crises in patients with diabetes. Diabetes Care 24:131–153
106. Bolli G et al (1979) Urinary excretion and plasma levels of norepinephrine and epinephrine during diabetic ketoacidosis. Acta Diabetol Lat 16:157–167
107. Porte D (1969) Sympathetic regulation of insulin secretion: its relation to diabetes mellitus. Arch Intern Med 123:252–260
108. Kreisberg RA (1978) Diabetic ketoacidosis: new concepts and trends in pathogenesis and treatment. Ann Intern Med 88:681–695

109. McRae JR et al (1985) Prostaglandin E2 metabolite levels during diabetic ketoacidosis. Diabetes 34:761–766
110. Johnson ME, Das NM, Butcher FR, Fain JN (1972) The regulation of gluconeogenesis in isolated rat liver cells by glucagon, insulin, dibutyryl cyclic adenosine monophosphate, and fatty acids. J Biol Chem 247:3229–3235
111. Rattigan S, Clark MG, Barrett EJ (1997) Hemodynamic actions of insulin in rat skeletal muscle: evidence for capillary recruitment. Diabetes 46:1381–1388
112. Laakso M, Edelman SV, Brechtel G, Baron AD (1992) Impaired insulin-mediated skeletal muscle blood flow in patients with NIDDM. Diabetes 41:1076–1083
113. Laakso M, Edelman SV, Brechtel G, Baron AD (1990) Decreased effect of insulin to stimulate skeletal muscle blood flow in obese man. A novel mechanism for insulin resistance. J Clin Invest 85:1844–1852
114. Bingham EM et al (2002) The role of insulin in human brain glucose metabolism: an 18fluorodeoxyglucose positron emission tomography study. Diabetes 51:3384–3390
115. Donohoe RT, Benton D (1999) Cognitive functioning is susceptible to the level of blood glucose. Psychopharmacology 145:378–385
116. Benton D, Owens DS (1993) Blood glucose and human memory. Psychopharmacology 113:83–88
117. Martin PY, Benton D (1999) The influence of a glucose drink on a demanding working memory task. Physiol Behav 67:69–74
118. Benton D (1990) The impact of increasing blood glucose on psychological functioning. Biol Psychol 30:13–19
119. Donohoe RT, Benton D (2000) Glucose tolerance predicts performance on tests of memory and cognition. Physiol Behav 71:395–401
120. Nelson SR, Schulz DW, Passonneau JV, Lowry OH (1968) Control of glycogen levels in brain1. J Neurochem 15:1271–1279
121. Sickmann HM, Waagepetersen HS, Schousboe A, Benie AJ, Bouman SD (2010) Obesity and type 2 diabetes in rats are associated with altered brain glycogen and amino-acid homeostasis. J Cereb Blood Flow Metab 30:1527–1537
122. Heikkilä O et al (2010) Cerebellar glucose during fasting and acute hyperglycemia in nondiabetic men and in men with type 1 diabetes. Cerebellum 9:336–344
123. Rseaquist E, Tkac I, Damberg G, Thomas W, Gruetter R (2005) Brain glucose concentrations in poorly controlled diabetes mellitus as measured by high-field magnetic resonance spectroscopy. Metabolism 54:1008–1013
124. Heikkilä O et al (2010) Evidence for abnormal glucose uptake or metabolism in thalamus during acute hyperglycaemia in type 1 diabetes – a 1 H MRS study. Metab Brain Dis 25:227–234
125. Criego AB et al (2005) Brain glucose concentrations in patients with type 1 diabetes and hypoglycemia unawareness. J Neurosci Res 79:42–47
126. Lei H, Gruetter R (2006) Effect of chronic hypoglycaemia on glucose concentration and glycogen content in rat brain: a localized 13C NMR study. J Neurochem 99:260–268
127. Boyle PJ et al (1994) Adaptation in brain glucose uptake following recurrent hypoglycemia. Proc Natl Acad Sci USA 91:9352–9356
128. Káplár M et al (2009) Changes in cerebral blood flow detected by SPECT in type 1 and type 2 diabetic patients. J Nucl Med 50:1993–1998
129. Cosentino F et al (2009) Impact of fasting glycemia and regional cerebral perfusion in diabetic subjects: a study with technetium-99 m-ethyl cysteinate dimer single photon emission computed tomography. Stroke 40:306–308
130. Mårin P, Andersson B, Krotkiewski M, Björntorp P (1994) Muscle fiber composition and capillary density in women and men with NIDDM. Diabetes Care 17:382–386
131. Lillioja S et al (1987) Skeletal muscle capillary density and fiber type are possible determinants of in vivo insulin resistance in man. J Clin Invest 80:415–424
132. Lash JM, Sherman WM, Hamlin RL (1989) Capillary basement membrane thickness and capillary density in sedentary and trained obese Zucker rats. Diabetes 38:854–860
133. Hattersley AT, Tooke JE (1999) The fetal insulin hypothesis: an alternative explanation of the association of low birthweight with diabetes and vascular disease. Lancet 353:1789–1792
134. Šalković-Petrišić M, Riederer P (2010) Brain glucose transporter protein 2 and sporadic Alzheimer's disease. Transl Neurosci 1:200–206
135. Wijesuriya HC, Bullock JY, Faull RLM, Hladky SB, Barrand MA (2010) ABC efflux transporters in brain vasculature of Alzheimer's subjects. Brain Res 1358:228–238
136. Correia SC et al (2011) Insulin-resistant brain state: the culprit in sporadic Alzheimer's disease? Ageing Res Rev 10:264–273
137. Klepper J, Voit T (2002) Facilitated glucose transporter protein type 1 (GLUT1) deficiency syndrome: impaired glucose transport into brain – a review. Eur J Pediatr 161:295–304
138. Seidner G et al (1998) GLUT-1 deficiency syndrome caused by haploinsufficiency of the blood-brain barrier hexose carrier. Nat Genet 18:188–191
139. Wang D et al (2006) A mouse model for Glut-1 haploinsufficiency. Hum Mol Genet 15:1169–1179
140. Korytkowski MT et al (1998) Reduced β-adrenergic sensitivity in patients with type 1 diabetes and hypoglycemia unawareness. Diabetes Care 21: 1939–1943
141. Cryer PE (1993) Hypoglycemia unawareness in IDDM. Diabetes Care 16(Suppl 3):40–47

142. Veneman TF, Erkelens DW (1997) Clinical review 88: hypoglycemia unawareness in noninsulin-dependent diabetes mellitus. J Clin Endocrinol Metab 82:1682–1684
143. Klepper J, Diefenbach S, Kohlschütter A, Voit T (2004) Effects of the ketogenic diet in the glucose transporter 1 deficiency syndrome. Prostaglandins Leukot Essent Fatty Acids 70:321–327
144. Klepper J et al (2005) Seizure control and acceptance of the ketogenic diet in GLUT1 deficiency syndrome: a 2- to 5-year follow-up of 15 children enrolled prospectively. Neuropediatrics 36:302–308
145. Liu Z-H, Guan T-J, Chen Z-H, Li L-S (1999) Glucose transporter (GLUT1) allele (XbaI-) associated with nephropathy in non-insulin-dependent diabetes mellitus. Kidney Int 55:1843–1848
146. Ng DPK et al (2002) Minor effect of GLUT1 polymorphisms on susceptibility to diabetic nephropathy in type 1 diabetes. Diabetes 51:2264–2269
147. Pontiroli AE et al (1996) Genetic contribution of polymorphism of the GLUT1 and GLUT4 genes to the susceptibility to type 2 (non-insulin-dependent) diabetes mellitus in different populations. Acta Diabetol 33:193–197
148. Hodgkinson AD, Millward BA, Demaine AG (2001) Polymorphisms of the glucose transporter (GLUT1) gene are associated with diabetic nephropathy. Kidney Int 59:985–989
149. Baroni M et al (1992) Polymorphisms at the GLUT1 (HepG2) and GLUT4 (muscle/adipocyte) glucose transporter genes and non-insulin-dependent diabetes mellitus (NIDDM). Hum Genet 88:557–561
150. Um HS et al (2008) Exercise training acts as a therapeutic strategy for reduction of the pathogenic phenotypes for Alzheimer's disease in an NSE/APPsw-transgenic model. Int J Mol Med 22:529–539
151. Boyle PJ et al (1994) Diminished brain glucose metabolism is a significant determinant for falling rates of systemic glucose utilization during sleep in normal humans. J Clin Invest 93:529–535
152. Maquet P et al (1990) Cerebral glucose utilization during sleep-wake cycle in man determined by positron emission tomography and [18 F]2-fluoro-2-deoxy-d-glucose method. Brain Res 513:136–143
153. Nofzinger EA et al (2002) Human regional cerebral glucose metabolism during non rapid eye movement sleep in relation to waking. Brain 125:1105–1115
154. Shah GN, Giddings SJ, Mooradian AD (1997) Shortening of poly (A) tail of glucose transporter – one mRNA in experimental diabetes mellitus. Brain Res 754:213–220
155. Pifferi F et al (2005) (n-3) polyunsaturated fatty acid deficiency reduces the expression of both isoforms of the brain glucose transporter GLUT1 in rats. J Nutr 135:2241–2246
156. Pifferi F et al (2007) n-3 Fatty acids modulate brain glucose transport in endothelial cells of the blood-brain barrier. Prostaglandins Leukot Essent Fatty Acids 77:279–286
157. Ximenes da Silva A et al (2002) Glucose transport and utilization are altered in the brain of rats deficient in n-3 polyunsaturated fatty acids. J Neurochem 81:1328–1337
158. Freemantle E et al (2006) Omega-3 fatty acids, energy substrates, and brain function during aging. Prostaglandins Leukot Essent Fatty Acids 75: 213–220
159. Liu D, Pitta M, Mattson MP (2008) Preventing NAD+depletion protects neurons against excitotoxicity. Ann N Y Acad Sci 1147:275–282
160. Mayman CI, Gatfield PD, Breckenridge BM (1964) The glucose content of brain in anaesthesia. J Neurochem 11:483–487
161. Duelli R, Kuschinsky W (2001) Brain glucose transporters: relationship to local energy demand. News Physiol Sci 16:71–76
162. McEwen BS, Reagan LP (2004) Glucose transporter expression in the central nervous system: relationship to synaptic function. Eur J Pharmacol 490:13–24
163. Zeller K, Rahner-Welsch S, Kuschinsky W (1997) Distribution of Glut1 glucose transporters in different brain structures compared to glucose utilization and capillary density of adult rat brains. J Cereb Blood Flow Metab 17:204–209
164. Grolund KM et al (1996) Chronic seizures increase glucose transporter abundance in rat brain. J Neuropathol Exp Neurol 55:832–840
165. Burkhalter J, Fiumelli H, Allaman I, Chatton J-Y, Martin J-L (2003) Brain-derived neurotrophic factor stimulates energy metabolism in developing cortical neurons. J Neurosci 23:8212–8220
166. Sparrow B, Liu J, Wegner DM (2011) Google effects on memory: cognitive consequences of having information at our fingertips. Science 333:776–778
167. van de Ven KCC et al (2011) Effect of acute hypoglycemia on human cerebral glucose metabolism measured by 13 C magnetic resonance spectroscopy. Diabetes 60:1467–1473
168. Gruetter R et al (1992) Direct measurement of brain glucose concentrations in humans by 13C NMR spectroscopy. Proc Natl Acad Sci USA 89:1109–1112
169. Oddo M et al (2008) Impact of tight glycemic control on cerebral glucose metabolism after severe brain injury: a microdialysis study. Crit Care Med 36:3233–3238
170. Kumagai AK, Kang YS, Boado RJ, Pardridge WM (1995) Upregulation of blood-brain barrier GLUT1 glucose transporter protein and mRNA in experimental chronic hypoglycemia. Diabetes 44:1399–1404
171. McCall AL et al (1986) Chronic hypoglycemia increases brain glucose transport. Am J Physiol Endocrinol Metab 251:E442–E447
172. Sutherland GR et al (1999) A mobile high-field magnetic resonance system for neurosurgery. J Neurosurg 91:804–813
173. Bays HE (2004) Current and investigational antiobesity agents and obesity therapeutic treatment targets. Obesity 12:1197–1211
174. Sandoval DA, Obici S, Seeley RJ (2009) Targeting the CNS to treat type 2 diabetes. Nat Rev Drug Discov 8:386–398

13 Beating Around the "Wrong" Bush?

It is time now to address the most important question as to why T2D is largely incurable or irreversible. There are three possible reasons for the apparent irreversibility. It is possible that the pathophysiological changes are irreversible by nature. Alternatively, we know what needs to be done in order to cure diabetes, but we do not have the right drugs or the technology to do it. The third possibility is that we do not know what the real cause is, and therefore, we are trying to hit a target that has nothing much to do with the disease. I am going to argue below that the third one is true. All the antidiabetic drugs currently in use and the new candidates in various stages of R&D are hitting at the wrong target and therefore are bound to be ineffective in principle. The problem does not lie in not having the technology to mend what is wrong. The problem lies in identifying what is really wrong. If insulin resistance is central to the pathophysiology of T2D, then efficient insulin-sensitizing drugs are available. But insulin-sensitizing drugs do not cure diabetes, and therefore, we need to suspect whether insulin resistance is really at the root of the pathophysiology of diabetes. If we knew that insulin resistance was the key to T2D but did not have any drug to reduce insulin resistance, that condition would have been comparable to beating *around* the bush. But the emerging picture is not this. We are unable to cure diabetes not because we do not have insulin-sensitizing drugs. We have them and we also know that exercises are insulin sensitizing. But still medicine has failed to prevent or cure diabetes. This situation is worse than beating around the bush, and I will call it beating around the *wrong* bush. It is not very difficult to see why I say this if we look at some of the recent evidence accumulating on the broader horizon of insulin action and insulin sensitivity across the animal world.

Wrong Bush 1: Insulin Resistance

The protein insulin and its equivalents are widespread in the animal world including vertebrates and invertebrates. Since impairment of insulin signaling or insulin resistance is believed to be central to a cluster of deadly diseases in humans, it would be useful to look at what happens when there is impairment of insulin signaling in other animals. And here lies a very fundamental paradox. In widely different taxa of animals, including both invertebrate and vertebrates, impairment of insulin signaling has no adverse effects on health; on the contrary, it increases life span. This is one very fundamental and perplexing fact reviewed very well by Kaletsky et al. [1].

Over 20 years ago, it was discovered that mutations in *daf-2* and *age-1* double the life span of a nematode worm model *Caenorhabditis elegans* [2, 3]. Subsequent cloning of these genes uncovered the importance of the insulin/insulin-like growth factor-1 (IGF-1) signaling pathway in aging regulation. *daf-2* encodes the only insulin/IGF-1 receptor expressed in worms, and *age-1* is the catalytic subunit of the downstream phosphoinositide 3-kinase (PI3K) [4, 5]. Mutations in

M. Watve, *Doves, Diplomats, and Diabetes: A Darwinian Interpretation of Type 2 Diabetes and Related Disorders*, DOI 10.1007/978-1-4614-4409-1_13, © Springer Science+Business Media New York 2013

these genes interfere with the normal insulin signaling pathway and therefore are equivalents to insulin resistance in humans. These "insulin-resistant" worms have a substantially longer life span. In addition to the overall life span extension, worms bearing either of these mutations, i.e., the insulin-resistant worms, are highly resistant to oxidative stress [6], hypoxia [7], heat [8], heavy metals [9], and bacterial pathogens [10].

The insulin/insulin-like growth factor signaling (IIS) pathway also acts as a food and stress sensor during development. When food is abundant, worms develop rapidly and uninterrupted through four larval stages to reach adulthood. If worms develop in food-limited or overcrowded conditions, they enter an alternative long-lived larval state called dauer in which reproductive maturity is delayed but resistance to environmental challenges increases. Favorable nutritional conditions promote exit from dauer which is again mediated by changes in insulin signaling pathway and development to reproductive adults [11]. The *daf-2* mutants are dauer phenotype by default [12]. Both the long-lived and dauer phenotypes of *daf-2* worms are dependent on the downstream forkhead transcription factor DAF-16 [2, 12, 13]. Briefly speaking, signaling through DAF-2 activates PI3K, which leads to the phosphorylation of DAF-16 and its inactivation by nuclear exclusion [14]. In the absence of IIS or in *daf-2* or *age-1* mutants, DAF-16 enters the nucleus and enacts a transcriptional program that doubles worm life span [10, 15–17]. Loss of *daf-2* or *age-1* also slows the age-related declines; thus, the insulin signaling pathway regulates longevity through its modulation of processes and pathways leading to aging.

This is not a worm story alone. Insulin signaling is an evolutionarily conserved pathway that regulates life span across many species. *Drosophila melanogaster* has a single insulin-like receptor (*dInR*) that, when mutated, extends life span in a manner that is dependent on its *daf-16* homolog *dFOXO* or simply *foxo* [18]. Overexpression of *dFOXO* in the fat body [19] or mutation of the insulin receptor substrate protein *chico* extends life span in the fly [20, 21]. Mammals encode several *daf-2* homologs including IGF-1 receptor (IGF-1R) and insulin receptor (IR)-A [22]. Despite these differences in insulin receptor expression, the functional consequences are similar, as reduced insulin/IGF signaling extends life span in multiple mammalian species [23, 24]. Longevity in dogs is inversely proportional to body size [25]. Interestingly, a single nucleotide polymorphism in IGF-1 is associated with small size and long life, suggesting that IIS influences aging in dogs [26]. Heterozygous IGF-1R knockout mice are long lived [27], and adipose tissue-specific insulin receptor knockout (FIRKO) mice also exhibit increases in life span [28, 29].

In mammals, there are some counterexamples where low insulin production and insulin sensitivity are associated with long life. Naked mole rats, known for their exceptionally long life spans, are insulin sensitive [30]. Recent studies suggest that IIS and aging are linked processes in mammals and specifically humans as well. But there appears to be a contradiction in the direction of the association. It is well known that insulin resistance is associated with a large cluster of disorders, many of which are fatal and therefore should reduce life span. Insulin sensitivity is a common feature of centenarians. Mutations in a human *daf-16* homolog, *FOXO3a*, are linked to increased longevity [31–33]. Long-lived men who are homozygous for the *FOXO3a GG* genotype also display greater insulin sensitivity [31]. Therefore it may appear that insulin sensitivity rather than insulin resistance is the key to longevity in humans. We may conclude that sometime in evolution, the relationship might have turned upside down.

But unfortunately this inference is not so straightforward. Other pieces of evidence point to the possibility that similar to worms and flies, insulin resistance increases life span in humans too. IGF-1R mutations are highly represented in populations of centenarians [34] indicating impaired IIS associated with long life. Of particular importance is the gene called *Klotho* known to be a longevity gene in mammals [35]. *Klotho*, the gene coding for a transmembrane enzyme, was discovered in 1997. Overexpression of this gene increases life span, and mutants with impaired expression age prematurely. It was therefore named after one of the three Greek

Moirae or fate called Clotho that weaves the thread of life. It was soon discovered that *Klotho* increases life span by inducing insulin resistance. *Klotho* induces adipocyte differentiation [36], lipogenesis [37], and insulin resistance [38, 39]. Contrasting with this stands out the fact that *Klotho* increases life span substantially. There are two different paradoxes here. First is that although almost every model of longevity extension involves insulin signaling, the direction appears confusing. In some examples longevity is associated with increased insulin sensitivity and in other models insulin resistance. The emerging solution to the paradox is that it is reduction in insulin signaling at a subcellular level that increases life span. This can be achieved either by lower levels of insulin or by reducing the sensitivity to insulin. If this turns out to be true, no real paradox is left since in long-lived phenotypes that are insulin sensitive such as naked mole rat, insulin production is indeed very low.

The concept of insulin resistance increasing life span is actually theoretically very sound at both proximate and ultimate levels. We have argued before that insulin resistance is a mechanism of modifying reproductive strategies to adapt to the demography and environment. Typical *r* reproductive strategy is insulin sensitive, and *K* reproductive strategy is insulin resistant. As a broad generalization across species, *K*-selected species are generally long lived as compared to r-selected species. We do not know whether this would be true for reproductive strategies within a species as well. But theoretically it should and therefore insulin-resistant individuals should be long lived as expected by life history theories. It is interesting to note that although the *Klotho* protein is essential for fertility, as demonstrated by infertility of *Klotho* knockouts [40], *Klotho* overexpression actually reduces fecundity [41]. Also note that placenta produces *Klotho* protein [42] which by inducing insulin resistance can draw more glucose through the placenta. Therefore *Klotho* appears to be the ideal *K* reproduction mechanism, one that increases longevity, reduces fecundity, and increases transplacental investment in the fetus through insulin resistance.

Insulin is a growth hormone during developmental period [43, 44]. Insulin has modulating effects on reproductive strategies [45–48], and insulin is an important modulator of life span as seen above. All three are important parameters of life history evolution. This means that insulin is the chief mediator of life history strategies. Being a fine-tuner of aggressive behavior, it also serves the function of selecting behavioral strategies for an individual. Aggression cannot be separated from life history strategies. Presence of predators is well known to shape life history strategies. Having aggressive competitors would play a role similar to predators. Since both predation and aggression increase the probability of death, life histories will tend more towards *r* and away from *K*. Therefore, insulin's role in aggression cannot be separated from its role in life history strategies. This appears to be the main role of insulin conserved across simple to complex animal taxa. Insulin's role in glucose homeostasis is only a small part of this bigger role. Just because insulin's role in glucose homeostasis was discovered earlier, medicine believed that that was the primary function of insulin. An evolutionary perspective shows that this is a grossly incomplete and biased view. Understanding insulin's function on a broader canvas can certainly deepen our understanding about insulin and thereby give us greater insights into diabetes too. Without understanding life history strategies and life sustenance strategies, it is impossible to understand insulin. Traditional medicine has been doing precisely this mistake so far and that must have contributed to the overall failure to prevent, control, or cure diabetes.

This leads us to the other paradox. How is insulin resistance associated with increased life span when it leads to so many deadly disorders? Here, once again, *Klotho* shows the way out. The paradox gets resolved when we look at the other metabolic and endocrine effects of *Klotho*. *Klotho* has anti-inflammatory effect [49], proangiogenic action [50], proendothelial function [51], and antioxidant activity [52]. We have seen earlier and will elaborate once again below that inflammatory changes, endothelial dysfunction, angiogenesis dysfunction, and oxidative damage are the main pathophysiological mechanisms of T2D. The prevalent perception is that insulin resistance or hyperglycemia is the driver of these pathological effects. Since in diabetes insulin resistance and/or hyperglycemia is invariably coupled with its pathophysiological mechanisms, the association has been assumed to be mandatory. A fundamental conceptual revision that *Klotho* has necessitated is that the association is not mandatory. Here is a mechanism in the form of *Klotho* protein that can decouple obesity and insulin resistance from the true pathological mechanisms of systemic inflammation, endothelial and angiogenesis dysfunction, and oxidative damages. Having thus decoupled, insulin resistance appears to increase longevity.

The inference that one can draw from *Klotho* research is that insulin resistance is not sufficient to drive the pathophysiological mechanisms of T2D. The real pathological drivers are different. In fact insulin resistance may be good for health if there is a way to decouple it from the true pathological drivers. *Klotho* is certainly an effective decoupler, but perhaps there can be other mechanisms too. Once we know that such a decoupling is possible in principle, researchers can specifically look for other mechanisms. The decoupling also means that if the pathophysiological mechanisms are not targeted, targeting insulin resistance will be ineffective as a strategy for the treatment of T2D. It may appear to normalize blood sugar temporarily but is most unlikely to cure diabetes because insulin resistance is not central to the pathophysiology of diabetes at all. Insulin resistance is the target for the mainstream pharmacological R&D in search of a wonder drug against T2D. A generation of insulin-sensitizing drugs is already in the market and being used commonly, but T2D remains an "incurable" disorder. This is because all our efforts are not only beating *around* the bush but beating around the *wrong* bush. Although insulin resistance is central to the behavioral transition from soldier to diplomat, it is not likely to be the root cause of T2D and related pathology, and therefore, targeting insulin resistance will never cure diabetes.

Wrong Bush 2: Hyperglycemia

The orthodox paradigm has revolved only around glucose and insulin and as a result may have completely misled medicine for decades together. Plasma glucose is an important parameter in diagnosis, and it will remain so for some more time. However, doubts can be raised about its value as an indicator of the effect of treatment and a predictor of pathological complications. First of all one should differentiate between hyperglycemia and diabetes. Hyperglycemia is only one of the symptoms of diabetes. The current treatment regime has the sole aim of normalizing plasma glucose, and therefore, it does not really treat diabetes. All that it can control is hyperglycemia. There is a strong prevalent belief that controlling sugar levels are sufficient to prevent all types of diabetic complications. Evidence is not completely in favor, and therefore, this postulate needs to be critically reexamined as we have discussed earlier (Chap. 3).

The paradigm that hyperglycemia is the root of all pathophysiology of T2D is an old one and has been there with us for several decades. Interestingly, in spite of decades of research, the mechanisms by which glucose can bring about wound-healing problems, peripheral neuropathy, endothelial dysfunction, and erectile dysfunction have never been clearly and completely worked out. All that we have are some incompletely worked out mechanisms, some other hypotheses that were never put to test, and some prevalent beliefs that are not even worth calling hypotheses. As a good example of the latter, it was believed that increased sugar encourages bacterial growth, and therefore, wound healing is difficult and infections are more common in diabetes. However, honey has been traditionally used as a wound-healing agent, and a number of careful studies have reproducibly demonstrated the wound-healing properties of honey [53, 54]. Here is an example of high sugar concentration not interfering in wound healing. Why should sugar interfere in wound healing only in diabetes? But this is a story that is popularly believed in the layman's circle, and even some practicing physicians seem to subscribe to it though fortunately not many researchers.

Glucose is proposed to cause a series of pathological changes through proposed mechanisms that we saw in Chap. 2. This is the current mainstream thinking. But the *Klotho* story suggests something else. Impaired *Klotho* pathway can give rise to a phenotype which is lean, insulin sensitive, and normoglycemic but shows rapid age-related degenerative changes including oxidative damage, chronic systemic inflammation, and vascular problems similar to those in diabetes. Since the pathological components of T2D such as inflammatory changes, oxidative excess, and endothelial and angiogenesis dysfunctions can be demonstrably decoupled from glucose homeostasis mechanisms, they are likely to have origins independent of glucose levels. This could imply that these pathological processes could also be ameliorated without worrying about glycemic control.

In order to elaborate on the possibility of glucose-independent origins of diabetic pathology, I will put together several pieces of evidence for the same set of pathological changes being caused by behavior-driven pathways. These alternative pathways work independent of sugar but are not incompatible with the sugar-driven pathways. We will see the alternative pathways first and then think of any convergence or contradiction between the behavior-driven and glucose-driven pathways.

This new landscape of behavior-driven pathology is more complex than the orthodox picture but much more realistic and evidence based. Although there are still some loose ends left where more research is needed, it is much more detailed than the orthodox one. Here the root of all pathophysiology lies in brain and behavior, not blood sugar. The brain and endocrine systems of a soldier-deficient or supernormal diplomat phenotype send independent signals that bring about simultaneous changes in (1) sex and aggression signals; (2) insulin secretion; (3) fat accumulation; (4) immune redistribution; (5) disinvestment from wound healing and angiogenesis mechanisms; (6) deficiencies of EGF, NGF, BDNF, and many other growth factors; and (7) deficiency of erythropoietin (Fig. 13.1). We have already seen the origin of these alterations at both ultimate and proximate levels, but a brief reiteration would be appropriate. Long-term absence of aggressive encounters reduces levels of aggression signals including testosterone. Higher insulin levels are needed by the brain to provide for the greater demand for cognitive functions. This is mediated by the parasympathetic stimulation of pancreas. High insulin activity induces fat accumulation. The reduced frequency as well as reduced anticipation of injuries leads to disinvestment from wound healing and angiogenesis mechanisms as well as accumulating deficiency of erythropoietin and growth factors. Reduced injuries along with increased levels of adipokines mediate redistribution of macrophages. The seven primary changes are partially independent but have interactions and mechanisms reinforcing each other. For example, testosterone deficiency facilitates adiposity and arrests EGF synthesis. Adipose tissue facilitates altered distribution of macrophages and so on. However I will still call all seven processes primary since they can also originate independently in the absence of others.

These half a dozen primary processes stemming independently from a common neurobehavioral origin trigger a chain of processes downstream (Fig. 13.1). Downstream to hyperinsulinemia, signals from adipose tissue, pancreas, and perhaps brain induce compensatory insulin resistance. Disinvestment from angiogenesis reduces capillary density which also reduces glucose uptake by muscle and other tissues contributing to insulin resistance. Immune redistribution leads to vascular and systemic inflammatory changes as well as oxidative excesses. Deficiency of EGF and other growth factors leads to progressive β cell loss [55, 56]. Reduced insulin levels lead to disinvestment from muscle on the one hand and increased plasma glucose on the other. As a result of disinvestment from muscle, soldier behavior becomes increasingly impossible, and this reinforces diplomat behavior completing the circle and thereby stabilizing the state.

The emerging alternative picture can explain each of the diabetic complications with a sound logic and substantial support from available

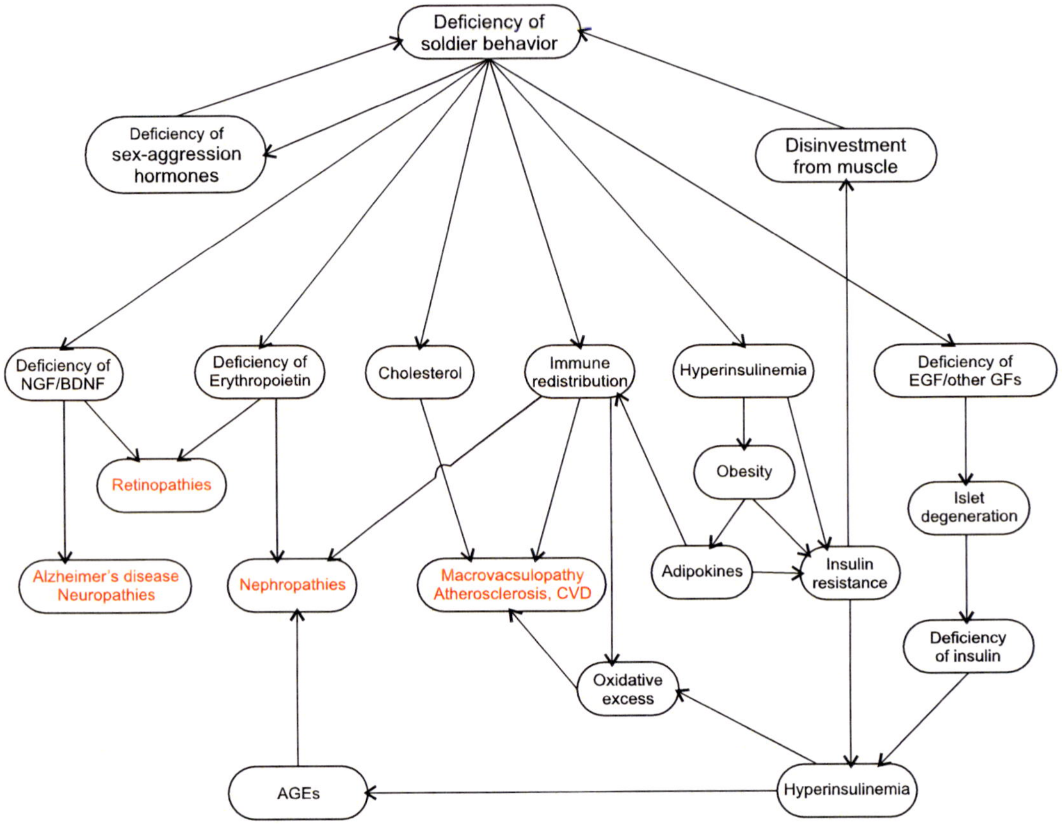

Fig. 13.1 Pathophysiology of diabetic complications: The new paradigm suggests a different route to diabetic complications that incorporates the effects of hyperglycemia but does not assume hyperglycemia to be the root cause of all complications. The alternative pathways account for many of the horse and cart paradoxes and are substantially supported by evidence. This line of thinking can potentially bring about major shifts in diabetes management and treatment (see text for detailed explanation)

evidence, although these hypotheses still remain to be tested directly.

1. Beta cell degeneration: The new paradigm suggests three independent causes for β cell degeneration. First is behavior-induced deficiency of growth factors needed for β cell regeneration. The effect of aggression on EGF secretion is well known [57]. The role of EGF in β cell growth has also been demonstrated [55, 56]; therefore, it is likely that EGF deficiency accumulating due to chronic aggression suppression is the main cause of progressive β cell degeneration as argued earlier. A little later, I will elaborate more on other growth factors required for β cell synthesis and how behavior can alter the supply of these growth factors too. The other possible cause is oxidative damage. Beta cells are unusually sensitive to oxidative damage [58–60], and Rashidi et al. think that the higher sensitivity of β cells to oxidative damage is an evolutionary adaptation [59, 60]. When the altered distribution and chronic activation of macrophages lead to oxidative excess, β cells are among the first to suffer. With altered aggression there is deficiency of sex hormones too. Testosterone and estrogens have antioxidant action, and estrogens have been shown to preserve islet function [61]. A third possible reason is that the reduced glut-1 levels chronically suppress insulin production

following which β cells experience disuse atrophy. The three prospective causes are mutually compatible and might have differential contributions to the islet pathology.

2. Macrovascular pathologies: The vascular pathologies are partially the result of immune redistribution where monocyte–macrophages accumulate within blood vessels as a result of loss of aggression and injury proneness (see Chap. 8 for details). The accumulation of macrophages in vessel walls is further aggravated by cholesterol which is also a marker of nonaggressive behavior. The other component in vascular pathology is NO deficiency which is also likely to be behavior induced. Endothelial NO synthase-deficient mice lose aggression [62] indicating that eNOS is involved somewhere in the aggression pathway itself. Since aggression involves violent muscle activity and also anticipates injuries, a fine control on vasoconstriction is needed to ensure sufficient blood flow to muscle and simultaneously minimize bleeding in case of an injury. It is crucial therefore that aggression should not be initiated unless the NO function is normal (see Chap. 7 for details). Such a behavioral checkpoint would be crucial for initiating aggression. It is likely therefore that loss of aggression may degenerate NO regulation. Sex hormones also have effects on NO-related mechanism. Estradiol induces calcium-dependent eNOS [61, 63, 64]. Testosterone has NO dependent as well as independent vasodilatory effects [65–74]. Therefore deficiency of sex and aggression signals could be a significant contributor to vascular pathology. Through the possible synergy of these mechanisms, endothelial dysfunction can precipitate independent of blood glucose. In the course of insulin resistance syndrome, signs of endothelial dysfunction appear much before diabetic hyperglycemia. Some have suggested that endothelial dysfunction is antecedent and causal rather than consequential [75]. However these processes are not incompatible with the traditional notion of hyperglycemia-induced oxidative excess at the endothelial level. Hyperglycemia may further hasten and worsen the process. But hyperglycemia does not appear to be essential for macrovascular pathology since the latter is not specific to diabetes, and tight glycemic control does not appear to be helpful in restoring endothelial function and reversing macrovascular pathology in all trials (see Chap. 3).

3. Degenerative changes in central and peripheral nervous system: In the alternative interpretation, diabetic neuropathy is a direct result of deficiency of NGF and other neurotrophins which arises from aggression deficiency. Aggression stimulates both EGF and NGF secretion by the salivary glands [76]. EGF and NGF together stimulate BDNF expression [77], and BDNF has strong neuroprotective effects [78]. Testosterone has also been shown to have protective action against oxidative stress-induced nerve damages [79, 80].

Alzheimer's disease appears to have multiple components. Traditionally amyloid deposits have been blamed for AD [81, 82]. The mechanism of deposition as well as its exact role in the etiology of T2D is debated. But there are alternative possibilities compatible with the new interpretation. Glut-1 downregulation is associated with AD [83]. Insulin deficiency in the brain following β cell loss or sympathetic suppression of insulin production is likely to play a causative role [84, 85], and insulin administration has ameliorating effects in AD [85–90]. NGF and BDNF deficiencies are also likely to be involved [91–99]. Along with BDNF, TrkB expression is also downregulated in AD [98]. Physical activity and aggression have been shown to be important stimulators of TrkB expression [100, 101] suggesting that deficiency of physical activity and aggression may play a direct role in AD. In an evolutionary perspective, deficiency of such hunter-gatherer behaviors is suggested to be important in the etiology of AD [102]. Being open to these alternative pathways may speed up the process of understanding AD.

4. Nephropathy: As compared to macrovascular pathologies, microvascular pathologies appear to be more tightly linked with hyperglycemia. Unlike atherosclerosis and CVD which are not specific to diabetes, microvascular pathologies are rarely seen in the absence of diabetes. In contrast with macrovascular pathologies, the effect of tight glycemic control on nephropathy is more consistently positive across studies. However sugar alone does not seem to account for all the pathology since tight glycemic control does not completely eliminate nephropathy. The risk is reduced by 10–30 % in different studies (see Chap. 3). This suggests that apart from glucose, there could be other pathophysiological mechanisms active in diabetic nephropathy. The glucocentric view does not explain certain patterns. Erythropoietin deficiency has been shown to precede nephropathy [103], and this remains unaccounted for in the glucocentric pathophysiological pathway.

   We can have a look then at the alternative picture. It is agreed that abnormal angiogenesis marks histological changes in diabetic nephropathy [104, 105]. Macrophage accumulation has been shown to be an early event in DN. Erythropoietin deficiency anemia also commonly precedes DN. Based on this we can synthesize the possible process. Erythropoietin deficiency anemia creates pockets of hypoxia which stimulate angiogenesis. Macrophages are important mediators of angiogenesis; therefore, wherever there is greater congregation of macrophages, excessive angiogenesis is more likely. Since in T2D macrophages accumulate in highly vascular tissues such as kidney and lungs, they are more prone to microangiopathy. Microangiopathy of lungs is common in diabetes for similar reasons, although fortunately serious complications are rare [106, 107]. A compelling evidence for early and fundamental involvement of macrophage accumulation in kidneys is the finding that MCP-1 knockouts or macrophage scavenger receptor A-deficient mice are resistant to diabetic nephropathy [108, 109]. Along with adipose tissue, kidneys are also involved in mediating redistribution of macrophages. Kidneys produce chemokines that alter innate immune cell dynamics in the insulin-resistant state. In this process excess macrophages accumulate in the kidneys. This excessive macrophage accumulation triggers a supernormal angiogenesis response on being stimulated by erythropoietin deficiency-mediated pocketed hypoxia. However aggressive behavior strengthens angiogenetic pathways, and in deficiency of aggression, there can be partial disinvestment from angiogenesis mechanisms. Therefore although angiogenesis is stimulated in pockets, the angiogenetic mechanisms are dysfunctional. This leads to excessive but defective angiogenesis. This abnormal angiogenesis underlies diabetic nephropathy.
5. Retinopathy: Diabetic retinopathy may also arise from growth factor deficiency combined with erythropoietin deficiency as follows. The role of EGF and NGF in preventing retinal degeneration is well demonstrated. In tissues where there is no direct blood supply, oxygen needs to be transported by cell-to-cell diffusion. Subacute degenerative changes in the retina may impair cell-to-cell diffusion of oxygen creating pockets of hypoxia. Erythropoietin deficiency anemia can increase the chances of localized hypoxia. The resulting hypoxia stimulates abnormal angiogenesis which is the main process underlying diabetic retinopathy [110, 111]. In retinopathy it is hard to visualize how, if at all, sugar plays a role.
6. Impaired wound healing: Wound healing is a complex process and involves a large number of signaling pathways. Many of them may be affected in diabetes. The factors that are known to be affected are NGF and EGF availability, reduced macrophage density in the subcutaneous tissues [112], and decreased angiogenesis response [113]. All three appear to be independent of sugar levels.

All the glucose-independent pathophysiological pathways proposed above are evidence based in the sense that every component of these processes has been demonstrated in the cited references. So far this work was piecemeal, and I have tried to put the pieces together to construct a coherent picture. Admittedly evidence for every step is not necessarily evidence for a ladder. But this is where more research is needed.

Agreeably there is evidence for hyperglycemia-driven pathology too. These pathways mainly involve alterations in NO expression, oxidative damages, and AGE-RAGE pathways (see Chap. 2 for details). If two alternative pictures explain the same pathology, we need some ways to resolve between the two. For example, oxidative state is implicated in both but the proposed origin of free radicals is different. This can possibly help us in resolving between the two hypotheses. By the glucose-driven view, oxidative state originates from glucose loading of cell respiratory mechanisms. Glucose loading is unlikely to happen in insulin-resistant cells but can happen in insulin-independent cells. On the other hand macrophage-driven oxidative stress will be distributed according to the distribution of macrophages. This suggests a differential testable prediction. If glucocentric hypothesis is true, insulin-independent tissues such as brain and endothelium should generate almost all the free radicals. If the macrophage hypothesis is true, origin of free radicals should be in tissues where macrophages accumulate. Macrophages accumulate in adipose tissue and vascular endothelium. Since vascular endothelium is common for both, it does not help in resolving. Adipose tissue-related predictions are in the opposite directions and therefore would be useful. Since adipose tissue is insulin dependent, it will not experience intracellular glucose excess and thereby a respiratory burst leading to free radical generation if the glucose hypothesis is correct. However since adipose tissue attracts macrophages in large numbers, it will have macrophage-generated free radicals if the alternative hypothesis is correct. If adipose tissues in T2D generate little free radicals, glucocentric hypothesis would be supported, and if adipose tissue generates larger amounts of free radicals, the macrophage hypothesis gets greater support. As we know today adipose tissue does generate substantial oxidative excess from an early stage and arresting oxidative stress in adipose tissue prevents shift of the adipokine imbalance towards proinflammatory state [114, 115]. Therefore oxidative stress in adipose tissue seems to matter, but a sound quantitative experimental comparison across different tissues would be more helpful. I expect in reality both the pathways could be contributing to pathology, but it is important to evaluate their relative roles since that can have important clinical implications.

There is another possible way in which the behavior-driven and glucose-driven pathways could be interacting. The hyperglycemia-driven pathways lead to oxidatively damaged proteins and cells, accumulation of AGEs, and AGE-driven protein damage. However, there is a rate at which damaged proteins and cells are being replaced in normal life. The effects of these damages need to be weighed against the rate of replacement of damaged components. In diabetes there is evidence that the rate of replacement is reduced. For example, the rate of formation of endothelial progenitor cells is reduced in diabetes [116]. This overall reduction is likely to be behavior driven since testosterone [117–119], estrogens [120–123], erythropoietin [124–129], and other behavioral signal molecules play a role in proliferation and mobilization of endothelial progenitor cells. EGF and other growth factors are needed for tissue regeneration. Testosterone, oxytocin, and some other behavioral signals promote protein synthesis. Therefore it is likely that these pathways stimulated by physical activity and aggression may effectively compensate the hyperglycemia-induced damages by replacing damaged cell components, cells, or tissues. Similarly even if we assume that hyperglycemia is the only source of oxidative radicals, antioxidant activity triggered by aggression can take care of it at least partially. Since aggression anticipates injuries and macrophages activated by injuries produce free radicals to kill bacteria, it is necessary to increase antioxidant capacity in anticipation. Therefore aggression should strengthen antioxidant defenses. Not surprisingly testosterone and estradiol have antioxidant actions [130–133], and physical exercise and aggression appears to have some testosterone-independent effect on antioxidant capacity [134–136]. If this is true it may be possible to prevent and perhaps reverse the hyperglycemia-induced pathophysiological processes by aggressive exercise even without tight glycemic control. This is a theoretical possibility currently but is testable and needs to be put to test sooner.

The two most important possible implications of the alternative pathophysiological pathways are that (1) even if raised blood sugar is demonstrated to cause some of the pathological effects, that is not sufficient to say that controlling sugar will eliminate them. Since the rise in blood sugar is associated with disinvestment from muscle and thereby loss of aggression, one would see an apparent effect of raised sugar on the behavior-driven pathophysiological mechanisms. But now, if sugar is controlled but behavior remains the same, we may not expect a mitigation of the pathophysiology. (2) If behavior is corrected appropriately, then even if sugar levels remain moderately higher, pathology can be prevented.

Partial support to the latter exists in literature on the effects of exercises. Exercises can bring about improvements in a number of functions without, before or independent of, improvement in glycemic control. Exercise increases growth factor and neurotrophin levels [137–142], improves angiogenesis [143], increases antioxidant enzymes in muscle [135, 144, 145], and enhances endothelial function [146], all independent of blood sugar levels.

The emerging alternative picture implies the need to drift away from glucocentric thinking. This is the basic inferential difference between the two models. In the orthodox model, since all pathologies are rooted in hyperglycemia, treating hyperglycemia is all that one needs to do. This is not so in the alternative picture. According to the alternative picture too, hyperglycemia can lead to pathological effects since it affects behavior and the behavioral change can result into downstream pathologies. However, in this model, controlling sugar will not mitigate complications if the behavior remains unchanged. On the other hand if the behavior is corrected or supplemented appropriately, as we will discuss in the coming chapter, pathological effects can be prevented even without normalizing glucose levels.

I have no intention to state that it is not important to control sugar, but the important point is that sole emphasis on sugar control is not sufficient. To be conservative, treatment aimed towards moderate control of glucose should continue until we have a proven alternative effective therapy against the true pathological processes of diabetes. At the same time research should focus on understanding and arresting the true pathological mediators of diabetes.

Finally we need to address the question of "incurability" of T2D here. The major depressive factor of diabetes for a patient is the understanding that he or she is a patient for life. One can never come out of diabetes. What one can do at the most is to prevent it from escalating. Unfortunately this too does not work in all cases. Glycemic control keeps on deteriorating slowly even in patients with good compliance to treatment.

A disease can be incurable for two reasons. One being that the disease causes some irreversible changes. There are many examples of this. Mammals have lost retinal regeneration capacity which lower vertebrates have. Therefore, in mammals, retinal damage is irreversible. In poliomyelitis, the viral infection can be cured but paralysis caused by the infection is irreversible and leads to lifetime disability. Is diabetes incurable because of some such irreversible changes?

It has been believed for a long time that β cell loss is irreversible. Therefore, by the popular belief, insulin resistance can be reversed, but once the progressive β cell loss begins, it cannot be reversed. However, unlike retina, we have not lost the capacity to regenerate β cells. There have been clear and reproducible demonstrations that β cells have good regeneration capacity, and this can happen by two alternative means. The β cells themselves can replicate or they can be generated from the acinar and ductal endothelial cells [147–152]. This means that as long as pancreatic tissue exists, β cells can regenerate. There is some evidence that even in autoimmune type 1 diabetes, β cells are continuously being regenerated [153, 154], although the rate of autoimmune destruction supersedes the rate of regeneration. In streptozotocin-treated rats, at least partial regeneration of β cells has been successfully brought about under the influence of EGF and gastrin [155, 156]. Therefore it is unlikely that β cell loss is irreversible. Perhaps we have failed to identify the appropriate internal environment needed for the growth of β cells. Although research on in vitro propagation of β cells for generation of functional islets from stem cells has not given us viable therapy options, it played an important role in recognizing the complex growth factor requirement for β cell growth.

Apart from EGF, a large number of other growth factors are now known to be required for β cell growth. The list may not be complete as yet, but interestingly, all the factors listed so far also have active roles in wound healing too. Insulin, IGF1, and IGF2 are important in β cell proliferation and preventing apoptosis [157–161], both IGFs are active in wound healing and also synthesized locally at the wound site [162–165]. Insulin itself has a beneficial action in wound healing [166]. Other growth factors that are involved in both β cell regeneration and wound healing are PDGF and FGF [167–169], HGF [148, 170–174], NGF [175–177], growth hormone [178–180], prolactin and placental lactogen [147, 180, 181], PTHrP [147, 182–184], gastrin [155, 156, 185], GRP [186], glucagon-like peptide [187–189], activin [190, 191], and betacellulin [148, 192]. Not only are these factors required in wound healing, most of them are synthesized systemically or locally at the site of wound after injury [172]. For the factors that are synthesized locally, some spillover in circulation is inevitable. It makes sense for the wound-healing growth factors to help β cell function since insulin itself is needed for wound healing. This raises the possibility that a chronically non-injury-prone lifestyle creates a deficiency of one or more of these factors affecting β cell regeneration. Interestingly two immunosuppressive agents that were used in the Edmonton protocol to prevent immune damage to transplanted β cells have now been shown to prevent natural β cell regeneration [150]. This is very likely to be related to the suppression of wound-healing mechanisms that the two agents are known to bring about [193–196]. Therefore the causes of impairment of wound healing and failure of effective β cell regeneration in diabetes appear to have much in common. This raises the possibility that the factors that would normalize β cell regeneration in diabetes will be identified in the near future, and a growth factor cocktail might turn out to be effective. But it is equally or perhaps more likely that behavioral intervention to stimulate growth factors and wound-healing mechanisms may work better. It is already known that two behaviors, namely, aggression and adventure, that anticipate injuries induce secretion of some of the growth factors. There are no studies on whether there is behavioral stimulation of all other growth factors, but one should expect it logically. If that is true, the right cocktail could be internally generated, and islet degeneration might be substantially reversed. If β cell loss can be reversed, then except for extremely advanced stages of diabetic complications, particularly retinopathy and nephropathy, there do not appear to be any truly irreversible changes involved in diabetes.

If irreversibility is not the cause of "incurability" of T2D, the only other possibility is that we have failed to detect the root cause of diabetes and treat it. The so-called incurability is because of beating around the wrong bush. This implies that if we attack the root cause with the right approach, we may successfully achieve complete cure. The current treatment regime is not only unsuccessful in curing or reversing diabetes, it is increasingly failing to maintain glycemic control as well. Any treatment measure currently does not have even 50 % success rate in the long run, although almost every drug has a brief honeymoon period of dramatic response in the beginning.

It would be possible in future to use a network model for testing and strengthening the alternative interpretation of the pathophysiological processes. This would need an extension of the model described in Appendix II. Currently the pathways leading to complications are not included in the model. It would be good to wait for more empirical data before doing so. But the network approach could prove to be a powerful tool in the near future to develop insights into the complex pathophysiology of diabetic complications.

We can appreciate here that at this stage, the new hypothesis takes a quantum leap and this leap is worth calling a paradigm shift. So far I have been using the words new hypothesis, new interpretation, or new theory, but now I will take the freedom to call it the new paradigm. Recent evidence is challenging two most fundamental assumptions behind the old theory, that of insulin resistance being central to diabetes and that of hyperglycemia being the root cause of all diabetic pathology. I will reiterate the evidence for the abolition of these two statements:

1. We learn from tissue-specific insulin receptor knockout models that muscle insulin resistance does not give rise to hyperglycemia or hyperinsulinemia and liver insulin resistance does not lead to sustained hyperglycemia (Chap. 3).
2. Gluconeogenesis in the liver parenchymal cells in vitro in the presence and absence of insulin differs only by about 20% (Chap. 12).
3. The IR-RII models and its possible variations do not give all the four typical characteristics of the altered glucose tolerance curve (Chap. 12).
4. The altered GTT and diabetic hyperglycemia are better explained by alternative brain- and behavior-dependent mechanisms that are independent of insulin resistance (Chap. 12).
5. Insulin resistance is not obligately coupled with pathophysiological mechanisms of diabetes as demonstrated by the *Klotho* action. *Klotho* overexpression leads to insulin resistance without any accompanying pathology. In *Klotho*-deficient mutants, many pathophysiological defects typical of T2D are seen without insulin resistance and hyperinsulinemia (this chapter, above).
6. Hyperglycemia-driven pathways do not explain all the pathophysiology and mechanisms of diabetic complications (this chapter, above).
7. Alternative pathways independent of glucose leading to pathophysiology and complications of diabetes are demonstrable (Chaps. 7, 8, and this chapter, above).
8. Aggressive normalization of plasma glucose does not normalize physiology and fails to *eliminate* the risk of diabetic complications (Chap. 3). Although the advantages of intensive glucose-lowering therapy reduce the risk of microvascular complications with statistical significance, the effect size is very small.

These evidences are sufficient to suspect whether insulin resistance and hyperglycemia are central to the pathology of T2D. If the suspicion turns out to be true, it demands not only a new hypothesis but a paradigm shift and an alternative paradigm that is logically more sound and compatible with available evidence that is emerging now. There are still some missing links in the new paradigm. The picture is not yet complete. But if we have given several decades for the classical paradigm to fail or succeed and noted that it has largely failed on all the fronts of preventing, controlling, or curing diabetes, we need to give sufficient time and research inputs to the alternative paradigm as well and see how it works. If the alternative paradigm is correct, T2D would certainly be curable if appropriate behavioral corrections are applied in time and for sufficiently long duration.

Now we have an answer to a question left unanswered from Chap. 6 as to whether it is soldier deficiency or supernormal diplomat behavior that matters more. From the discussion in this chapter, it should be clear that although diplomat behavior could be important in inducing an HIIR state, the true pathophysiological mechanisms stem from a deficiency of soldier behavior. There is no doubt that a diplomat state pushes the systems towards insulin resistance, but there is nothing wrong in diplomat behavior or the insulin-resistant state. It is an adaptation after all. What turns the health conditions bad is the supernormal deficiency of soldier behavior. Therefore adequate supplementation of the deficient behavior is likely to be effective in prevention and reversal of T2D and related disorders. Trying to reverse insulin resistance could be both unnecessary and fruitless but would certainly happen eventually with behavioral supplementation.

References

1. Kaletsky R, Murphy CT (2010) The role of insulin/IGF-like signaling in *C. elegans* longevity and aging. Dis Model Mech 3:415–419
2. Kenyon C, Chang J, Gensch E, Rudner A, Tabtiang R (1993) A *C. elegans* mutant that lives twice as long as wild type. Nature 366:461–464
3. Friedman DB, Johnson TE (1988) Three mutants that extend both mean and maximum life span of the nematode, *Caenorhabditis Elegans*, define the Age-1 gene. J Gerontol 43:B102–B109
4. Kimura KD, Tissenbaum HA, Liu Y, Ruvkun G (1997) Daf-2, an insulin receptor-like gene that regulates longevity and diapause in *Caenorhabditis elegans*. Science 277:942–946
5. Morris JZ, Tissenbaum HA, Ruvkun G (1996) A phosphatidylinositol-3-OH kinase family member regulating longevity and diapause in *Caenorhabditis elegans*. Nature 382:536–539
6. Honda Y, Honda S (1999) The daf-2 gene network for longevity regulates oxidative stress resistance and Mn-superoxide dismutase gene expression in *Caenorhabditis elegans*. FASEB J 13:1385–1393
7. Scott BA, Avidan MS, Crowder CM (2002) Regulation of hypoxic death in *C. elegans* by the insulin/IGF receptor homolog DAF-2. Science 296:2388–2391
8. Lithgow GJ, Walker GA (2002) Stress resistance as a determinate of *C. elegans* lifespan. Mech Ageing Dev 123:765–771
9. Barsyte D, Lovejoy DA, Lithgow GJ (2001) Longevity and heavy metal resistance in daf-2 and age-1 long-lived mutants of *Caenorhabditis elegans*. FASEB J 15:627–634
10. Garsin DA et al (2003) Long-lived *C. elegans* daf-2 mutants are resistant to bacterial pathogens. Science 300:1921
11. Cassada RC, Russell RL (1975) The dauer larva, a post-embryonic developmental variant of the nematode *Caenorhabditis elegans*. Dev Biol 46:326–342
12. Riddle DL, Swanson MM, Albert PS (1981) Interacting genes in nematode dauer larva formation. Nature 290:668–671
13. Ogg S et al (1997) The fork head transcription factor DAF-16 transduces insulin-like metabolic and longevity signals in *C. elegans*. Nature 389:994–999
14. Lee RY, Hench J, Ruvkun G (2001) Regulation of *C. elegans* DAF-16 and its human ortholog FKHRL1 by the daf-2 insulin-like signaling pathway. Curr Biol 11:1950–1957
15. Garigan D et al (2002) Genetic analysis of tissue aging in *Caenorhabditis elegans*: a role for heat-shock factor and bacterial proliferation. Genetics 161:1101–1112
16. Huang C, Xiong C, Kornfeld K (2004) Measurements of age-related changes of physiological processes that predict lifespan of *Caenorhabditis elegans*. Proc Natl Acad Sci USA 101:8084–8089
17. Schmeissner PJ et al (2002) Stochastic and genetic factors influence tissue-specific decline in ageing *C. elegans*. Nature 419:808–14
18. Tatar M et al (2001) A mutant drosophila insulin receptor homolog that extends life-span and impairs neuroendocrine function. Science 292:107–110
19. Hwangbo DS, Gersham B, Tu M-P, Palmer M, Tatar M (2004) Drosophila dFOXO controls lifespan and regulates insulin signalling in brain and fat body. Nature 429:562–566
20. Clancy DJ et al (2001) Extension of life-span by loss of CHICO, a drosophila insulin receptor substrate protein. Science 292:104–106
21. Böhni R et al (1999) Autonomous control of cell and organ size by CHICO, a Drosophila homolog of vertebrate IRS1-4. Cell 97:865–875
22. Benyoucef S, Surinya KH, Hadaschik D, Siddle K (2007) Characterization of insulin/IGF hybrid receptors: contributions of the insulin receptor L2 and Fn1 domains and the alternatively spliced exon 11 sequence to ligand binding and receptor activation. Biochem J 403:603–613

23. Barbieri M, Bonafè M, Franceschi C, Paolisso G (2003) Insulin/IGF-I-signaling pathway: an evolutionarily conserved mechanism of longevity from yeast to humans. Am J Physiol Endocrinol Metab 285:E1064–E1071
24. Hsieh C-C, DeFord JH, Flurkey K, Harrison DE, Papaconstantinou J (2002) Implications for the insulin signaling pathway in Snell dwarf mouse longevity: a similarity with the *C. elegans* longevity paradigm. Mech Ageing Dev 123:1229–1244
25. Greer KA, Canterberry SC, Murphy KE (2007) Statistical analysis regarding the effects of height and weight on life span of the domestic dog. Res Vet Sci 82:208–214
26. Sutter NB et al (2007) A single IGF1 allele is a major determinant of small size in dogs. Science 316: 112–115
27. Holzenberger M et al (2003) IGF-1 receptor regulates lifespan and resistance to oxidative stress in mice. Nature 421:182–187
28. Blüher M, Kahn BB, Kahn CR (2003) Extended longevity in mice lacking the insulin receptor in adipose tissue. Science 299:572–574
29. Klöting N, Blüher M (2005) Extended longevity and insulin signaling in adipose tissue. Exp Gerontol 40: 878–883
30. Buffenstein R (2005) The naked mole-rat: a new long-living model for human aging research. J Gerontol A Biol Sci Med Sci 60:1369–1377
31. Willcox BJ et al (2008) FOXO3A genotype is strongly associated with human longevity. Proc Natl Acad Sci USA 105:13987–13992
32. Flachsbart F et al (2009) Association of FOXO3A variation with human longevity confirmed in German centenarians. Proc Natl Acad Sci USA 106:2700–2705
33. Anselmi CV et al (2009) Association of the FOXO3A locus with extreme longevity in a southern Italian centenarian study. Rejuvenation Res 12:95–104
34. Suh Y et al (2008) Functionally significant insulin-like growth factor I receptor mutations in centenarians. Proc Natl Acad Sci USA 105:3438–3442
35. Kuro-o M (2010) Klotho. Pflugers Arch 459:333–343
36. Chihara Y et al (2006) Klotho protein promotes adipocyte differentiation. Endocrinology 147:3835–3842
37. Mori K et al (2000) Disruption of Klotho gene causes an abnormal energy homeostasis in mice. Biochem Biophys Res Commun 278:665–670
38. Bartke A (2006) Long-lived Klotho mice: new insights into the roles of IGF-1 and insulin in aging. Trends Endocrinol Metab 17:33–35
39. Unger RH (2006) Klotho-induced insulin resistance: a blessing in disguise? Nat Med 12:56–57
40. Kuro-o M et al (1997) Mutation of the mouse klotho gene leads to a syndrome resembling ageing. Nature 390:45–51
41. Kurosu H et al (2005) Suppression of aging in mice by the hormone Klotho. Science 309:1829–1833
42. Wang Y, Sun Z (2009) Current understanding of Klotho. Ageing Res Rev 8:43–51
43. Travers JP, Pratten MK, Beck F (1989) Effects of low insulin levels on rat embryonic growth and development. Diabetes 38:773–778
44. Akashi M et al (1991) Effects of insulin and myo-inositol on embryo growth and development during early organogenesis in streptozocin-induced diabetic rats. Diabetes 40:1574–1579
45. Nakayama Y, Yamamoto T, Abé SI (1999) IGF-I, IGF-II and insulin promote differentiation of spermatogonia to primary spermatocytes in organ culture of newt testes. Int J Dev Biol 43:343–347
46. Brüning JC et al (2000) Role of brain insulin receptor in control of body weight and reproduction. Science 289:2122–2125
47. Sun LL et al (2010) Effect of insulin on oogenesis from mouse fetal germ cells in a serum-free 3D culture system. Reprod Biomed Online 20:11–25
48. Nef S et al (2003) Testis determination requires insulin receptor family function in mice. Nature 426:291–295
49. Maekawa Y et al (2009) Klotho suppresses TNF-α-induced expression of adhesion molecules in the endothelium and attenuates NF-κB activation. Endocrinology 35:341–346
50. Fukino K et al (2002) Regulation of angiogenesis by the aging suppressor gene Klotho. Biochem Biophys Res Commun 293:332–337
51. Maekawa Y et al (2011) Klotho protein diminishes endothelial apoptosis and senescence via a mitogen-activated kinase pathway. Geriatr Gerontol Int 11: 510–516
52. Rakugi H et al (2007) Anti-oxidative effect of Klotho on endothelial cells through cAMP activation. Endocrine 31:82–87
53. Efem SEE (1988) Clinical observations on the wound healing properties of honey. Brit J Surg 75: 679–681
54. Bergman A, Yanai J, Weiss J, Bell D, David MP (1983) Acceleration of wound healing by topical application of honey: an animal model. Am J Surg 145:374–376
55. Bouwens L (2006) Beta cell regeneration. Curr Diabetes Rev 2:3–9
56. Miettinen P, Ormio P, Hakonen E, Banerjee M, Otonkoski T (2008) EGF receptor in pancreatic β-cell mass regulation. Biochem Soc Trans 36: 280–285
57. Belsare PV et al (2010) Metabolic syndrome: aggression control mechanisms gone out of control. Med Hypotheses 74:578–589
58. Modak MA, Parab PB, Ghaskadbi SS (2009) Pancreatic islets are very poor in rectifying oxidative DNA damage. Pancreas 38:23–29
59. Rashidi A, Kirkwood TBL, Shanley DP (2009) Metabolic evolution suggests an explanation for the weakness of antioxidant defences in β-cells. Mech Ageing Dev 130:216–221
60. Rashidi A, Kirkwood TBL, Shanley DP (2009) On the surprising weakness of pancreatic B-cell antioxidant

defences: an evolutionary perspective. Evol Biol 109–125
61. Goetz RM et al (1999) Estradiol induces the calcium-dependent translocation of endothelial nitric oxide synthase. In Evolutionary Biology: Concept, modeling and application, Berlin-Heidelberg: Springer, pp 109–126
62. Demas GE et al (1999) Elimination of aggressive behavior in male mice lacking endothelial nitric oxide synthase. J Neurosci 19:RC30
63. Weiner CP et al (1994) Induction of calcium-dependent nitric oxide synthases by sex hormones. Proc Natl Acad Sci USA 91:5212–5216
64. Hayashi T et al (1995) Estrogen increases endothelial nitric oxide by a receptor mediated system. Biochem Biophys Res Commun 214:847–855
65. Akishita M et al (2007) Low testosterone level is an independent determinant of endothelial dysfunction in men. Hypertens Res 30:1029–1034
66. Littleton-Kearney M, Hurn PD (2004) Testosterone as a modulator of vascular behavior. Biol Res Nurs 5:276–285
67. Jones RD, Pugh PJ, Jones TH, Channer KS (2003) The vasodilatory action of testosterone: a potassium-channel opening or a calcium antagonistic action? Brit J Pharmacol 138:733–744
68. Worboys S, Kotsopoulos D, Teede H, McGrath B, Davis SR (2001) Evidence that parenteral testosterone therapy may improve endothelium-dependent and -independent vasodilation in postmenopausal women already receiving estrogen. J Clin Endocrinol Metab 86:158–161
69. Jones RD, Pugh PJ, Jones TH, Channer KS (2003) The vasodilatory action of testosterone: a potassium-channel opening or a calcium antagonistic action? Brit J Pharmacol 138:733–744
70. Álvarez E, Cairrão E, Morgado M, Morais C, Verde I (2010) Testosterone and cholesterol vasodilation of rat aorta involves L-type calcium channel inhibition. Adv Pharmacol Sci 2010:1–10
71. Yildiz O, Seyrek M (2007) Vasodilating mechanisms of testosterone. Exp Clin Endocrinol Diabetes 115:1–6
72. Tep-areenan P, Kendall DA, Randall MD (2002) Testosterone – induced vasorelaxation in the rat mesenteric arterial bed is mediated predominantly via potassium channels. Brit J Pharmacol 135:735–740
73. Ding AQ, Stallone JN (2001) Testosterone-induced relaxation of rat aorta is androgen structure specific and involves K+ channel activation. J Appl Physiol 91:2742–2750
74. Tep-areenan P, Kendall DA, Randall MD (2003) Mechanisms of vasorelaxation to testosterone in the rat aorta. Eur J Pharmacol 465:125–132
75. Tooke JE, Hannemann MM (2000) Adverse endothelial function and the insulin resistance syndrome. J Intern Med 247:425–431
76. Lakshmanan J (1986) Aggressive behavior in adult male mice elevates serum nerve growth factor levels. Am J Physiol 250:386–392
77. Tirassa P, Triaca V, Amendola T, Fiore M, Aloe L (2003) EGF and NGF injected into the brain of old mice enhance BDNF and ChAT in proliferating subventricular zone. J Neurosci Res 72:557–564
78. Mattson MP, Maudsley S, Martin B (2004) BDNF and 5-HT: a dynamic duo in age-related neuronal plasticity and neurodegenerative disorders. Trends Neurosci 27:589–594
79. Túnez I et al (2007) Effect of testosterone on oxidative stress and cell damage induced by 3-nitropropionic acid in striatum of ovariectomized rats. Life Sci 80:1221–1227
80. Kinderman N, Jones K (1993) Testosterone enhancement of the nerve cell body response to injury: evidence using in situ hybridization and ribosomal DNA probes. J Neurosci 13:1523–1532
81. Selkoe DJ (1989) Molecular pathology of amyloidogenic proteins and the role of vascular amyloidosis in Alzheimer's disease. Neurobiol Aging 10: 387–395
82. Mann DM, Jones D, Prinja D, Purkiss MS (1990) The prevalence of amyloid (A4) protein deposits within the cerebral and cerebellar cortex in Down's syndrome and Alzheimer's disease. Acta Neuropathol 80:318–327
83. Liu Y, Liu F, Iqbal K, Grundke-Iqbal I, Gong C-X (2008) Decreased glucose transporters correlate to abnormal hyperphosphorylation of tau in Alzheimer disease. FEBS Lett 582:359–364
84. Steen E et al (2005) Impaired insulin and insulin-like growth factor expression and signaling mechanisms in Alzheimer's disease—is this type 3 diabetes? J Alzheimers Dis 7:63–80
85. Wang X et al (2010) Insulin deficiency exacerbates cerebral amyloidosis and behavioral deficits in an Alzheimer transgenic mouse model. Mol Neurodegener 5:46
86. Kern W et al (2001) Improving influence of insulin on cognitive functions in humans. Neuroendocrinology 74:270–280
87. Zhao W-Q, Chen H, Quon MJ, Alkon DL (2004) Insulin and the insulin receptor in experimental models of learning and memory. Eur J Pharmacol 490:71–81
88. Correia SC et al (2011) Insulin-resistant brain state: the culprit in sporadic Alzheimer's disease? Ageing Res Rev 10:264–273
89. Benedict C et al (2004) Intranasal insulin improves memory in humans. Psychoneuroendocrinology 29:1326–1334
90. Kern W, Born J, Fehm HL (2002) Role of insulin in Alzheimer's disease: approaches emerging from basic animal research and neurocognitive studies in humans. Drug Dev Res 56:511–525
91. Olson L (1993) NGF and the treatment of Alzheimer's disease. Exp Neurol 124:5–15
92. Salehi A, Delcroix J-D, Swaab DF (2004) Alzheimer's disease and NGF signaling. J Neural Transmission 111:323–345

93. Tuszynski MH et al (2005) A phase 1 clinical trial of nerve growth factor gene therapy for Alzheimer disease. Nat Med 11:551–555
94. Scott SA, Crutcher KA (1994) Nerve growth factor and Alzheimer's disease. Rev Neurosci 5:179–211
95. Higgins GA, Mufson EJ (1989) NGF receptor gene expression is decreased in the nucleus basalis in Alzheimer's disease. Exp Neurol 106:222–236
96. Phillips HS et al (1991) BDNF mRNA is decreased in the hippocampus of individuals with Alzheimer's disease. Neuron 7:695–702
97. Kunugi H et al (2001) A novel polymorphism of the brain-derived neurotrophic factor (BDNF) gene associated with late-onset Alzheimer's disease. Mol Psychiatry 6:83–86
98. Ferrer I et al (1999) BDNF and full-length and truncated TrkB expression in Alzheimer disease. Implications in therapeutic strategies. J Neuropathol Exp Neurol 58(7):729–739
99. Narisawa-Saito M et al (1996) Regional specificity of alterations in NGF, BDNF and NT-3 levels in Alzheimer's disease—Abstract—UK PubMed Central. Neuroreport 7:2925–2928
100. Widenfalk J, Olson L, Thorén P (1999) Deprived of habitual running, rats downregulate BDNF and TrkB messages in the brain. Neurosci Res 34: 125–132
101. Fiore M, Amendola T, Triaca V, Alleva E, Aloe L (2005) Fighting in the aged male mouse increases the expression of TrkA and TrkB in the subventricular zone and in the hippocampus. Behav Brain Res 157:351–362
102. Brito GN (2009) Exercise and cognitive function: a hypothesis for the association of type II diabetes mellitus and Alzheimer's disease from an evolutionary perspective. Diabetol Metab Syndr 1:7–7
103. Bosman DR, Winkler AS, Marsden JT, Macdougall IC, Watkins PJ (2001) Anemia with erythropoietin deficiency occurs early in diabetic nephropathy. Diabetes Care 24:495–499
104. Nakagawa T (2007) Uncoupling of the VEGF-endothelial nitric oxide axis in diabetic nephropathy: an explanation for the paradoxical effects of VEGF in renal disease. Am J Physiol Renal Physiol 292:1665–1672
105. Yamamoto Y et al (2004) Tumstatin peptide, an inhibitor of angiogenesis, prevents glomerular hypertrophy in the early stage of diabetic nephropathy. Diabetes 53:1831–1840
106. Sandler M (1990) Is the lung a 'target organ' in diabetes mellitus? Arch Intern Med 150:1385–1388
107. Goldman MD (2003) Lung dysfunction in diabetes. Diabetes Care 26:1915–1918
108. Chow FY et al (2006) Monocyte chemoattractant protein-1 promotes the development of diabetic renal injury in streptozotocin-treated mice. Kidney Int 69:73–80
109. Usui HK et al (2007) Macrophage scavenger receptor-a–deficient mice are resistant against diabetic nephropathy through amelioration of microinflammation. Diabetes 56:363–372
110. Aiello LP et al (1994) Vascular endothelial growth factor in ocular fluid of patients with diabetic retinopathy and other retinal disorders. N Engl J Med 331:1480–1487
111. Suzuma K et al (2005) Vitreous levels of angiopoietin 2 and vascular endothelial growth factor in patients with proliferative diabetic retinopathy. Am J Ophthalmol 139:476–481
112. Maruyama K et al (2007) Decreased macrophage number and activation lead to reduced lymphatic vessel formation and contribute to impaired diabetic wound healing. Am J Pathol 170:1178–1191
113. Galiano RD et al (2004) Topical vascular endothelial growth factor accelerates diabetic wound healing through increased angiogenesis and by mobilizing and recruiting bone marrow-derived cells. Am J Pathol 164:1935–1947
114. Kurata A et al (2006) Blockade of Angiotensin II type-1 receptor reduces oxidative stress in adipose tissue and ameliorates adipocytokine dysregulation. Kidney Int 70:1717–1724
115. Chevillotte E, Giralt M, Miroux B, Ricquier D, Villarroya F (2007) Uncoupling protein-2 controls adiponectin gene expression in adipose tissue through the modulation of reactive oxygen species production. Diabetes 56:1042–1050
116. Loomans CJM et al (2004) Endothelial progenitor cell dysfunction. Diabetes 53:195–199
117. Foresta C et al (2008) Androgens stimulate endothelial progenitor cells through an androgen receptor – mediated pathway. Clin Endocrinol 68:284–289
118. Foresta C et al (2006) Reduced number of circulating endothelial progenitor cells in hypogonadal men. J Clin Endocrinol Metab 91:4599–4602
119. Foresta C et al (2009) Effect of vardenafil on endothelial progenitor cells in hypogonadotrophic hypogonadal patients: role of testosterone treatment. Clin Endocrinol 71:412–416
120. Fadini GP et al (2009) Effects of androgens on endothelial progenitor cells in vitro and in vivo. Clin Sci (Lond) 117:355–364
121. Strehlow K et al (2003) Estrogen increases bone marrow–derived endothelial progenitor cell production and diminishes neointima formation. Circulation 107:3059–3065
122. Iwakura A et al (2003) Estrogen-mediated, endothelial nitric oxide synthase–dependent mobilization of bone marrow–derived endothelial progenitor cells contributes to reendothelialization after arterial injury. Circulation 108:3115–3121
123. Imanishi T, Hano T, Nishio I (2005) Estrogen reduces endothelial progenitor cell senescence through augmentation of telomerase activity. J Hypertens 23:1699–1706
124. Bahlmann FH et al (2003) Endothelial progenitor cell proliferation and differentiation is regulated by

erythropoietin rapid communication. Kidney Int 64:1648–1652

125. Heeschen C et al (2003) Erythropoietin is a potent physiologic stimulus for endothelial progenitor cell mobilization. Blood 102:1340–1346
126. Bahlmann FH et al (2004) Erythropoietin regulates endothelial progenitor cells. Blood 103:921–926
127. Westenbrink BD et al (2007) Erythropoietin improves cardiac function through endothelial progenitor cell and vascular endothelial growth factor mediated neovascularization. Eur Heart J 28: 2018–2027
128. George J et al (2005) Erythropoietin promotes endothelial progenitor cell proliferative and adhesive properties in a PI 3-kinase-dependent manner. Cardiovasc Res 68:299–306
129. Satoh K et al (2006) Important role of endogenous erythropoietin system in recruitment of endothelial progenitor cells in hypoxia-induced pulmonary hypertension in mice. Circulation 113:1442–1450
130. Ahlbom E, Prins GS, Ceccatelli S (2001) Testosterone protects cerebellar granule cells from oxidative stress-induced cell death through a receptor mediated mechanism. Brain Res 892:255–262
131. Mooradian AD (1993) Antioxidant properties of steroids. J Steroid Biochem Mol Biol 45:509–511
132. Shwaery GT, Vita JA, Keaney JF (1998) Antioxidant protection of LDL by physiologic concentrations of estrogens is specific for 17-β-estradiol. Atherosclerosis 138:255–262
133. Ayres S, Tang M, Ravi Subbiah MT (1996) Estradiol-17[β] as an antioxidant: some distinct features when compared with common fat-soluble antioxidants. J Lab Clin Med 128:367–375
134. Isaksson C et al (2011) Aggression, but not testosterone, is associated to oxidative status in a free-living vertebrate. Behaviour 148:713–731
135. Powers SK, Ji LL, Leeuwenburgh C (1999) Exercise training-induced alterations in skeletal muscle antioxidant capacity: a brief review. Med Sci Sports Exerc 31:987–997
136. Venditti P, Di Meo S (1996) Antioxidants, tissue damage, and endurance in trained and untrained young male rats. Arch BiochemBiophys 331:63–68
137. Konradsen L, Nexø E (1988) Epidermal growth factor in plasma, serum and urine before and after prolonged exercise. Regul pepts 21:197–203
138. Oliff HS, Berchtold NC, Isackson P, Cotman CW (1998) Exercise-induced regulation of brain-derived neurotrophic factor (BDNF) transcripts in the rat hippocampus. Mol Brain Res 61:147–153
139. Cotman CW, Engesser-Cesar C (2002) Exercise enhances and protects brain function. Exerc Sport Sci Rev 30:75–79
140. Cotman CW, Berchtold NC, Christie L-A (2007) Exercise builds brain health: key roles of growth factor cascades and inflammation. Trends Neurosci 30:464–472
141. Nexø E, Hansen MR, Konradsen L (1988) Human salivary epidermal growth factor, haptocorrin and amylase before and after prolonged exercise. Scand J Clin Lab Invest 48:269–273
142. Gustafsson T, Puntschart A, Kaijser L, Jansson E, Sundberg CJ (1999) Exercise-induced expression of angiogenesis-related transcription and growth factors in human skeletal muscle. Am J Physiol Heart Circ Physiol 276:H679–H685
143. Swain RA et al (2003) Prolonged exercise induces angiogenesis and increases cerebral blood volume in primary motor cortex of the rat. Neurosci 117: 1037–1046
144. McAardle A, Jackson MJ (2000) Exercise, oxidative stress and ageing. J Anatomy 197:539–541
145. Ji LL (1993) Antioxidant enzyme response to exercise and aging. Med Sci Sports Exerc 25:225–231
146. Clarkson P et al (1999) Exercise training enhances endothelial function in young men. J Am Coll Cardiol 33:1379–1385
147. Yesil P, Lammert E (2008) Islet dynamics: a glimpse at β cell proliferation. Histol Histopathol 23: 883–895
148. García-Ocaña A et al (2001) Using β-cell growth factors to enhance human pancreatic islet transplantation. J Clin Endocrinol Metab 86:984–988
149. Trucco M (2005) Regeneration of the pancreatic β cell. J Clin Invest 115:5–12
150. Nir T, Melton DA, Dor Y (2007) Recovery from diabetes in mice by β cell regeneration. J Clin Invest 117:2553–2561
151. Bonal C, Avril I, Herrera PL (2008) Experimental models of β-cell regeneration. Biochem Soc Trans 36:286–289
152. Oyama K et al (2006) Spontaneous recovery from hyperglycemia by regeneration of pancreatic β-cells in Kir6.2G132S transgenic mice. Diabetes 55:1930–1938
153. Meier JJ, Bhushan A, Butler AE, Rizza RA, Butler PC (2005) Sustained β cell apoptosis in patients with long-standing type 1 diabetes: indirect evidence for islet regeneration? Diabetologia 48:2221–2228
154. Meier JJ et al (2006) Direct evidence of attempted β cell regeneration in an 89-year-old patient with recent-onset type 1 diabetes. Diabetologia 49: 1838–1844
155. Suarez-Pinzon WL, Lakey JRT, Brand SJ, Rabinovitch A (2005) Combination therapy with epidermal growth factor and gastrin induces neogenesis of human islet β-cells from pancreatic duct cells and an increase in functional β-cell mass. J Clin Endocrinol Metab 90:3401–3409
156. Brand SJ et al (2002) Pharmacological treatment of chronic diabetes by stimulating pancreatic β-cell regeneration with systemic co-administration of EGF and gastrin. Pharmacol Toxicol 91:414–420
157. Ueki K et al (2006) Total insulin and IGF-I resistance in pancreatic [β] cells causes overt diabetes. Nat Genet 38:583–588
158. Xuan S et al (2002) Defective insulin secretion in pancreatic β cells lacking type 1 IGF receptor. J Clin Invest 110:1011–1019

159. Hennige AM et al (2003) Upregulation of insulin receptor substrate-2 in pancreatic β cells prevents diabetes. J Clin Invest 112:1521–1532
160. Van Haeften TW, Twickler TB (2004) Insulin – like growth factors and pancreas β cells. Eur J Clin Invest 34:249–255
161. Petrik J, Arany E, McDonald TJ, Hill DJ (1998) Apoptosis in the pancreatic islet cells of the neonatal rat is associated with a reduced expression of insulin-like growth factor ii that may act as a survival factor. Endocrinol 139:2994–3004
162. Blakytny R, Jude EB, Martin Gibson J, Boulton AJM, Ferguson MW (2000) J. Lack of insulin – like growth factor 1 (IGF1) in the basal keratinocyte layer of diabetic skin and diabetic foot ulcers. J Pathol 190:589–594
163. Kratz G, Lake M, Gidlund M (1994) Insulin like growth factor-1 and –2 and their role in the re-epithelialisation of wounds; interactions with insulin like growth factor binding protein type 1. Scand J Plastic Reconstruct Surg Hand Surg 28:107–112
164. Suh DY, Hunt TK, Spencer EM (1992) Insulin-like growth factor-I reverses the impairment of wound healing induced by corticosteroids in rats. Endocrinology 131:2399–2403
165. Tsuboi R, Shi C-M, Sato C, Cox GN, Ogawa H (1995) Co-administration of insulin-like growth factor (IGF)-I and IGF-binding protein-1 stimulates wound healing in animal models. J Investig Dermatol 104:199–203
166. Pierre EJ et al (1998) Effects of insulin on wound healing. J Trauma 44:342–345
167. Greenhalgh DG, Sprugel KH, Murray MJ, Ross R (1990) PDGF and FGF stimulate wound healing in the genetically diabetic mouse. Am J Pathol 136: 1235–1246
168. Swenne I, Heldin C-H, Hill DJ, Hellerstrom C (1988) Effects of platelet-derived growth factor and somatomedin-c/insulin-like growth factor i on the deoxyribonucleic acid replication of fetal rat islets of langerhans in tissue culture. Endocrinology 122:214–218
169. Swenne I (1992) Pancreatic B-cell growth and diabetes mellitus. Diabetologia 35:193–201
170. Neuss S, Becher E, Wöltje M, Tietze L, Jahnen – Dechent W (2004) Functional expression of HGF and HGF receptor/c – met in adult human mesenchymal stem cells suggests a role in cell mobilization, tissue repair, and wound healing. Stem Cells 22:405–414
171. Chandrasekher G, Kakazu AH, Bazan HEP (2001) HGF- and KGF-induced activation of PI-3K/p70 S6 kinase pathway in corneal epithelial cells: its relevance in wound healing. Exp Eye Res 73:191–202
172. Wilson SE, Chen L, Mohan RR, Liang Q, Liu J (1999) Expression of HGF, KGF, EGF and receptor messenger RNAs following corneal epithelial wounding. Exp Eye Res 68:377–397
173. Garcia-Ocaña A et al (2000) Hepatocyte growth factor overexpression in the islet of transgenic mice increases B cell proliferation, enhances islet mass, and induces mild hypoglycemia. J Biol Chem 275:1226–1232
174. Vasavada RC et al (2006) Growth factors and β cell replication. Int J Biochem Cell Biol 38:931–950
175. Polak M et al (1993) Nerve growth factor induces neuron-like differentiation of an insulin-secreting pancreatic β cell line. Proc Natl Acad Sci USA 90:5781–5785
176. Li AK, Koroly MJ, Schattenkerk ME, Malt RA, Young M (1980) Nerve growth factor: acceleration of the rate of wound healing in mice. Proc Natl Acad Sci USA 77:4379–4381
177. Matsuda H et al (1998) Role of nerve growth factor in cutaneous wound healing: accelerating effects in normal and healing-impaired diabetic mice. J Exp Med 187:297–306
178. Herndon DN, Barrow RE, Kunkel KR, Broemeling L, Rutan RL (1990) Effects of recombinant human growth hormone on donor-site healing in severely burned children. Ann Surg 212:424–431
179. Gilpin DA, Barrow RE, Rutan RL, Broemeling L, Herndon DN (1994) Recombinant human growth hormone accelerates wound healing in children with large cutaneous burns. Ann Surg 220:19–24
180. Nielsen JH et al (1992) The role of growth hormone and prolactin in β cell growth and regeneration. Adv Exp Med Biol 321:9–17 (discussion 19–20)
181. Corbacho A, Martinez G, Clapp C (2002) Roles of prolactin and related members of the prolactin/growth hormone/placental lactogen family in angiogenesis. J Endocrinol 173:219–238
182. Blomme EAG, Zhou H, Kartsogiannis V, Capen CC, Rosol TJ (1999) Spatial and temporal expression of parathyroid hormone-related protein during wound healing. J Invest Dermatol 112:788–795
183. Bostrom MP et al (2000) Parathyroid hormone-related protein analog RS-66271 is an effective therapy for impaired bone healing in rabbits on corticosteroid therapy. Bone 26:437–442
184. Cozar-Castellano I et al (2006) Molecular control of cell cycle progression in the pancreatic β-cell. Endocr Rev 27:356–370
185. Schmassmann A, Reubi JC (2000) Cholecystokinin-B/gastrin receptors enhance wound healing in the rat gastric mucosa. J Clin Invest 106:1021–1029
186. Yamaguchi Y et al (2002) Gastrin – releasing peptide, a bombesin – like neuropeptide, promotes cutaneous wound healing. Dermatol Surg 28: 314–319
187. Bulut K et al (2004) Glucagon-like peptide 2 improves intestinal wound healing through induction of epithelial cell migration in vitro–evidence for a TGF-[β]-mediated effect. Regul pepts 121: 137–143
188. Stoffers DA (2004) The development of B-cell mass: recent progress and potential role of GLP-1. Horm Metab Res 36:811–821
189. Xu G, Stoffers DA, Habener JF, Bonner-Weir S (1999) Exendin-4 stimulates both β-cell replication

and neogenesis, resulting in increased β-cell mass and improved glucose tolerance in diabetic rats. Diabetes 48:2270–2276

190. Li L, Yi Z, Seno M, Kojima I (2004) Activin A and betacellulin: effect on regeneration of pancreatic β-cells in neonatal streptozotocin-treated rats. Diabetes 53:608–615

191. Hübner G, Hu Q, Smola H, Werner S (1996) Strong induction of activin expression after injury suggests an important role of activin in wound repair. Dev Biol 173:490–498

192. Barrientos S, Stojadinovic O, Golinko MS, Brem H, Tomic – Canic M (2008) Perspective article: growth factors and cytokines in wound healing. Wound Repair Regenerat 16:585–601

193. Schäffer MR, Fuchs N, Proksch B, Bongartz M, Beiter T, Becker HD (1998) Tacrolimus impairs wound healing: a possible role of decreased nitric oxide synthesis. Transplantation 65(6):813–818

194. Guilbeau J (2002) Delayed wound healing with sirolimus after liver transplant. Ann Pharmacother 36:1391–1395

195. Dean PG, Lund WJ, Larson TS, Prieto M, Nyberg SL, Ishitani MB, Kremers WK, Stegall MD (2004) Wound-healing complications after kidney transplantation: a prospective randomized comparison of sirolimus and tacrolimus. Transplantation 77:1555–1561

196. Knight RJ et al (2007) Risk factors for impaired wound healing in sirolimus – treated renal transplant recipients. Clin Transplant 21:460–465

14 Behavioral Deficiencies and Behavioral Supplementation

At the core of the new paradigm of T2D is the concept that behavioral mechanisms are at the root of and are the drivers of all the endocrine, metabolic, immunological, and neurological changes observed in T2D and its complications. A simple expectation that arises from this concept is that there should be detectable behavioral changes accompanying or actually preceding diabetes. Also one should be able to observe behavioral differences between diabetics and age-matched nondiabetic controls. This has not been looked at seriously because hardly anyone suspected so. The other reason is that the behavior of diabetics is always perceived as normal, and there is no doubt that is it generally "normal" in the sense that it is not what psychiatry calls a disorder. Psychiatry looks at behavioral disorders and is generally not interested in behavioral differences within the "normal and healthy" range of behaviors. Psychologists and ethologists, on the other hand, who are interested in "normal" behavior do not study diabetes. As a result whether diabetics differ in any types of behaviors or personality within the normal range of behaviors was a question almost never addressed so far. The word diabetic personality has occasionally been used and debated [1–3], but this literature does not talk about the kind of personality traits the new paradigm expects. But differences do exist and I will cite one example.

An interesting two-minute game called ultimatum game involves two-player bargaining. The two players are unknown to each other. One player (player 1) plays the role of allocator, and the other player (player 2) plays the role of recipient. Player 1 is promised some money and is asked to divide it between himself and the other player. The rules stipulate that player 1 must make an offer, and player 2 can either accept the offer or reject it. If player 2 accepts the offer, player 1 receives the promised money and will have to give the offered fraction to player 2. If player 2 rejects the offer, none of them gets any money.

If we assume that each player plays to maximize his benefits, the rational strategies are simple. For player 2, accepting and rejecting are the only two options. Since rejection is bound to give zero returns, accepting any nonzero offer is the only rational strategy. Assuming player 1 knows what is rational for player 2, offering minimum nonzero amount would be the most rational strategy for player 1. However, it is observed that most people do not go by this rationale. In a large number of studies across age groups and cultures, the modal offer is typically 50%, and the mean offer lies between 40 and 50% of the total amount. Although the game is simple to perform and analyze, the interpretation of the results is complex. The surprising agreement about roughly equal sharing has been commonly interpreted as a result

M. Watve, *Doves, Diplomats, and Diabetes: A Darwinian Interpretation of Type 2 Diabetes and Related Disorders*, DOI 10.1007/978-1-4614-4409-1_14, © Springer Science+Business Media New York 2013

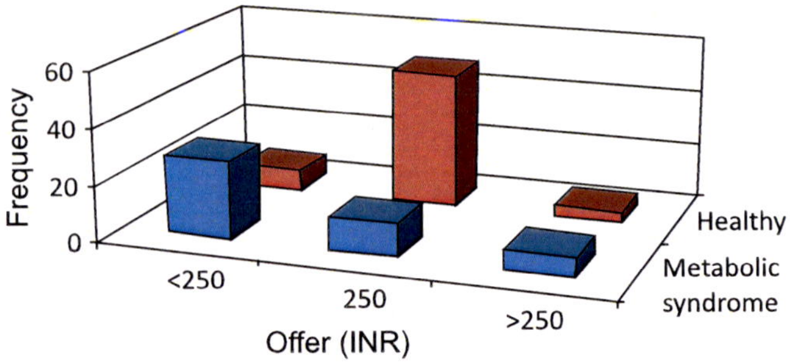

Fig. 14.1 Behavior of people with metabolic syndrome disorders in ultimatum game: Frequencies of ultimatum offers by people with metabolic syndrome (MS) disorders are depicted in comparison with healthy controls: Data are divided in three categories, namely, modal (50±5%), below modal, and above modal class. MS disorders give significantly lower offers (χ^2=32.01, df=2, $p<0.001$)

of an innate human tendency to appreciate fair and to retaliate unfair decisions [4]. This interpretation is certainly attractive and generally widely agreed. However the demonstration that serotonin [5–7] and testosterone [8, 9] levels affect ultimatum offers has raised other possibilities too. Both serotonin and testosterone are associated with social dominance hierarchy and aggression in diametrically opposite ways. Manipulating brain serotonin levels can change the dominant status and behavior of an individual [10, 11]. It is possible therefore that the ultimatum game offers reflect social hierarchical behavior in some way and not fairness alone. Social factors related to status and hierarchy have also been reported as risk factors for metabolic syndrome [12, 13]. Both serotonin and testosterone play important roles in metabolic syndrome. Chronically elevated serotonin signaling in the hypothalamus induces peripheral insulin resistance [14, 15]. Testosterone levels of diabetics are typically low, and testosterone has a protective role against many pathological consequences of metabolic syndrome [16–18] with the possible exception of advanced stages of nephropathy. Since serotonin is negatively associated with ultimatum offers and testosterone is positively associated [5–9], one may expect that diabetics may give lower offers in ultimatum game.

In a comparative study of people with metabolic syndrome (MS) and age-matched controls, as compared to the healthy control group, the MS group deviated significantly from the mode, much of the deviation being towards the left. The mean offer by the MS group (Rs. 202.55) was lower than the control group (Rs. 241.05), and frequency below 50% offer was significantly greater than the control group (χ^2=32.01, df=2, $p<0.001$) (Fig. 14.1). The comparison of diabetics and nondiabetics showed similar pattern, and the difference was highly significant (χ^2=28.42, df=2, $p<0.001$). For other disorders the sample sizes were too small to allow a meaningful statistical test. There was a weak negative correlation between BMI and ultimatum game offers in the pooled data which was nonsignificant.

Logistic regression with presence of at least one of the metabolic syndrome disorders as a dependent binary variable and including sex, occupation, BMI, and ultimatum offers revealed that the effects of sex and occupation were nonsignificant, whereas BMI (coefficient β=0.088, Wald=3.09, p=0.078) and ultimatum offer (coefficient β=−0.004, Wald=3.56, p=0.059) were marginally significant with overall predictability of the model being 62.4%. Exclusion of BMI from the regression rendered the effect of ultimatum offer significant (Wald=5.12, p=0.02) with predictability improving to 68.6%. On exclusion of ultimatum offer, the effect of BMI was nonsignificant and predictability declined to 56.78%. Considering T2D alone the patterns were very similar. Only ultimatum offer significantly predicted T2D in logistic regression (coefficient β=−0.004, Wald=4.42, p=0.035), whereas BMI, sex, and occupation did not show significant effects, the predictability of the model being 65.52%. In cross-sectional data ultimatum game offer was a better predictor of diabetes than BMI.

Ultimatum game is generally considered as a "fairness" game. However it would be unfair to jump to the conclusion that people with metabolic syndrome are "unfair" to others. This is because the fairness interpretation of ultimatum game itself can be questioned. People may deviate from the economically rational low offers owing to a number of possible alternative reasons apart from fairness: (1) The offers may represent a valuation of relative social ranking with anonymous person being given a default equal ranking; (2) A high offer may be viewed as a costly signal intended to advertise one's own status and generosity; or (3) It may be driven by a hidden prediction of repeated and reciprocal interactions. All these explanations can be grouped as social status related explanations as opposed to the economic rationality. It can be perhaps generalized that economic rationality prompts low offers, and social rank-related factors prompt offers substantially higher than the economically rational ones. On this scale diabetics appear to be more inclined towards economic rationality than social rank- or social justice-related factors. We have seen earlier that social ranks play an important role in determining metabolic state. The hormones which have previously been shown to affect economic game behavior are serotonin and testosterone, both of which are known to play a role in social dominance hierarchy [10, 11]. Therefore a plausible explanation of lower hits by diabetics is likely to be related to the social hierarchy factors. It makes sense for a diplomat to be economically rational and care less about aggressive social hierarchical struggle. The results of the survey are therefore compatible with the behavioral origin paradigm of metabolic syndrome. On the other hand there is no a priori reason why "fairness" should be affected in metabolic syndrome disorders.

Of much potential interest is the result that ultimatum game offers are good predictors of T2D and other disorders in a cross-sectional sample. At least in our sample ultimatum offers were a better predictor of diabetes than BMI. This may not be surprising because of two reasons. On the one hand in Indian population, insulin resistance is not necessarily associated with very high BMI [19], and on the other hand, the inadequacy of obesity alone in explaining insulin resistance syndrome is increasingly being recognized.

This small study should therefore stimulate further studies along three paths. One would be to test the robustness of the association of ultimatum game offers with metabolic syndrome cross-culturally, the other to design and standardize a set of tests to cover a wider variety of behaviors that could be markers of metabolic syndrome, and the third to test whether the set of behavioral differences can predict the development of metabolic syndrome disorders in longitudinal studies.

A number of other behavioral differences between diabetics and nondiabetics are expected but have not been tested. These include physical risk taking, readiness for adventure, and tolerance to physical discomfort. Specific situational tests need to be designed where there can be two alternative solutions possible for solving a given problem, one involving physical aggression and the other involving social manipulation. The prediction of course is that diabetics would avoid the soldier solution and prefer a diplomat solution.

If these tests demonstrate a behavioral difference, the question of cause–effect relationship would still remain. Are the behavioral differences caused by diabetes or are these behaviors a risk factor for diabetes? This question can only be answered through longitudinal studies where a series of behavioral studies are performed on a cohort, and the cohort is then monitored for several years to see who develops insulin resistance or related phenomenon.

Leaving the question to researchers of whether diabetics differ from nondiabetics in behavior and whether behavioral differences precede or follow diabetes, I would now proceed to reframe the central concept of the new paradigm of T2D in a slightly different perspective and explore the possibility that the concept is much broader and goes beyond T2D and encompasses a wider range of disorders, particularly the ones that have become common suddenly with modern urban lifestyle. The broader concept can be called behavioral deficiency disorders. The mainframe argument of behavioral deficiency disorders goes in the following sequence:

1. We evolved as hunter-gatherers, and a number of behaviors evolved with us as adaptive behaviors for a hunter-gatherer life.
2. Every behavior is linked with some neuroendocrine pathway(s).
3. Therefore chronic deficiency of particular behaviors can affect specific neuroendocrine pathways affecting its downstream links and thereby eventually leading to endocrine, metabolic, immunological, and other functional disorders.
4. Just as dietary deficiencies are best treated with dietary supplementation, the best treatment for behavioral deficiencies would be behavioral supplementation.

Traditional medicine identifies many different types of dietary deficiencies, but the concept of behavioral deficiencies is nonexistent in current medicine. Much is talked about Paleolithic diet, but the importance of Paleolithic behavior remains largely unrecognized. This should not be too surprising because of two main reasons. One is that epidemiologically behavioral deficiency disorders appear to have become common fairly recently. Most of the disorders that we will list below as possible behavioral deficiency disorders have suddenly become common throughout the world in the last 2–3 generations, and their prevalence is rapidly on the rise. This is because it is only the modern urban life which has created these deficiencies. So it is only recently that any serious research could focus on these. In comparison, dietary deficiencies have been with us for a few thousand years, perhaps beginning with agriculture. As a result research on dietary deficiencies has a much longer history. Also the concept of relation of diet with health is very old although specific vitamins and other essential nutrient deficiencies and their effects were discovered only in the last two centuries. The other reason for the failure to identify behavioral deficiency disorders is the Cartesian body–mind division implicit in medicine. Behavior is thought to be more related to the mind than the body. The "mind" is rarely a part of research of biochemists. Psychiatry deals with the mind and brain, but it becomes active only when there is a perceived "behavioral disorder." The behavioral disorders handled by psychiatry are rather extreme cases and have always been treated as "defects." As a result the behavior–biochemistry connection in normal life has been largely neglected. Behavioral ecologists on the other hand could have been quicker to make these connections, but they never consider medicine as their field. This segmentation of science and the extremely myopic view in each segment have prevented science from identifying the logical and obvious possibility of behavioral deficiency disorders.

I have so far discussed T2D, hypertension, hypercholesterolemia, atherosclerosis, and CVD typical of the modern lifestyle as behavioral deficiency disorders, although I did not use this word so explicitly before. All the details that were discussed in the last few chapters can be summarized in this new phrase: behavioral deficiency disorder. We have seen the pathways by which deficiency of soldier behaviors can lead to some of the typical modern "lifestyle" disorders. This may not be a complete list. This is only a beginning of a new concept which we may expect to develop further with more research inputs.

The list of behavioral deficiencies that I can currently identify can be summarized by the acronym PARALISIS, summarizing deficiencies of (1) physical aggression; (2) agility and rapid action; (3) romantic love, sex, and reproductive functions; (4) adventure; (5) exposure to the little creatures around; (6) injury proneness; (7) serenity and solitude; (8) intellectually intensive activities; and (9) sun, heat, and other skin exposures.

1. Physical aggression: Through the entire book, we have been looking at what the deficiency of physical aggression can cause. Almost the entire pathophysiological picture of T2D can be explained by deficiency of physical aggression. But how and when did we develop this deficiency and why are its effects apparent only in the modern urban lifestyle?

   It is not difficult to perceive that until modern times, the importance and frequency of aggressive neuromotor actions have not changed in human history in spite of going through major transitions such as hunter-gatherer to agricultural and agricultural to urban life. Unlike many popular perceptions, hunter-gatherer life is not full of violence. Most hunter-gatherer societies are generally peace loving with occasional intergroup tensions and rarely actual wars. The main aggressive neuromotor actions are not because of wars, they are most frequent during day-to-day foraging. Both hunting and gathering have aggressive neuromotor actions. Aggression in hunting does not need any elaboration, but gathering exercises also involve digging, cutting, throwing, and other aggressive actions. Other day-to-day maintenance activities such as cutting wood for fire, mending houses, and making tools also involve a number of aggressive actions. This did not change with agriculture. Even in a peaceful agricultural society, aggressive actions are involved in digging, cutting, driving animal-drawn carts, controlling animal herds, and driving away birds and nuisance animals like foxes and jackals. Even in urban life, as long as there was the use of firewood, wood cutting was necessary. Manual cutting, grating, dough mixing, and other cooking preparations did involve aggressive neuromotor actions. Whether good or bad, punishing a child with a quick spank was a routine. The games children played were more of rough and tumble type. Social signaling among friends could easily involve some aggressive actions like a forceful stroke on the back or a mild punch on the arm.

All these mild and minimum aggressive actions are also lost in the modern mechanized, mannerized, automated, and comfortable urban life. Modern life is not free of violence, but modern violence too is aggression deficient. Pulling the trigger of a revolver and pressing a button of a remote control bomb are examples of nonaggressive violence. Even spanking our own children has vanished from our lives now. The nature of games that children play has also changed substantially. Parents are overanxious that their kids do not engage in aggression even while playing. The disappearance of aggression at an early age may be one of the reasons why T2D and other disorders have started appearing at an earlier age now. Aggressive actions have almost disappeared from social signaling as well since they are being considered increasingly uncivilized. This is where extreme deficiency of aggression accumulated over decades of life might start showing its effects. If this is true, then participation in even mildly aggressive exercises or sports would be able to prevent most of the modern lifestyle disorders. Unfortunately sports have become a specialist and highly competitive activity. It has not remained a frequent and casual activity. Parents like their kids to participate in some sport only to excel. While excelling is good, the other side of the coin is that it discourages a larger proportion of individuals who do not excel. So a very small proportion of youngsters actively engage in sports, others remain passive spectators.

2. Agility and rapid action: We talked a lot about aggression since there are substantial data on the physiological effects of aggression. That does not mean that aggression is the only or the most important behavioral deficiency. A number of others are indicated here about which there is suggestive evidence, but researchers have mostly neglected them mainly owing to the lack of a paradigm that recognizes their importance. But the difference between a soldier and diplomat lifestyle is not restricted to the presence and absence of aggression. There are a number of other components of soldier life that may be equally important.

   Rapid action is one such possibility. Although very little research has gone into it, there is a demonstration that hyperinsulinemia is associated with downregulation of quick preattentive neuronal responses and enhancement of some of the slower cognitive responses [20]. This is clearly a soldier–diplomat difference. Rapid and complex nerve–muscle coordination (NMC) action is required in hunting, fighting, or explosive sports. The cerebellum is responsible for such coordination. There is some evidence indicating that the neurodegenerative changes in the aged individuals are more due to cerebellar degeneration than cortical degeneration [21, 22]. This might mean that deficiency of complex NMCs is the ultimate cause of neurodegenerative disorders. This is currently only a speculation but it looks promising and warrants research.

3. Romantic love, sex, and reproduction: Although romantic love, sex, and reproduction have very different connotations, they form a continuum, and therefore, here we group them together. Romantic love increases NGF levels, and NGF deficiency is one of the important contributors to degenerative nerve disorders. Affectionate body contact behavior stimulates oxytocin release which is believed to be a stress-protecting hormone. We have noted before the many protective effects of male and female sex hormones against oxidative damage, apoptosis, and many of the diabetic complications [23–26]. Mating is shown to upregulate BDNF in the brain [27]. In females, EGF levels go up in pregnancies and remain higher than baseline after pregnancy as well [28]. Thus there are many indications that the trio of romantic love, sex, and reproduction has a number of beneficial effects on health. It is also known that castration induces obesity and insulin resistance in rats [29, 30]. Many contraceptive measures are known to have components of metabolic syndrome as their side effects [31, 32]. But whether the altered nature of love, sex, and their relevance to reproduction in the modern society has anything to do with the cluster of modern diseases is an open question so far largely ignored by the scientific community.
4. Adventure: Voluntarily participating and enjoying an activity involving potential physical risk is adventure. The words "physical risk" are important since it necessarily excludes metabolically risky behaviors such as alcoholism or smoking. It also excludes social or business risks. Adventure is another behavior differentiating soldier life from diplomat life. Again there is little research on the metabolic or endocrine effects of adventure except the demonstration that adventure triggers NGF secretion. NGF can effectively prevent neurodegenerative disorders, and therefore, there can be a strong connection between adventure and long-term health. Interestingly, in one study, not only participation in adventure but even the thought of participating in adventure triggered salivary NGF secretion [33]. Unfortunately this study does not estimate other growth factors. NGF is most likely to have an adaptive role in adventure since adventure anticipates injuries and NGF is necessary for regeneration and healing. By the same logic, adventure should also trigger other growth factors such as EGF. Since both aggression and adventure anticipate injuries, many of their immunologic and metabolic effects could be similar. But this has not been tested so far. Again this is not at all difficult to test, just that the paradigm was absent, and so researchers did not think of looking at it.
5. Little creatures around: One major deficiency of modern urban life is contact with a wide variety of small and large creatures including microbes, animals, and plants. This contact certainly affects immunity and may also influence many other aspects of physiology.

Growing evidence suggests a role of intestinal bacterial flora in obesity [34–37]. This parallels the effects of intestinal microorganisms on brain and behavior. Although surprising at the first glance, there is good evidence that intestinal bacteria affect emotional behavior and central GABA receptor expression [38]. It would not be surprising if it is not a one-way traffic. I would expect behavior to affect intestinal flora too since behavior affects immunity in many ways (Chaps. 8 and 11), and mucosal immunity is likely to be a critical determinant of the resident flora. Thus, the 4 Bs, namely, brain, behavior, BMI, bacteria, could be in a complex interacting loop that we have not sufficiently understood as yet, but there is strong evidence pointing in that direction.

Allergic diseases including asthma or hay fever are less common in a farm environment than in an urban environment [39–44], and this is most likely to be an effect of early life exposure to common microorganisms associated with soil and animals. Early life overexposure to antibiotics also increases the risk of asthma [45]. There is a widely discussed and debated "hygiene hypothesis" which states that early life exposure to a diversity of microorganisms is protective against allergic diseases, and the current increase in the incidence of allergies and asthma is a product of growing in a too hygienic environment. Substantial data support the hygiene hypothesis, although the mechanisms are debated [46–50]. Even type 1 diabetes is included under the hygiene hypothesis by some [51]. There are possible links with type 2 diabetes as well. Arthropod bites and stings have anti-inflammatory properties. I suspect them to be immune redistributors rather than being anti-inflammatory. Frequent exposure to insect bites is likely to drive the innate immune cells peripheral thus ameliorating one of the main pathophysiological processes of T2D (see Chap. 8).

Why do I call deficiency of exposure to the natural biota a behavioral deficiency rather than an environmental factor? The reason is simple. Soil and its microbial and insect flora are generally abundant everywhere even in urban environments. However, people's attitude to look at it is widely different. Playing in mud could be a routine for farmers' children, while in urban culture, typically, mud is treated as dirt and children are discouraged from playing in it. This is a behavioral rather than environmental difference.

6. Injury proneness: We have already argued before that anticipation of injuries by aggression or adventure stimulates EGF and NGF secretion [52–55]. In addition there can be proneness to injuries independent of both aggression and adventure. Simply walking through a dense forest or thorn scrub that involves neither, one can get many minor pricks and scratches on the skin. Traditional cooking also involves frequent minor injuries. These minor injuries attract innate immune cells of the body and help normalize their distribution within the body. They also stimulate the expression of PPARs and some of the growth factors that have effects on insulin and insulin sensitivity. A major change in the frequency of minor injuries characterizes modern urban life. Again we do not have studies to evaluate to what extent this actually matters.
7. Solitude and serenity: We have seen that population density and its perception affect aggression, social behavior, and reproductive behavior. In a hunter-gatherer society, one is exposed to the wilderness, vast expanses of forest, grassland, deserts, or the sea very frequently and often alone. This happens very rarely in modern urban life. The feeling of being in a serene environment has a number of perceivable effects on feelings and mood, and therefore, it is likely that it has many effects on the endocrine states as well. This appears to have never been studied since the paradigm was nonexistent. There is some literature on loneliness where it is perceived as an entirely negative emotion and a stress [56]. The endocrine effects of an enjoyable and tranquil state of solitude have seldom been studied. I would expect that solitude upregulates adiponectin since adiponectin is a marker of *r* reproductive strategy. But there can be many other pro-health effects of solitude, and it remains an unexplored area of research.
8. Intellectually intensive activities: A lot has been talked about the importance of physical exercises. I would like to suggest here that intellectual exercises are equally important. As there has been a reduction in total physical activity with modern urban lifestyle, there is also a reduction in brain activity. A substantial part of the brain work is reduced with reduction in complex nerve muscle coordination activity. The typical hunter- or soldier-related brain activities such as mental mapping and spatial cognition are also reduced substantially. In the soldier to diplomat transition, it is expected that this loss is made up by increase in cognitive activities. In modern life reliance on electronic devices has taken off some of the cognitive activities as well. A reduction in the total brain activity is likely to result into downregulation of glut-1 in the brain capillaries leading to hyperglycemia. It is also possible

that NGF and BDNF are downregulated in an inactive brain. These possibilities have not been tested.

9. Sun, heat, and other climatic exposures: The story of sunlight and vitamin D is well known. Wide cultural differences exist in the perception about exposure to sunlight. Generally people living in cold climates welcome a sunny day and enjoy sunbathing. In tropical countries, there is a reverse trend. In tropical urban environments, avoiding direct exposure to sun has reached such an extreme that there is a paradoxical deficiency of vitamin D in the sunniest places [57–60]. This again is clearly a behavioral rather than environmental deficiency.

The above list of PARALISIS deficiencies is perhaps incomplete. For many, the nature of evidence is currently only suggestive, and therefore, some of them remain only speculations. But speculations stimulate research and therefore need to be respected, simultaneously understanding their limitations. The question to ask further is that if these are the deficiencies, what are the possible disorders caused by these deficiencies and what are the pathways connecting them. We will obviously not talk about T2D and CVD anymore. A number of other diseases that can be primarily behavioral deficiency disorders are as follows:

1. Cancer: It is well known that cancer cells arise by somatic mutations. However, a single mutation may not be sufficient to make a cell malignant. More than one mutation is involved in carcinogenesis. Since mutation is a rare event, occurrence of many specific mutations that turn a cell cancerous must be extremely unlikely. This is like an old evolutionary debate. A complex structure such as the eye cannot arise by a single mutation. There need to be a series of mutations before a fully functional eye can evolve. This example was used as a procreationist or intelligent design argument. Richard Dawkins and other Darwinians counterargued saying that even intermediate stages in eye development would serve as a primitive eye with a small selective advantage. A small selective advantage can increase the frequency of the primitive eye phenotypes. This forms the platform for further rare beneficial mutations. Since selection is coupled with mutations, the impossible-looking combination of rare mutations can actually materialize. It is likely that the case with cancer is similar. A common feature of a range of carcinomas is abnormal expression of EGF receptor (EGF-R) [61–63]. EGF is an important growth factor for tissue regeneration [64, 65]. As long as EGF levels are normal, the tissue regeneration process will be normal. If a deficiency of EGF arises, the normal cell will have a reduced growth rate, and on this background, a mutant with abnormal expression of EGF-R or autoactivation of EGF-R will get a selective advantage. Since EGF deficiency can arise due to behavioral deficiency, one of the causes of cancer can be behavioral deficiency. A strong support to the behavioral origins of cancer is given by the mice experiments in which mice living in an enriched environment showed reduced tumor growth as compared to mice in traditional cage, and this difference could not be accounted for by total physical activity alone [66]. These experiments monitored BDNF levels and correlated them with tumor suppression. Although these experimenters did not monitor EGF or NGF levels, other experiments have demonstrated that EGF and NGF together stimulate BDNF [67]. It is likely therefore that the difference in the mice in conventional cage versus enriched environment is driven by behavioral differences. Perhaps the link between behavior and cancer goes beyond EGF involving other growth factors and their receptors. The behavior-induced altered distribution of macrophages can affect angiogenesis mechanisms which are known to be important in tumor growth. In short there is sufficient evidence to suspect a link between behavior and cancer, and researchers need to explore this possibility seriously.

2. Proliferative skin disorders: Two of the proliferative skin disorders, namely, acanthosis nigricans and psoriasis, share two common features. Both involve hyperproliferation of skin tissue and both have abnormal expression

of EGF-R [68–70]. Here both have a fundamental similarity with cancers in being proliferative disorders and having altered EGF-R. Therefore it is likely that similar to cancer, both primarily arise from behavior-induced EGF deficiency. Acanthosis nigricans is associated with diabetes [71, 72] which itself arises from loss of aggression, and there are data showing association of psoriasis with reduced aggression as well [73].

3. Sex and reproductive system disorders: The anovulatory condition polycystic ovary syndrome (PCOS) is known to be linked with insulin resistance and therefore presumably has the same behavioral etiology. In many nonhuman primates, subordinate females have suppressed reproduction [74–81]. The mechanisms involved in the suppression of ovulation involve abnormal expression of estrogens, LH, and FSH. Abnormal levels of estradiol, LH, and FSH characterize human PCOS as well. The subordination-induced suppression of ovulation in primates cannot be explained by "stress" mechanisms alone [75, 80]; there appear to be specific endocrine pathways carrying out this function. Since social status can change with time, the subordination-induced suppression is reversible. Therefore it makes sense to be thrifty and store energy for future reproductive opportunities. Therefore thrift should be associated with suppression of reproduction. PCOS is likely to be the human counterpart of this behavior-induced ovulation suppression, although it is not exactly parallel. Since social subordination is related to insulin resistance, some underlying mechanisms of insulin resistance and PCOS are likely to be common.

   Another anovulatory disorder of a later age, dysfunctional uterine bleeding (DUB), is also frequently associated with impaired glucose tolerance and therefore is likely to have similar behavioral origins [82]. In men erectile dysfunction (ED) has strong positive association with diabetes. Testosterone and NO have a direct role to play in both ED and T2D, and therefore, ED needs to be investigated as a behavioral deficiency disorder. Currently a wild speculation, but it might be worth exploring the possibility of homosexuality also arising out of early life behavioral deficiencies. Homosexuality in some nonhuman primates has been shown to be more commonly initiated by subordinate females indicating the importance of social subordination.

4. Frailty syndrome, osteopenia, osteoporosis, and sarcopenia: All these appear to be disorders connected with progressive disinvestment from bone and/or muscle. We have been talking about such behavior-induced disinvestment mechanisms in this book. There are strong reasons to suspect behavioral origins of these disorders too. Bone metabolism is finely modulated by serotonin and dopamine in such a way that an aggressive personality would be accompanied by high bone density, and aggression suppression would erode it [83, 84]. Glucocorticoids also chronically weaken bones [85–87]. Lack of adequate exposure to sunlight leads to deficiency of vitamin D and thereby weakens bones. Giving more calcium does not resolve the problem. In the most likely case, behavioral intervention would be effective in reversing most low bone density conditions. Same can be said about disorders involving progressive loss of muscle. Clinical trials for behavioral intervention need to be undertaken to test the possibility.
5. Chronic fatigue syndrome (CFS), depression, and other disorders involving brain monoamines: Currently the etiology of CFS is unknown. We have seen in Chap. 7 that the central fatigue signals are activated much before energy exhaustion and this is adaptive. Serotonin is involved in the central fatigue mechanisms, and higher levels of serotonin activity could trigger the central fatigue signal faster. An extreme case of this could be CFS. In CFS, there is increased brain serotonin activity and decreased dopamine activity [88–90]. DHEA is also low [91, 92]. CFS is distinct from depression in having low HPA activity and cortisol levels. In a large number of disorders linked with brain monoamines including depression, the possibility of behavioral deficiencies needs to be tapped since brain

monoamines, which appear to have a central role in these disorders, have the most direct link with behavioral strategies.

6. Chronic inflammatory conditions: A large number of inflammatory disorders are becoming increasingly common with the modern lifestyle. The possibility that they arise from behavior-induced immune redistribution (see Chap. 8) needs to be explored seriously.

If behavioral deficiencies are important contributing factors to some of these disorders, we should be able to treat them by "behavioral supplements" just as dietary deficiencies are treated with pills of vitamins, calcium, iron, or whatever. The behavioral pills can take the form of specific games, activities, and exercises, specifically designed for supplementing the specific deficient behavior. The importance of exercise in preventing and controlling diabetes is well accepted. But the current perception of exercise is only a means of burning extra calories. Exercises are much more than calorie drains. They can make up for specific behavioral deficiencies, and therefore, the precise nature of exercise is crucial in deciding its effect. Exercises need to be specifically designed to supplement one or more of the deficiencies diagnosed. Depending upon personal history, the need can be widely different in different individuals.

The exercises commonly recommended for the prevention and control of T2D and its complications today are grossly inadequate to supplement most of the components of PARALISIS set of deficiencies. Treadmill, jogging, cycling, and swimming are exercises deficient in aggression, quick unanticipated actions, and complex nerve muscle coordination. Also the psychological components of adventure, perceived proneness to injuries, and discomfort are missing. The precise nature of recommended exercises would also depend upon age and current condition of a patient. This is going to be a specialist job, and once the principles are established, expertise will be built up eventually.

The ultimate argument is quite simple then. Playing the right kinds of games should be able to cure diabetes and a list of possibly behavioral deficiency disorders. The right game would be highly personalized and would depend upon the specific deficiencies identified for that person and the appropriate corrective measure for it, suitably modified for the given age and condition of the patient. This, I can imagine, would sound ridiculous to many readers. Diabetes is a complex disorder that a battery of drugs developed over decades of research have been unable to control, leave alone cure!! Will playing games cure diabetes?

There has been a similar story in the history of medicine. Scurvy was known as a disease since Hippocrates' time, i.e., about 400 BC, and was a common cause of morbidity and even death among sailors and pirates. A Scottish surgeon in the British Royal Navy, James Lind, first proved with experiments that he described in his 1753 book, *A Treatise of the Scurvy*, that it could be treated with citrus fruit. Lind backed his claim by carefully designed experiments. The design of these experiments was historical and formed the backbone of experimental medicine for the future. The evidence that oranges could cure scurvy was strong. But for everyone, this was a ridiculous statement. An intractable disease that affected the most robust, sturdy, and brave soldiers and was not cured by any medicine for centuries together would be gone in no time by eating oranges and lemons? That cannot be true!! In fact James Lind's advice was not implemented by the Royal Navy for ages to come. Interestingly nobody seems to have experimentally challenged or disproved what Lind said. It was simply ignored. At last in 1932, vitamin C and its direct role in scurvy were discovered, and scurvy ceased to be a problem in no time. It appears that the diabetes story will repeat history. If T2D is truly a behavioral deficiency disorder, simply playing the right kinds of games that make up for the behavioral deficiencies will be sufficient to prevent, control, reverse, and perhaps even cure type 2 diabetes and a number of other disorders. But this would sound ridiculous to most physicians and patients, and for decades, together, they may keep on relying on expensive medicines which essentially do not work. But this time we can avoid making the same mistake since we are much better equipped with scientific tools. There was a gap of almost two centuries between the realization that oranges work and working out the biochemistry behind it. This time, at least, some of the pathways linking behavior to endobolism are already known. If the scientific community focuses on adequate research efforts on the existing gaps and works out the complete pathways by which behavioral deficiencies and behavioral supplementation works, the delay and thereby loss to several generations can be avoided.

References

1. Palmer RW (1958) The diabetic personality. J Indiana State Med Assoc 51:1399–1400
2. Dunn SM, Turtle JR (1981) The myth of the diabetic personality. Diabetes Care 4:640–646
3. Iványi J, Gyimesi A, Hanyecz V, Kállai-Szabó K (1988) Personality examinations in individuals with insulin-dependent diabetes mellitus. Acta Diabet Lat 25:109–116
4. Thaler RH (1988) Anomalies: the ultimatum game. J Econ Perspect 2:195–206
5. Emanuele E, Brondino N, Bertona M, Re S, Geroldi D (2008) Relationship between platelet serotonin content and rejections of unfair offers in the ultimatum game. Neurosci Lett 437:158–161
6. Crockett MJ, Clark L, Tabibnia G, Lieberman MD, Robbins TW (2008) Serotonin modulates behavioral reactions to unfairness. Science 320:1739
7. Emanuele E, Bertona M, Re S, Brondino N (2009) Human economic and financial behavior: the serotonergic hypothesis. Bio Hypotheses 2:109–110
8. Burnham TC (2007) High-testosterone men reject low ultimatum game offers. Proc Biol Sci 274:2327–2330
9. Zak PJ et al (2009) Testosterone administration decreases generosity in the ultimatum game. PLoS One 4:e8330
10. Edwards DH, Kravitz EA (1997) Serotonin, social status and aggression. Curr Opin Neurobiol 7: 812–819
11. Larson ET, Summers CH (2001) Serotonin reverses dominant social status. Behav Brain Res 121:95–102
12. Agardh EE et al (2003) Work stress and low sense of coherence is associated with type 2 diabetes in middle-aged Swedish women. Diabetes Care 26:719–724
13. Sapolsky RM (2005) The influence of social hierarchy on primate health. Science 308:648–652
14. Luo S, Meier AH, Cincotta AH (1998) Bromocriptine reduces obesity, glucose intolerance and extracellular monoamine metabolite levels in the ventromedial hypothalamus of Syrian hamsters. Neuroendocrinology 68:1–10
15. Luo S, Luo J, Cincotta AH (1999) Chronic ventromedial hypothalamic infusion of norepinephrine and serotonin promotes insulin resistance and glucose intolerance. Neuroendocrinology 70:460–465
16. Holmäng A, Björntorp P (1992) The effects of testosterone on insulin sensitivity in male rats. Acta Physiol Scand 146:505–510
17. Spark RF (2007) Testosterone, diabetes mellitus, and the metabolic syndrome. Curr Urol Rep 8:467–471
18. Muller M, Grobbee DE, den Tonkelaar I, Lamberts SWJ, van der Schouw YT (2005) Endogenous sex hormones and metabolic syndrome in aging men. J Clin Endocrinol Metab 90:2618–2623
19. Yajnik CS (2004) Obesity epidemic in India: intrauterine origins? Proc Nutr Soc 63:387–396
20. Kern W et al (2001) Improving influence of insulin on cognitive functions in humans. Neuroendocrinology 74:270–280
21. Hogan MJ et al (2011) Cerebellar brain volume accounts for variance in cognitive performance in older adults. Cortex 47:441–450
22. Iacopino AM, Christakos S (1990) Specific reduction of calcium-binding protein (28-kilodalton calbindin-D) gene expression in aging and neurodegenerative diseases. Proc Natl Acad Sci USA 87:4078–4082
23. Ayres S, Tang M, Ravi Subbiah MT (1996) Estradiol-17[β] as an antioxidant: Some distinct features when compared with common fat-soluble antioxidants. J Lab Clin Med 128:367–375
24. Behl C, Widmann M, Trapp T, Holsboer F (1995) 17-[β] Estradiol protects neurons from oxidative stress-induced cell death in vitro. Biochem Biophys Res Commun 216:473–482
25. Shwaery GT, Vita JA, Keaney JF (1998) Antioxidant protection of LDL by physiologic concentrations of estrogens is specific for 17-β-estradiol. Atherosclerosis 138:255–262
26. Ahlbom E, Prins GS, Ceccatelli S (2001) Testosterone protects cerebellar granule cells from oxidative stress-induced cell death through a receptor mediated mechanism. Brain Res 892:255–262
27. Franklin TB, Perrot-Sinal TS (2006) Sex and ovarian steroids modulate brain-derived neurotrophic factor (BDNF) protein levels in rat hippocampus under stressful and non-stressful conditions. Psychoneuroendocrinology 31:38–48
28. Brown MJ, Zogg JL, Schultz GS, Hilton FK (1989) Increased binding of epidermal growth factor at preimplantation sites in mouse uteri. Endocrinology 124:2882–2888
29. Hausberger FX, Hausberger BC (1966) Castration-induced obesity in mice. Eur J Endocrinol 53: 571–583
30. Georgiev IP et al (2009) Evaluation of insulin resistance in obese castrated New Zealand white rabbits. Rev Med Vet 160:335–340
31. Baird DT, Glasier AF (1993) Hormonal contraception. N Engl J Med 328:1543–1549
32. Foussard-blanpin O, Paillot-renaud P, Bruneau-bigot A (1984) Oral contraception: failures and risks. Lyon Pharm 35:385–393
33. Aloe L et al (1994) Emotional stress induced by parachute jumping enhances blood nerve growth factor levels and the distribution of nerve growth factor

receptors in lymphocytes. Proc Natl Acad Sci USA 91:10440–10444
34. DiBaise JK et al (2008) Gut microbiota and its possible relationship with obesity. Mayo Clin Proc 83:460–469
35. Cope K, Risby T, Diehl AM (2000) Increased gastrointestinal ethanol production in obese mice: implications for fatty liver disease pathogenesis. Gastroenterology 119:1340–1347
36. Bajzer M, Seeley RJ (2006) Physiology: obesity and gut flora. Nature 444:1009–1010
37. Cani PD et al (2007) Metabolic endotoxemia initiates obesity and insulin resistance. Diabetes 56:1761–1772
38. Bravo JA et al (2011) Ingestion of lactobacillus strain regulates emotional behavior and central GABA receptor expression in a mouse via the vagus nerve. Proc Natl Acad Sci USA 108:16050–16055
39. Ernst P, Cormier Y (2000) Relative scarcity of asthma and atopy among rural adolescents raised on a farm. Am J Respir Crit Care Med 161:1563–1566
40. Riedler J, Eder W, Oberfeld G, Schreuer M (2000) Austrian children living on a farm have less hay fever, asthma and allergic sensitization. Clin Exp Allergy 30:194–200
41. Leynaert B et al (2001) Does living on a farm during childhood protect against asthma, allergic rhinitis, and atopy in adulthood? Am J Respir Crit Care Med 164:1829–1834
42. Kilpeläinen M, Terho EO, Helenius H, Koskenvuo M (2000) Farm environment in childhood prevents the development of allergies. Clin Exp Allergy 30:201–208
43. Ehrenstein V et al (2000) Reduced risk of hay fever and asthma among children of farmers. Clin Exp Allergy 30:187–193
44. Adler A, Tager I, Quintero DR (2005) Decreased prevalence of asthma among farm-reared children compared with those who are rural but not farm-reared. J Allergy Clin Immunol 115:67–73
45. Kozyrskyj AL, Ernst P, Becker AB (2007) Increased risk of childhood asthma from antibiotic use in early life. Chest 131:1753–1759
46. Wills-Karp M, Santeliz J, Karp CL (2001) The germless theory of allergic disease: revisiting the hygiene hypothesis. Nat Rev Immunol 1:69–75
47. Liu AH, Murphy JR (2003) Hygiene hypothesis: fact or fiction? J Allergy Clin Immunol 111:471–478
48. Guarner F et al (2006) Mechanisms of disease: the hygiene hypothesis revisited. Nat Clin Pract Gastroenterol Hepatol 3:275–284
49. Weiss ST (2002) Eat dirt—the hygiene hypothesis and allergic diseases. N Engl J Med 347:930–931
50. Matricardi PM, Bonini S (2000) High microbial turnover rate preventing atopy: a solution to inconsistencies impinging on the Hygiene hypothesis? Clin Exp Allergy 30:1506–1510
51. Gale E (2002) A missing link in the hygiene hypothesis? Diabetologia 45:588–594
52. Lakshmanan J (1986) Aggressive behavior in adult male mice elevates serum nerve growth factor levels. Am J Physiol Endocrinol Metabol 250:E386–E392
53. Nexo E, Hollenberg M, Bing J (1981) Aggressive behavior in mice provokes a marked increase in both plasma epidermal growth factor and renin. Acta Physiol Scand 111:367–71
54. Nexø E, Olsen PS, Poulsen K (1984) Exocrine and endocrine secretion of renin and epidermal growth factor from the mouse submandibular glands. Regul Pept 8:327–334
55. Sánchez O, Viladrich M, Ramírez I, Soley M (2007) Liver injury after an aggressive encounter in male mice. Am J Physiol Regul Integr Comp Physiol 293:R1908–1916
56. Wawrzyniak AJ, Whitemn MCP (2011) Perceived stress, loneliness, and interaction with fellow students does not affect innate mucosal immunity in first year university students. Jpn Psychol Res 53:121–132
57. Sachan A et al (2005) High prevalence of vitamin D deficiency among pregnant women and their newborns in northern India. Am J Clin Nutr 81: 1060–1064
58. Marwaha RK et al (2005) Vitamin D and bone mineral density status of healthy schoolchildren in northern India. Am J Clin Nutr 82:477–482
59. Harinarayan CV et al (2007) High prevalence of low dietary calcium, high phytate consumption, and vitamin D deficiency in healthy south Indians. Am J Clin Nutr 85:1062–1067
60. Holick MF, Chen TC (2008) Vitamin D deficiency: a worldwide problem with health consequences. Am J Clin Nutr 87:1080S–1086S
61. Suo Z et al (2002) EGFR family expression in breast carcinomas. c-erbB-2 and c-erbB-4 receptors have different effects on survival. J Pathol 196:17–25
62. Paez JG et al (2004) EGFR mutations in lung cancer: correlation with clinical response to gefitinib therapy. Science 304:1497–1500
63. Lee JW et al (2005) Somatic mutations of egfr gene in squamous cell carcinoma of the head and neck. Clin Cancer Res 11:2879–2882
64. Grotendorst GR, Soma Y, Takehara K, Charette M (1989) EGF and TGF α are potent chemoattractants for endothelial cells and EGF like peptides are present at sites of tissue regeneration. J Cell Physiol 139:617–623
65. Maas-Szabowski N, Stärker A, Fusenig NE (2003) Epidermal tissue regeneration and stromal interaction in HaCaT cells is initiated by TGF-α. J Cell Sci 116:2937–2948
66. Cao L et al (2010) Environmental and genetic activation of a brain-adipocyte BDNF/leptin axis causes cancer remission and inhibition. Cell 142:52–64
67. Tirassa P, Triaca V, Amendola T, Fiore M, Aloe L (2003) EGF and NGF injected into the brain of old mice enhance BDNF and ChAT in proliferating subventricular zone. J Neurosci Res 72:557–564
68. Haase I, Hunzelmann N (2002) Activation of epidermal growth factor receptor//erk signaling correlates with suppressed differentiation in malignant *Acanthosis nigricans*. J Invest Dermatol 118: 891–893

69. Nanney LB, Yates RA, King LE Jr (1992) Modulation of epidermal growth factor receptors in psoriatic lesions during treatment with topical EGF. J Invest Dermatol 98:296–301
70. King LE Jr, Gates RE, Stoscheck CM, Nanney LB (1990) Epidermal growth factor/transforming growth factor α receptors and psoriasis. J Invest Dermatol 95:10S–12S
71. Stuart CA et al (1998) *Acanthosis nigricans* as a risk factor for non-insulin dependent diabetes mellitus. Clin Pediatr 37:73–79
72. Grandhe N, Bhansali A, Dogra S, Kumar B (2005) *Acanthosis nigricans*: relation with type 2 diabetes mellitus, anthropometric variables, and body mass in Indians. Postgrad Med J 81:541–544
73. Niemeier V, Fritz J, Kupfer J, Gieler U (1999) Aggressive verbal behaviour as a function of experimentally induced anger in persons with psoriasis. Eur J Dermatol 9:555–558
74. Abbott DH (1987) Behaviourally mediated suppression of reproduction in female primates. J Zool 213:455–470
75. Abbott DH, McNeilly AS, Lunn SF, Hulme MJ, Burden FJ (1981) Inhibition of ovarian function in subordinate female marmoset monkeys (*Callithrix jacchus jacchus*). J Reprod Fert 63:335–345
76. Abbott DH, Saltzman W, Schultz-Darken NJ, Tannenbaum PL (1998) Adaptations to subordinate status in female marmoset monkeys. Comp Biochem Physiol C Pharmacol Toxicol Endocrinol 119:261–274
77. Wasser SK, Barash DP (1983) Reproductive suppression among female mammals: implications for biomedicine and sexual selection theory. Quart Rev Biol 58:513–538
78. Barrett J, Abbott DH, George LM (1990) Extension of reproductive suppression by pheromonal cues in subordinate female marmoset monkeys, *Callithrix jacchus*. J Reprod Fertil 90:411–418
79. Saltzman W, Schultz-Darken NJ, Wegner FH, Wittwer DJ, Abbott DH (1998) Suppression of cortisol levels in subordinate female marmosets: reproductive and social contributions. Horm Behav 33:58–74
80. Abbott DH, Saltzman W, Schultz-darken NJ, Smith TE (1997) Specific neuroendocrine mechanisms not involving generalized stress mediate social regulation of female reproduction in cooperatively breeding marmoset monkeys. Ann N Y Acad Sci 807: 219–238
81. Shively CA, Clarkson TB (1994) Social status and coronary artery atherosclerosis in female monkeys. Arterioscler Thromb 14:721–726
82. Benjamin F (1960) Glucose tolerance in dysfunctional uterine bleeding and in carcinoma of endometrium. Brit Med J 1:1243–1246
83. Warden SJ, Bliziotes MM, Wiren KM, Eshleman AJ, Turner CH (2005) Neural regulation of bone and the skeletal effects of serotonin (5-hydroxytryptamine). Mol Cell Endocrinol 242:1–9
84. Bliziotes M, Gunness M, Eshleman A, Wiren K (2002) The role of dopamine and serotonin in regulating bone mass and strength: studies on dopamine and serotonin transporter null mice. J Musculoskelet Neuronal Interact 2:291–295
85. Chyun YS, Kream BE, Raisz LG (1984) Cortisol decreases bone formation by inhibiting periosteal cell proliferation. Endocrinology 114:477–480
86. Dennison E et al (1999) Profiles of endogenous circulating cortisol and bone mineral density in healthy elderly men. J Clin Endocrinol Metabol 84: 3058–3063
87. Dietrich JW, Canalis EM, Maina DM, Raisz LG (1979) Effects of glucocorticoids on fetal rat bone collagen synthesis in vitro. Endocrinology 104:715–721
88. Georgiades E et al (2003) Chronic fatigue syndrome: new evidence for a central fatigue disorder. Clin Sci 105:213
89. Sharpe M, Hawton K, Clements A, Cowen P (1997) Increased brain serotonin function in men with chronic fatigue syndrome. BMJ 315:164–165
90. Cleare AJ (2003) The neuroendocrinology of chronic fatigue syndrome. Endocr Rev 24:236–252
91. Harlow B, Signorello L, Hall J, Dailey C, Komaroff A (1998) Reproductive correlates of chronic fatigue syndrome. Am J Med 105:94S–99S
92. Scott LV, Salahuddin F, Cooney J, Svec F, Dinan TG (1999) Differences in adrenal steroid profile in chronic fatigue syndrome, in depression and in health. J Affect Disord 54:129–137

15 Where Do We Go From Here?

A famous musician was once asked how he visualized music after 20 years. He replied, "If I knew, I will be playing it today!"

Similar to music, the path of research and development is always difficult to predict. I started an inquiry into the evolutionary origins of type 2 diabetes as a curiosity exercise in evolutionary biology without a slightest idea that it may lead to reinterpretation of the whole phenomenon and that it may have radical implications for clinical practice too. The pursuit of a small but interesting question and some recombination of preexisting ideas led me a long way to challenge the long-standing conceptual foundation of T2D itself. Now I will make an attempt to predict where research would lead us in the near future being completely aware that it is just too difficult to make such predictions.

The Old Versus New Paradigm

Before going to the future, it will be good to summarily state the new paradigm of T2D once again in the form of a logical sequence of statements:

1. A plurality of behavioral strategies can coexist in a population, and hawk–dove or soldier–diplomat is such a coexisting dichotomy seen in animal and human populations respectively. The difference is mainly in physical strength and aggression (soldier characteristic) versus cognitive abilities and social manipulation (diplomat characteristic).
2. The two strategies are accompanied by a wide range of physiological adaptations to support the respective behavioral repertoire. In brief, the soldier strategy is marked by strong muscle and bone, more subcutaneous distribution of body fat, low serotonergic activity, low cholesterol and cortisol, low plasma insulin, insulin sensitivity, and protein anabolic bias in metabolism. The diplomat strategy is accompanied by disinvestment from muscle and bone, more visceral accumulation of fat, high serotonergic activity, high plasma levels of cholesterol, cortisol and insulin, insulin resistance, and lipid anabolic bias in metabolism.
3. Insulin, cholesterol, leptin, and cortisol have important roles in learning, cognitive brain functions, and behavioral modifications. Higher levels of these molecules are involved in aggression suppression, physical risk aversion, and enhancement of cognitive functions, in effect suppression of soldier and facilitation of diplomat behavior.
4. Aggression anticipates injuries, and a number of mechanisms of combating injuries are activated in anticipation. These include migration of innate immune cells towards the periphery and production of growth factors such as EGF and NGF that are involved in wound healing and regeneration of tissue. Deficiency of physical aggression results in more central distribution of innate immune cells leading to low-grade chronic systemic inflammation and chronic deficiency of many growth factors.

M. Watve, *Doves, Diplomats, and Diabetes: A Darwinian Interpretation of Type 2 Diabetes and Related Disorders*, 319
DOI 10.1007/978-1-4614-4409-1_15, © Springer Science+Business Media New York 2013

5. Crowding, a perception of population density, can affect reproductive strategies, aggressive behavior, and energy reserves and thereby overall metabolic and endocrine balance. Crowding is identified as a new potential risk factor for obesity, T2D, and related disorders.
6. Most of the complications of T2D and related disorders are due to growth factor deficiency, altered dynamics of macrophages and other innate immune cells, oxidative excess, and angiogenesis dysfunction, all of which can arise from behavioral signals independent of hyperglycemia. Hyperglycemia may play a role in aggravating some of these effects, but it is unlikely to be the sole driver of diabetic complications.
7. Beta cell degeneration is most likely to be due to deficiency of EGF and other growth factors. Beta cells have a high regeneration capacity, and islet degeneration in diabetes is likely to be reversible if we identify the right internal environment required for the regeneration. This would include EGF, gastrin, and possibly a number of other growth factors.
8. The association of insulin resistance with growth factor deficiency, chronic inflammation, oxidative excess, and endothelial and angiogenesis dysfunction does not appear to be obligate. Insulin resistance can be decoupled from these mechanisms as shown by the *Klotho* function, and if so decoupled, insulin resistance can promote health and longevity instead of a series of deadly disorders. Therefore insulin resistance may not be central to the pathophysiology of T2D. This necessitates a deviation from the obesity–glucose–insulin-centered paradigm of T2D and expansion of our understanding to a more accommodative picture that simultaneously accounts for all the changes in a wide spectrum of organs and systems.
9. Along with the insulin–glucagon-driven peripheral mechanisms of glucose homeostasis, central mechanisms appear to play a more important role than previously believed, and there are indications that the rate of glucose transport from the brain capillaries may have a crucial role in determining plasma glucose levels. More research efforts are needed to understand glucose homeostasis in the brain and the interaction between central and peripheral mechanisms in glucose homeostasis. The key to glucose homeostasis is most likely to be found in the brain.

The new paradigm is thus radically different from the orthodox one in that:

1. Obesity is not considered to be central to T2D and related disorders. It is only one of the risk factors. Obesity is associated with suppression of the soldier set of behaviors with two-way causality, and that is what mediates the association between obesity and metabolic syndrome.

2. Insulin sensitivity and insulin secretion are affected by a large number of signal molecules including sex hormones, myokines, osteocalcin, BDNF, CCK, endorphins, HISS, serotonin, dopamine, and melatonin among them. The classical view considers only FFAs, triglycerides, and adipokines to affect insulin sensitivity, whereas the new paradigm accounts for the action of almost all the known molecules at proximate and ultimate levels simultaneously.
3. T2D is not primarily or mainly about insulin and glucose homeostasis. Alterations in insu-

lin and glucose are only a part of the wide variety of changes that take place simultaneously in the body.

4. All the pathophysiological mechanisms are not driven by hyperglycemia alone, although it may aggravate many of them. As a corollary, glycemic control alone may not be sufficient to prevent all the complications of T2D.
5. All pathophysiology of T2D originates from brain and behavior rather than diet and energy imbalance.
6. T2D is not an incurable disorder. With the exception of advanced stages of complications, the baseline alterations in the body systems are not irreversible. Therefore, potentially there should be ways to reverse or "cure" T2D.
7. A reversal or cure is unlikely to be brought about by medication. Since the origin of the disorders is in behavioral deficiencies, only behavioral supplementation can mediate a reversal. Even complete cure may be possible.

In Chap. 3 we noted five major horse and cart paradoxes that existed on the platform of the orthodox theory. Somewhere in the book, there is a solution to each of the five paradoxes, but let me explicitly relate the solutions once again here to emphasize that the new paradigm is logically sounder and resolves almost all the unresolved paradoxes.

1. Cart pulls the horse: Insulin resistance first or hyperinsulinemia? This paradox is successfully resolved by the new paradigm with a clear-cut answer that hyperinsulinemia comes first and the hypoglycemic effects of it are avoided by compensatory peripheral insulin resistance. In low birth weight infants, insulin is necessary as a growth hormone to promote catch-up growth. Independent of this, on adopting a diplomat livelihood, insulin is necessary to facilitate cognitive functions for which increased insulin secretion is a primary adaptation. High levels of indigenous insulin are backed up by built-in mechanisms to induce insulin resistance. The orthodox paradigm believed in insulin resistance being primary leading to compensatory hyperinsulinemia. But no mechanism appears to exist by which insulin resistance can stimulate higher insulin secretion without ongoing hyperglycemia. See Chaps. 3 and 7 for more discussion.
2. Horse moves, cart left behind: In a hyperinsulinemic and insulin-resistant state, if insulin secretion is artificially suppressed, blood sugar should rise if the orthodox view is correct. Experiments show that this does not happen. On suppressing insulin production, insulin resistance comes down and blood sugar is not affected. This paradox is partly resolved by the solution to the first paradox itself. If insulin resistance is a reaction to insulin levels, then reducing insulin would reduce resistance levels automatically. But this is only a partial solution. The other question as to why this does not happen in diabetes needs a separate solution. This is provided by the brain glucose dynamics. If the brain is short of the required sugar levels, it suppresses insulin production, facilitates glucagon production, and stimulates liver glucose production thereby increasing blood sugar until desired brain sugar levels are reached. Insulin, glucagon, liver, and muscle glucose uptakes are only tools that the central mechanisms manipulate. In effect, the blood sugar level is decided by the brain and only executed by insulin, glucagon, and insulin sensitivity. If one of them is changed, others adjust themselves to achieve the desired level. In diabetes, the desired level is increased, and therefore, following insulin suppression, insulin resistance does not drop sufficiently to keep sugar levels to "normal." See Chaps. 3 and 12 for details.
3. Horse stops, cart keeps moving: Reducing insulin resistance does not reduce blood sugar: In an early diabetic who has insulin resistance, hyperinsulinemia, and hyperglycemia, if insulin resistance is reduced by exercise or by insulin sensitizing drug, insulin levels drop rapidly but sugar levels do not. This paradox also has the same solution as above. Since desired glucose levels are decided by the brain sugar dynamics, when insulin resistance is

forcibly reduced, insulin levels reduce to such an extent that the blood sugar level remains the same. This is an important point that can help us resolve between the old and new interpretations. But currently, data are inadequate here, and more research inputs are needed for a finer scale resolution of the events when insulin resistance is reversed.

4. Cart starts moving before the horse: By the classical paradigm, raised blood sugar causes oxidative stress leading to endothelial dysfunction. However, early signs of endothelial dysfunction appear before overt diabetes and often without diabetes. So hyperglycemia could not be *the* or *the only* horse. On the other hand, vascular pathology can explain hyperglycemia in the new paradigm. Vascular pathology has been shown to reduce blood flow in the brain. If we include reduction in glucose transport across the brain capillaries in vascular defects, the two together cause a reduction in steady-state levels of brain glucose. This is sufficient to increase blood sugar by neuronal manipulation of insulin, glucagon, and liver. See Chaps. 3, 12, and 13 for details.

   Somewhat similar is the case of β cell dysfunction and glucotoxicity. Glucotoxicity is unlikely to be the horse since without cell dysfunction, blood sugar is unlikely to rise. Here again, it is possible that the answer lies in brain glucose dynamics which needs to be explored seriously. If glucotoxicity is "real," the sequence of events could be that hyperglycemia is caused by altered brain glucose dynamics which results in glucotoxicity leading to eventual β cell loss.

5. Half cart forward, half cart backward: Paradoxical angiogenesis: Research papers on wound healing in diabetes have often blamed impaired angiogenesis for wound healing problems. On the other hand, diabetic nephropathy and retinopathy appear to be caused by excessive angiogenesis. Is angiogenesis reduced or increased in diabetes? The glucocentric paradigm does not explain why and how angiogenesis is affected in diabetes and further why it is affected in opposite directions in different parts of the same body. It is unlikely to be caused by raised blood sugar since the same blood supplies all organs. The new paradigm has a solution in the form of altered macrophage distribution in the body. Since macrophages are important initiators of angiogenetic pathways, wherever there is greater density of macrophages, there is hyperangiogenesis, and where it is less, there is deficient angiogenesis. See Chaps. 8 and 13 for details.

Does the new theory fulfill the expectations from an evolutionary theory? At the beginning of Chap. 4, I had expressed my expectations from a theory of evolutionary origins of diabetes. Does the emerging theory fulfill these expectations? What stand does it take on each of these issues? Let us see one by one. (1) Why a sudden rise? The sudden rise in the population in 1–2 generations is because of altered physical and social behavior. Behavioral deficiencies developed and reached an extreme only in the modern urban life, and therefore, these disorders are specific to modern life. (2) Polymorphism: When alternative behavioral strategies have negative frequency dependence, stable polymorphism is achieved. This need not always happen at the genetic level. The same principle may work at the phenotypic level too. (3) The strong association between birth weight and metabolic syndrome is explained by the new theory by the developmental constraints laid by intrauterine growth retardation. Since the growth retardation is unequal and brain is spared, diplomat strategy is a better option after IUGR. (4) An evolutionary theory should not stop at explaining the origins of obesity but also explain why obesity is associated with insulin resistance and its consequences with both proximate and ultimate components of reasoning. The emerging theory is unique in this respect since all the prior theories have talked about why a tendency to accumulate fat may be adaptive but do not proceed to explain the downstream effects of obesity, insulin resistance, and other pathophysiological processes. The new theory deals with why insulin resistance is adaptive independent of obesity, why obesity is associated with (but not the only cause of) insulin resistance, and why insulin resistance is associated with (but not necessarily causal to other pathophysiological processes. Almost every aspect of the multitude of changes in the body is accounted for by the new theory at both ultimate and proximate level simultaneously. (5) We have seen above how almost every paradox has a plausible solution in the light of the neurobehavioral way of interpreting diabetes. (6) The new theory has raised the possibility that T2D is actually not irreversible and cure may be reasonably possible by treating behavioral deficiencies. Only carefully designed long-term studies will be able to put this claim to test. (7) Testable predictions are abundant all over the book, and many of them are simple and practicable. Researchers with different set of expertise need to work in concert to test these ideas. (8) Lastly, beyond being an evolutionary theory, it has many implications which may change clinical practice in the future if careful trials support the possible implications. Thus the neurobehavioral origins appear to fulfill all the expectations from an evolutionary theory of T2D, but more work is certainly needed to fill in the gaps in the picture and ultimately come out with strategies that will help the current-generation diabetics and prevent diabetes in the next generation.

A few questions that were not explicitly answered so far need to be answered now. First is about the age dependence of T2D. Readers might have worked out the answer already since it is implicit in Chaps. 5–9. To state it more explicitly, aggression, sex, and reproduction, the three main determinants of metabolic states, are certainly affected by age. The optimum strategies for sex and reproduction change with changing marital status, parity status, and reproductive capacity. Accompanying the change in reproductive strategy and capacity, a change in physical aggression is natural. Risk-taking behavior should decrease after having offspring at various stages of dependence. In humans, grandparental status induces further behavioral changes. It is no surprise therefore that behavioral as well as metabolic states change simultaneously with age. The rate of change is dependent on lifestyle though. In women, menopause is a major transition, and the risk of obesity, CVD, and T2D increase after menopause is consistent with this logic. Even more dramatic is the metabolic change following hysterectomy. Loss of sex hormones is an important risk factor. In men, although there is no well-defined condition comparable to menopause, a slow andropause-like condition marked by loss of testosterone is a well-recognized risk factor. It is perceivable therefore that loss of sex and aggression hormones may not be only correlated to the increased risk of T2D and related disorders, it may have a causal role.

The other possible FAQ is about the apparent heritability of diabetes. Although attempts to find a gene for T2D have failed and GWA studies have demonstrated the near impossibility of obesity and T2D being genetic, one cannot deny familial tendency in the susceptibility to T2D. This is likely to be due to epigenetic to some extent, but the possibility of extragenetic or nongenetic biological inheritance cannot be denied. For example, maternal EGF has an important role in fetal development [1–4]. We have already seen that EGF secretion is behaviorally triggered. This raises the possibility that maternal behavior affects her growth hormone levels which influences the developmental pattern of the fetus. We know that fetal development has lifetime programming effects. This can influence behavioral strategies of the progeny and when it reproduces its hormonal and growth factor levels further influence the third-generation phenotype. The EGF case is interesting because EGF specifically affects branching morphogenesis of fetal submandibular glands that are the sources of adult EGF secretion [5]. This can make EGF deficiency transgenerational. Therefore a transgenerational biological inheritance without involving genetic or epigenetic mechanisms is possible which we may call extragenetic or nongenetic inheritance. It is not impossible that there are other mechanisms with similar heritable programming effects. This inheritance, unlike true genetic inheritance, can be changed by environment or behavior within a generation or two. Such a mechanism could have evolved to support adaptive behavioral programming for social status since social hierarchies are at least partially heritable. Good correlation exists between maternal and progeny social status, and therefore, transgenerational transmission of behavioral strategies through fetal programming would be adaptive. Further, behavioral traits are passed on culturally from parents to offspring, which does not need a biological mechanism but may have biological effects. A vertical cultural inheritance mimics genetic inheritance in certain aspects. Therefore there are a number of alternative possible mechanisms that lead to familial tendencies that are not genetic. The implication of understanding that obesity and T2D are not genetic is extremely important. Blaming genes for T2D reduces the motivation for controlling or reversing the condition. When patients are given the impression that your diabetes is genetic, it is implied that it is inevitable and needs to be accepted as it is. If it is not genetic, the implication is that it can be reversed since behavioral patterns can be at least partially changed by conscious efforts.

There is one more question that is not adequately answered by either the old or the new school. That is regarding why there is a need for acute-phase insulin response (AIR). Since there are neural mechanisms involved in inducing AIR [6], it would be safe to conclude that AIR must have been a response evolved for some specific function. But what function did it evolve for? There would be a temptation to argue that it serves the function of keeping the GTT "normal." It needs to be realized that if AIR was not there, the "normal" would have been different. Absence of AIR leads to an increase in the area under the curve of both glucose and late-phase insulin, but there is no evidence that this by itself leads to any pathological consequences. Therefore we may have considered that curve as "normal" if AIR had not evolved. There are more riddles associated with AIR for which I have no definitive answers currently. Since AIR is launched even before absorption of digested food begins, an acute insulin response should lead to a transient hypoglycemia but that does not seem to happen for some unknown reason. I suspect that AIR evolved for behavioral rather than metabolic reasons. Insulin affected rodent behavior in open-field test [7] as well as elevated plus maze test [8]. Animals administered insulin avoided being in the open and exploring. This is a sensible response from a behavioral ecology point of view. During hunting or foraging, risk-taking and exploring may be needed. The hormonal composition of the body therefore needs to be that of a risk-taker. However once some rich food is obtained, it is better to hide and be inconspicuous to avoid aggressive competitors. This is particularly true for species that can carry food to a safer place and eat. This needs a sudden change in mood, and the acute insulin response may be instrumental in mediating this change. Feeding-induced corticosteroids serve a similar purpose. If AIR evolved to serve a behavioral purpose, I presume mechanisms will have evolved to avoid accompanying hypoglycemia.

I am sure there will be more unanswered questions than the ones that have been answered so far. Certainly the new school has answered more questions than the old one, and this process has not ended.

The new paradigm is not "complete" at this stage. We can at the best say that it looks far more promising than the old one, but a number of gaps in data can still be visualized. This is not surprising. Research is driven by hypotheses. If a line of thinking does not exist, experimenters are most unlikely to collect any data to support or reject it. Very surprising is the fact that although completely new, the theory already has much evidence in support. One way available evidence can be used is to know all the existing evidence and then try to make a coherent picture out of it. But this is not how this concept evolved. Rather it evolved like the way experimental research evolves. One has a hypothesis, based on which testable predictions are defined, and then experiments are designed and performed to test the predictions. During the evolution of this theory, I made certain predictions and then looked into literature to see whether there were any data already available to test that prediction, and to my surprise, most frequently, data already existed and the predictions turned out to be true. To cite a few examples, in the very beginning of this effort, when I learned that insulin-resistant mothers have larges babies, I interpreted it as an increased investment in a fetus and following the theory of *r* and *K* selection predicted that by some mechanism the offspring number must be coming down. It could be reduced ovulation or failure of implantation or something else. At this time, I was completely unaware of the link of PCOS to insulin resistance. I was amazed to find from literature that anovulatory disorders like PCOS are commonly associated with insulin resistance. One of my students attempted a meta-analysis carefully going through literature with an expectation that all molecular signals positively correlated with aggression should have antiobesity and antidiabetic action and vice versa. This association turned out to be highly significant [9]. One of the antiobesity antidiabetes signals for which we did not find any relation to aggression was adiponectin. Here I thought if adiponectin was not related to aggression, it must be related to regulating *r* and *K* strategies of reproduction. Specifically it should increase fecundity and decrease transplacental nutrient flow. To my great amazement, some experimenters had looked at these relations, and both of them turned out to be true (Chap. 9). Thus the journey of the theory so far has happened with a hypothesis testing approach. But a large number of predictions are yet to be tested, and here I will briefly outline what I feel are the important predictions that need to be tested and also the areas where data are conspicuously missing.

It is likely that some of the data needed already exists in unpublished studies. I am aware of some unpublished results that support the new paradigm. They remained unpublished because either the investigators did not understand the relevance of the results or they found it hard to publish. One such study showed that the insulin resistance of youngsters doing nonaggressive exercise was not different from those doing no exercise. A third group doing aggressive exercise was substantially insulin sensitive as compared to both no exercise and nonaggressive exercise groups. In another study, diabetics with poor glycemic control were found to be more extrovert and socially dependent as compared to those with excellent glycemic control. The latter were significantly introvert and socially independent. A study of maternal and early childhood nutrition in relation to cognitive development reveals that after correcting for birth weight, children with higher insulin levels are better at certain cognitive tasks as compared to the ones with low insulin levels and higher insulin sensitivity. A study in *Drosophila* demonstrates that crowding in the larval stages gives rise to adults that are less aggressive than those coming from uncrowded environments. When I started publishing and talking about my hypotheses, I found people saying, "Oh, we did observe something like this, but we did not (or could not) publish it." If an experimental result does not make sense in the prevalent paradigm, it is most unlikely to be published. My small sample size might be an indication that there are many more unpublished results that are of relevance to and make sense in the light of the new paradigm. We can expect that they would be published sooner or later. Much more research is nevertheless needed to fill in the existing gaps.

Major Areas of Data Voids

1. The multidimensionality of aggression: Aggression is a very complex phenomenon, and all dimensions of it are not yet quite explored. Throughout this book, we have used aggression more or less as a black box. The reason why all aspects of the aggression syndrome are not yet clear is that aggression has been handled by behavioral ecologists, psychologists, physiologists, and neurobiologists in different ways and with different perspectives. In physiology and neurobiology experiments, often, aggression is reduced to simple stereotyped behaviors such as a resident–intruder test. In real life, there are several different contexts of aggression, and accordingly, there are several types of aggressions. All types and all contexts of aggressions are unlikely to have identical effects. Although some aggression-related physiological changes would be common, there can be substantial difference in a few others. Our understanding of such subtleties is still in the infancy. I made an attempt in Chap. 7 to account for some of the complexities of aggression physiology, but reality is bound to be much more variable and complex.

A particularly worrying information void is female aggression mechanisms. Most of our knowledge related to the physiology of aggression is based on male aggression. Aggression researchers have not paid sufficient attention to female aggression. This bias needs to be removed as fast as possible. Female aggression and dominance are important in deciding female social ranks. Maternal aggression is a female specific form of aggression. But unlike the testosterone story of male aggression where there is abundant literature, female aggression is largely a literature void.

2. Interaction between diet, behavior, peripheral, and central mechanisms affecting insulin sensitivity and other metabolic alterations: It is adequately clear by now that neither peripheral nor central mechanisms alone are likely to give us a complete understanding of insulin resistance syndrome. There is evidence that peripheral changes such as glucose transporter levels that were thought to originate peripherally are under neuronal control [10], and the central systems take cues continuously from peripheral systems. There have been some indents into the interaction between peripheral and central mechanisms [11–17], but so far, there have not been enough insights into how the CNS modulates insulin sensitivity and other aspects of metabolism. Diet is closely linked to behavior, but this area is largely neglected so far [18]. The behavior-centric paradigm is likely to give some insights into how and why neurobehavioral mechanisms should be involved. There is a cross talk between neuronal mediators of insulin sensitivity and those involved in anxiety, exploratory behavior, aggression, and other behaviors. A study of such interactions is likely to give many leads in research, and this I presume would be an active and most fruitful field of research in the near future.
3. Although there have been demonstrations that insulin, cholesterol, and leptin enhance cognitive function, so far, there are no studies on the mechanism by which they do so. Not many researchers so far have taken interest in it, again because a paradigm that expects insulin to play a role in cognitive brain functions did not exist. Similar are the questions as to how cholesterol suppresses aggression or corticosteroids suppress risk-taking behavior. Since corticosteroids are dubbed stress hormones, the research on behavioral correlates of corticosteroids has got locked into a narrow vision basin, and the true expanse of adaptive roles they play in behavioral fine-tuning remains unexplored.
4. For decades, it has been assumed that hyperglycemia is the cause of all pathophysiological processes leading to diabetic complications. Despite decades of research, the precise mechanisms are not yet known, although a number of pathways have been suggested. One of the reasons for the slow progress could be the belief that glucose drives everything. There is no doubt that some possible mechanisms by which glucose causes or aggravates the pathological processes have been demonstrated. But the picture is highly incomplete. I have suggested here an array of parallel mechanisms that originate from behavior and are independent of glucose levels. The two pictures are not mutually incompatible, and both may operate synergistically to multiply the effects. Nevertheless there is a need to evaluate the relative roles of hyperglycemia-induced and behavioral deficiency-driven mechanisms in the pathophysiology of diabetes. This has important consequences in deciding whether the current focus on glycemic control is both necessary and sufficient in management of diabetes or there needs to be a radical change in the approach.
5. I suspect population density to affect behavior and physiology to a considerable extent. Although there is evidence that crowding has a number of effects on behavior, reproductive strategies, and physiology, we do not know to what extent crowding contributes to the current epidemic of obesity and metabolic syndrome. It is not difficult to look at epidemiological correlates of crowding, obesity, and insulin resistance. There are some data in support of the crowding hypothesis, but there are a number of confounding factors

(Chap. 9). There is a need for studies that carefully control for confounding factors and specifically look at the effects of crowding on obesity and insulin resistance syndrome parameters.

6. Glucose dynamics in the brain appears to have a lot of potential of giving insights into glucose homeostasis in the body. This was an almost entirely unexplored field until recently because of methodological difficulties. Some of the modern technological solutions have offered efficient ways out. In the near future, focusing on brain glucose dynamics can lead a long way in taking us to the root causes of diabetic hyperglycemia.

Are There Any Inconvenient Truths or Paradoxes in the New Paradigm?

I would be surprised if the new paradigm is free of problems. Although I do not perceive any serious paradox comparable to the series of paradoxes in the old paradigm, there are certain areas where some components of the theory stand on slippery grounds and need evidence to either support or reject/replace them. The ones that I can visualize currently are the following. These are already discussed and possible solutions suggested, but emphasizing them once again will not be out of place:

1. Glut-1 mutant not hyperglycemic: The effects of glut-1 downregulation on hyperglycemia is a promising hypothesis in spite of being more speculative currently because it gives logical solutions to many unresolved puzzles. However the strong evidence against it is that gut-1-deficient mutants are not hyperglycemic. The possible solution is also not very difficult to visualize and is discussed in sufficient details in Chap. 12. Admittedly this is a weakness in the hypothesis of brain-driven hyperglycemia. But it does not make sense to discard this hypothesis without investigating, since there are perceivable reasons for the apparent contradiction. Also the brain-driven hyperglycemia model can work independent of glut-1 as seen in Chap. 12. Therefore there is no need to give up this hypothesis even if the proposed role of glut-1 fails to get support. Further experimental probe is the only possible approach to the problem.
2. Testosterone in females: The aggression hormone testosterone has a number of beneficial effects in males. But high levels of testosterone could be problematic in females. High testosterone in females is associated with hirsutism, PCOS, and related disorders. However, this does not really challenge the central hypothesis. The high-testosterone hirsute, or PCOS females are not known to be excessively physically aggressive [19]; on the contrary, some studies find inhibited aggression in hirsute women [20]. The relationship of testosterone to female aggression is complex. For example, in mice, testosterone administration in the prenatal phase increased maternal aggression [21] but testosterone during lactation reduced it [22]. Physical aggression is central to our hypothesis, not testosterone. Testosterone is not the only aggression hormone as seen in Chap. 7. Presumably the endocrinology of female aggression is more complex and involves a number of different signal molecules. Also there can be a number of different causes for rise in testosterone in females, and all may not have the same effects. Therefore high testosterone associated with PCOS and insulin resistance in females is not contradictory to the new paradigm but needs better understanding of why testosterone rises in these conditions.

   Some of the functions of testosterone are done through alternative routes in females. For example, testosterone stimulates EGF synthesis, and since females have low testosterone activity, they have lower levels of EGF as well. But at least some of the roles of EGF appear to be played by or compensated by estrogens. For example, estrogens have a direct facilitating role in wound healing [23–26] as well as β cell protection and insulin secretion [27, 28].

I am sure skeptics will find more possible problems in the theory, and it is important to encourage skepticism and address these problems as and when they are pointed out. The orthodox paradigm exists with us for several decades now and is still left with many paradoxes and unexplained areas. This picture is unlikely to improve since I think some of the basic tenets of the classical paradigm are wrongly perceived. Now we need to give at least a few decades for the upcoming paradigm as well. If adequate research inputs are given for testing the new paradigm and filling the current information voids, it will most probably give a clear and transparent understanding of the complex phenomena. But in case it fails to do so, doors would be open to reject it and look for more novel interpretations. This would certainly be a better approach than sticking to the old paradigm and trying to overlook the glaring gaps and paradoxes existing with it.

Testing the New Ideas

What can we do to test the new set of ideas? On almost every page of this book, there is some new suggestion or hypothesis or speculation along with its testable prediction that awaits experimental or epidemiological tests. Therefore testing the new theory in its entirety is a massive job. But that is in fact a justification for writing a book at this stage. The decision to write a book at this stage was not easy. Today's science is a lot more apprehensive about making speculations. While some components of the new theory have substantial evidence in support, certain other aspects are still largely speculative. Would it not have been a better idea to test the speculative components of the theory with experiments or with epidemiological data and then propose the theory? This would certainly have improved the acceptability and impact of the theory. But this could have come only at a great time cost and a potential loss to the community at large. As readers would have already visualized and as reemphasized below, testing the theory would involve a large body of research involving a wide range of researchers including cell and molecular biologists, physiologists, systems biologists, geneticists and epigeneticists, behavioral ecologists, behavioral neurobiologists, evolutionary biologists, and pharmaceutical and clinical researchers. Therefore no single laboratory would be able to take up the challenge. Even if there existed such a lab or institute, it would have taken several decades of work. Now if we assume that the theory obtained adequate support and was then published, it would be a loss to a generation or two of diabetic patients. On the other hand, if I am successful in attracting the attention of different research groups throughout the globe, the necessary research would materialize in a much shorter time, and the resulting implementation at preventive and treatment level will be available to people more rapidly.

To test the new interpretation of T2D and related disorders, we need experiments that differentiate it from the old one. While trying to do so, it is important to realize that the new paradigm includes some of the important elements of the old one. In fact, the gluco-lipo-centric paradigm can be said to be a small subset of the new and broader paradigm. Therefore the already-existing or any new evidence in support of the old theories is not evidence against the new one. For example, we still do not know the precise pathways through which obesity alters insulin sensitivity. But suppose such a pathway is found, it is not evidence against the new paradigm. Obesity and behavioral strategies are interdependent and correlated, and therefore, any demonstration of effect of obesity does not rule out the effects of behavior and vice versa. I have defined above the precise differences between the two interpretations, and any attempts towards comparative tests of the two should focus on these specific differences.

Although differentially testable components of the theory are very large and will be found on almost every page of the book, I would list here a few lines of research that I think should be followed with priority in order to test, support, or falsify the new paradigm. As said earlier, we need researchers from almost every field of biology. The suggested lines are classified below according to the broad categories of research fields:

A. At fundamental level:

(a) Challenges for physiologists:

1. Aggressive versus nonaggressive exercises: The positive effects of exercise on physiology are undisputed. The new thought is about the component of aggression in exercises and the specific physiological effects of this component. Studies addressing this issue would be the most important of all kinds of studies in testing the central idea behind my thinking and its actual clinical implications too. Designing such studies is not as easy as it may look at first glance. The crucial part of it is in separating the effects of aggression from the energetic effects of exercise. For this, it is necessary to compare groups of volunteers with no exercise, with nonaggressive exercise, and with aggressive exercise. Two unpublished preliminary studies including one by my lab have already indicated that aggressive exercises have an efficient insulin sensitizing action, whereas nonaggressive exercises have marginal effects. However, these studies did not precisely measure the calorie component of exercise. In both the studies, the nonaggressive exercises were more energy intensive than aggressive exercises. Therefore the greater observed effect of aggressive exercise is unlikely to be caused by greater energy expenditure. Nevertheless, care needs to be taken to see that the nonaggressive and aggressive exercises are calorically equivalent. There is also a need for parameters to measure the aggression component of an exercise which is possible to do by monitoring changes in aggression-related signal molecules such as testosterone, EGF, or CCK. Such a comparison would reveal whether calorie burning matters or the aggression component of exercise matters. The parameters to be monitored in such studies should not be restricted to insulin sensitivity but should also include inflammatory markers, many growth factors, sex and aggression hormones, subcutaneous macrophage density, endothelial function, endothelial progenitor cells, and oxidative state. For example, one specific testable prediction could be that aggressive exercises would alter the adipokines expression in adipose cells independent of total fat content, and there would be a quantitative difference in aggressive and nonaggressive exercises in this effect. The two can also have differential detectable methylation patterns in muscle tissue that can be studied in a muscle biopsy specimen. My prediction is that aggressive exercise will have much more dramatic effects on the true pathophysiological processes reflected by oxidative state, growth factor levels, angiogenesis, endothelial function, and macrophage distribution than the effects on glucose levels.
2. I have speculated on other possible behavioral deficiencies such as adventure, agility, and so on (Chap. 14). There already exists much data on the effects of aggression. But for others, there are little preexisting data. The effect of each of them needs to be studied separately and in combination with a design similar to the above.
3. One of the main arguments has been about the behavior-induced alterations in the distribution of macrophages in the body. My interpretation is that it is not a generalized inflammatory response that marks metabolic syndrome. Rather it is the altered distribution of inflammatory cells that cause tissue-specific inflammatory patterns. Although there is evidence that in obesity and diabetes there is a reduction in the density of macrophages in the peripheral tissue and accumulation of them in vascular tissue, adipose tissue, and kidneys in the long run, we still do not have a vision of their whole-body distribution and dynamics. Quantitative studies on the dynamics of macrophages and other innate immune

cells brought about by behavioral and metabolic signals will not only be crucial in testing our hypotheses but also in developing a basic understanding of the pathophysiology of systemic inflammatory state. Studying the dynamics of immune redistribution is one part of the research, the other is to determine the causal factors of the redistribution. It is important to determine whether it is driven by hyperglycemia or is driven by behavioral signals. A simple way to test it would be to test whether the macrophage distribution can be normalized by aggressive exercises and epidermal injury-like simulation without altering plasma glucose. If this works, it would be a great support to the new hypothesis. A better animal experimental idea would be to subject a group of animals to aggressive encounters but simultaneously subject them with frequent hyperglycemic conditions by infusing glucose and study the changes in macrophage distribution in comparison with appropriate controls. It would also pay to see whether good glycemic control alone can normalize macrophage distribution.

4. Diet and behavior: A large number of studies address how diet affects physiology. From our perspective, it is equally important to study how diet affects behavior and whether there are any adaptive correlates between the effects of diet on behavior and on physiology. Fewer studies have looked at the effects of diet on behavior, and perhaps, no study addresses the three-way interaction between diet behavior and metabolism. For example, high-fat diet has been shown to suppress aggression in some studies. Tryptophan and glucose also suppress aggression. This makes a logical connection between increased cholesterol levels induced by high-fat diet since cholesterol is negatively correlated with aggression. An important question of relevance in the near future is how the arrows of causation are placed. Does high-fat diet simultaneously and independently reduce aggressive behavior on the one hand and increase cholesterol on the other or whether increase in cholesterol reduces aggression or whether aggression suppression is necessary first to increase cholesterol. Appropriate interventional studies can address these problems.

   I would like to point out a serious problem that has been largely ignored so far. There are a large number of experiments and studies that demonstrate the effects of diet on metabolism. However, to the best of my knowledge, none of them controls for behavior. If diet affects behavior too, the important question is what the effects of diet are independent of behavior. Since diet has been central in the classical paradigm but relatively peripheral in the new one, it would be important to test to what extent diet affects physiology if behavior is meticulously maintained constant. This is particularly difficult to achieve since a significant dietary change is bound to bring about many subtle changes in behavior, and matching behavior across diet groups is not a trivial job since there are many subtle aspects of behavior. If we focus on selected behavioral parameters, it would be possible to match behaviors across diet groups. Such experiments can be conclusive in only one way. If one does not find a significant effect of diet after controlling for behavior, it can be concluded that the change is brought about by behavior and not diet. But if one finds a difference across behavior-matched diet groups, inference can be difficult since all behaviors cannot be matched simultaneously. Nevertheless, in spite of the limitation, such a set of experiments will be extremely valuable.

(b) Challenges for cell, molecular, and genome biologists:

5. Beta cell dynamics: There is no dearth of research on the β cell. But all the research is limited by a basin of relatively narrow

vision. It has been assumed all the time that any difference in the normal behavior of β cells has to be a "defect." People have not been open to alternatives such as it might be an adaptive or plastic change in its behavior. For example, proinsulin may have some role to play in the body, and a higher level of proinsulin may be serving some purpose not related to glucose homeostasis. The role of insulin in cognitive functions was not imagined by the earlier generation of insulin researchers. We know it today. Similarly proinsulin may turn out to have a role in cognition or sexual or social behavior or something else not even imagined today. Unless we can rule out alternative possibilities, it is not good to make haste in drawing conclusions about β cell "defects." Beta cell degeneration may be evolutionarily programmed for an adaptive purpose as suggested by Rashidi et al. [29, 30]. If we understand the adaptive basis of tissue programming, we will certainly be closer to solutions.

Currently two areas need immediate inputs. One is the role of neuronal inputs in β cell function. It is important to test whether the decreased response of β cells to glucose is accompanied by increased response to autonomic inputs. The other is the role of growth factors and other subtle conditions required for the in vivo regeneration of β cells. If we are able to identify and manipulate these conditions in vivo, there is no need for stem cells, and if we are unable to know the crucial factors in β cell survival and regeneration, stem cells will be useless in the long run anyway.

6. Genome biology of behavior: GWA studies have served a great useful purpose, that of exposing the myth of genetic origins of obesity and diabetes. This is a very important step which will now enable us to come out of a thinking trap and move ahead. Information on statistically significant associations is also important in identifying all potential players in the next-generation era of phenomics elaborated below. But as a prerequisite to enter that era, we need to look at genome-wide associations of subtle behavioral patterns. I am not pointing to psychiatric disorders here. I mean variations in the normal range of human behavior and personalities. This is a potentially important branch of science yet to reach even infancy but is expected to have a tremendous potential in understanding behavior and health.
7. Behavioral epigenetics: Perhaps much more important than behavioral genomics would be behavioral epigenetics. The concept of fetal programming in obesity and diabetes is well known, but it accounts for only a minority of diabetics. Fetal programming is suspected to have a strong epigenetic component, although a clear picture is yet to emerge. I suspect that apart from IUGR, there is likely to be behavioral programming taking place at various developmental stages, much of it currently being pushed under the carpet of stress effects. Specific behaviors and developing personalities are likely to program specific gene expression patterns, and this is currently a big information void.

(c) Challenges for evolutionary, organismal, and systems biologists:

8. Phenomics: Over the 1990s and early 2000s, human genome project raised large expectations. The genomic era and its limitations have contributed much to our understanding which paves the way to a much more difficult era of phenomics [31]. Phenomics largely consists of how a genome interacts with a vast array of social and ecological challenges. Genome biology was handled by molecular biologists and informaticians. Phenomics should be a challenge to ecologists, ethologists, and evolutionary

and organismal biologists simultaneously with cell, molecular, and systems biologists. But currently, there appears to be a huge gap between the two ends, and it would perhaps take several decades for the two ends to meet.

One possible approach is to look for ordination and clustering in phenomic space that consists of multidimensional information on an organism's genome, developmental history, metabolism, behavior, social circle, and ecological relations. This is an unimaginable huge task; several orders of magnitude bigger than genomics but some small preparatory steps in this direction can be perceived. One quick example is an endocrine space mapping. In the classical approach, people have related one hormone to one behavior such as testosterone to aggression. In reality, this is too naïve. For today's science, it should be possible to have data on several hormones and look at the clusters that emerge in the multidimensional space. In the most likely situation, these clusters will be signatures of personalities as well as links between behavior and metabolism. This is certainly doable with the current state of the art.

Another highly ambitious project can be suggested which needs team efforts from almost all fields of animal biology. A colony of rodents or any other suitable social mammalian species is allowed to develop in a semi-wild environment during which the behavioral profile of every individual is carefully noted. This should include diet, feeding behavior, aggressive or agonistic behavior, risk-taking, response to challenges, sexual behavior, maternal behavior, and play behavior. After sufficient social stabilization of the colony, the entire colony can be biopsied and if necessary even sacrificed to study holistically every parameter that one can think of. This would include morphometry, body composition, hormone and growth factor levels, metabolome, proteome, tissue-specific gene expression profiles and methylome, and also a number of brain-related parameters. Such an experiment is likely to generate the kind of data needed for a futuristic phenomic approach.

9. Mechanisms of life history strategies: Evolutionary biologists have been talking about life history strategies for over four decades. Every organism has to strive to reach the optimum distribution of resources between growth, reproduction, and longevity since the resources are finite and there are inevitable trade-offs between the three major activities of life. Life history strategies can differ across as well as within species. It is the within-species choice of strategies that matters in the present context. So far, almost all the theoretical as well as empirical work in life history strategies is at an organismal level. Not much work has gone in discovering the mechanisms that sense the environment, make a choice of strategy, and bring about the necessary changes in body chemistry to execute and support the strategy. I am suggesting here that insulin, leptin, cholesterol, and adiponectin are all important players that execute the life history strategy. There is still a substantial gap in the proximate and ultimate levels in the area of life history strategies. Evolutionary biologists need to look at both the levels carefully. On the one hand, they need to look at whether there are broader scale correlations between population density, aggression, reproductive strategies, lipid metabolism, insulin signaling, and longevity. Some data are no doubt available, and I have based my arguments on the available fragments of data, but more work is needed to fill in the glaring gaps and make a coherent picture.

10. Unstressing stress: The whole area of stress research needs radical rethinking. Only evolutionary biology can form a sound foundation on which the reinterpretations can be based. The reinterpretations should take the form of finding appropriate adaptive responses to specific challenges and the inevitable trade-offs associated with these responses. The earlier we get rid of ambiguous concepts like "stress hormone," the faster will be the progress of phenomic biology.

B. At clinical level:

(a) Challenges for epidemiologists and clinical researchers:

11. Profession/occupation-specific incidence: It is generally agreed that people doing sedentary jobs are more prone to obesity, T2D, and related disorders. However, among sedentary jobs, whether the actual nature of the job makes a difference is not clear. It is difficult to test this from published data since apart from physical activity, no other details of job profile are generally given. The new paradigm has suggested a possibility that people in those kinds of professions that need social manipulation will be more insulin resistant and will have high cholesterol or high cortisol. Apart from job profiles, personalities should also matter. Individuals who have greater skills in social manipulation are likely to be more prone to these conditions. Careful epidemiological studies that take sufficient care of the confounding factors would be illuminating in answering a question raised earlier, whether it is the deficiency of soldier components of behavior alone that matters or components of diplomat behavior have a direct role to play.

12. Social deficiency and diplomat state: If insulin-resistant phenotype is associated with social manipulations, individuals deficient in social skills should be less prone to metabolic syndrome disorders. Autism is an example of social deficiencies. A possibility is that autistic individuals should be less prone to metabolic syndrome. There is a predominant research as well as publication bias against negative associations between disorders. Positive associations quickly attract attention of epidemiologists, but negative associations are seldom published. Here is a theoretical prediction of a negative association which is not difficult to test. Some data about negative association between autism and cholesterol [32–35] or low cortisol levels in autistics in spite of stress conditions [36–39] already exists. Other associations need to be checked.

13. Differentiating between peripheral versus central components in glucose homeostasis at the clinical level: Since hyperglycemia can be caused either by peripheral or central mechanisms, it would be important to develop diagnostic procedures for identifying the predominant cause of hyperglycemia for a given patient. I have suggested certain approaches in Chap. 12. Painstaking standardization and population studies would be required before these approaches can be clinically useful. But diagnosing the cause of hyperglycemia as peripheral or central can substantially improve treatment strategies in a personalized manner.

14. Perhaps even more urgent than that is the need to develop and standardize other markers for follow-up. We have relied too much on glucose, and glucose does not reflect all pathophysiology of T2D. According to the new paradigm, it should be possible to ameliorate a number of diabetes-related pathologies without the need for normalizing sugar. On the other

hand, normalizing sugar may fail to prevent all pathological processes. Therefore we need to have indicators directly reflecting the mediators of pathologies. The possible candidates are EGF-NGF and other growth factors, BDNF, sex hormones, inflammatory markers, and oxidative stress markers. If an improvement in these markers is seen, glucose need not be normalized, since these markers are more directly linked to the multiple pathologies than glucose. If there is reduction in the rate of glucose transport in the brain, alteration in the brain glucose dynamics, and central shift in the homeostatic mechanisms, the normal range of plasma glucose itself can shift up and it would be difficult to decide whether higher fasting glucose represents an elevated normal or a pathological increase. Therefore rather than relying on glucose alone, it would be prudent to look for other parameters directly reflecting the pathophysiological pathways. Currently the information on the other suggested parameters is too scanty to allow any useful interpretations.

(b) Challenges for drug discovery and development researchers:

15. Novel antidiabetic strategies: Since diabetes has been viewed with an extremely myopic vision, drug development also has not gone beyond glycemic control. Perhaps by a stroke of fortune or perhaps because glycemic control was not possible without mitigating some root pathologies such as endothelial dysfunction, drugs like metformin or some of the glitazones happen to have independent simultaneous actions on endothelial function [40–43]. But since tight glycemic control is unable to prevent all complications in the long run, drug discovery efforts to directly target the true pathological processes, at times even ignoring glycemic control, need to be undertaken.

The concept of lifelong dependence on any drug is completely flawed and is an indication of the failure of these drugs to attack the root cause. By the new paradigm, only behavioral correction can show long-term effects towards reversal or complete cure for diabetes. However medication can provide some short-term aids. For example, along with aggressive exercises, supplementary medication with something like myostatin inhibitors may help improve muscle strength and in turn myokine production.

There are two other areas where pharma R&D should focus in the near future. One is identifying the set of growth factors to facilitate β cell regeneration and look for enhancers of glut-1 expression in the brain capillaries.

Clinical Implications of the Upcoming Paradigm

We can imagine possible changes in clinical practice at two levels. One could be immediate and the other if and when substantial research supports the new paradigm and recommends radical changes in clinical practice. At the first level, it would be good to be conservative and without giving up traditional practice try to introduce some elements suggested by the new paradigm, which are almost without any risk. This would include changing the recommended exercises while continuing on the traditional line of medication. The outlook of exercises should change radically. Exercises are not meant for calorie burning and weight loss. Exercises should be intended to make up for the behavioral deficiencies of aggression, adventure, agility, and complex nerve muscle coordination. The exact nature of the games takes different forms in different cultures. Apart from modern games such as football, tennis, badminton, cricket, and baseball which could be good anyway, every culture has a set of traditional games whose usefulness should be tested. Generally the main theme of sports is to mimic the process of hunting or battle in some way or the other. The main neuromotor acts involved in Stone Age hunting or combat are involved in most sports activities as well. They include chasing, grabbing, hitting, kicking, throwing, dodging, and escaping in some way or the other. Thus the very basic concept of sport is in favor of the new paradigm. To some extent, many types of dances also mimic hunting or warlike movements. Traditionally, in most societies, there are occasions on which such games or dances are performed by everyone. Unfortunately, in modern societies, sports became a competitive affair because of which it became an activity of young and highly competitive individuals representing a small sector of the society and others became mere spectators. This is a perverted version of the original concept of sports that is leading to today's state of behavioral deficiencies. In the highly competitive era of sports, even a good player retires from the game in his or her 30s, and old people just do not play any sports. There is a need to come out of this mind-set and realize that active sports that mimic hunter-warrior behavior are more essential for middle to early old-age people, because this is the age at which behavioral deficiencies start exerting their effects on health and physiology.

The current perception of exercise for weight loss is causing more damage than being useful to the society. This is because lean people remain under the false belief that they do not have to exercise. A substantial proportion of people in many societies are lean diabetics, and the false impression of exercising for weight loss discourages them from exercising. On the other hand, a substantial proportion of people are unable to reduce their weight by exercise and get frustrated and give up. They need to be told that the benefits of exercises are independent of weight loss. Appropriate exercises would normalize macrophage distribution, increase EGF, NGF, BDNF, myokines, and a number of other signal molecules directly without involving fat tissue either way. Weight loss is a relatively minor by-product of exercise.

The effects of these kinds of exercises on diabetes would be visible in terms of reduction in inflammatory markers and HOMA index, but it may not normalize blood sugar rapidly since blood sugar appears to be determined by brain glucose dynamics about which we know little today. One of the reasons why diabetics get frustrated with exercise soon is because it does not normalize sugar in a short run. But the effect on HOMA can be shown very quickly. Unfortunately assays for EGF and other growth factors are not yet available at the diagnostic level. But if there is substantial awareness among physicians and a need is felt, they will become available in the near future. Exercise brings about a rise in EGF, testosterone, BDNF, and other markers more reliably. These markers are likely to matter more for preventing diabetic complications than sugar. Demonstrating alterations in these parameters before achieving glycemic control can be encouraging for patients. This much can be incorporated in diabetes management practices very rapidly since it does not involve any new drug and no change in medication is recommended. All that is recommended is playing games, which does not need FDA approval or any ethics committee clearance.

Let us also talk about the other level of clinical implications, although this will need much more time and research to get acceptance. Now for the time being, I will assume that there are sufficient research inputs on the new paradigm and most of the major arguments are supported. The implications at this stage will be radical. We can speculate on what diabetes clinical practice would be after accepting the new paradigm, although there is a long way to reach this stage:

1. Blood sugar may still be used as a diagnostic tool, but it will no more remain important in the follow-up of the treatment.
2. Behavioral therapy in the form of activities, games, exercises, advice, and counseling will begin from day one after diagnosis and will remain the main line of treatment.
3. Insulin sensitizing drugs will no more be recommended for lifetime or indefinite use. They may be used for a short while as a secondary supplement to behavioral treatment.
4. The efficiency of treatment will be judged by novel and standardized lab tests for inflammatory markers, growth factors, sex hormones, and angiogenesis markers. Glucose levels will no more be used as first indicators of success of the treatment.

5. Simultaneously therapy for normalizing glut-1 levels will begin parallel to the above in the form of behavioral therapy consisting of a series of exercises for memory, spatial skills, nerve muscle coordination, etc., to increase brain glucose metabolism which will stimulate upregulation of BBB glucose transport. This can be supplemented by next-generation antihyperglycemic drugs that could be glut-1 agonists or glut-1-specific transcription enhancers. This will ultimately normalize blood sugar which can remain normal with continued behavioral exercises but without any medication. This can be the definition of a case of "cured" diabetes.
6. Insulin will not be used for treatment unless there is advanced β cell degeneration. Decision to administer insulin will be taken based on levels of indigenous insulin production, not based on glucose levels.
7. Effective growth factor cocktail may be developed by then to enhance endogenous β cell regeneration in the case of excessive β cell degeneration at diagnosis itself. Both 6 and 7 will become rare since in T2D it is rare to find extreme islet loss at first diagnosis. If appropriate behavioral therapy begins soon, β cell degeneration would be an extremely rare case in type 2 diabetes.

This is no doubt a remote dream, but dreams are needed to drive research. Human genome sounded an impossible dream which suddenly materialized in an amazingly short time. Medieval sailors needed the North Star to navigate. No one reached the North Star, but without the North Star, so many would have aimlessly drifted and died. My North Star is not so impossible to reach.

References

1. Lindqvist P, Grennert L, Maršál K (1999) Epidermal growth factor in maternal urine—a predictor of intrauterine growth restriction? Early Hum Dev 56:143–150
2. Kamei Y et al (1999) Maternal epidermal growth factor deficiency causes fetal hypoglycemia and intrauterine growth retardation in mice: possible involvement of placental glucose transporter GLUT3 expression. Endocrinology 140:4236–4243
3. Cain MP, Kramer SA, Tindall DJ, Husmann DA (1994) Alterations in maternal epidermal growth factor (EGF) effect testicular descent and epididymal development. Urology 43:375–378
4. Popliker M et al (1987) Onset of endogenous synthesis of epidermal growth factor in neonatal mice. Dev Biol 119:38–44
5. Kashimata M et al (2000) The ERK-1/2 signaling pathway is involved in the stimulation of branching morphogenesis of fetal mouse submandibular glands by EGF. Dev Biol 220:183–196
6. Ahrén B, Holst JJ (2001) The cephalic insulin response to meal ingestion in humans is dependent on both cholinergic and noncholinergic mechanisms and is important for postprandial glycemia. Diabetes 50:1030–1038
7. Kopf SR, Baratti CM (1999) Effects of posttraining administration of insulin on retention of a habituation response in mice: participation of a central cholinergic mechanism. Neurobiol Learn Mem 71:50–61
8. Dhingra D, Parle M, Kulkarni SK (2003) Effect of combination of insulin with dextrose, D(−) fructose and diet on learning and memory in mice. Indian J Pharmacol 35:151–156
9. Belsare PV et al (2010) Metabolic syndrome: aggression control mechanisms gone out of control. Med Hypotheses 74:578–589
10. Schaan BD et al (2005) Sympathetic modulation of the renal glucose transporter GLUT2 in diabetic rats. Auton Neurosci 117:54–61
11. Knauf C et al (2008) Brain glucagon-like peptide 1 signaling controls the onset of high-fat diet-induced insulin resistance and reduces energy expenditure. Endocrinology 149:4768–4777
12. Knauf C et al (2005) Brain glucagon-like peptide-1 increases insulin secretion and muscle insulin resistance to favor hepatic glycogen storage. J Clin Invest 115:3554–3563
13. Darleen S (2008) CNS GLP-1 regulation of peripheral glucose homeostasis. Physiol Behav 94:670–674
14. Klockener T et al (2011) High-fat feeding promotes obesity via insulin receptor/PI3K-dependent inhibition of SF-1 VMH neurons. Nat Neurosci 14: 911–918
15. Könner AC et al (2007) Insulin action in AgRP-expressing neurons is required for suppression of hepatic glucose production. Cell Metab 5:438–449
16. Lautt WW (2004) A new paradigm for diabetes and obesity: the hepatic insulin sensitizing substance (HISS) hypothesis. J Pharmacol Sci 95:9–17
17. Chowdhury KK, Legare DJ, Lautt WW (2011) Insulin sensitization by voluntary exercise in aging rats is mediated through hepatic insulin sensitizing substance (HISS). Exp Gerontol 46:73–80
18. Bosch G, Beerda B, van der Poel AFB, Hendriks WH (2007) Properties of food that may modulate canine and feline behaviour. *Engormix.com* (June 21, 2007). http://en.engormix.com/
19. Albert DJ, Walsh ML, Jonik RH (1993) Aggression in humans: what is its biological foundation? Neurosci Biobehav Rev 17:405–425

20. Lundberg U et al (1983) Hirsute women with elevated androgen levels: psychological characteristics, steroid hormones, and catecholamines. J Psychosomat Obstet Gynecol 2:86–93
21. Mann MA, Svare B (1983) Prenatal testosterone exposure elevates maternal aggression in mice. Physiol Behav 30:503–507
22. Bruce S (1980) Testosterone propionate inhibits maternal aggression in mice. Physiol Behav 24:435–439
23. Ashcroft GS et al (1997) Estrogen accelerates cutaneous wound healing associated with an increase in TGF-[β]1 levels. Nat Med 3:1209–1215
24. Ashcroft GS, Greenwell-Wild T, Horan MA, Wahl SM, Ferguson MWJ (1999) Topical estrogen accelerates cutaneous wound healing in aged humans associated with an altered inflammatory response. Am J Pathol 155:1137–1146
25. Ashcroft GS et al (2003) Estrogen modulates cutaneous wound healing by downregulating macrophage migration inhibitory factor. J Clin Invest 111:1309–1318
26. Hardman MJ, Emmerson E, Campbell L, Ashcroft GS (2008) Selective estrogen receptor modulators accelerate cutaneous wound healing in ovariectomized female mice. Endocrinology 149:551–557
27. Contreras JL, Smyth CA, Bilbao G, Young CJ, Thompson JA, Eckhoff DE (2002) 17[β]-Estradiol protects isolated human pancreatic islets against proinflammatory cytokine-induced cell death: molecular mechanisms and islet functionality. Transplantation 74:1252–1259
28. Matute ML, Kalkhoff RK (1973) Sex steroid influence on hepatic gluconeogenesis and glycogen formation. Endocrinology 92:762–768
29. Rashidi A, Kirkwood TBL, Shanley DP (2009) Metabolic evolution suggests an explanation for the weakness of antioxidant defences in β-cells. Mech Ageing Dev 130:216–221
30. Rashidi A, Kirkwood TBL, Shanley DP (2009) On the surprising weakness of pancreatic B-cell antioxidant defences: an evolutionary perspective. In: Pontarotti P (ed) Evolutionary biology. Springer, Heidelberg, pp 109–125
31. Houle D (2009) Colloquium paper: numbering the hairs on our heads: the shared challenge and promise of phenomics. Proc Natl Acad Sci USA 107:1793–1799
32. Tierney E et al (2006) Abnormalities of cholesterol metabolism in autism spectrum disorders. Am J Med Genet B Neuropsychiatr Genet 141B: 666–668
33. Aneja A, Tierney E (2008) Autism: the role of cholesterol in treatment. Int Rev Psychiatry 20:165–170
34. Sikora DM, Pettit-Kekkel K, Penfield J, Merkens LS, Steiner RD (2006) The near universal presence of autism spectrum disorders in children with Smith–Lemli–Opitz syndrome. Am J Med Genet A 140A:1511–1518
35. Clark-Taylor T (2004) Is autism a disorder of fatty acid metabolism? Possible dysfunction of mitochondrial β-oxidation by long chain acyl-CoA dehydrogenase. Med Hypotheses 62:970–975
36. Corbett BA, Schupp CW, Levine S, Mendoza S (2009) Comparing cortisol, stress, and sensory sensitivity in children with autism. Autism Res 2:39–49
37. Marinović-Ćurin J et al (2007) Slower cortisol response during ACTH stimulation test in autistic children. Eur Child Adolesc Psychiatry 17:39–43
38. Tordjman S et al (1997) Plasma β endorphin, adrenocorticotropin hormone, and cortisol in autism. J Child Psychol Psychiatr 38:705–715
39. Marinovic Curin J et al (2010) Lower cortisol and higher ACTH levels in individuals with autism. J Autism Dev Disord 33:443–448
40. Jager J et al (2005) Effects of short term treatment with metformin on markers of endothelial function and inflammatory activity in type 2 diabetes mellitus: a randomized, placebo controlled trial. J Intern Med 257:100–109
41. Katakam PVG, Ujhelyi MR, Hoenig M, Miller AW (2000) Metformin improves vascular function in insulin-resistant rats. Hypertension 35:108–112
42. Majithiya JB, Balaraman R (2006) Metformin reduces blood pressure and restores endothelial function in aorta of streptozotocin-induced diabetic rats. Life Sci 78:2615–2624
43. Diamanti-Kandarakis E et al (2005) Metformin administration improves endothelial function in women with polycystic ovary syndrome. Eur J Endocrinol 152:749–756

Epilogue: I May be Wrong After All!!

After thinking for over a decade, and challenging and debating with myself, I am completely convinced that T2D and a variety of other disorders originate in brain and behavior, not so much in diet and metabolism. Behavioral deficiency is the key to a number of modern disorders, and if we understand this, solutions are going to be simple. Nevertheless I keep the possibility open that I have actually got into a thinking trap from where it is really difficult to get out. But I would like someone to try and convince me that it is so. I tried doing it myself and it did not work. There was only one ground on which I could make myself suspect that I am wrong and that is I am not trained as a mathematician but I am using mathematical models, I am not a trained as a systems biologist but I am using systems approach, I am not trained as a physiologist but I am making physiological arguments, and I have no background in medicine and I am criticizing current antidiabetic medicine. Therefore, the possibility remains that I have made erroneous arguments out of inadequate knowledge of the field. I would be glad to correct myself if anyone points out specific problems in my way of thinking.

As emphasized repeatedly, the book is mainly intended to stimulate rethinking and research. It is not intended to and not every reader is expected to be convinced right away. I can visualize four possible reactions to this book:

1. Uncritical rejection: This could perhaps be the commonest reaction from the field of medicine, although I do not expect it to be too common in the field of basic biology. Most practicing diabetologists are less worried about the evolution of the disorder and the fundamental biological processes behind it. Their main interest is to manage their patients in the best possible way under all given constraints. For this they have to have a set of assumptions which they rarely challenge although they are very much aware that what they see in real life is very different from the text book picture. Practicing medicine is a field of minimum risk tolerance. In the current era of human rights and awareness, a doctor can be easily sued for a wrong decision. Therefore doctors are generally risk averters. There is a safety factor in following traditional practice. Gone are the days when an Edward Jenner could inject an 8-year-old boy with the ooze from blisters of a milkmaid and then challenge him directly with small pox, or an Alexander Fleming could inject a crude culture extract from a fungus by lumber puncture into the csf of a meningitis patient and save him. The current social and legal constraints have made practicing doctors extreme risk averters and as a result they are least open to novel thinking. Needless to say one would always find exceptions.

   There is another interesting behavioral aspect to uncritical rejection. An uncritical rejection is more often decided by who said it rather than what is being said. As a student of behavioral evolution, I find this phenomenon extremely interesting. It is demonstrated in Japanese macaques that if a monkey innovates

M. Watve, *Doves, Diplomats, and Diabetes: A Darwinian Interpretation of Type 2 Diabetes and Related Disorders*, DOI 10.1007/978-1-4614-4409-1, © Springer Science+Business Media New York 2013 something and finds it useful others imitate it. But the rate of imitation is not decided by how useful the idea is. It is decided by who invented it. One coming from a high ranking monkey is quickly mimicked. One coming from a lower order monkey is not. This is a primate tendency and the scientific primate is perhaps not an exception. I am a smaller monkey and therefore I should not expect rapid acceptance of these ideas. I expect uncritical rejection to be common and it will be an interesting observation for the ethologist in me. But from the point of view of progress in research this is the least welcome of all possible reactions.

2. Uncritical acceptance: Uncritical acceptance is no more welcome than uncritical rejection. The scientific community has made a major mistake of not challenging the existing diabetes paradigm for several decades in spite of so many inconvenient truths accumulating over many decades. One should not repeat the same mistake again, and therefore an uncritical acceptance is strongly discouraged.
3. Quasi-critical rejection: The process by which we make decisions about something being right or wrong is somewhat counterintuitive but extremely interesting. Most often in the human mind a decision is made first and then reasons and justifications for the decisions sought later. This is perhaps an innate human tendency hard but not impossible to override. Therefore it could be a common reaction to decide first that no, this is weird! It can't be true. The current thinking is reasonably alright and there is no need for any alternative thinking. Having decided this already, one only needs reasons to justify the stand and then fishes out only those points that can be criticized. This is what I would like to call quasi-critical rejection. This could be no better than uncritical rejection and needs to be ignored at its best.
4. Experimental challenge approach: The most desirable reaction would be to say "Hey, there are so many testable predictions made. Let us put them to experimental challenge first." This is the most or rather the only desirable approach. I hope the research minded readers take such an approach and at least some of them are motivated to perform experiments to evaluate the new paradigm vis-a-vis the old one.

My dream does not end here. If this story achieves even a moderate success it would be a strong example of the importance of Darwin in the field of medicine. The ultimate success of the story will be when students of medicine and biomedical researchers read modern evolutionary theory and behavioral ecology first and learn to complete the bridge on the river why with simultaneous use of ultimate and proximate reasoning. Medicine will no more be blind then. It will be truly insightful and standing on a sound foundation. More than having attempted a new interpretation of diabetes, I am more interested in this outcome. I am only an undergraduate teacher after all!!

Appendix I: Genes/Molecules that are Associated with Aggression and also Associated with Some Component of Metabolic Syndrome

	Gene/molecule	Association with aggression	Association with metabolic syndrome
1	Acetylcholine	Aggression increases with gene overexpression [1]	Insulin resistance decreases with overexpression [2]
2	ACTH (adenocorticotropic hormone)	ACTH reduced aggression [3]	Promotes insulin-resistant and inflammatory state [4]
3	Activin (TGF β superfamily)	Transgenic mice overexpressing activin E showed aggressive behavior in resident–intruder tests [5]	Facilitates β cell regeneration [6]. A high follistatin/activin ratio contributes to the pathophysiology of PCOS [7]. Activin A plays a role in human adipogenesis [8]
4	ADORA 1 (adenosine receptor)	Mice lacking adenosine receptors are aggressive but less exploratory [9]	Adenosine receptors involved in regulation of systemic inflammatory responses [10]. Interact with D2 Dopamine receptor [11]. Increased signaling in diabetes. Selective $A_{2B}R$ blockers reduce insulin resistance [12]
5	α MSH (melanocyte stimulating hormone)	Releases olfactory aggression signals in rodents [13]. Increased aggression when the receptor *mc5r*gene is overexpressed [14], and decreased aggression in knockouts [15]	α MSH decreases intra-abdominal fat and enhances the actions of insulin [16, 17]
6	Androgen receptors	Mutation that fails to produce the long form of the androgen receptor (AR) is not aggressive [18, 19]	Shorter allele polymorphism associated with PCOS [20, 21]. Length polymorphism modulates body fat mass and serum concentrations of insulin in men [22] Association between T and IR is modified by the CAG repeat polymorphism within the AR [23]
7	AVP Arginine Vasopressin (*avpl* gene)	Strong role in aggression in zebrafish [24]. In mice, overexpression of gene seen in dominant individuals [19]	Plasma levels increased in uncontrolled diabetes as well as hypoglycemia [25]. Obese men may have an altered pituitary response to combined CRH/AVP stimulation. ACTH hyper-responsiveness after CRH/AVP stimulation also appears to be related to hyperinsulinemia [26]. AVP causes insulin release from β cells [27]. Elevated in diabetes [28]

(continued)

(continued)

	Gene/molecule	Association with aggression	Association with metabolic syndrome
8	Aromatase	Aromatase inhibition increased aggression and aromatase activity in the bed nucleus of the stria terminalis was negatively correlated with aggression [29]. Changes in central aromatase activity may play a role in determining how social experiences affect whether an individual engages in aggressive behavior [30]	Aromatase mutant showed MS characterized by abdominal obesity and hyperinsulinemia. This reinforces congenital estrogen deprivation in the role of fat accumulation and distribution in men [31]. Features of PCOS seen in patients with aromatase deficiency [32]. Short-term aromatase inhibition affected glucose metabolism, lipid metabolism, and leptin secretion [33]
9	BDNF (brain-derived neurotropic factor)	BDNF deficiency leads to aggressive phenotype and heterozygote knockouts showed increased aggression [34]. Mice with global deletion of one BDNF allele or with forebrain-restricted deletion of both alleles show elevated aggression [35]. Aggressive encounters and particularly defeat response upregulate BDNF	BDNF is insulin sensitizing [36]. BDNF regulates food intake and glucose metabolism in rodent obese diabetic models [37]. Hyperphagia, severe obesity, and impaired cognitive function with chromosomal inversion in region having *BDNF* gene [38]
10	Beta 2 microglobulin	Homozygous knockout mice for the gene *b2m* show decreased aggression [39]	Elimination of cell surface MHC class I expression with disrupted *B2m* gene blocked autoimmune diabetes in NOD/Lt mice [40]. Beta cell disruption by transgene overexpression involved a defect in insulin secretion [41]. Activity synergistic with IGF I [42]
11	Caffeine	Aggressive behavior was increased following higher doses of caffeine. Non-monotonic effect of caffeine on aggression [43]	Consumption of high amounts of coffee or equivalent doses of pure caffeine reduces hyperglycemia and decreases fat mass. Some effects of caffeine in diabetic animals are mediated by blockade ofA2BRs [12]. Caffeine can alter insulin sensitivity in healthy humans, possibly as a result of elevated plasma epinephrine levels [44]. High coffee consumption associated with better glucose tolerance and a substantially lower risk of T2D [45]
12	CamK2 (calcium-dependent calmodulin protein kinase-2)	α-CaMKII homozygous knockout mice display reduced aggression. Heterozygotes increased aggression [46]	Involved in insulin exocytosis from the β cells [47]. Responsible for insulin signaling and caused *glut*-4 translocation to the plasma membrane and increase of glucose uptake [48, 49]
13	Cb1 Endocannabinoid receptor	CB1 knockout mice show increased aggression [50]	Has a role in fat accumulation. Cb1 blockade has been used in the treatment of obesity [51]. Substrate for insulin receptor kinase [52]
14	CCK (cholecystokinin)	Increased aggression in mice with upregulation of CCK response [53]	Lack of CCK-A receptors in OLETF rats resulted in hyperphagia and obesity [54]. CCK 8 exerts antidiabetic action [55]. Reduces postprandial hyperglycemia [56]. Women with PCOS have reduced CCK secretion. Impaired CCK secretion played a role in binge eating and overweight in women with PCOS [57]
15	Cholesterol	The negative association between cholesterol and physical aggression is quite robust [58, 59]. In contrast with nonphysical aggression there is a positive association [59]	High plasma cholesterol is strongly associated with MS [60, 61]

	Gene/molecule	Association with aggression	Association with metabolic syndrome
16	COMT (catechol-O-methyltransferase)	Homozygous knockouts have no difference in aggression levels, but heterozygous knockouts show increased aggression [62]	Genetic polymorphisms in *COMT* involved in hypertension in the Japanese population [63] are is associated with abdominal obesity and blood pressure in men [64]. Has a role in body fat regulation and modification in effect of exercise on fat loss in postmenopausal women [65]. Its role in PCOS is debated [66]
17	CRH (CRF)/ Urocortin	Intracerebroventricular injections of CRF significantly inhibited maternal aggression [67]	Elevation of CRF associated with a reduction of food intake and body weight gain in normal and obese animals [68]. CRH acts synergistically with AVP and causes insulin release from β cells [27]. CRH slightly affects basal plasma levels of insulin and glucagon in mice [69]. Urocortin causes suppression of feeding in rodents [70]
18	DHEA (dehydroepi-androsterone)	Expression of animal aggression in nonbreeding season when gonadal T synthesis is low [71]	Protects against fat accumulation and muscle insulin resistance in rats [72–74]. Has anti-inflammatory properties [75]. Exogenous application increases wound healing [76]
19	Dopamine	Positively associated with aggression [77, 78]	Diabetes associated with decreased dopamine activity [79]. In mice, the use of dopamine receptor antagonist was associated with weight gain and insulin resistance, whereas agonists of the D1 and D2 dopamine receptor isoforms decrease food intake and improve insulin sensitivity [80]
20	EGF (epidermal growth factor)	Secreted in saliva in response to injury or aggressive acts [81, 82]	Physiologically significant improvement in glucose tolerance achieved by stimulation of β cell regeneration with EGF administration [83]. Implicated to have a role in PCOS [84] and a role in β cell replication [85]. EGF mimicked effects of insulin and increased the tyrosine phosphorylation and activation of IRS-1 and IRS-2; thus, it can augment the downstream signaling of insulin in insulin-resistant states like T2D [86]
21	Endorphin	Endorphin levels were negatively associated with aggression scores in rodents [87]. Has inhibitory effect on aggression and reproductive behavior in birds [88]	Beta endorphin associated with overeating and obesity in mice, rats, and humans [89, 90]. Involved in exercise-induced insulin sensitivity [91]
22	Endothelin	Heterozygous ET-1 knockout mice exhibit diminished aggressive and autonomic responses toward intruders [92]	Insulin signaling upregulated endothelin production [93, 94] and endothelin-induced insulin resistance as well as insulin secretion [95–97]. Levels are generally increased in obesity and diabetes. Endothelin-1 was elevated in women with PCOS [98], and may play a role in endothelial dysfunction associated with PCOS [99]
23	Epinephrine	Overexpression of PNMT that catalyzes the formation of epinephrine from norepinephrine caused increase in aggression [100]	Increases insulin resistance [101, 102]

(continued)

(continued)

	Gene/molecule	Association with aggression	Association with metabolic syndrome
24	Estrogen Esr-1 and Esr- 2 (estrogen receptor)	Knockouts for both the receptors together show decreased aggression [103, 104]. Male mice with targeted disruption of the gene encoding the α-isoform of the estrogen receptor (ERαKO) display reduced aggression in several testing situations [104, 105]. Conversely, the β-isoform null mice (ERβKO) exhibit normal or increased aggression depending on social experience [103]. ERαKO females exhibit increased levels of aggression toward other female mice relative to wild-type females [104, 105]	In postmenopausal women MS is associated with lack of estrogen [106]. Estrogen use in American Indian postmenopausal women related to deterioration of glucose tolerance [107]. Estrogen induces insulin resistance in women using contraceptive [108]. Esr-1 contributed to T2D and CVD risk via pleiotropic effects, leading to insulin resistance, a poor lipid profile, and obesity [109]. Introns 1 and 2 of ESR1 gene contain functionally important regions related to T2D risk [110]. Esr located in the hypothalamus serve as a master switch to control food intake, energy expenditure, and body fat distribution [111]
25	Fat and triglycerides	Negatively associated with aggression [112]	Plasma triglyceride levels are associated with insulin resistance [113]
26	FGF1R (fibroblast growth factor receptor)	Mutation of the gene *fgfr1a* increased aggression [114]	It ameliorates insulin and leptin resistance, enhances fat oxidation, and suppresses de novo lipogenesis in liver [115]. *FGFR1* influences adipose tissue and the hypothalamic control of appetite [116]
27	*Fyn* (Fyn tyrosine kinase)	Homozygous knockouts show decreased aggression [117]	*FRK* polymorphisms found associated with overweight/obesity in Korean population [118]. Fyn tyrosine kinase gene was associated with increased hyperglycemia in mice [119]
28	GABA (gamma-amino butyric acid)	In *gad2*(GABA)gene knockouts aggression decreased [120]	GABA is intricately involved in food intake regulation and thereby obesity [121, 122]. It had a positive relation with obesity [123]
29	*GDi* (GDP dissociating inhibitor)	Homozygous knockouts for the gene show decreased aggression [124]	*GDi* retention on adipocyte membranes was associated with complete arrest of insulin-induced translocation of *glut*-4 glucose transporters onto plasma membrane [125]. Overexpression of *GDi* significantly inhibited glucose-induced insulin secretion from β cells [126]
30	b-geo 22 *Gtrgeo22* (gene trap ROSA)	Homozygous mutants show decreased inter-male aggression [127]	The mice having this mutation show reduced body fat [127]
31	GH (growth hormone)	GH administration increases isolation-induced aggression [128, 129]	GH secretion is profoundly blunted in obesity [130]. Deficiency is characterized by central obesity [131–133]
32	Ghrelin	Ghrelin levels are negatively associated with cholesterol and violent individuals have low cholesterol and leptin along with elevated ghrelin [134]	Lower levels associated with T2D, insulin resistance, and hypertension [135, 136]
33	Glucocorticoids	Low basal levels associated with increased aggression [71, 137, 138]	Contributes to increased insulin resistance [139–141], raised blood pressure, and elevated plasma glucose [141]. N363S variant in the glucocorticoid receptor associated with obesity [142]. They had chronic effect to increase central obesity, and acute effects to enhance lipolysis [141]

	Gene/molecule	Association with aggression	Association with metabolic syndrome
34	Glucose	Low levels of plasma glucose known to stimulate aggression [143–145]	Fasting levels of plasma glucose are higher in T2D. T2D is characterized by hyperglycemia
35	GnRh (gonadotropin releasing hormone)	Higher expression of this hormone in dominant animals including elephants [146], antelopes [147], and squirrels [148]	Insulin may regulate reproductive function by stimulating GnRH gene expression [149]. Long-term suppression by GnRH analogue leads to increase in fat and insulin resistance [150]
36	Histamine	In *hrh* gene homozygous knockouts aggression decreases [151]	Leads to reduction in obesity and food intake [152–154]
37	IGF-1 (insulin-like growth factor-1)	Expression associated with increased level of aggression in fish [155]. Dominant pudu males had higher IGF-1 levels than their subordinate pen mates [156]	IGF 1 treatment reduced obesity and hyperinsulinemia [157]. IGF-1 reduced hyperphagia, obesity, hyperleptinemia, and hyperinsulinemia in rats [157, 158]. Liver IFG-1-deficient mice showed insulin resistance [159]
38	IL-6 (interleukin-6)	IL-6 overexpression in mice decreases aggression [160]. It was associated with the presence of aggressive behavior in domestic dogs [161]	Knockouts develop obesity and centrally acting IL-6 had antiobesity effects in rodents [162]. Raised IL-6 levels were found in obese patients [163] and in obese women with PCOS [164]. Has a role in insulin resistance in humans and may act in concert with other cytokines that also are upregulated in adipose cells in insulin resistance [165]
39	Kiss1peptin (kisspeptin or metastin)	Exogenous administration of kisspeptin stimulates the GnRH system [166]. Higher levels of expression in dominant males [167]	The GPR54 (Kiss1 receptor)/KISS1 system is expressed in the endocrine pancreas, where it influences β cell secretory function [168]
40	Leptin	Levels are negatively associated with aggression [58, 169]	Although leptin regulates hyperphagia and obesity, plasma levels go up in obesity [170], believed to be accompanied with leptin resistance [171]. Plasma leptin levels are associated with increased insulin levels [171]
41	MAOA (monoamine oxidase A)	Point mutation in humans causes increase in aggression in human males [172]. Homozygous knockouts in mice for the gene causes increase in aggression [173]	Increased serum MAOA concentration was also registered in persons with diabetes mellitus [174]
42	Melatonin	Subcutaneous injections of melatonin increase aggression [175]	Administration of this hormone associated with a reduction in hyperinsulinemia, hyperleptinemia, and hypertriglyceridemia [176, 177] and acts as a regulator of obesity [177]
43	mTOR (the mammalian target of rapamycin)	mTOR is activated by exercise [178–180] and testerone [181, 182], which are main components of aggression	Increased mTOR activity reduces insulin action, thought to be a crucial link in insulin resistance [183, 184]. But action of T on mTOR is insulin sensitizing
44	Myostatin	Myostatin limits muscle growth and increasing levels of myostatin are thought to be responsible for age-induced sarcopenia. The aggression hormone T suppresses myostatin [185]	Increases insulin resistance [186]. Myostatin inhibitors prevent the development of obesity and T2D [187]
45	N-CAM (neural cell adhesion molecule)	Both homozygous and heterozygous knockouts of *Ncam gene* show increase in aggression [188]	Promotes insulin signaling and adipocyte differentiation of adult stem cells [189]. NCAM(–/–) mice exhibit impaired glucose tolerance and basal hyperinsulinemia [190]
46	NEP (neutral endopeptidase)	Homozygous knockout mice show increased aggression [191]	NEP plays a role in regulating nerve function in insulin-deficient diabetes and diet-induced obesity [192]

(continued)

(continued)

	Gene/molecule	Association with aggression	Association with metabolic syndrome
47	NPY (neuropeptide Y)	Interactions between (NPY) and serotonergic neurons provided link between aggressive behavior in mammals and its link with feeding [193]	NPY and leptin regulated body fat mass and food intake [194]. Intracerebroventricular infusion of NPY Receptor produced hyperphagia, accompanied by increased adipose tissue weight, hyperinsulinemia, and hyperleptinemia [195]. In the absence of NPY, mice are less obese [196, 197] and are less severely affected by diabetes [197]
48	NGF (nerve growth factor)	High serum β NGF levels found in aggressive mice [198]. Basal serum NGF levels of lactating females were higher than those of virgin females but did not increase significantly above base after an aggressive encounter with a male or a female conspecific [199]	NGF levels altered in obesity, T2D, and the MS in women [200]. NGF involved in the development of cardiovascular disease in patients with MS [201]. NFG deficiency blamed for diabetic neuropathy [202]
49	NO (nitrous oxide)	Homozygous knockouts of nNOS show increase in aggression [203]. Heterozygote knockouts of eNOS show reduced aggression [204]. NO also appears to affect aggressive behavior via 5-HT [19]	Reduced bioavailability of NO characterizes obesity and related disorders [205]. eNOS knockout mice were hypertensive and had fasting hyperinsulinemia, hyperlipidemia, and insulin resistance [206]. eNOS gene polymorphisms have significant association with T2D [207]. The effects of nNOS are sex specific in mice
50	Osteocalcin	Testosterone stimulates osteocalcin production and vice versa [208, 209]	Osteocalcin also has insulin sensitizing and antiobesity actions and anti-inflammatory actions [210, 211]. Induces β cells to release insulin and also induces the release of adiponectin [210]
51	Oxytocin	Homozygous knockouts show increase [212] or decrease [213] in aggression	Effects of oxytocin on obesity, hypertension, and insulin resistance are also mixed: it facilitates lipogenesis but has antiobesity effects; it facilitates liver gluconeogenesis but still reduces insulin resistance [214–225]
52	PEPCK (phosphoenol pyruvate carboxykinase)	Overexpression in muscle increased endurance making muscle almost immune to fatigue. Animals with such increased muscle strength increase their aggression levels substantially [226]	Animals overexpressing in muscle have high levels of intramuscular triglycerides but are highly insulin sensitive [226, 227]. Overexpression of this enzyme in liver leads to liver insulin resistance and upregulation of gluconeogenesis [228, 229]. Overexpression in adipose tissue leads to increased accumulation of fat, not accompanied by higher levels of FFAs in blood. Also in spite of obesity, there was no change in insulin levels or insulin sensitivity [230]
53	Progastrin-Gastrin	Transgenic mice overexpressing progastrin showed increased aggression [53]	Central obesity and insulin resistance are increased in mice lacking gastrin expression [231]
54	Prolactin	Causes postpartum aggression [232]. Higher levels of prolactin were associated with high aggressiveness [233]	Obese patients have higher basal levels of serum prolactin than those of healthy subjects [234]
55	Rgs2 (regulator of G protein signaling 2)	The *RGS2*$^{-/-}$ mice are less aggressive than their heterozygous littermates [235]	Polymorphism associated with hypertension [236]. Knockouts associated with hypertensive phenotype [237]
56	5-HT (serotonin)	Low serotonin levels are associated with impulsive aggression or violent suicide [78, 238, 239]	Serotonin 5-HT2C receptor knockouts are obese [240]. 5-HT responsible for hyperinsulinemia, insulin resistance, and glucose intolerance [241]. Chronically elevated levels in ventromedial hypothalamus induce insulin resistance [241, 242]. Reciprocally hyperinsulinemia and insulin resistance result in elevated 5-HT [243]

	Gene/molecule	Association with aggression	Association with metabolic syndrome
57	SF-1 (steroidogenic factor-1)	CNS-specific knockouts of SF-1 had significantly more risk avoiding behavior [244]	Global knockout mice lacking SF-1 exhibit delayed-onset obesity [244, 245]
58	Sildenafil	Aggressive behavior is a possible side effect [246–248]	It has insulin sensitizing effect [246]
59	*Slc6a2* (norepinephrine receptor)	Knockouts show increased aggression [249]	Polymorphism associated with hypertension [250] and with weight loss treatment success [251]
60	Substance P	This has a complex role in aggression. It is vital for territorial aggression and defensive aggression [252, 253] but also regulates aggression via NK-1 [132]	Substance P antagonist prevents weight gain and fat accumulation in mice [254]. Stimulative effect on glucose-induced insulin secretion from β cells under normobaric oxygen, and depressed insulin secretion under high baric incubation [255]
61	Testosterone	Higher levels of T cause greater aggression. Dominant males have higher level of T [256, 257]	Obesity and MS have been associated with lower levels of T [258, 259]. Low T levels and impaired mitochondrial function promote insulin resistance in men [260] and women [261]. T administered to female rats was followed by insulin resistance [262]. Insulin stimulates human ovarian T production in PCOS [263]
62	TNF-α (tumor necrosis factor α)	Hostility and aggression in men were associated with an increased expression of TNF-α [264]	The absence of in knockouts improved insulin sensitivity and reduced levels of FFAs in plasma [265]. TNF-α increased in obesity and associated with insulin resistance [266] and diabetes that often accompany obesity [267]. In PCOS, TNF-*α* correlated more strongly with indices of insulin resistance [164]. Higher levels found in PCOS patients [268]
63	Trpc (receptor potential channel)	Homozygous knockouts show decreased aggression [269]	In MS, abnormally elevated adrenal *TRPC* expression may underlie increased plasma epinephrine and heart rate. The excess of plasma catecholamines and increased heart rate are risk factors for cardiovascular disease [270]. Involved in serotonin [271] and leptin [272, 273] signaling
64	Tryptophan	Supplementation reduces and depletion increases aggression [274–277]	Long-term feeding of surplus dietary TRP inhibits both baseline adrenocortical and sympathetic nervous system activities; it induces insulin resistance for both glucose and amino acids [278]
65	VGF (nerve growth factor) inducible	Homozygous knockouts mice show decreased aggression [279]	Homozygous mutants are small, hypermetabolic, hyperactive, and infertile, with markedly reduced leptin levels and fat stores and altered hypothalamic POMC [279]. VGF mutant mice have increased insulin sensitivity. VGF deficiency lowered circulating glucose and insulin levels in several murine models of obesity that are also susceptible to adult-onset diabetes mellitus. Ablation of Vgf in ob/ob mice decreased circulating glucose and insulin levels but did not affect adiposity [280]
66	Vitamin D	Disruption of the vitamin D receptor gene results in loss of aggression in mice [281, 282]	Vitamin D deficiency is a significant contributor to insulin resistance and Vitamin D3 supplementation was an effective insulin sensitizer [283, 284]

T testosterone, *MS* metabolic syndrome, *PCOS* polycystic ovarian syndrome, *T2D* type 2 diabetes

From the list of 66 genes or molecules above it can be seen that 37 molecules show a negative association between aggression and metabolic syndrome components (molecules that have a pro-aggression effect or are elevated during aggression are antiobesity or insulin sensitizing and anti-aggression molecules are pro-obesity or insulin resistance), 14 show a positive association, and 15 have contradicting, complex, or non-monotonic associations. Going by the numbers alone there appears to be a significant negative association between physical aggression and metabolic syndrome.

References

1. Pucilowski O, Eichelman B, Overstreet DH, Rezvani AH, Janowsky DS (1990) Enhanced affective aggression in genetically bred hypercholinergic rats. Neuropsychobiology 24:37–41
2. Xie H, Lautt WW (1996) Insulin resistance caused by hepatic cholinergic interruption and reversed by acetylcholine administration. Am J Physiol Endocrinol Metab 271:E587–E592
3. Brain PF, Nowell NW, Wouters A (1971) Some relationships between adrenal function and the effectiveness of a period of isolation in inducing intermale aggression in albino mice. Physiol Behav 6:27–29
4. Iwen KAH et al (2008) Melanocortin crosstalk with adipose functions: ACTH directly induces insulin resistance, promotes a pro-inflammatory adipokine profile and stimulates UCP-1 in adipocytes. J Endocrinol 196:465–472
5. Sekiyama K et al (2009) Abnormalities in aggression and anxiety in transgenic mice overexpressing activin E. Biochem Biophys Res Commun 385:319–323
6. Li L, Yi Z, Seno M, Kojima I (2004) Activin A and betacellulin. Diabetes 53:608–615
7. Eldar-Geva T (2001) Follistatin and activin A serum concentrations in obese and non-obese patients with polycystic ovary syndrome. Hum Reprod 16:2552–2556
8. Zaragosi L-E et al (2010) Activin A plays a critical role in proliferation and differentiation of human adipose progenitors. Diabetes 59:2513–2521
9. Giménez-Llort L et al (2002) Mice lacking the adenosine A1 receptor are anxious and aggressive, but are normal learners with reduced muscle strength and survival rate. Eur J Neurosci 16:547–550
10. Ohta A, Sitkovsky M (2001) Role of G-protein-coupled adenosine receptors in downregulation of inflammation and protection from tissue damage. Nature 414:916–920
11. Kamiya T, Saitoh O, Yoshioka K, Nakata H (2003) Oligomerization of adenosine A2A and dopamine D2 receptors in living cells. Biochem Biophys Res Commun 306:544–549
12. Figler RA et al (2011) Links between insulin resistance, adenosine A2B receptors, and inflammatory markers in mice and humans. Diabetes 60:669–679
13. Nowell NW, Thody AJ, Woodley R (1980) The source of an aggression-promoting olfactory cue, released by α-melanocyte stimulating hormone, in the male mouse. Peptides 1:69–72
14. Gonzalez MI, Vaziri S, Wilson CA (1996) Behavioral effects of [α]-MSH and MCH after central administration in the female rat. Peptides 17:171–177
15. Morgan C, Thomas RE, Cone RD (2004) Melanocortin-5 receptor deficiency promotes defensive behavior in male mice. Horm Behav 45:58–63
16. Obici S et al (2001) Central melanocortin receptors regulate insulin action. J Clin Invest 108:1079–1085
17. Huszar D et al (1997) Targeted disruption of the melanocortin-4 receptor results in obesity in mice. Cell 88:131–141
18. Olsen K (1983) Genetic determinants of sexual differentiation. In: Balthazart J, Prove E, Gilles R (eds) Hormones and behavior in higher vertebrates. Springer, Heidelberg, p 139
19. Nelson RJ, Chiavegatto S (2001) Molecular basis of aggression. Trends Neurosci 24:713–719
20. Ibáñez L et al (2003) Androgen receptor gene CAG repeat polymorphism in the development of ovarian hyperandrogenism. J Clin Endocrinol Metab 88:3333–3338
21. Jääskeläinen J et al (2008) Androgen receptor gene CAG repeat length in women with metabolic syndrome. Gynecol Endocrinol 24:411–416
22. Zitzmann M, Gromoll J, von Eckardstein A, Nieschlag E (2003) The CAG repeat polymorphism in the androgen receptor gene modulates body fat mass and serum concentrations of leptin and insulin in men. Diabetologia 46:31–39
23. Möhlig M et al (2006) The androgen receptor CAG repeat modifies the impact of testosterone on insulin resistance in women with polycystic ovary syndrome. Eur J Endocrinol 155:127–130
24. Filby AL, Paull GC, Hickmore TF, Tyler CR (2010) Unravelling the neurophysiological basis of aggression in a fish model. BMC Genomics 11:498
25. Zerbe RL, Vinicor F, Robertson GL (1979) Plasma vasopressin in uncontrolled diabetes mellitus. Diabetes 28:503–508
26. Pasquali R et al (1999) ACTH and cortisol response to combined corticotropin releasing hormone-arginine vasopressin stimulation in obese males and its relationship to body weight, fat distribution and parameters of the metabolic syndrome. Int J Obesity 23:419–424
27. O'Carroll A-M, Howell GM, Roberts EM, Lolait SJ (2008) Vasopressin potentiates corticotropin-releasing hormone-induced insulin release from mouse pancreatic β-cells. J Endocrinol 197:231–239
28. Bankir L, Bardoux P, Ahloulay M (2001) Vasopressin and diabetes mellitus. Nephron 87:8–18
29. Trainor BC, Bird IM, Marler CA (2004) Opposing hormonal mechanisms of aggression revealed through

short-lived testosterone manipulations and multiple winning experiences. Horm Behav 45:115–121
30. Trainor B, Kyomen H, Marler C (2006) Estrogenic encounters: how interactions between aromatase and the environment modulate aggression. Front Neuroendocrinol 27:170–179
31. Maffei L et al (2007) A novel compound heterozygous mutation of the aromatase gene in an adult man: reinforced evidence on the relationship between congenital oestrogen deficiency, adiposity and the metabolic syndrome. Clin Endocrinol 67:218–224
32. Belgorosky A et al (2003) Hypothalamic-pituitary-ovarian axis during infancy, early and late prepuberty in an aromatase-deficient girl who is a compound heterocygote for two new point mutations of the CYP19 gene. J Clin Endocrinol Metab 88:5127–5131
33. Lapauw B, T'Sjoen G, Mahmoud A, Kaufman JM, Ruige JB (2009) Short-term aromatase inhibition: effects on glucose metabolism and serum leptin levels in young and elderly men. Eur J Endocrinol 160:397–402
34. Lyons WE (1999) Brain-derived neurotrophic factor-deficient mice develop aggressiveness and hyperphagia in conjunction with brain serotonergic abnormalities. Proc Natl Acad Sci USA 96: 15239–15244
35. Ito W, Chehab M, Thakur S, Li J, Morozov A (2011) BDNF-restricted knockout mice as an animal model for aggression. Genes Brain Behav 10:365–374
36. Duan W, Guo Z, Jiang H, Ware M, Mattson MP (2003) Reversal of behavioral and metabolic abnormalities, and insulin resistance syndrome, by dietary restriction in mice deficient in brain-derived neurotrophic factor. Endocrinology 144:2446–2453
37. Nakagawa T et al (2000) Brain-derived neurotrophic factor regulates glucose metabolism by modulating energy balance in diabetic mice. Diabetes 49:436–444
38. Gray J et al (2006) Hyperphagia, severe obesity, impaired cognitive function, and hyperactivity associated with functional loss of one copy of the brain-derived neurotrophic factor (BDNF) gene. Diabetes 55:3366–3371
39. Loconto J et al (2003) Functional expression of murine V2R pheromone receptors involves selective association with the M10 and M1 families of MHC Class Ib molecules. Cell 112:607–618
40. Serreze DV, Leiter EH, Christianson GJ, Greiner D, Roopenian DC (1994) Major histocompatibility complex class I-deficient NOD-B2mnull mice are diabetes and insulitis resistant. Diabetes 43:505–509
41. Allison J et al (1991) Overexpression of β 2-microglobulin in transgenic mouse islet β cells results in defective insulin secretion. Proc Natl Acad Sci USA 88:2070–2074
42. Centrella M, McCarthy TL, Canalis E (1989) Beta 2-microglobulin enhances insulin-like growth factor I receptor levels and synthesis in bone cell cultures. J Biol Chem 264:18268–18271
43. Wilson JF et al (2000) Effects of low doses of caffeine on aggressive behavior of male rats. Psychol Rep 86:941–946
44. Keijzers GB, De Galan BE, Tack CJ, Smits P (2002) Caffeine can decrease insulin sensitivity in humans. Diabetes Care 25:364–369
45. van Dam RM, Hu FB (2005) Coffee consumption and risk of type 2 diabetes. JAMA: J Am Med Assoc 294:97–104
46. Chen C, Rainnie D, Greene R, Tonegawa S (1994) Abnormal fear response and aggressive behavior in mutant mice deficient for α-calcium-calmodulin kinase II. Science 266:291–294
47. Yamamoto H, Matsumoto K, Araki E, Miyamoto E (2003) New aspects of neurotransmitter release and exocytosis: involvement of Ca2+/calmodulin-dependent phosphorylation of synapsin i in insulin exocytosis. J Pharmacol Sci 93:30–34
48. Wright D et al (2004) A role for calcium/calmodulin kinase in insulin stimulated glucose transport. Life Sci 74:815–825
49. Illario M et al (2009) Calcium-calmodulin-dependent kinase II (CaMKII) mediates insulin-stimulated proliferation and glucose uptake. Cell Signal 21:786–792
50. Martin M, Ledent C, Parmentier M, Maldonado R, Valverde O (2002) Involvement of CB1 cannabinoid receptors in emotional behaviour. Psychopharmacology (Berl) 159:379–387
51. Viveros MP, de Fonseca FR, Bermudez-Silva FJ, McPartland JM (2008) Critical role of the endocannabinoid system in the regulation of food intake and energy metabolism, with phylogenetic, developmental, and pathophysiological implications. Endocr Metabol Immune Disord Drug Targ 8:220–230
52. Saltiel AR, Kahn CR (2001) Insulin signalling and the regulation of glucose and lipid metabolism. Nature 414:799–806
53. Li Q, Deng X, Singh P (2007) Significant increase in the aggressive behavior of transgenic mice overexpressing peripheral progastrin peptides: associated changes in CCK2 and serotonin receptors in the CNS. Neuropsychopharmacol 32:1813–1821
54. Moran TH, Katz LF, Plata-Salaman CR, Schwartz GJ (1998) Disordered food intake and obesity in rats lacking cholecystokinin A receptors. Am J Physiol 274:R618–625
55. Ahrén B, Holst JJ, Efendic S (2000) Antidiabetogenic action of cholecystokinin-8 in type 2 diabetes. J Clin Endocrinol Metab 85:1043–1048
56. Liddle RA et al (1988) Physiological role for cholecystokinin in reducing postprandial hyperglycemia in humans. J Clin Invest 81:1675–1681
57. Hirschberg AL, Naessén S, Stridsberg M, Byström B, Holtet J (2004) Impaired cholecystokinin secretion and disturbed appetite regulation in women with polycystic ovary syndrome. Gynecol Endocrinol 19:79–87
58. Golomb BA (1998) Cholesterol and violence: is there a connection? Ann Intern Med 128:478–487
59. Hillbrand M (1999) Cholesterol and aggression. Aggress Violent Behav 4:359–370
60. Ford ES, Giles WH, Dietz WH (2002) Prevalence of the metabolic syndrome among US adults. JAMA: J Am Med Assoc 287:356–359

61. Aso Y et al (2005) Metabolic syndrome accompanied by hypercholesterolemia is strongly associated with proinflammatory state and impairment of fibrinolysis in patients with type 2 diabetes synergistic effects of plasminogen activator inhibitor-1 and thrombin-activatable fibrinolysis inhibitor. Diabetes Care 28:2211–2216
62. Gogos JA (1998) Catechol-O-methyltransferase-deficient mice exhibit sexually dimorphic changes in catecholamine levels and behavior. Proc Natl Acad of Sci USA 95:9991–9996
63. Kamide K et al (2007) Association of genetic polymorphisms of ACADSB and COMT with human hypertension. J Hypertension 25:103–110
64. AnnerbrinkKetal(2008)CatecholO-methyltransferase val158-met polymorphism is associated with abdominal obesity and blood pressure in men. Metabolism 57:708–711
65. Tworoger SS et al (2004) The effect of CYP19 and COMT polymorphisms on exercise-induced fat loss in postmenopausal women. Obes Res 12:972–981
66. Hill LD et al (2012) Catechol-O-methyltransferase (COMT) single nucleotide polymorphisms and haplotypes are not major risk factors for polycystic ovary syndrome. Mol Cell Endocrinol 350:72–77
67. Gammie SC, Negron A, Newman SM, Rhodes JS (2004) Corticotropin-releasing factor inhibits maternal aggression in mice. Behav Neurosci 118:805–814
68. Heinrichs SC et al (1996) Corticotropin-releasing factor-binding protein ligand inhibitor blunts excessive weight gain in genetically obese Zucker rats and rats during nicotine withdrawal. Proc Natl Acad Sci USA 93:15475–15480
69. Karlsson S, Ahrén B (1988) Effects of corticotropin-releasing hormone on insulin and glucagon secretion in mice. Acta Endocrinol 117:87–92
70. Spina M et al (1996) Appetite-suppressing effects of urocortin, a CRF-related neuropeptide. Science 273:1561–1564
71. Soma KK, Scotti M-AL, Newman AEM, Charlier TD, Demas GE (2008) Novel mechanisms for neuroendocrine regulation of aggression. Front Neuroendocrinol 29:476–489
72. Cleary MP, Zisk JF (1986) Anti-obesity effect of two different levels of dehydroepiandrosterone in lean and obese middle-aged female Zucker rats. Int J Obes 10:193–204
73. Han DH, Hansen PA, Chen MM, Holloszy JO (1998) DHEA treatment reduces fat accumulation and protects against insulin resistance in male rats. J Gerontol A Biol Sci Med Sci 53:B19–24
74. Villareal DT, Holloszy JO (2004) Effect of DHEA on abdominal fat and insulin action in elderly women and men. J Am Med Assoc 292:2243–2248
75. Lopez-Marure R, Huesca-Gomez C, MdJ I-S, Zentella A, Perez-Mendez O (2007) Dehydroepiandrosterone delays LDL oxidation in vitro and attenuates several oxLDL-induced inflammatory responses in endothelial cells. Inflam Allergy Drug Targ (Formerly 'Curr Drug Targ Inflam Allergy) 6:174–182
76. Mills SJ, Ashworth JJ, Gilliver SC, Hardman MJ, Ashcroft GS (2005) The sex steroid precursor DHEA accelerates cutaneous wound healing via the estrogen receptors. J Invest Dermatol 125:1053–1062
77. Ferrari PF, van Erp AMM, Tornatzky W, Miczek KA (2003) Accumbal dopamine and serotonin in anticipation of the next aggressive episode in rats. Eur J Neurosci 17:371–378
78. Seo D, Patrick CJ, Kennealy PJ (2008) Role of serotonin and dopamine system interactions in the neurobiology of impulsive aggression and its comorbidity with other clinical disorders. Aggress Violent Behav 13:383–395
79. Murzi E et al (1996) Diabetes decreases limbic extracellular dopamine in rats. Neurosci Lett 202: 141–144
80. Cincotta AH, Tozzo E, Scislowski PW (1997) Bromocriptine/SKF38393 treatment ameliorates obesity and associated metabolic dysfunctions in obese (ob/ob) mice. Life Sci 61:951–956
81. Nexø E, Hollenberg MD, Bing J (1981) Aggressive behaviour in mice provokes a marked increase in both plasma epidermal growth factor and renin. Acta Physiol Scand 111:367–371
82. Sanchez O, Viladrich M, Ramirez I, Soley M (2007) Liver injury after an aggressive encounter in male mice. Am J Physiol Regulat Integrat Comp Physiol 293:R1908–R1916
83. Brand SJ et al (2002) Pharmacological treatment of chronic diabetes by stimulating pancreatic β-cell regeneration with systemic co-administration of EGF and gastrin. Pharmacol Toxicol 91:414–420
84. Giovanni Artini P et al (2007) Growth factors and folliculogenesis in polycystic ovary patients. Expert Rev Endocrinol Metabol 2:215–223
85. Bouwens L, Rooman I (2005) Regulation of Pancreatic B-Cell Mass. Physiol Rev 85:1255–1270
86. Gogg S, Smith U (2002) Epidermal growth factor (EGF) and TGF α mimic the effects of insulin in human fat cells and augment down-stream signaling in insulin resistance. J Biol Chem 277:36046–36051
87. Tordjman S et al (2003) Aggression and the three opioid families (endorphins, enkephalins, and dynorphins) in mice. Behav Genet 33:529–536
88. Kotegawa T, Abe T, Tsutsui K (1997) Inhibitory role of opioid peptides in the regulation of aggressive and sexual behaviors in male Japanese quails. J Exp Zool 277:146–154
89. Giugliano D et al (1992) Physiological elevations of plasma β-endorphin alter glucose metabolism in obese, but not normal-weight, subjects. Metabo 41:184–190
90. Margules DL, Moisset B, Lewis MJ, Shibuya H, Pert CB (1978) β-Endorphin is associated with overeating in genetically obese mice (ob/ob) and rats (fa/fa). Science 202:988–991
91. Su CF et al (2005) Mediation of β-endorphin in exercise-induced improvement in insulin resistance in obese Zucker rats. Diabetes Metab Res Rev 21:175–182
92. Kurihara Y et al (2000) Role of endothelin-1 in stress response in the central nervous system. Am J Physiol Regul Integr Comp Physiol 279:R515–521

93. Ferri C et al (1995) Insulin stimulates endothelin-1 secretion from human endothelial cells and modulates its circulating levels in vivo. J Clin Endocrinol Metab 80:829–835
94. Hu RM, Levin ER, Pedram A, Frank HJ (1993) Insulin stimulates production and secretion of endothelin from bovine endothelial cells. Diabetes 42:351–358
95. Ottosson-Seeberger A, Lundberg JM, Alvestrand A, Ahlborg G (1997) Exogenous endothelin-1 causes peripheral insulin resistance in healthy humans. Acta Physiol Scand 161:211–220
96. Wilkes JJ, Hevener A, Olefsky J (2003) Chronic endothelin-1 treatment leads to insulin resistance in vivo. Diabetes 52:1904–1909
97. Juan C-C et al (1996) Endothelin-1 induces insulin resistance in conscious rats. Biochem Biophys Res Commun 227:694–699
98. Diamanti-Kandarakis E, Spina G, Kouli C, Migdalis I (2001) Increased endothelin-1 levels in women with polycystic ovary syndrome and the beneficial effect of metformin therapy. J Clin Endocrinol Metab 86:4666–4673
99. Wenner MM, Taylor HS, Stachenfeld NS (2011) Endothelin B receptor contribution to peripheral microvascular function in women with polycystic ovary syndrome. J Physiol (Lond) 589:4671–4679
100. Sørensen DB et al (2005) PNMT transgenic mice have an aggressive phenotype. Horm Metab Res 37:159–163
101. Lee ZSK et al (2001) Urinary epinephrine and norepinephrine interrelations with obesity, insulin, and the metabolic syndrome in Hong Kong Chinese. Metabolism 50:135–143
102. Deibert DC, DeFronzo RA (1980) Epinephrine-induced insulin resistance in man. J Clin Invest 65:717–721
103. Ogawa S et al (2000) Abolition of male sexual behaviors in mice lacking estrogen receptors α and β (α β ERKO). Proc Natl Acad Sci USA 97:14737–14741
104. Ogawa S, Lubahn DB, Korach KS, Pfaff DW (1997) Behavioral effects of estrogen receptor gene disruption in male mice. Proc Natl Acad Sci USA 94:1476–1481
105. Ogawa S et al (1998) Roles of estrogen receptor-α gene expression in reproduction-related behaviors in female mice. Endocrinology 139:5070–5081
106. Carr MC (2003) The emergence of the metabolic syndrome with menopause. J Clin Endocrinol Metabol 88:2404–2411
107. Zhang Y et al (2002) The effect of estrogen use on levels of glucose and insulin and the risk of type 2 diabetes in American Indian postmenopausal women. Diabetes Care 25:500–504
108. Godsland IF et al (1992) Insulin resistance, secretion, and metabolism in users of oral contraceptives. J Clin Endocrinol Metab 74:64–70
109. Gallagher CJ et al (2007) Association of the estrogen receptor-α gene with the metabolic syndrome and its component traits in African–American families: the insulin resistance atherosclerosis family study. Diabetes 56:2135–2141
110. Gallagher CJ et al (2007) Investigation of the estrogen receptor-α gene with type 2 diabetes and/or nephropathy in African–American and European–American populations. Diabetes 56:675–684
111. Musatov S (2007) *et al* Silencing of estrogen receptor α in the ventromedial nucleus of hypothalamus leads to metabolic syndrome. Proc Natl Acad Sci USA 104:2501–2506
112. Scheffel A (1996) Serum cholesterol, triglycerides, HDL and LDL in aggressive elderly patients with dementia. Psychiatr Pol 30:159–170
113. Koyama K, Chen G, Lee Y, Unger RH (1997) Tissue triglycerides, insulin resistance, and insulin production: implications for hyperinsulinemia of obesity. Am J Physiol Endocrinol Metab 273:E708–E713
114. Norton WHJ et al (2011) Modulation of Fgfr1a signaling in zebrafish reveals a genetic basis for the aggression–boldness syndrome. J Neurosci 31:13796–13807
115. Coskun T et al (2008) Fibroblast growth Factor 21 corrects obesity in mice. Endocrinology 149:6018–6027
116. Jiao H et al (2011) Genetic association and gene expression analysis identify FGFR1 as a new susceptibility gene for human obesity. J Clin Endocrinol Metab 96:E962–E966
117. Miyakawa T, Yagi T, Takao K, Niki H (2001) Differential effect of Fyn tyrosine kinase deletion on offensive and defensive aggression. Behav Brain Res 122:51–56
118. Jung MY, Kim BS, Kim YJ, Koh IS, Chung J-H (2008) Assessment of relationship between Fyn-related kinase gene polymorphisms and overweight/obesity in Korean population. Korean J Physiol Pharmacol 12:83
119. Nadler ST et al (2000) The expression of adipogenic genes is decreased in obesity and diabetes mellitus. Proc Natl Acad Sci USA 97:11371–11376
120. Stork O et al (2000) Postnatal development of a GABA deficit and disturbance of neural functions in mice lacking GAD65. Brain Res 865:45–58
121. Coscina DV, Nobrega JN (1984) Anorectic potency of inhibiting GABA transaminase in brain: studies of hypothalamic, dietary and genetic obesities. Int J Obes 8(Suppl 1):191–200
122. Fisler JS, Shimizu H, Bray GA (1989) Brain 3-hydroxybutyrate, glutamate, and GABA in a rat model of dietary obesity. Physiol Behav 45:571–577
123. Tong Q, Ye C-P, Jones JE, Elmquist JK, Lowell BB (2008) Synaptic release of GABA by AgRP neurons is required for normal regulation of energy balance. Nat Neurosci 11:998–1000
124. D'Adamo P (2002) Deletion of the mental retardation gene Gdi1 impairs associative memory and alters social behavior in mice. Hum Mol Genet 11:2567–2580

125. Chinni SR, Brenz M, Shisheva A (1998) Modulation of GDP dissociation inhibitor protein membrane retention by the cellular redox state in adipocytes. Exp Cell Res 242:373–380
126. Kowluru A, Veluthakal R (2005) Rho guanosine-diphosphate–dissociation inhibitor plays a negative modulatory role in glucose-stimulated insulin secretion. Diabetes 54:3523–3529
127. Campbell PK et al (2002) Mutation of a novel gene results in abnormal development of spermatid flagella, loss of intermale aggression and reduced body fat in mice. Genetics 162:307–320
128. Matte AC (1981) Growth hormone and isolation-induced aggression in wild male mice. Pharmacol Biochem Behav 14(Suppl 1):85–87
129. Weltman A, Weltman JY, Veldhuis JD, Hartman ML (2001) Body composition, physical exercise, growth hormone and obesity. Eat Weight Disord 6:28–37
130. Scacchi M, Pincelli AI, Cavagnini F (1999) Growth hormone in obesity. Int J Obes 23:260–271
131. Rizza RA, Mandarino LJ, Gerich JE (1982) Effects of growth hormone on insulin action in man. Mechanisms of insulin resistance, impaired suppression of glucose production, and impaired stimulation of glucose utilization. Diabetes 31:663–669
132. Jose Augusto SB et al (2002) Familial isolated growth hormone deficiency is associated with increased systolic blood pressure, central obesity and dislipidemia. J Clin Endocrinol Metab 87(5): 2018–2023
133. Krag MB et al (2007) Growth hormone-induced insulin resistance is associated with increased intramyocellular triglyceride content but unaltered VLDL-triglyceride kinetics. Am J Physiol Endocrinol Metab 292:E920–E927
134. Atmaca M et al (2006) Serum ghrelin and cholesterol values in suicide attempters. Neuropsychobiology 54:59–63
135. Pöykkö SM et al (2003) Low plasma ghrelin is associated with insulin resistance, hypertension, and the prevalence of type 2 diabetes. Diabetes 52:2546–2553
136. Tschöp M et al (2001) Circulating ghrelin levels are decreased in human obesity. Diabetes 50:707–709
137. McBurnett K, Lahey BB, Rathouz PJ, Loeber R (2000) Low salivary cortisol and persistent aggression in boys referred for disruptive behavior. Arch Gen Psychiatry 57:38–43
138. Sever'ianova LA (1981) Role of ACTH and corticosteroids in the aggressive-defensive behavior of rats]. Fiziol Zh SSSR Im I M Sechenova 67:1117–1122
139. Andrews RC, Walker BR (1999) Glucocorticoids and insulin resistance: old hormones, new targets. Clin Sci 96:513–523
140. Christiansen JJ et al (2007) Effects of cortisol on carbohydrate, lipid, and protein metabolism: studies of acute cortisol withdrawal in adrenocortical failure. J Clin Endocrinol Metab 92:3553–3559
141. Macfarlane DP, Forbes S, Walker BR (2008) Glucocorticoids and fatty acid metabolism in humans: fuelling fat redistribution in the metabolic syndrome. J Endocrinol 197:189–204
142. Lin RCY, Wang XL, Dalziel B, Caterson ID, Morris BJ (2003) Association of obesity, but not diabetes or hypertension, with glucocorticoid receptor N363S variant. Obesity 11:802–808
143. Andrade ML, Benton D, Brain PF, Ramirez JM, Walmsley SV (1988) A reexamination of the hypoglycemia-aggression hypothesis in laboratory mice. Int J Neurosci 41:179–186
144. Benton D (1988) Hypoglycemia and aggression: a review. Int J Neurosci 41:163–168
145. Bolton R (1976) Hostility in fantasy: A further test of the hypoglycemia-aggression hypothesis. Aggressive Behav 2:257–274
146. De Nys HM et al (2010) Vaccination against GnRH may suppress aggressive behaviour and musth in African elephant (Loxodontaafricana) bulls–a pilot study. J S Afr Vet Assoc 81:8–15
147. Penfold LM et al (2002) Case studies in antelope aggression control using a GnRH agonist. Zoo Biol 21:435–448
148. Millesi E, Hoffmann IE, Steurer S, Metwaly M, Dittami JP (2002) Vernal changes in the behavioral and endocrine responses to GnRH application in male European ground squirrels. Horm Behav 41:51–58
149. Kim HH, DiVall SA, Deneau RM, Wolfe A (2005) Insulin regulation of GnRH gene expression through MAP kinase signaling pathways. Mol Cell Endocrinol 242:42–49
150. Tascilar ME, Bilir P, Akinci A, Kose K, Akçora D, Inceoglu D, Fitöz SO (2011) The effect of gonadotropin-releasing hormone analog treatment (leuprolide) on body fat distribution in idiopathic central precocious puberty. Turk J Pediatr 53:27–33
151. Yanai K (1998) Behavioural characterization and amounts of brain monoamines and their metabolites in mice lacking histamine H1 receptors. Neuroscience 87:479–487
152. Yoshimatsu H, Chiba S, Tajima D, Akehi Y, Sakata T (2002) Histidine suppresses food intake through its conversion into neuronal histamine. Exp Biol Med 227:63–68
153. Malmlof K et al (2005) Influence of a selective histamine H3 receptor antagonist on hypothalamic neural activity, food intake and body weight. Int J Obes Relat Metab Disord 29:1402–1412
154. Masaki T et al (2004) Involvement of hypothalamic histamine H1 receptor in the regulation of feeding rhythm and obesity. Diabetes 53:2250–2260
155. Vera Cruz EM, Brown CL (2007) The influence of social status on the rate of growth, eye color pattern and insulin-like growth factor-I gene expression in Nile tilapia, *Oreochromisniloticus*. Horm Behav 51:611–619
156. Bartos L, Reyes E, Schams D, Bubenik G, Lobos A (1998) Rank dependent seasonal levels of IGF-1, cortisol and reproductive hormones in male pudu

(Pudupuda). Comp Biochem Physiol A Mol Integr Physiol 120:373–378
157. Vickers MH, Ikenasio BA, Breier BH (2001) IGF-I treatment reduces hyperphagia, obesity, and hypertension in metabolic disorders induced by fetal programming. Endocrinology 142:3964–3973
158. O'Connell T, Clemmons DR (2002) IGF-I/IGF-binding protein-3 combination improves insulin resistance by GH-dependent and independent mechanisms. J Clin Endocrinol Metab 87:4356–4360
159. Yakar S et al (2001) Liver-specific igf-1 gene deletion leads to muscle insulin insensitivity. Diabetes 50:1110–1118
160. Alleva E et al (1998) Behavioural characterization of interleukin-6 overexpressing or deficient mice during agonistic encounters. Eur J Neurosci 10:3664–3672
161. Re S, Zanoletti M, Emanuele E (2008) Association of inflammatory markers elevation with aggressive behavior in domestic dogs. J Ethology 27:31–33
162. Wallenius V et al (2002) Interleukin-6-deficient mice develop mature-onset obesity. Nature Med 8:75–79
163. Roytblat L et al (2000) Raised interleukin-6 levels in obese patients. Obesity 8:673–675
164. Vgontzas AN et al (2006) Plasma interleukin 6 levels are elevated in polycystic ovary syndrome independently of obesity or sleep apnea. Metabolism 55:1076–1082
165. Rotter V, Nagaev I, Smith U (2003) Interleukin-6 (IL-6) induces insulin resistance in 3T3-L1 adipocytes and is, like IL-8 and tumor necrosis factor-α, overexpressed in human fat cells from insulin-resistant subjects. J Biol Chem 278:45777–45784
166. Castellano JM et al (2006) Ontogeny and mechanisms of action for the stimulatory effect of kisspeptin on gonadotropin-releasing hormone system of the rat. Mol Cell Endocrinol 257–258:75–83
167. Grone BP, Maruska KP, Korzan WJ, Fernald RD (2010) Social status regulates kisspeptin receptor mRNA in the brain of *Astatotilapia burtoni*. Gen Comp Endocrinol 169:98–107
168. Hauge-Evans AC et al (2006) A role for kisspeptin in islet function. Diabetologia 49:2131–2135
169. Kaplan JR, Manuck SB, Shively C (1991) The effects of fat and cholesterol on social behavior in monkeys. Psychosom Med 53:634–642
170. Klein KO (1998) Effect of obesity on estradiol level, and its relationship to leptin, bone maturation, and bone mineral density in children. J Clin Endocrinol Metab 83:3469–3475
171. Widjaja A et al (1997) UKPDS 20: plasma leptin, obesity, and plasma insulin in type 2 diabetic subjects. J Clin Endocrinol Metab 82:654–657
172. Brunner H, Nelen M, Breakefield X, Ropers H, van Oost B (1993) Abnormal behavior associated with a point mutation in the structural gene for monoamine oxidase A. Science 262:578–580
173. Cases O et al (1995) Aggressive behavior and altered amounts of brain serotonin and norepinephrine in mice lacking MAOA. Science 268:1763–1766
174. Nilsson SE, Tryding N, Tufvesson G (2009) serum monoamine oxidase (mao) in diabetes mellitus and some other internal diseases. Acta Med Scand 184:105–108
175. Jasnow AM, Huhman KL, Bartness TJ, Demas GE (2002) Short days and exogenous melatonin increase aggression of male Syrian hamsters (Mesocricetusauratus). Horm Behav 42:13–20
176. Nishida S, Segawa T, Murai I, Nakagawa S (2002) Long-term melatonin administration reduces hyperinsulinemia and improves the altered fatty-acid compositions in type 2 diabetic rats via the restoration of Delta-5 desaturase activity. J Pineal Res 32:26–33
177. Prunet-Marcassus B et al (2003) Melatonin reduces body weight gain in sprague dawley rats with diet-induced obesity. Endocrinology 144:5347–5352
178. Bodine SC (2006) mTOR signaling and the molecular adaptation to resistance exercise. Med Sci Sports Exerc 38:1950–1957
179. Dreyer HC et al (2008) Leucine-enriched essential amino acid and carbohydrate ingestion following resistance exercise enhances mTOR signaling and protein synthesis in human muscle. Am J Physiol Endocrinol Metab 294:E392–E400
180. Bolster DR et al (2003) Immediate response of mammalian target of rapamycin (mTOR)-mediated signalling following acute resistance exercise in rat skeletal muscle. J Physiol 553:213–220
181. Wu Y, Bauman WA, Blitzer RD, Cardozo C (2010) Testosterone-induced hypertrophy of L6 myoblasts is dependent upon Erk and mTOR. Biochem Biophys Res Commun 400:679–683
182. Altamirano F et al (2009) Testosterone induces cardiomyocyte hypertrophy through mammalian target of rapamycin complex 1 pathway. J Endocrinol 202:299–307
183. Zick Y (2005) Ser/Thr phosphorylation of IRS proteins: a molecular basis for insulin resistance. Sci STKE 2005:pe4
184. Khamzina L, Veilleux A, Bergeron S, Marette A (2005) Increased activation of the mammalian target of rapamycin pathway in liver and skeletal muscle of obese rats: possible involvement in obesity-linked insulin resistance. Endocrinology 146:1473–1481
185. Kovacheva EL, SinhaHikim AP, Shen R, Sinha I, Sinha-Hikim I (2010) Testosterone supplementation reverses sarcopenia in aging through regulation of myostatin, c-Jun NH2-terminal kinase, notch, and Akt signaling pathways. Endocrinology 151:628–638
186. Hittel DS, Berggren JR, Shearer J, Boyle K, Houmard JA (2009) Increased secretion and expression of myostatin in skeletal muscle from extremely obese women. Diabetes 58:30–38
187. McPherron AC, Lee S-J (2002) Suppression of body fat accumulation in myostatin-deficient mice. J Clin Invest 109:595–601
188. Stork O, Welzl H, Cremer H, Schachner M (1997) Increased intermale aggression and neuroendocrine

response in mice deficient for the neural cell adhesion molecule (NCAM). Eur J Neurosci 9:1117–1125
189. Yang HJ et al (2011) A novel role for neural cell adhesion molecule in modulating insulin signaling and adipocyte differentiation of mouse mesenchymal stem cells. J Cell Sci 124:2552–2560
190. Olofsson CS et al (2009) Impaired insulin exocytosis in neural cell adhesion molecule–/– mice due to defective reorganization of the submembrane F-actin network. Endocrinology 150:3067–3075
191. Chiavegatto S (2006) Using mouse models to unravel aggressive behavior. In: Canli T (ed) Biology of personality and individual differences. The Guilford Press, New York, NY, pp 385–406
192. Davidson E et al (2009) The roles of streptozotocin neurotoxicity and neutral endopeptidase in murine experimental diabetic neuropathy. Exp Diabetes Res 2009:431980
193. Emeson RB, Morabito MV (2005) Food fight: the NPY-serotonin link between aggression and feeding behavior. Sci STKE 2005:pe12
194. Wang Q et al (1997) Interactions between leptin and hypothalamic neuropeptide Y neurons in the control of food intake and energy homeostasis in the rat. Diabetes 46:335–341
195. Mashiko S et al (2003) Characterization of neuropeptide Y (NPY) Y5 receptor-mediated obesity in mice: chronic intracerebroventricular infusion of d-Trp34NPY. Endocrinology 144:1793–1801
196. Palmiter RD, Erickson JC, Hollopeter G, Baraban SC, Schwartz MW (1998) Life without neuropeptide Y. Recent Prog Horm Res 53:163–199
197. Erickson JC, Hollopeter G, Palmiter RD (1996) Attenuation of the obesity syndrome of ob/ob mice by the loss of neuropeptide Y. Science 274:1704–1707
198. Lakshmanan J (1986) Aggressive behavior in adult male mice elevates serum nerve growth factor levels. Am J Physiol 250:E386–E392
199. Alleva E, Aloe L, Cirulli F, Della Seta D, Tirassa P (1996) Serum NGF levels increase during lactation and following maternal aggression in mice. Physiol Behav 59:461–466
200. Bulló M, Peeraully MR, Trayhurn P, Folch J, Salas-Salvadó J (2007) Circulating nerve growth factor levels in relation to obesity and the metabolic syndrome in women. Eur J Endocrinol 157:303–310
201. Chaldakov GN et al (2004) Neurotrophin presence in human coronary atherosclerosis and metabolic syndrome: a role for NGF and BDNF in cardiovascular disease? Prog Brain Res 146:279–289
202. Apfel SC et al (1998) Recombinant human nerve growth factor in the treatment of diabetic polyneuropathy. Neurology 51:695–702
203. Nelson RJ et al (1995) Behavioural abnormalities in male mice lacking neuronal nitric oxide synthase. Nature 378:383–386
204. Demas GE et al (1999) Elimination of aggressive behavior in male mice lacking endothelial nitric oxide synthase. J Neurosci 19:RC3
205. Gruber H-J et al (2008) Obesity reduces the bioavailability of nitric oxide in juveniles. Int J Obes 32:826–831
206. Duplain H et al (2001) Insulin resistance, hyperlipidemia, and hypertension in mice lacking endothelial nitric oxide synthase. Circulation 104:342–345
207. Monti LD et al (2003) Endothelial nitric oxide synthase polymorphisms are associated with type 2 diabetes and the insulin resistance syndrome. Diabetes 52:1270–1275
208. Oury F, Sumara G, Sumara O, Ferron M, Chang H, Smith CE, Hermo L, Suarez S, Roth BL, Ducy P, Karsenty G (2011) Endocrine regulation of male fertility by the skeleton. Cell 144:796–809
209. Karsenty G, Ferron M (2012) The contribution of bone to whole-organism physiology. Nature 481:314–320
210. Ferron M, Hinoi E, Karsenty G, Ducy P (2008) Osteocalcin differentially regulates β cell and adipocyte gene expression and affects the development of metabolic diseases in wild-type mice. Proc Natl Acad Sci USA 105:5266–5270
211. Saleem U, Mosley TH, Kullo IJ (2010) Serum osteocalcin is associated with measures of insulin resistance, adipokine levels, and the presence of metabolic syndrome. Arteriosclerosis Thrombosis Vasc Biol 30:1474–1478
212. Winslow JT et al (2000) Infant vocalization, adult aggression, and fear behavior of an oxytocin null mutant mouse. Horm Behav 37:145–155
213. DeVries AC, Young WS III, Nelson RJ (2003) Reduced aggressive behaviour in mice with targeted disruption of the oxytocin gene. J Neuroendocrinol 9:363–368
214. Holt-Lunstad J, Birmingham WA, Light KC (2008) Influence of a "warm touch" support enhancement intervention among married couples on ambulatory blood pressure, oxytocin, α amylase, and cortisol. Psychosom Med 70:976–985
215. Light KC, Grewen KM, Amico JA (2005) More frequent partner hugs and higher oxytocin levels are linked to lower blood pressure and heart rate in premenopausal women. Biol Psychol 69:5–21
216. Stock S, Bremme K, Uvnas-Moberg K (1993) Is oxytocin involved in the deterioration of glucose tolerance in gestational diabetes? Gynacol Obstet Invest 36:81–86
217. Altszuler N, Hampshire J (1981) Oxytocin infusion increases plasma insulin and glucagon levels and glucose production and uptake in the normal dog. Diabetes 30:112–114
218. Camerino C (2009) Low sympathetic tone and obese phenotype in oxytocin-deficient mice. Obesity (Silver Spring) 17:980–984
219. Augert G, Exton JH (1988) Insulin and oxytocin effects on phosphoinositide metabolism in adipocytes. J Biol Chem 263:3600–3609
220. Braun T, Hechter O, Rudinger J (1969) "Insulin-like" action of oxytocin: evidence for separate oxytocin-sensitive and insulin-sensitive systems in fat cells. Endocrinology 85:1092–1096

221. Mirsky IA, Perisutti G (1961) The insulin-like action of oxytocin on adipose tissue. Biochim Biophys Acta 50:603–604
222. Tolson KP et al (2010) Postnatal Sim1 deficiency causes hyperphagic obesity and reduced Mc4r and oxytocin expression. J Neurosci 30:3803–3812
223. Takayanagi Y et al (2008) Oxytocin receptor-deficient mice developed late-onset obesity. Neuroreport 19:951–955
224. Kublaoui BM, Gemelli T, Tolson KP, Wang Y, Zinn AR (2008) Oxytocin deficiency mediates hyperphagic obesity of Sim1 haploinsufficient mice. Mol Endocrinol 22:1723–1734
225. Stock S, Granström L, Backman L, Matthiesen AS, Uvnäs-Moberg K (1989) Elevated plasma levels of oxytocin in obese subjects before and after gastric banding. Int J Obes 13:213–222
226. Novak CM et al (2009) Endurance capacity, not body size, determines physical activity levels: role of skeletal muscle PEPCK. PLoS One 4:e5869
227. Hanson RW, Hakimi P (2008) Born to run; the story of the PEPCK-Cmus mouse. Biochimie 90:838–842
228. Valera A, Pujol A, Pelegrin M, Bosch F (1994) Transgenic mice overexpressing phosphoenolpyruvate-carboxykinase develop non-insulin-dependent diabetes mellitus. Proc Natl Acad Sci USA 91:9151–9154
229. Sun Y et al (2002) Phosphoenolpyruvate carboxykinase overexpression selectively attenuates insulin signaling and hepatic insulin sensitivity in transgenic mice. J Biol Chem 277:23301–23307
230. Franckhauser S et al (2002) Increased fatty acid re-esterification by PEPCK overexpression in adipose tissue leads to obesity without insulin resistance. Diabetes 51:624–630
231. Cowey SL et al (2005) Abdominal obesity, insulin resistance, and colon carcinogenesis are increased in mutant mice lacking gastrin gene expression. Cancer 103:2643–2653
232. Kellner R, Buckman MT, Fava M, Fava GA, Mastrogiacomo I (1984) Prolactin, aggression and hostility: a discussion of recent studies. Psychiatr Dev 2:131–138
233. Gleason P, Michael S, Christian J (1981) Prolactin-induced aggression in female. Behav Neural Biol 33:243–248
234. Grimaldi F et al (1990) Changes in secretion of prolactin in obesity. Minerva Endocrinol 15:267–271
235. Oliveira-dos-Santos AJ et al (2000) Regulation of T cell activation, anxiety, and male aggression by RGS2. Proc Natl Acad Sci USA 97:12272–12277
236. Sartori M et al (2008) RGS2 C1114G polymorphism and body weight gain in hypertensive patients. Metabolism 57:421–427
237. Heximer SP et al (2003) Hypertension and prolonged vasoconstrictor signaling in RGS2-deficient mice. J Clin Invest 111:445–452
238. Cleare AJ, Bond AJ (1997) Does central serotonergic function correlate inversely with aggression? A study using d-fenfluramine in healthy subjects. Psychiatr Res 69:89–95
239. Linnoila VM, Virkkunen M (1992) Aggression, suicidality, and serotonin. J Clin Psychiatr 53(suppl):46–51
240. Nonogaki K, Strack AM, Dallman MF, Tecott LH (1998) Leptin-independent hyperphagia and type 2 diabetes in mice with a mutatedserotonin 5-HT2C receptor gene. Nature Med 4:1152–1156
241. Luo S, Luo J, Cincotta AH (1999) Chronic ventromedial hypothalamic infusion of norepinephrine and serotonin promotes insulin resistance and glucose intolerance. Neuroendocrinology 70:460–465
242. Rattigan S, Clark MG, Barrett EJ (1999) Acute vasoconstriction-induced insulin resistance in rat muscle in vivo. Diabetes 48:564–569
243. Singh RB, Pella D, Mechirova V, Otsuka K (2004) Can brain dysfunction be a predisposing factor for metabolic syndrome? Biomed Pharmacother 58(Suppl 1):S56–68
244. Zhao L et al (2008) Central nervous system-specific knockout of steroidogenic factor 1 results in increased anxiety-like behavior. Mol Endocrinol 22:1403–1415
245. Majdic G et al (2002) Knockout mice lacking steroidogenic factor 1 are a novel genetic model of hypothalamic obesity. Endocrinology 143:607–614
246. Ayala JE et al (2007) Chronic treatment with sildenafil improves energy balance and insulin action in high fat–fed conscious mice. Diabetes 56: 1025–1033
247. Milman HA, Arnold SB (2002) Neurologic, psychological, and aggressive disturbances with sildenafil. Ann Pharmacother 36:1129–1134
248. Hotchkiss AK et al (2005) Aggressive behavior increases after termination of chronic sildenafil treatment in mice. Physiol Behav 83:683–688
249. Haller J et al (2002) Behavioral responses to social stress in noradrenaline transporter knockout mice: effects on social behavior and depression. Brain Res Bull 58:279–284
250. Ono K et al (2003) Epidemiological evidence of an association between SLC6A2 gene polymorphism and hypertension. Hypertens Res 26:685–689
251. Spraggs CF et al (2005) Pharmacogenetics and obesity: common gene variants influence weight loss response of the norepinephrine/dopamine transporter inhibitor GW320659 in obese subjects. Pharmacogenet Genomics 15:883–889
252. Felipe CD et al (1998) Altered nociception, analgesia and aggression in mice lacking the receptor for substance P. Nature 392:394–397
253. Gregg TR, Siegel A (2001) Brain structures and neurotransmitters regulating aggression in cats: implications for human aggression. Prog Neuropsychopharmacol Biol Psychiatry 25:91–140
254. Karagiannides I et al (2008) Substance P as a novel anti-obesity target. Gastroenterology 134:747–755
255. Fu XW, Sun AM (1989) Stimulative effect of substance P on insulin secretion from isolated rat islets under normobaric oxygen incubation. Zhongguo Yao Li Xue Bao 10:69–73

256. van Bokhoven I et al (2006) Salivary testosterone and aggression, delinquency, and social dominance in a population-based longitudinal study of adolescent males. Horm Behav 50:118–125
257. Albert DJ, Jonik RH, Walsh ML (1992) Hormone-dependent aggression in male and female rats: experiential, hormonal, and neural foundations. Neurosci Biobehav Rev 16:177–192
258. HolmäNg A, BjöRntorp P (1992) The effects of testosterone on insulin sensitivity in male rats. Acta Physiol Scand 146:505–510
259. Stewart TM et al (2009) Associations between andrological measures, hormones and semen quality in fertile Australian men: inverse relationship between obesity and sperm output. Hum Reprod 24:1561–1568
260. Pitteloud N et al (2005) Relationship between testosterone levels, insulin sensitivity, and mitochondrial function in men. Diabetes Care 28:1636–1642
261. Polderman KH, Gooren LJ, Asscheman H, Bakker A, Heine RJ (1994) Induction of insulin resistance by androgens and estrogens. J Clin Endocrinol Metab 79:265–271
262. Holmang A, Svedberg J, Jennische E, Bjorntorp P (1990) Effects of testosterone on muscle insulin sensitivity and morphology in female rats. Am J Physiol Endocrinol Metab 259:E555–E560
263. Nestler JE et al (1998) Insulin stimulates testosterone biosynthesis by human thecal cells from women with polycystic ovary syndrome by activating its own receptor and using inositolglycan mediators as the signal transduction system. J Clin Endocrinol Metab 83:2001–2005
264. Suarez EC, Lewis JG, Kuhn C (2002) The relation of aggression, hostility, and anger to lipopolysaccharide-stimulated tumor necrosis factor (TNF)-α by blood monocytes from normal men. Brain Behav Immun 16:675–684
265. Uysal KT, Wiesbrock SM, Marino MW, Hotamisligil GS (1997) Protection from obesity-induced insulin resistance in mice lacking TNF-α function. Nature 389:610–614
266. Hofmann C et al (1994) Altered gene expression for tumor necrosis factor-α and its receptors during drug and dietary modulation of insulin resistance. Endocrinology 134:264–270
267. Hotamisligil GS, Shargill NS, Spiegelman BM (1993) Adipose expression of tumor necrosis factor-α: direct role in obesity-linked insulin resistance. Science 259:87–91
268. Sayin NC et al (2003) Elevated serum TNF-α levels in normal-weight women with polycystic ovaries or the polycystic ovary syndrome. J Reprod Med 48:165–170
269. Leypold BG (2002) Altered sexual and social behaviors in trp2 mutant mice. Proc Natl Acad Sci USA 99:6376–6381
270. Hu G, Oboukhova EA, Kumar S, Sturek M, Obukhov AG (2009) Canonical transient receptor potential channels expression is elevated in a porcine model of metabolic syndrome. Mol Endocrinol 23:689–699
271. Sohn J-W et al (2011) Serotonin 2C receptor activates a distinct population of arcuate pro-opiomelanocortin neurons via TRPC channels. Neuron 71:488–497
272. Qiu J, Fang Y, Rønnekleiv OK, Kelly MJ (2010) Leptin excites proopiomelanocortin neurons via activation of TRPC channels. J Neurosci 30:1560–1565
273. Qiu J, Fang Y, Bosch MA, Rønnekleiv OK, Kelly MJ (2011) Guinea pig kisspeptin neurons are depolarized by leptin via activation of TRPC channels. Endocrinology 152:1503–1514
274. Winberg S, Øverli Ø, Lepage O (2001) Suppression of aggression in rainbow trout (Oncorhynchusmykiss) by dietary L-tryptophan. J Exp Biol 204:3867–3876
275. Höglund E, Bakke MJ, Øverli Ø, Winberg S, Nilsson GE (2005) Suppression of aggressive behaviour in juvenile Atlantic cod (Gadusmorhua) by l-tryptophan supplementation. Aquaculture 249:525–531
276. DeNapoli JS, Dodman NH, Shuster L, Rand WM, Gross KL (2000) Effect of dietary protein content and tryptophan supplementation on dominance aggression, territorial aggression, and hyperactivity in dogs. J Am Vet Med Assoc 217:504–508
277. Chamberlain B, Ervin FR, Pihl RO, Young SN (1987) The effect of raising or lowering tryptophan levels on aggression in vervet monkeys. Pharmacol Biochem Behav 28:503–510
278. Koopmans SJ, Ruis M, Dekker R, Korte M (2009) Surplus dietary tryptophan inhibits stress hormone kinetics and induces insulin resistance in pigs. Physiol Behav 98:402–410
279. Hahm S et al (1999) Targeted Deletion of the Vgf Gene Indicates that the Encoded Secretory Peptide Precursor Plays a Novel Role in the Regulation of Energy Balance. Neuron 23:537–548
280. Watson E et al (2005) VGF ablation blocks the development of hyperinsulinemia and hyperglycemia in several mouse models of obesity. Endocrinology 146:5151–5163
281. Kalueff AV et al (2006) Behavioural anomalies in mice evoked by "Tokyo" disruption of the Vitamin D receptor gene. Neurosci Res 54:254–260
282. Kalueff A, Lou Y, Laaksi I, Tuohimaa P (2004) Increased grooming behavior in mice lacking vitamin D receptors. Physiol Behav 82:405–409
283. Holick MF (2007) Vitamin D deficiency. N Engl J Med 357:266–281
284. Borissova AM, Tankova T, Kirilov G, Dakovska L, Kovacheva R (2003) The effect of vitamin D3 on insulin secretion and peripheral insulin sensitivity in type 2 diabetic patients. Int J Clin Pract 57:258–261
285. Goto K, Miyauchi T, Homma S, Ohshima N (1992) Calcitonin gene-related peptide in the regulation of cardiac function. Ann N Y Acad Sci 657:194–203

Appendix II: Network Model of Type 2 Diabetes

Since the levels of a large number of molecules are altered in T2D and a large number of signals affect insulin sensitivity, glucose homeostasis, endothelial function, angiogenesis, oxidative state, and other pathophysiological processes of T2D, unless we develop a perspective of all the players and their interaction, our understanding of T2D will not be complete. Figure II.1 represents a systems view of how the different players from a variety of systems and organs interact with each other to form a complex network.

A simplest possible model based on the complex network consists of only three states for each player, namely, normal (0), upregulated (+1), and downregulated (−1). If the state of one player is changed, it changes the state of others through positive (denoted by clockwise solid lines with pointed arrows) or negative links (anticlockwise dotted lines with blunted ends). The model starts with a default state of zero or normal for all the players. User can perturb the system by changing one player, let us say food intake to 1 (upregulated), acutely (only for 1 time unit) or chronically (for a defined length or permanently) and observe the effect on the entire network. For the network dynamics, the state of each player at time t is examined and upregulation or downregulation signals are given to other players through the positive or negative links. For each player all such signals at a given time t are collected, and based upon the net positive or negative value, its state at time $t+1$ altered. The system is run over a long time to observe whether it ends in any stable state. By changing the nature and duration of perturbations the basins of attractions, if any, can be identified and their critical causal pathways determined.

The interdependence of links: For example, insulin action depends upon another variable, insulin sensitivity. This is incorporated by conditional operation of the link, i.e.

ins at state 1 will downregulate *gng* only if *ina* is 0 or 1 but not if *ina* is −1.

Although currently the model works in a qualitative mode with only three possible values for each variable, if reliable quantitative data are available, one or more of the links can be made quantitative. Making a complete quantitative model would need decades of research to generate necessary empirical data and large computational power.

Limitations of the model: The model as depicted here does not include all diabetic complications. It also does not include all food intake regulation molecules. Some players whose actions are overlapping are clubbed together, e.g., all pro-inflammatory adipokines are shown as a single player.

Integration of the model with empirical data: Empirically it is difficult to get data on the players in different tissues. Particularly for tissues like brain, it is currently impossible. However, all factors that circulate through blood can be sampled and characterized easily. Therefore, the inner circle in the figure is the possible interface to interaction with empirical data on humans. If the model predicts some stable basins of attraction, each of these basins should have their own

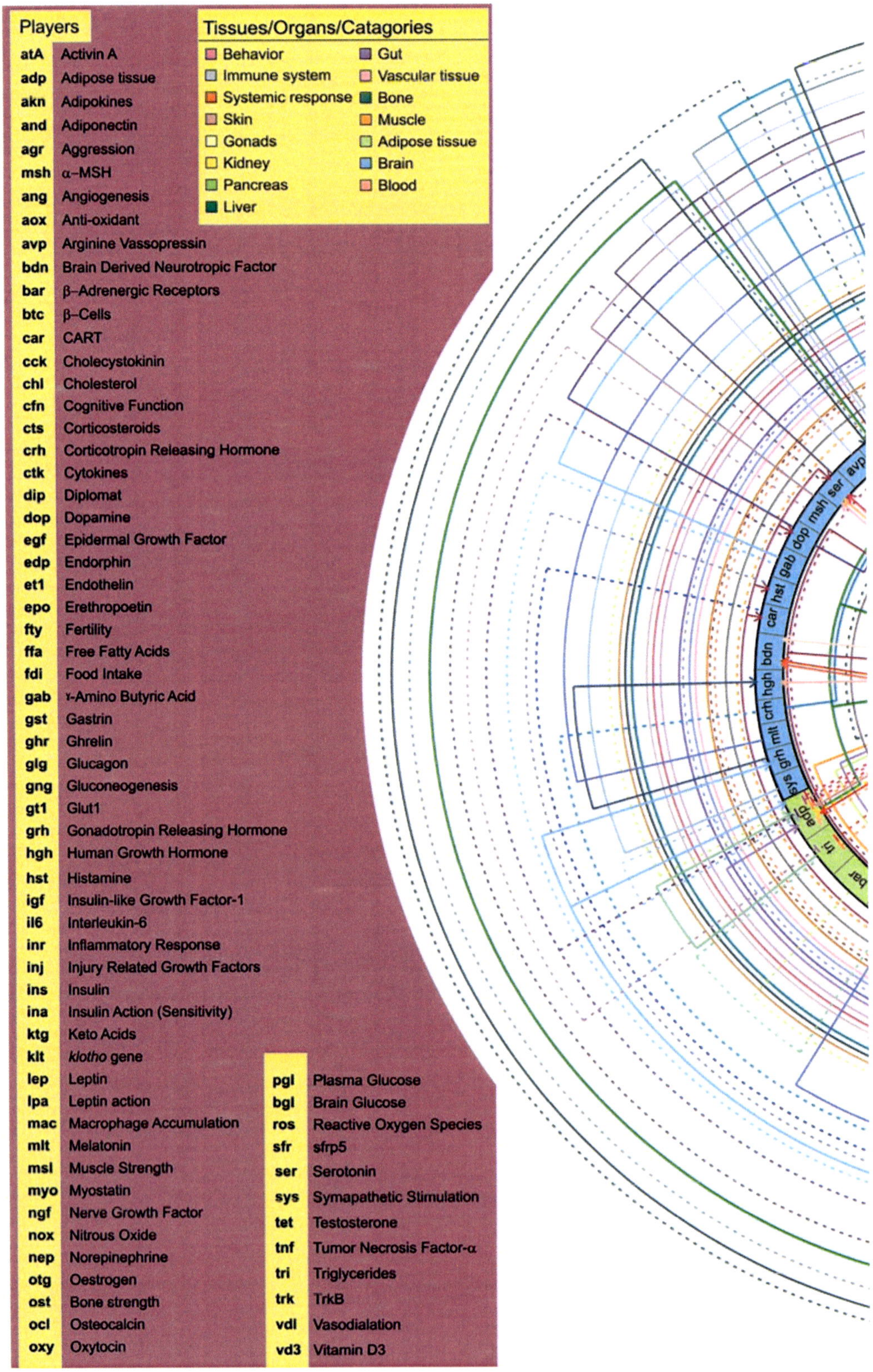

Fig. II.1 A whole body–brain behavior systems view of all known regulatory interactions involved in T2D and related conditions. The network was constructed by literature search on all factors that affect fat accumulation, insulin sensitivity, insulin response, and beta cell function. They include factors recognized by the classical theory as well as those revealed by recent research. Further all factors affecting these influences were searched. The figure depicts all the players classified organ

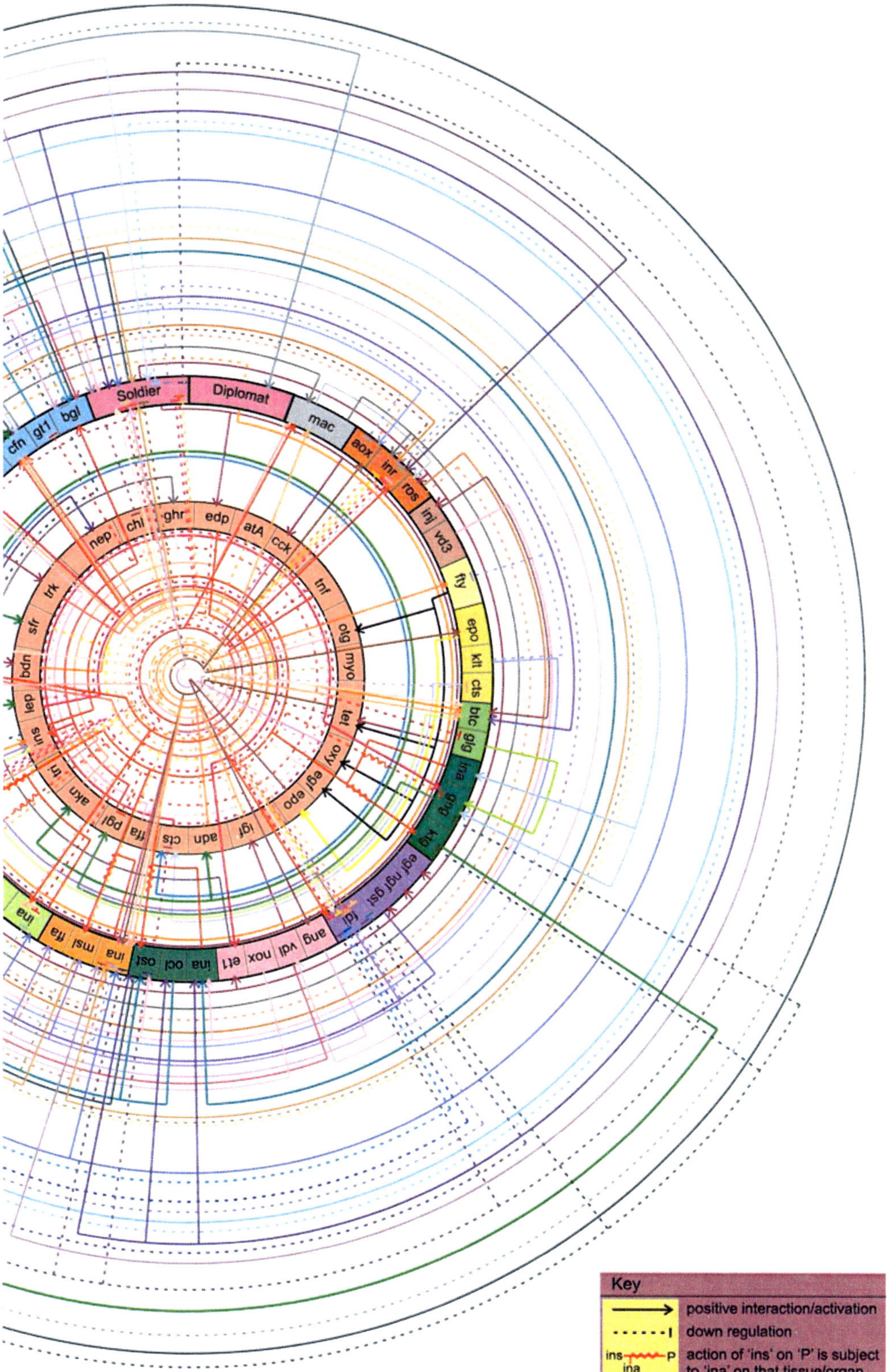

wise (*outer circle*, each organ denoted by a different color), and those that circulate in blood (*inner circle*). Upregulation interactions are shown by clockwise *solid lines* with *pointed arrows*. Downregulation is shown by anticlockwise *dotted lines* with *blunted ends*. The colors of the *arrows* correspond to the organ of their origin signatures detectable in the inner circle parameters. Data on these parameters in health as well as in T2D and related conditions could be subject to clusters in the multidimensional space, and clusters emerging from such analysis should correspond to the predicted signatures. Such correspondence can be used to validate the model.

The model is far from being complete as of today. We may have missed some links in our search and some links may not have been discovered as yet. Therefore the results should be treated as tentative. However, even at this stage the results are stable and clear. A complex system can easily enter a chaotic state. However, in spite of a large number of players and complexity of the network the system, as per the model stands today, never enters a chaotic state. This can be taken as an indication of a highly structured system evolved for switching between two or more stable states. It can also be an indication that a substantial part of the real network is included in the model. Some of the preliminary results relevant to the synthesis in this book are as follows:

1. Increasing food intake acutely or chronically leads to an insulin-resistant state along with change in behavior but without obesity. If food intake is increased for threshold duration or more, the state is maintained even if user control on food intake is removed. This seems to happen owing to the behavioral positive feedback loops.
2. If all the direct and indirect links of food intake to behavior are disabled, insulin-resistant state is not sustained and insulin sensitivity is reestablished as soon as user control on food intake is removed.
3. Obesity does not stabilize without independently downregulating adiponectin even if high food intake is sustained.
4. Suppression of adiponectin alone is able to induce obesity, insulin resistance, and T2D even if food intake is not perturbed by the user.
5. Suppression of aggression alone is sufficient to cause not only insulin-resistant state, but also a diabetic state with β cell degeneration. This can happen even if food intake and obesity is forcibly frozen at normal level.

In the long run the model can be used to establish causality in spite of the complexity of the system and correlational nature of empirical data. Also deleting some of the links selectively the vitality or redundancy of every link can be examined. The network can be used to judge the short- and long-term effects of any intervention or therapy and also their possible "side effects."

We believe that this model is only a beginning. Much refinement in our understanding of diabetes is yet to come. The future of complex disease research perhaps lies in network models of this type.

Appendix III: Model for the Effect of Population Density on Aggression

The model is based on the classical hawk and dove model. We will describe the hawk–dove model first for general readers and then go to modifications of the model to incorporate the possible effects of population density on the outcome.

The Classical Hawk and Dove Model

The standard payoff matrix of a Hawk–dove game is

	H (HAWK)	D (DOVE)
H (HAWK)	½ $(V-C)$	V
D (DOVE)	0	$V/2$

where V is the reward after winning and C is the cost of getting injured.

Next, we assume an infinite population of individuals, each adopting hawk or dove strategy, pairing off at random. Before the contest, all individuals have a fitness, W_0.

Let p = frequency of H strategists in the population, $W(H)$, $W(D)$ = fitness of H and D strategists, respectively, $E(H, D)$ = payoff to individual adopting H against a D opponent (and a similar notation for other strategy pairs).

Then,

If each individual engages in one contest,

$$W(H) = W_0 + pE(H,H) + (1-p)E(H,D)$$

$$W(D) = W_0 + pE(D,H) + (1-p)E(D,D).$$

An evolutionary stable strategy (ESS) is a strategy which when adopted by all the players in the population will not be invaded by any other rare mutant strategy. So strategy I will be an ESS when a rare mutant strategy J will get less payoff than the resident strategy I. Thus, $E(I, I) > E(J, I)$. However, if $E(I, I) = E(J, I)$ then I is still an ESS if $E(I, J) > E(J, J)$. D is not an ESS because $E(D, D) < E(H, D)$.

A population of doves can be invaded by a hawk mutant. H is an ESS if ½ $(V-C) > 0$ or $V > C$, i.e., to say, if it is worth risking injury to obtain the resource, H is the only stable strategy. But if $V < C$, neither H nor D is an ESS. Then, we can have an individual that can play sometimes H and sometimes D. Strategy I is thereafter defined as "play H with probability P, and D with probability $(1-P)$." Now, if I is a mixed ESS which includes, with nonzero probability, the pure strategies A, B, C... Then,

$$E(A,I) = E(B,I) = E(C,I),\ldots = E(I,I).$$

Because if $E(A, I) > E(B, I)$, then definitely, it would pay to adopt A more often than B, and in that eventuality, I would not be an ESS. Hence, if I is an ESS, the expected payoffs to the various strategies composing I must be equal. Now, if I is the ESS of hawk–dove game, we have

$$E(H,I) = E(D,I)$$
$$= pE(H,H) + (1-p)E(H,D) = pE(D,H) + (1-p)E(D,D)$$
$$= p \times \frac{1}{2}(V-C) + (1-p) \times V = p \times 0 + (1-p) \times \frac{V}{2}$$
$$\therefore p = \frac{V}{C}$$
$$\therefore \frac{p}{(1-p)} = \frac{V/C}{1-V/C} = \frac{V}{C-V}.$$

Thus if all the individuals in a population adopt the strategy such that they play hawk with a probability $p = V/C$ and play dove otherwise. Alternatively, we can say that the population at equilibrium will have $p = V/C$ hawks and $(1-p) = 1-(V/C)$ doves and the proportion of hawks to doves will be $V/(C-V)$.

Adding Interpersonal Knowledge to the Hawk–Dove Game

We now make our hawks more smart and sophisticated. Why should they always fight and get injured? They should be smart enough to make a judgment of the chances of winning versus chances of getting injured and avoid fight when chances of getting injured are higher. This actually happens in animal societies. There is a social hierarchy established according to the size, power, and possibly certain other strengths of individuals. Accordingly when an individual comes across an opponent that is higher up in the hierarchy, he is most likely to retreat. But when faced with a weaker individual it is beneficial to be aggressive. One can therefore make a simple thumb rule, behave as a hawk when facing a weaker individual and play a dove when facing someone higher up. For this to work, there needs to be knowledge and memory of the relative strengths of other individuals.

How would having such a knowledge change the payoffs of the hawk–dove game?

For examining this we will assume that doves always behave as doves. Hawks behave as hawks when facing an opponent known to be weaker and doves facing a stronger one. Assuming that with a probability Q one has adequate prior knowledge of the opponents relative strength, the payoff matrix becomes

	H (HAWK)	*D* (DOVE)
H (HAWK)	$(QV/2)+(1-Q)(V-C)/2$	V
D (DOVE)	0	$V/2$

Using the modified payoffs, the new p can be calculated as

$$E(H,I) = E(D,I)$$
$$= pE(H,H) + (1-p)E(H,D) = pE(D,H) + (1-p)E(D,D)$$
$$= p \times \frac{QV}{2} + (1-Q)\frac{1}{2}(V-C) + (1-p) \times V = p \times 0 + (1-p) \times \frac{V}{2}.$$

This condition is satisfied when

$$p = \frac{V}{C-Q}.$$

Thus using prior knowledge to avoid risky conflicts increases the frequency of hawks in the population. This is because hawks can avoid actual fights with known higher ups in ranking and also get a clear win with known lower-downs. This saves the cost of fights and injuries. Thus social knowledge increases the frequency of hawks. However, in the new scenario two hawks need not always fight. As a result although the proportion of hawks goes up, the frequency of fights goes down. Fights will happen only when two hawks meet and there are no clearly established prior ranks amongst the two. The frequency of aggression in the society will now be given by

$$A = p^2 \times (1 - Q).$$

Thus although Q increases the proportion of hawks, it may actually decrease the frequency of fights in the society.

The Effect of Crowding on Hawk–Dove Behavior

We attempt here to build a theoretical framework to show the effect of population density on aggression. Crowding has three potential effects:

(a) The value of the reward per fight will decrease with increasing population.
(b) With higher population density the mean time between two fights will decrease and therefore there will be less time for healing of injuries. As a result the cost of injuries will add up.
(c) As the population increases, it becomes difficult to learn, know, and remember each potential opponent's physical strength and as a result Q will tend to decrease. A decreased value of Q might increase A at a given p.

Intuitively factors (a) and (b) above would tend to decrease aggression with increasing population density, whereas (c) might have the opposite effect. We will now incorporate the above three factors in the hawk and dove model to check the effect of increasing population density on aggression, i.e., the change in p as well as in A. For a working definition, the population X is defined as the number of individuals that a focal individual can potentially interact in lifetime.

(a) When there is an increase in X, either the reward gets divided or increased number of fights are needed to get the same reward. Therefore the reward per fight is expected to decrease with increasing population. We make a simplistic assumption that the reward is inversely proportional to the population size.

Since $p = V/C$ with decline in V, there will be decrease in p, i.e., the frequency of hawk strategists in the population and therefore a decrease in the frequency of aggression in the population.

(b) With increase in population density, there will be an increase in the number or frequency of conflicts amongst the individuals constituting the population. The classical hawk dove game assumes that the payoff of each encounter is independent of others. This is certainly not true for injuries, particularly if the time interval between two fights is small. We make a limiting assumption that an individual is completely exhausted after a defeat but recovers with time following a saturation curve:

$$W = \frac{(W_0 \times T)}{(K + T)},$$

where K is a constant, inversely related to rate of healing; T is the time; W_0 is the maximum fitness; and W is the realized fitness.

The realized integrated cost of injuries (C) is given by

$$C = W_0 - W$$
$$= W_0 - \frac{(W_0 \times T)}{(K + T)}.$$

Assuming mean time between two combats varies inversely with population, we can write time in terms of population as

$$T = \frac{K1}{X},$$

where $K1$ is the proportionality constant reflecting ecological factors determining frequency of aggressive encounters.

$$C = W_0 - \frac{(W_0 \times T)}{(K + T)}$$
$$= W_0 \left[1 - \frac{T}{K + T} \right]$$
$$= W_0 \left[\frac{KX}{KX + K1} \right].$$

Hence, noting the variation of both benefit as well as cost with increase in population density, there seems to be a decrease in the overall level of aggression in the population.

Taking only (a) and (b) there is a clear reduction in aggression with increasing population. The third factor (c) is likely to have a complex effect and therefore it is necessary to consider its effect first in isolation and then in integration with (a) and (b).

(c) When the population is very small, each individual has a higher probability Q of knowing the strength of its opponents but when there is an increase in population, this probability tends to decline as then it would not be practically possible for each individual to keep track of the strength of most of its opponents. Q is expected to decrease with X such that (see Figs. III.1 and III.2 for the logic behind this form of equation)

$$Q = \frac{K_2^n}{X^n + K_2^n},$$

where $K2$ reflects the mean learning/memory capacity of an individual and the power n decides the variance around this mean such that larger power means smaller variance. An approximate intuitive interpretation of $K2$ would be the number of individuals about which an average individual can have knowledge.

Using the new population dependent values of V, C, and Q, we can calculate p as well as A as

$$p = \frac{V}{C - Q}$$

$$p = \left[\frac{\frac{R}{X}}{W_0 \left[\frac{KX}{KX + K1} \right] - \frac{K_2^n}{X^n + K_2^n}} \right]$$

and

$$A = p^2 \times (1 - Q),$$

$$A = \left[\frac{\frac{R}{X}}{W_0 \left[\frac{KX}{KX + K1} \right] - \frac{K_2^n}{X^n + K_2^n}} \right]^2 \times \left[1 - \frac{K_2^n}{X^n + K_2^n} \right].$$

Numerical solutions show that although the actual shape of the curve is complex and parameter dependent, both A and p decrease monotonically with population density. At small X the effect on A predominates that on p. In fact p may remain 1 for a substantial range of X. However as X increases beyond a threshold, the effect on p predominates and a rapid decrease in p keeps aggression at a very low level (see chapter 9, Fig. 9.4).

Appendix IV: Glucose Homeostasis Model

This is an exploratory theoretical model of glucose homeostasis which is meant to test how different mechanisms of homeostasis might interact and which mechanisms give patterns compatible with known observed patterns. The model can be used to make qualitative predictions about the behavior of different components of the system which will help us in understanding the role of each of the components and test various alternative hypotheses explaining diabetic hyperglycemia. The model is not intended to make quantitative predictions since there are no empirical estimates for a number of parameters used in the model. Nevertheless it can be useful for qualitative answers to questions such as whether altered beta cell responsivity can or cannot lead to delayed insulin curve, whether hyperglycemia is possible without beta cell degeneration, and so on. The possible/impossible types of conclusions can also lead us a long way to an insightful understanding of glucose homeostasis.

Assumptions: The model is based on the interaction between peripheral and central mechanisms of glucose homeostasis depicted in Fig. 12.1 (Chap. 12). Two distinct control circuits exist with a partial overlap in the mechanisms involved. One is a peripheral circuit that regulates plasma glucose through direct glucose-mediated modulation of pancreatic hormones insulin and glucagon. The other is a central circuit that senses brain glucose and can regulate brain glucose with feedback control exerted through autonomic control of liver and endocrine pancreas.

Level 1

In the first level of the model we take four variables, namely, levels of plasma glucose, brain glucose, plasma insulin, and glucagon and all relationships are assumed to be linear. Accordingly the model can be written as a set of four equations (see Table IV.1 for symbols used).

(1) Change in plasma glucose=(Glucose assimilated from gut)+(Glucose produced in liver)−(Glucose uptake by muscle and other tissues)−(Glucose transported to brain)−(Glucose converted into glycogen and fat)

$$\dot{P}_g = gut_g + \frac{K_l \cdot C}{I \cdot P_g \cdot \mathrm{IS}} - MU \cdot I \cdot IS \cdot P_g - \frac{glut-1 \cdot P_g}{B_g} - \frac{K_s P_g \cdot I \cdot IS}{C} \qquad \text{IV.1}$$

Table IV.1 Equation parameters

1	A	Amylin: plasma levels and activity (arbitrary units)
2	B_g	Concentration of brain glucose (mg/dL)
3	BU	Glucose utilization by the brain tissue (mg/dL/*t*)
4	C	Glucagon: plasma levels and activity (arbitrary units)
5	D_a	Constant for amylin degradation (t^{-1})
6	D_g	Constant for glucagon degradation (t^{-1})
7	D_i	Constant for insulin degradation (t^{-1})
8	*glut*–1	Activity of *glut*–1 transporter protein across the BBB (arbitrary units)
9	gut_g	Glucose available from gut per unit plasma volume
10	I	Insulin: plasma levels and activity (μIU/ml)
11	IS	Insulin sensitivity
12	K_5	Constant for insulin secretion under peripheral control (t^{-1})
13	K_5'	Constant for insulin secretion under central control (t^{-1})
14	K_6	Constant for glucagon secretion under peripheral control (t^{-1})
15	K_6'	Constant for glucagon secretion under central control (t^{-1})
16	K_a	Saturation constant of amylin action on insulin sensitivity
17	K_g	Glucose concentration at half maximal rate of insulin secretion (mg/dL)
18	K_i	Insulin concentration at half the maximum muscle uptake (μIU/ml)
19	K_l	Rate constant for liver glucose production: peripheral control [(mg/dL)^2/*t*]
20	K_l'	Rate constant for liver glucose production: central control [(mg/dL)^2/*t*]
21	K_s	Rate constant for energy storage
22	MU	Glucose pick up from plasma by muscle (mg/dL/min)
23	n	A parameter determines the shape of sigmoid curve
24	P_g	Concentration of plasma glucose (mg/dL)
25	z	Amylin to insulin ratio in β cells

(2) Change in brain glucose = (Glucose transport from plasma to brain) – (Utilization of glucose by brain tissue)

$$\dot{B}_g = \frac{P_g \cdot glut-1}{B_g} - \text{BU} \cdot B_g \qquad \text{IV.2}$$

(3) Change in plasma insulin = (Secretion of insulin) – (Degradation of insulin)

$$\dot{I} = K_5 \cdot P_g - D_i \cdot I \qquad \text{IV.3}$$

(4) Change in plasma glucagon = (Secretion of glucagon) – (Degradation of glucagon)

$$\dot{C} = \frac{K_6}{P_g} - D_g \cdot C \qquad \text{IV.4}$$

The set of four equations is used for numerical simulations over time. Parameters can be chosen to give a realistic set of steady-state plasma and brain glucose, insulin, and glucagon. For example if we assume an arbitrary unit for BU, the value for *glut*–1 can be derived from (D.2) using known normal steady-state values of P_g (80–90 mg/dL) and B_g (20–25 mg/dL). Similarly since half-life of insulin in plasma is known, D_i can be derived from data and K_5 calculated to give normal steady-state insulin. These values can then be fed back in (D.1) to determine other parameters. Thus although a large number of parameters are used, not more than 2–3 need to be arbitrarily chosen. This restricts the parameter space for operating the model.

The above model depicts the classical pathway driven by peripheral mechanisms. If we assume instead that only the central mechanisms work, i.e., brain glucose rather than plasma glucose gives the feedback for insulin and glucagon production as well as for liver glucose production, the equations will take the form:

(5) $$\dot{P}_g = gut_g + \frac{K_l' \cdot C}{I \cdot B_g \cdot \mathrm{IS}} - MU \cdot I \cdot IS \cdot P_g - \frac{glut - 1 \cdot P_g}{B_g} - \frac{K_s P_g \cdot I \cdot IS}{C} \quad \text{IV.5}$$

(6) $$\dot{B}_g = \frac{P_g \cdot glut - 1}{B_g} - \mathrm{BU} \cdot B_g \quad \text{IV.6}$$

(7) $$\dot{I} = K_5' \cdot B_g - D_i \cdot I \quad \text{IV.7}$$

(8) $$\dot{C} = \frac{K_6'}{B_g} - D_g \cdot C \quad \text{IV.8}$$

It can be noted that there are two possible ways in which central control affects glucose homeostasis. One is by giving a positive feedback for insulin production and negative one for glucagon. The other is by directly influencing liver glucose production. In reality either or both can operate simultaneously. If we use the set of (IV.5), (IV.6), (IV.3), and (IV.4) central control on liver glucose production is incorporated but central control of pancreas is not. If we use the set of (IV.1), (IV.2), (IV.7), and (IV.8) central control of pancreas is incorporated but central control of liver glucose production is not. If (IV.5), (IV.6), (IV. 7), and (IV.8) are used central control on both liver glucose production and pancreas is incorporated in the model.

Steady state

All these models give a steady-state fasting condition and the steady-state levels of glucose and hormones are determined by the parameters. Since the normal fasting values of glucose insulin and glucagon are known, the parameters can be back calculated as above.

Perturbations

The standing levels of P_g, B_g, I, and C may change when certain perturbations are introduced in the system. Such perturbations include changes in glucose utilization by muscle and brain as well as glucose assimilation from the gut. Perturbations are introduced in the model by changing the values of BU, MU, or gut_g at certain time for either a short or sustained duration.

There are four possible ways in which the central and peripheral regulation can interact:

1. One in which only the peripheral mechanisms work, i.e., (IV.1)–(IV.4).
2. One in which only the central mechanisms work, i.e., (IV.5)–(IV.8).
3. Both the mechanisms work simultaneously, i.e., both P_g and B_g give a feedback simultaneously but their relative importance can vary. For example, the insulin production term would be $(x \cdot K_5' \cdot B_g) + (1 - x) \cdot K_5 \cdot P_g$, where x goes from 0 to 1. At $x=0$ only the peripheral system is operating. At $x=1$ only the central system is operating. The intermediate values of x depict the relative emphasis on central and peripheral control.
4. The central control operates only when the brain glucose goes below a threshold, i.e., as per the classical model of counterregulatory response.

It is also possible to combine points (3) and (4) above, i.e., the system normally works with differential involvement of central and peripheral mechanisms ($0<x<1$), but if B_g goes below a threshold then x becomes 1 as long as B_g is below the threshold.

Level 2

The limitation of linear model of level 1 is that the levels of glucose and insulin cannot be in the normal range under both fasting and meal-stimulated

Fig. IV.1

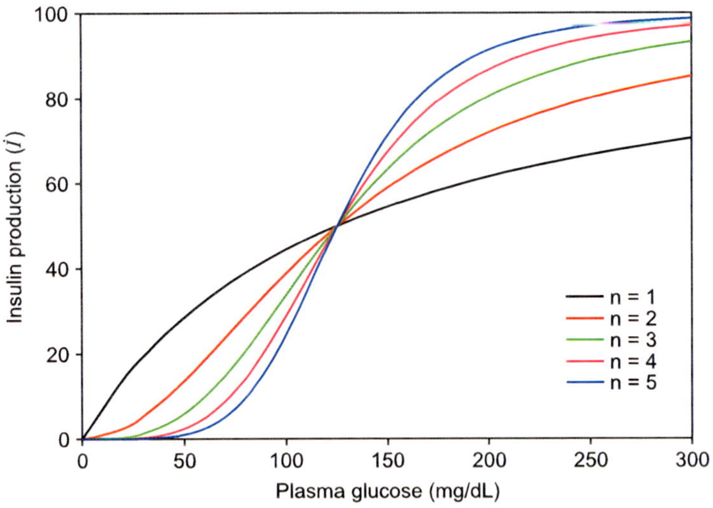

conditions at any values of parameters. For obtaining a normal OGTT curve for glucose and insulin simultaneously nonlinearity in insulin response to glucose and insulin-stimulated muscle glucose uptake is imperative. The nonlinearity incorporated in the model is as follows.

The insulin response to glucose is assumed to be a sigmoid response given by a modification of the Michelis–Menten equation of the form $K_5 \cdot P_g^n / P_g^n + K_g^n$. This gives a sigmoid shape with n deciding the sharpness of the rise as in Fig. IV.1. At $n=1$, the behavior is typical of a Michelis Menten curve.

The muscle response to insulin is similarly assumed to be saturating or sigmoid. Hence muscle response was represented by similar term. Including these modifications the (IV.5) and (IV.7) for example would change to

1.
$$\dot{P}_g = gut_g + \frac{K_l' \cdot C}{I \cdot B_g \cdot \mathrm{IS}} - MU \cdot IS \cdot P_g \frac{I^n}{I^n + K_i^n} - \frac{glut - 1 \cdot P_g}{B_g} - \frac{K_s P_g \cdot I \cdot IS}{C}. \qquad \text{IV.5}$$

2.
$$\dot{I} = \frac{K_5' \cdot B_g^n}{B_g^n + K_g^n} - D_i \cdot I \qquad \text{IV.7}$$

In the level 2 model it was possible to choose a set of parameter values that resulted in a normal fasting glucose and insulin levels as well as a normal OGTT curves for both. Because of the complex nature of the model and a large number of parameters whose empirical estimates are unavailable, there could be more than one set of parameters that can give normal fasting and OGTT. Therefore the set of parameters used (Table IV.2) may not be realistic and the model may not be used for making quantitative predictions. Nevertheless, the model can be used to analyze qualitative trends, make differential testable predictions, and differentiate between possible and impossible outcomes. In order to achieve this, the following "pathological" alterations are tested:

(1) Reduction in IS to simulate insulin resistance.
(2) Increase in K_g simulates reduced responsivity of β cells to glucose.
(3) Decreased K_5 simulates decreased beta cell capacity. Alternatively the equation term is unaltered, but a maximum limit on insulin

Table IV.2 Numerical values of parameter used for simulation results depicted in Figs. 12.8, 12.9, 12.10, 12.11, 12.12, 12.13 (units as above)

1	BU	0.1 at rest; at acute mental exercise 0.2 for 10 *t*; for sustained 0.13
2	D_a	0.105
3	D_g	0.05
4	D_i	0.35
5	*glut*–1	0.8
6	gut_g	0 at fasting; on feeding 20 which gradually comes down to 0 in 15 *t*
7	K_5	100
8	K_5'	100
9	K_6	100
10	K_6'	33
11	K_a	16
12	K_g	125
13	K_g'	40
14	K_i	25
15	K_l	30
16	K_l'	13
17	MU	0.1 at resting; at exercise 0.5 sustained for 10 *t*
18	n	5
19	z	0.1

At $t = 2$ min many of the parameters close to available empirical estimates

production rate (I_{max}) is imposed, i.e., insulin production rate = $\min\left(I_{max}, K_5 \cdot P_g^n / P_g^n + K_g^n\right)$.

(4) Decreased glucose supply to brain is simulated by decreased *glut*–1.

Incorporation of amylin: It was observed that with levels 1 and 2 models at certain parameter values the post-meal glucose shows a decline giving reactive hypoglycemia. This was sometimes followed by oscillations that gradually dampened. Amylin was incorporated to remove this effect. Amylin is co-secreted along with insulin by the β cells and at high concentrations it decreases insulin sensitivity. This was incorporated in the model by addition of plasma amylin levels as a variable:

$$\dot{A} = \frac{z \cdot K_5 \cdot P_g^n}{P_g^n + K_g^n} - D_a \cdot A.$$

Since amylin has a longer half-life than insulin $D_a < D$ throughout the model. Insulin sensitivity was assumed to be an inverse sigmoid function of amylin levels (Fig. IV.2) represented by the equation

$$\text{IS} = \frac{\text{IS}_0 \cdot K_a^n}{K_a^n + A^n}.$$

Limitations of the Model

1. We are considering the brain and the body as two separate boxes. For glucose homeostasis in the brain, utilization of glucose by different parts of the brain matters but has not been considered in this model. There are indications that there is differential utilization and different *glut*–1 levels in different parts of the brain, but currently data are inadequate to consider it in the model.
2. The conclusions made from the model are useful for qualitative predictions only.
3. Empirical estimates for many parameters are currently unavailable.

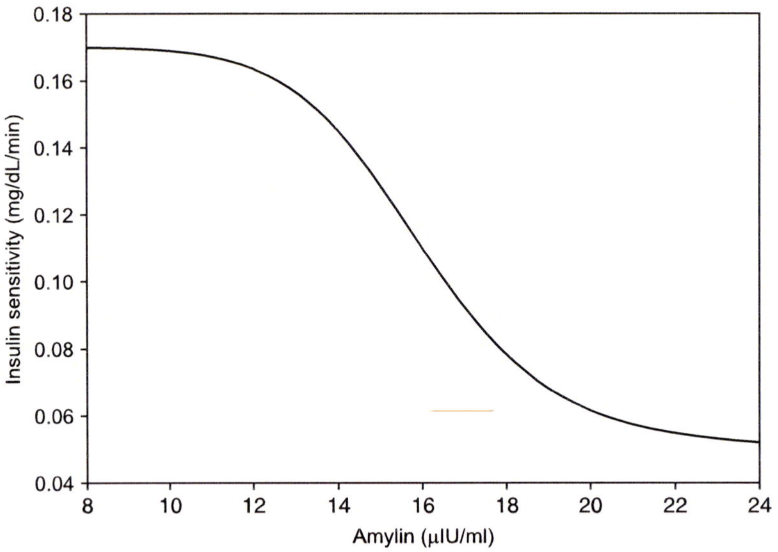

Fig. IV.2

4. The model assumes that fat stores of the body are not limiting. Stored energy and energy balance is not included in the model.

Results

The baseline parameters for all the variations of the model (i.e., only peripheral control, only central control, switch over control, and dual control with varying emphasis) were set to get roughly equal glucose levels at the fasting steady state and during GTT. However there were certain inevitable differences in the nature of the insulin and other curves in different models.

Performance of Central Versus Peripheral controls when perturbed (Fig. 12.2)

The peripheral control is always better at plasma glucose regulation against all types of perturbations, but it cannot prevent brain glucose from going down to dangerously low levels when brain glucose consumption increases in an acute or sustained fashion. The central mechanism is not as efficient as peripheral control in handling muscle activity perturbations in terms of both plasma glucose and brain glucose, but it is highly efficient in keeping brain glucose levels almost unchanged during sustained intensive brain activity and also can prevent brain glucose from reducing to dangerously low levels after acute mental stress. Dual control is intermediate in all the above characters. Switch-over control is good in handling muscle perturbation as well as short-term intensive brain activity, but may cause fluctuations, often chaotic, while handling sustained intensive brain activity.

These general trends in results are parameter and level invariant. The parameters used in the simulations represented in Fig. 12.2 were as in Table IV.2. The same baseline parameters are used in all the figures illustrated. Simulations were run exploring other sets of parameters whose results were qualitatively similar but not shown here.

Some other parameter independent qualitative trends were that:

1. The meal-stimulated insulin peak is always smaller in central control as compared to peripheral control model. This is because the amplitude of change in brain glucose is always smaller than the amplitude in plasma glucose.
2. Increased glucose consumption by brain over short or sustained duration always reduced plasma glucose marginally in the peripheral

control model, whereas it increased plasma glucose substantially in central control.
3. The HIIR state: It is interesting to note that any model based on any of the set of assumptions above was unable to give a compensatory response leading to HIIR on its own. That is introduction of insulin resistance by reducing IS did not automatically result in higher insulin production without causing hyperglycemia. Similarly a primary increase in insulin responsivity leading to hyperinsulinemia did not bring about a compensatory rise in insulin resistance such that glucose levels remained normal. This indicates that the normoglycemic hyperinsulinemic insulin-resistant state (HIIR) cannot be obtained by the mechanisms assumed in the model. There need to be some other compensatory mechanisms responsible for the HIIR state. To incorporate such a mechanism a new parameter *HIIR* is introduced in the insulin production and insulin resistance terms. *HIIR* is unity at normal insulin sensitivity. A change in *HIIR* simultaneously changes IS and K_5.

Effects of *Glut*-1 Downregulation

In diabetic rat models *glut*-1 is shown to be downregulated in the brain capillaries which would slow down the rate of transport of glucose from blood to brain. This is currently believed to be a consequence of hyperglycemia. We examine here the possible consequences of *glut*-1 downregulation on glucose homeostasis.

From (IV.2) it can be seen that *glut*-1 reduction will decrease the steady-state ratio of brain glucose to plasma glucose. It needs to be appreciated at the same time that lower *glut*-1 will reduce the amplitude of fluctuation of brain glucose. To study this we oscillate plasma glucose in a sine curve with a given amplitude and study the effects on oscillations in brain glucose with varying levels of *glut*-1 using (IV.2) only. With decreasing *glut*-1, both the mean and amplitude of oscillation of brain glucose decrease. It is possible to maintain the same mean brain glucose and reduce the amplitude of fluctuation by simultaneously decreasing *glut*-1 with a compensatory rise in the mean plasma glucose (Fig. 12.3). If maintenance of constant brain glucose is the homeostatic target, then theoretically reducing *glut*-1 and simultaneously increasing plasma glucose is the best possible solution.

Chronic Changes

The model considered so far depicts rapid changes in glucose and hormone levels. In addition to these changes, there are certain chronic changes known to take place in diabetes. A chronic change is incorporated in the model by taking a long time average and changing the target with a very small increment. For example:

1. If the plasma sugar averaged over time T is consistently high above a threshold, then *glut*-1 is reduced by a small Δ*glut*-1. Other chronic changes that were introduced alone or in different combinations in the model were.
2. If BU to MU ratio is consistently high then x increases by Δx.
3. If MU is consistently low IS decreases by ΔIS.
4. If the amplitude of oscillation of plasma glucose is high then *glut*-1 is decreased by Δ*glut*-1.

Incorporation of chronic change 1 in long-term dynamics can give rise to an autocatalytic cycle. High average blood sugar over time T slowly reduces *glut*-1. Low *glut*-1 results in reduced B_g which, if the central mechanisms are sufficiently active, stimulates liver glucose production and suppresses insulin. This raises plasma sugar further and reduces *glut*-1 further in the long run. Simulations show that this vicious cycle operates only if x is above a threshold. The threshold x is influenced by MU as well as BU in a complex way (Figs. 12.8, 12.9, and 12.10).

A similar vicious cycle is also possible if the chronic effect 4 is operative (results not shown).

Effect of Exogenous Insulin

To test the effect of exogenous insulin in the long-term model a constant is added to (IV.3) or (IV.7).

$$\dot{I} = K_5 \cdot P_g - D_i \cdot I + \varepsilon,$$

where ε is the contribution of exogenous insulin. The results show that exogenous insulin is effective in the long run if the system has not entered the vicious cycle. If the system is in a vicious cycle then a given dose ε_1 of insulin reduces mean plasma glucose in the short run but is unable to stop the progress of the vicious cycle owing to which the plasma glucose level slowly increases. A higher dose ε_2 is then needed to achieve the effect that ε_1 could achieve earlier (Fig. 12.11). This result is interesting because it matches with the frequently observed trend in diabetes treatment. Here the increase in the insulin dose required is not a result of increased insulin resistance but is due to reduced *glut*-1 levels in the BBB. This is a novel possibility that needs to be empirically tested.

The results of the model described above and in Chap. 12 raise a number of novel possibilities and alternative explanations for altered glucose dynamics in diabetes. They are testable and need to be subject to experimental verification. It is important because if they find experimental support they can alter the clinical practice radically.

Index

M. Watve, *Doves, Diplomats, and Diabetes: A Darwinian Interpretation of Type 2 Diabetes and Related Disorders*,
DOI 10.1007/978-1-4614-4409-1, © Springer Science+Business Media New York 2013

C

D

U

V

W